18.50/1

Karl Esser · Rudolf Kuenen

Genetics of Fungi

Translated by Erich Steiner

With 74 Figures

Springer Verlag New York Inc. 1967

Karl Esser
Dr. phil., o. Professor, Ruhr-Universität Bochum (Germany)

Rudolf Kuenen
Dr. rer. nat., Oberstudienrat, Köln (Germany)

Erich Steiner
Professor of Botany, University of Michigan
Ann Arbor (USA)

Cover illustration. Asci of *Sordaria macrospora* from a two-factor cross between the dark brown spored wild type ($lu^+ c^+$) and a white spored double mutant ($lu c$). The left ascus is a parental ditype ($lu^+ c^+$ and $lu c$). The ascus in the center is a recombination ditype for yellow and gray spores ($lu c^+$ and $lu^+ c$) respectively. The ascus on the right is a tetratype.

© by Springer-Verlag Berlin · Heidelberg 1967. Printed in Germany

Library of Congress Catalog Card Number 67—11496

The use of general descriptive names, trade names, trade marks, etc. in this publication, even if the former are not especially identified, is not to be taken as a sign that such names, as understood by the Trade Marks and Merchandise Marks Act, may accordingly be used freely by anyone.

Title-No. 1460

Preface

The significance of genetics in biology today stems to a considerable extent from the knowledge which has been obtained through the use of fungi as experimental objects. As a result of their short generation time, their ease of culture under laboratory conditions, and the possibility of identifying the four products of meiosis through tetrad analysis, the fungi have proven themselves in many ways superior to the classic genetic experimental material such as *Drosophila* and maize. Because they permit investigation of genetic fine structure as well as biochemical analysis of the function of the genetic material, the fungi can be used, just as the bacteria and bacteriophages, for molecular biological research. Further, the fungi, because of their simple organization, are suitable for investigation of the genetic and physiological bases of morphogenesis and of extrachromosomal inheritance.

This monograph is an attempt to summarize and interpret the results of genetic research on fungi. The reader should be reminded that review and interpretation of original research are inevitably influenced by the authors' own opinions.

An understanding of the basic principles of genetics is assumed. The volume is primarily for those interested in the problems of fungal genetics and is not intended to serve as a textbook of genetics. Nevertheless, for the general reader we have treated those special problems and methods, e.g. tetrad analysis, that are generally not included in genetic textbooks. A textbook presentation has been used for those areas, which because of their significance should be included here, but in which the fungi have been little or not at all used as research material (e.g. replication).

The subject matter has been arranged as follows: After a brief historical treatment of the significance of fungi as objects of research, a chapter entitled "Morphology" presents the particulars of the reproductive behavior of fungi, their classification, and the life histories of species which have been most important as genetic research material. In the following chapter, "Reproduction", the genetic bases of morphogenesis and of sexual reproduction are discussed. The fundamental

V

characteristics of the genetic material are treated in the chapters, "Replication", "Recombination", "Mutation", and "Function". In the last chapter, the phenomenon of extrachromosomal inheritance is discussed and interpreted. The interrelationships of these separate areas of genetic research are indicated by cross references which are organized into a subject matter index. Every chapter concludes with a summary of its main conclusions. Literature citations are given at the end of each chapter. To facilitate the survey of the main ideas, the less important details of the text, for example, description of methods and individual experiments, have been set in small type. If the reader wishes to glean only the general concepts, these sections can be omitted without losing the continuity of thought.

It is our pleasant duty to thank our colleagues and friends who have aided in this task through criticism and stimulating discussion. In particular we wish to express our thanks to D. M. BONNER, C. BRESCH, R. H. DAVIS, H. FAILLARD, R. W. KAPLAN, J. R. RAPER, P. STARLINGER, and W. STUBBE for the critical reading of particular chapters. We wish also to thank Mrs. I. TESCHE and Mrs. A. GEBAUER for their assistance in the typing and in the preparation of the index. Finally we are grateful to the publisher for his efforts to fulfill our wishes with regard to the production of this volume.

Köln, winter 1963   KARL ESSER   RUDOLF KUENEN

Preface to English Edition

The lively interest in the German edition of this work led us and the publisher to believe that an English translation was warranted. We are deeply indebted to ERICH STEINER for undertaking this task. For critical reviews of the individual chapters we wish to express our sincere thanks to R. H. DAVIS (Function), F. LINGENS (parts of Function), L. S. OLIVE (Reproduction), J. R. RAPER (Introduction, Morphology), D. R. STADLER (Replication, Recombination), and C. STERN (Mutation, Extra-chromosomal inheritance). We are also grateful for the comments on the text sent to us by readers of the German edition. These have been valuable and in so far as possible have been taken into consideration in the translation. Lists of culture collections and genetic newsletters have been included as an appendix.

We are fully aware that since the appearance of the German edition in 1965 many new investigations in all areas of fungal genetics have been published; these call for a reinterpretation of the experimental data at some points in the text and additional discussion at others. This is true particularly of the chapter on function, in which only small changes have been made. Because the translation was already in progress in 1965, it was impracticable to simultaneously undertake a revision. In order to compensate somewhat for this deficiency many of the papers which have appeared since the publication of the German edition are listed according to topic at the end of the bibliography for each chapter.

We wish also to acknowledge with sincere thanks the help of Miss R. FRINK, Mrs. A. GEBAUER and Mrs. CAROL NIMKE in the preparation of the index and the typing of the manuscript. Finally we wish to extend our thanks to the publisher for his unstinting efforts in bringing out the English edition.

Bochum and Köln, spring 1967   KARL ESSER   RUDOLF KUENEN

Translator's Note

One of the hazards of translation is the temptation to edit and alter the authors' presentation to reflect the views of the translator. Every effort has been made to avoid this pitfall; thus the ideas and views of the authors are not necessarily those of the translator. However, in order to arrive at a readable style a literal translation is often impossible and an interpretation of the authors' meaning and intent becomes essential. Errors in such interpretation are the complete responsibility of the translator.

Ann Arbor, spring 1967 ERICH STEINER

Contents

Introduction

B. O. DODGE, to whom we are indebted for the first fundamental studies of *Neurospora*, in a report in 1950 commenting upon the current status of genetic research on the fungi, said: *"The results as reported are such as would have been unbelievable ten years ago"* (DODGE, 1952). The same statement can be made again today, fifteen years later, for the fungi have demanded their place as a research material of choice, in the context of an exciting development of genetics that has led in recent years to an attack on problems at the molecular level. A brief view of the history of biology will show how the fungi have attained their present day significance in biological research after playing a secondary role for a long time.

Man has been aware of the fungi since ancient times. The macroscopically recognizable fruiting bodies of Basidiomycetes were familiar. The Greeks called the fungi *"mykes"* (THEOPHRASTUS) and appeared to have no doubt as to their vegetable nature in spite of the absence of *"stem, shoots, leaves, flowers, fruit, bark, pith, fibers, and veins"*. The word *"fungus"* occurs in PLINIUS; further, he mentions different kinds of fungi. It is assumed that the German word *"Pilz" (= fungus)*, used since the 16th century, is derived from the term *"boletus"* which goes back to PLINIUS, by way of the old German word *"buliz = piliz"* and the middle German *"bliz = bülez"*.

In the herbals of the Middle Ages, the fungi are frequently designated as *"Schwämme" (= sponges)*. This word is still used, especially in many dialects of South Germany.

Although THEOPHRASTUS assigned the fungi to the plant kingdom, many scholars of ancient times and the Middle Ages associated mystical ideas with these organisms.

HIERONYMUS BOCK thus wrote in his herbal (1552): *"Sponges are neither herbs, nor roots, nor flowers, nor seeds, but mere superfluous moisture of the earth, of the trees, of the decaying wood and other rotting things. Out of such moisture grow all the 'Tubera' and fungi. One can be certain that all the sponges described above (particularly those used for the table) grow best when it thunders and rains, says AQUINAS PONTA. Therefore, wise people paid particular attention and believed that the 'Tubera' (those which did not grow from seed) had a relationship with heaven. PORPHYRIUS also writes in this vein and says: the children of God are called fungi and Tubera because they arise from seeds, and are not born like other people."*

Even 200 years later, according to LINNAEUS (1767), the fungi, although plants without leaves and like the moss and fern, lacking a structure analogous to the fruit, nevertheless were spoken of either as *"flour or seed"* which in lukewarm water develop into small worms; these *"weave an infinitely delicate web to which they fix themselves immovably until they again swell up to form a sponge."*

Not all scientists had such erroneous conceptions about the fungi. H. v. HALLER, a contemporary of LINNAEUS, classified the fungi as a natural group in his system. It is worth noting that, in this period, the absence of anatomical and morphological differentiation of the fungi was stressed, but the heterotrophy which resulted from the lack of chlorophyll went unrecognized. Not until 1831 did DE CANDOLLE write that the fungi were *"unable to decompose carbon dioxide in sunlight."* [The literature cited up to this point was taken from MÖBIUS (1937).]

Nevertheless, apart from the speculations concerning the nature of the fungi, the *exact description and systematic classification of fungi* made steady progress during these centuries. In this connection, the publications of PERSOON (1801) and FRIES (1821—1832) should be mentioned. Concurrent with these systematically oriented investigations, the study of *fungal life histories* was initiated. The TULASNE brothers and DE BARY were productive in this area during the second half of the 19th century. The knowledge of the fungi held at this time was summarized in DE BARY's book *"Morphology and Physiology of the Fungi, Lichens, and Myxomycetes"* (1866). In this study mycological research based on precise observation was summarized, for the first time, in the form of a textbook.

The almost exclusively descriptive methods of biology of this period did not suggest either experimentation or inheritance studies with fungi. The *first genetic experiments with fungi* were not begun until the beginning of the present century, shortly after the rediscovery of MENDEL's work (BLAKESLEE, BURGEFF, EDGERTON, and their coworkers).

BLAKESLEE (1904, 1906) was the first to carry out crossing experiments and recognized that two mating types which are determined by hereditary factors exist in species of mucors (p. 50). BURGEFF (1912, 1914, 1915) was able, by systematic genetic analysis, to prove that a pair of alleles is responsible for the *plus* and *minus* mating types. The first crossing experiments with Eumycetes (*Glomerella*, and Ascomycetes) were made by EDGERTON (1912, 1914).

In spite of the success of these first genetic experiments with the mucoraceous Phycomycetes and the Ascomycete *Glomerella*, genetic research shifted its emphasis in the years following to the Basidiomycetes. It was KNIEP, above all, who, from 1919 to his early death at the beginning of the 1930's, studied the genetics of sexuality in this group. We are indebted to him for the first review of the genetics of fungi (KNIEP, 1928, 1929).

A major factor in the rapid development of fungal genetics was the work of DODGE on the Ascomycetes, particularly the study of the newly established genus *Neurospora* (SHEAR and DODGE, 1927).

DODGE was successful, not only in clarifying the development and sexual behavior of the three *Neurospora* species, *N. crassa*, *N. sitophila*, and *N. tetrasperma*, but also in working out suitable culture methods for these fungi. He recognized the species *N. crassa* and *N. sitophila* to be self-incompatible and to produce fruiting bodies only upon the crossing of *plus* and *minus* strains. Thus the foundation for the extensive investigations of LINDEGREN (1932—1934) was established, who worked out the formal genetics of *Neurospora* and first recognized the great advantage of tetrad analysis (p. 147ff.), that was made possible by the linear arrangement of the ascopores. [See BEADLE (1945) for a historical review of *Neurospora* genetics.]

In spite of this progress, the contribution of fungal genetics to fundamental genetic problems was relatively meagre. This suddenly changed, when BEADLE and TATUM in 1940, recognized that the Ascomycete *Neurospora crassa* was particularly suited for the study of the physiological effects of hereditary factors. This new direction of investigation, known as *biochemical genetics*, had previously been pursued to a limited degree in insects (KÜHN, BUTENANDT, EPHRUSSI, BEADLE; for literature see the chapter, *"Function"*). The *Neurospora* work, however, developed this field fully and led to the study of molecular genetics. As a result, our knowledge of the structure and function of genetic material has been significantly extended. Thus, the geneticist experiments with the fungi, not for their own sake, but rather, because the fungi, along with the bacteria and viruses, are suited for the solution of many of the problems of molecular genetics.

In conclusion, we would like to cite the book of FINCHAM and DAY (Fungal Genetics, 1963, 2nd edition 1965), which appeared shortly after the completion of this manuscript. The volume also treats the entire field of fungal genetics.

Literature

BEADLE, G. W.: Genetics and metabolism in *Neurospora*. Physiol. Rev. 25, 643—663 (1945).
BLAKESLEE, A. F.: Sexual reproduction in the Mucorineae. Proc. Amer. Acad. Arts Sci. 40, 205—319 (1904).
— Zygospore germination in the Mucorineae. Ann. Mycol. 4, 1—28 (1906).
BURGEFF, H.: Über Sexualität, Variabilität und Vererbung bei *Phycomyces nitens* KUNZE. Ber. dtsch. bot. Ges. 30, 679—685 (1912).
— Untersuchungen über Variabilität, Sexualität und Erblichkeit bei *Phycomyces nitens* KUNZE. I. Flora (Jena) 107, 259—316 (1914). II. Flora (Jena) 108, 353—448 (1915).
DE BARY, A.: Morphologie und Physiologie der Pilze, Flechten und Myxomyceten. Handbuch der physiologischen Botanik, Bd. 2. Leipzig 1866.
DODGE, B. O.: The fungi come into their own. Mycologia (N.Y.) 44, 273—291 (1952).
EDGERTON, C. W.: Plus and minus strains in an ascomycete. Science 35, 151 (1912).
— Plus and minus strains in the genus *Glomerella*. Amer. J. Bot. 1, 244—254 (1914).
FINCHAM, J. R. S., and P. R. DAY: Fungal genetics. Oxford 1963, 2. edit. 1965.
FRIES, E.: Systema Mycologicum, sistens fungorum ordines genera et species, huc usque cognitas, quas ad normam methodi naturalis determinavit, disposuit atque descripsit. Vol. 1—2, Lundae, ex Officina Berlingiana 1821—1823. Elenchus Fungorum, sistens commentarium in systema mycologium. Vol. 1—3, Sumtibus Ernesti Mauritzii, Gryphiswaldiae 1828—1832.
KNIEP, H.: Die Sexualität der niederen Pflanzen. Jena 1928.
— Vererbungserscheinungen bei Pilzen. Bibl. genet. 5, 371—475 (1929).
LINDEGREN, C. C.: The genetics of *Neurospora*. I. The inheritance of response to heat treatment. Bull. Torrey bot. Club 59, 81—102 (1932a).

1*

Introduction

LINDEGREN, C. C.: II. The segregation of the sex factors in the asci of
    N. crassa, *N. sitophila* and *N. tetrasperma*. Bull. Torrey bot. Club **59**,
    119—138 (1932b).
— III. Pure breed stocks and crossing-over in *N. crassa*. Bull. Torrey bot.
    Club **60**, 133—154 (1933).
— IV. The inheritance of *tan* versus normal. Amer. J. Bot. **21**, 55—65
    (1934a).
— V. Self-sterile bisexual heterokaryons. J. Genet. **28**, 425—435 (1934b).
— VI. Bisexual akaryotic ascospores from *N. crassa*. Genetica **16**, 315—320
    (1934c).
MÖBIUS, M.: Geschichte der Botanik. Jena 1937.
PERSOON, C. H.: Synopsis methodica Fungorum. Sistens enumerationem
    omnium huc usque detectarum specierum, cum brevibus descriptionibus
    nec non synonymis et observationibus selectis. Gottingiae apud Hen-
    ricium Dietrich 1801.
SHEAR, C. L., and B. O. DODGE: Life histories and heterothallism of the red
    bread mould fungi of the *Monilia sitophila* group. J. agric. Res. **34**,
    1019—1042 (1927).

Chapter I

Morphology

To provide a basis for the succeeding chapters, which describe specific aspects of fungal genetics, we wish to discuss the life cycles and taxonomy of the most significant experimental objects of genetic research. This discussion will be initiated with a description of the morphological characteristics and reproduction of these fungi. The genetic principles of reproduction will be treated in the following chapter.

For more detailed information the reader is referred to the following textbooks, e.g. WOLF and WOLF (1947); BESSEY (1950); ALEXOPOULOS (1962); TROLL (1959); INGOLD (1961); HARDER et al. (1962); GÄUMANN (1964); and the literature surveys of KNIEP (1928); GREIS (1943); OLIVE (1953); RAPER and ESSER (1964).

A. Basic concepts of fungal morphology

The fungi are not capable of assimilating carbon dioxide, because they lack chromatophores. They live as *parasites* or *saprophytes* in terrestrial and fresh water habitats, and more rarely in salt water. Almost without exception, the saprophytes can be readily cultured in the laboratory. Some parasitic species can also be cultured. Many species are not only *heterotrophic for carbon and nitrogen*, but also for various growth factors.

The vegetative plant body consists in general of branching, tube-like cells, which frequently contain more than one haploid nucleus. The walls of these filaments, called hyphae, are usually composed of chitin, less often of cellulose. The mass of *hyphae* is designated the *mycelium*.

Some Phycomycetes and Ascomycetes lack mycelia and grow as single cells. The Myxomycetes possess plant bodies lacking cell walls and are completely different in form from those of the other groups. Since the Myxomycetes have not been studied genetically to an appreciable extent, they will not be considered here.

Among the mycelial-forming fungi are species with hyphae segmented by cross walls; others have hyphae that lack cross walls. In the forms lacking cross walls, the entire mycelium is essentially a single multinucleate cell (coenocyte). Cross walls are formed in the latter organisms only to cut off reproductive organs. The species with septa must also be considered coenocytic, since their cross walls have a central perforation that allows cytoplasmic continuity and may permit the movement of nuclei between neighboring cells (SHATKIN and TATUM, 1959). Only in Basidiomycetes is the mycelium septated into cells comparable to those of the higher plants. The central perforation is surrounded by a protruding membrane and has the diameter of a plasmodesmon. Partial disintegration of the cross wall is necessary for nuclear migration (GIRBARDT, 1960; GIESY and DAY, 1965).

Hyphae of the higher fungi grow together to form a tissue-like mass in the production of fruiting bodies. Such *plectenchyma* consists of a thick web of hyphae, which upon superficial examination can be mistaken for a genuine tissue.

The fungi show great diversity with respect to their reproductive behavior. Along with species with asexual reproduction by spores and sexual reproduction by means of gametes, are those which lack one or the other, or both, of these reproductive means. All species that are unable to reproduce sexually are collectively designated *Fungi Imperfecti*.

Asexual propagation of aquatic fungi occurs by means of flagellated *zoospores*, which are produced within zoosporangia. Terrestrial forms, on the other hand, produce only non-motile spores, which, depending upon their method of formation, are variously called *sporangiospores*, *conidia*, or *oidia*.

Sporangiospores develop endogeneously within sporangia (Phycomycetes). Conidia are produced exogenously from specialized cells. Oidia arise by the fragmentation of hyphae into single spores. Conidia and oidia are typical of the Eumycetes. Under unfavorable conditions, aquatic as well as terrestrial fungi often form thick-walled resting spores (*gemmae* or *chlamydospores*) and these generally function as asexual spores.

Sexual reproduction is initiated through plasmogamy, which may occur in various ways:

Gametogamy. Fusion of morphologically or physiologically differentiated gametes.

Gametangiogamy. Copulation of morphologically or physiologically differentiated gametangia.

Oogamy. Fertilization of a female organ (oogonium) by male gametes or male nuclei.

Somatogamy. Fusion of vegetative cells which are not sexually differentiated (hyphae or conidia).

There are a few species in which no plasmogamy occurs: the multinucleate female sex organ develops into a fruiting body without being fertilized by male gametangia, gametes, or nuclei functioning as male gametes (apandry). Karyogamy is autogamous.

Between plasmogamy and karyogamy a *dikaryotic phase* of greater or shorter duration may occur, during which haploid paternal and maternal nuclei multiply vegetatively within the hyphae arising from plasmogamy. *Meiosis follows karyogamy immediately. Spores* arise, either immediately or following mitoses, from the nuclei produced by the reduction division. The *fungi are* thus predominantly *haplonts*. Exceptions are found among the yeasts, which are diplonts, and among some lower fungi, which exhibit an alteration of haploid and diploid generations.

There are a large number of fungi in which the sexual process does not take place between random male and female organs, cells or nuclei, but is possible only between particular combinations of different sexual partners. This phenomenon, known as *incompatibility*, leads to a situation in which a *hermaphroditic mycelium is no longer capable of carrying through its sexual life cycle* without the aid of a physiologically different, although morphologically identical, sexual partner. Hereditary factors are responsible for the physiological differences that determine the existence of mating types. The wide distribution of incompatibility among the fungi and the great significance that it holds for the reproductive behavior of these organisms will be discussed in more detail in the chapter, "*Reproduction*" (p. 54ff.).

Among the fungi that require two different mating types for sexual reproduction are a number of species with sexual behavior not characterized by incompatibility. Rather, in these species, plasmogamy involves the fusion of morphologically identical gametangia or gametes. Since the individual mating types cannot be considered hermaphroditic because of the absence of any morphological sexual differentiation, the absence of self-fertility in a *plus* or *minus* mycelium cannot be explained either by incompatibility or by dioecy. We have designated these species as physiologically dioecious (see further in the chapter, "*Reproduction*", p. 51ff.).

B. Systematics and morphology of the most important experimental objects of fungal genetics

The true fungi are divided into two classes. The *Phycomycetes* (lower fungi, algal fungi) possess either a *thallus lacking cross walls* or in some exceptional cases (Entomophthoraceae) septated hyphae. The *Eumycetes* (higher fungi) have septate hyphae. Before proceeding to a brief systematic treatment and a discussion of individual species which have significance for genetic research, a few technical comments are pertinent.

The diagrammatic representations of the life cycles will be complemented with figures which in part are taken from the text of WALTER (1961). In this scheme, the nuclear phases are designated as P = plasmogamy, K = karyogamy, M = meiosis. The haploid phase is characterized by a thin line, the dikaryotic phase between plasmogamy and karyogamy by a double line, and the diploid phase by a thick line. In species with two mating types, the figures show the *plus* mating type nuclei as black and the *minus* mating type nuclei as white. The more complicated mating system found most widely distributed among the Basidiomycetes, will be treated more extensively in Chapter II (Reproduction). In the present chapter, we have considered only the simplest case, that is, with two mating types.

If the development of closely related forms (different species of the same genus, as well as different genera of the same family) are similar, our presentation will portray and discuss only one form. The text will then mention the characteristics and peculiarities by which the other forms deviate.

At the end of this chapter (p. 30—31), the characteristics of the fungi of interest to the geneticist are summarized (Table I-1).

I. Class: Phycomycetes

The Phycomycetes include terrestrial as well as aquatic forms. Relatively few representatives of this group have been studied genetically. These belong primarily to the family Mucoraceae of the order Zygomycetales and grow saprophytically as the so-called "black molds" on plant and animal materials. As a representative example of this group we shall discuss the life cycle of *Phycomyces blakesleeanus* BURG., which under the synonym, *P. nitens* (KZE.) v. T. et le M., was the first fungus to be used for genetic research (BLAKESLEE, 1904, 1906a, b) (p. 2).

P. blakesleeanus propagates vegetatively by sporangiospores, which are mostly multinucleate. Sexual reproduction of this physiologically dioecious fungus is initiated by gametangiogamy, which occurs only between *plus* and *minus* haploid mating types.

Since *P. blakesleeanus* has also frequently served as an object of physiological investigations, there are many publications concerning this fungus, and most of them include information describing the life cycle. This literature has been most recently summarized by HARM (as yet unpublished). The publications of BURGEFF (1915, 1924, 1925, 1928), KEENE (1919), ORBAN (1919), BURGEFF and SEYBOLD (1927), WESENDONCK (1930), OORT (1931),

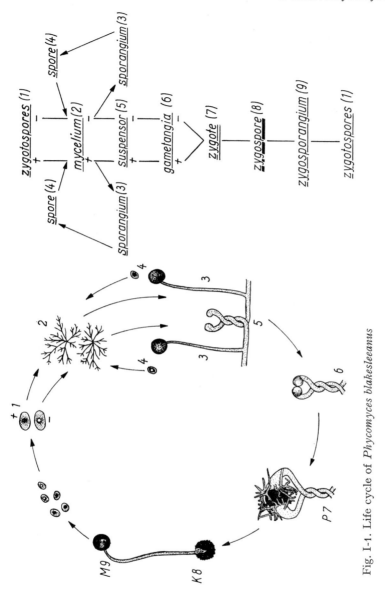

Fig. I-1. Life cycle of *Phycomyces blakesleeanus*

ROBBINS and KAVANAGH (1942), ROBBINS et al. (1942, 1943), SJÖWALL (1945), JOHANNES (1950), HESSELTINE (1955) are primarily concerned with this topic.

As seen in Fig. I-1, out of a *plus* or *minus zygotospore* (1) arises a mycelium (2), which in two days produces upright sporangiophores bearing sporangia. The sporangiophores reach a height of several cm. The sporangium (3) is cut off at the apex by septation; it may produce up to 10^4 multi-

9

nucleate spores (4), which are able to germinate to form new mycelia without undergoing a resting phase.

Sexual reproduction in *P. blakesleeanus* begins when hyphal tips of *plus* and *minus* mycelia meet. The tips twist around each other several times and grow above the substrate to reach a height of several millimeters. The hyphal tips, now called suspensors (5), cut off terminal gametangia (6) which may contain up to 100 nuclei. The two gametangia fuse to form a multinucleate zygote, around which develops a thick black wall (7). The wall then becomes overgrown with thin, branching, antler-like hyphae arising out of the suspensors and later turning black. It is assumed that during these events, the nuclei of the gametangia, and subsequently of the zygote, undergo numerous mitotic divisions. The majority of the haploid nuclei fuse in pairs, the remainder degenerating. The zygospore then enters a resting stage lasting from 4 to 6 months. After this dormant period, it produces a germ tube (8), at the end of which a zygosporangium (9) forms; this is similar in appearance to the asexual sporangium. Genetic and cytological investigations suggest that upon germination, only one diploid nucleus of the zygospore divides meiotically, while the remaining nuclei die. From these four meiotic products, two of each mating type, *plus* and *minus*, the several thousand 2- to 6-nucleate zygotospores of the zygosporangium arise. The percentage of germination of these spores, which do not require a resting stage, can be markedly increased through heat treatment.

Genetic experiments (p. 85) indicate the spore initial to contain only a single nucleus, which divides mitotically during maturation of the spore. In exceptional cases, however, more than one nucleus may be included in a young spore. When these nuclei carry different alleles for mating type, a heterokaryotic mycelium results (p. 86).

The sexual reaction is induced by sexual hormones, which are produced by both mating partners and diffuse through the substrate to their sites of action (p. 90).

The *remaining species of Mucoraceae*, of which *Mucor mucedo* (L.) FRES. has repeatedly been utilized for genetic experiments, are likewise mostly physiologically dioecious, although some are self-fertile, monoecious forms. They do not differ in any essential way from *P. blakesleeanus* in their life history. Differences concern primarily the habit of the gametangia and suspensors. For example, in *Mucor mucedo* the suspensors do not twist around one another and they lack the hyphal outgrowths, which are typical for *P. blakesleeanus*. In other species, there are marked differences between the *plus* and *minus* gametangia. In many cases, it is not possible to say whether this morphological difference constitutes true dioecy: The difference may be due to genes having morphogenetic effects but having no relation to sexual differentiation, i.e., genes able to appear in either mating type through recombination (p. 49). "Sexual hormones" have also been described and analyzed for *Mucor mucedo* (p. 89).

Allomyces arbuscula BUTLER and different species of *Blastocladiella* will be mentioned only briefly in our discussion (p. 47, 314, 318). They also belong to the *Phycomycetes* (order: Blastocladiales). Both genera are soil inhabitants and are with the exception of *B. emersonii* characterized by an alternation of generations. The haploid gametophyte gives rise to flagellated gametes. The zygote germinates to produce a sporophyte which is morphologically identical to the gametophyte. Following reduction division the sporophyte produces haploid zoospores which germinate to form gametophytes. *B. varabilis* HARDER and SÖRGEL and *B. stuebenii* COUCH and WHIFFEN are physiologically dioecious with isomorphic *plus* and *minus* gametophytes. The gametophyte of

A. arbuscula is monoecious and gives rise to anisogamous gametes; the sporophyte reproduces vegetatively by means of diploid zoospores. *B. emersonii* CANT. differs from all three of these in having no sexual cycle.

The life cycle of *Achlya ambisexualis* RAPER (order: Oomycetales) will be described in Chapter II (Reproduction) in connection with a discussion of the sexual hormones typical of these aquatic fungi.

Literature: Blastocladiella (CANTINO, 1951), *Allomyces* (summary of the literature see STUMM, 1958), *Achlya* (RAPER, 1940).

II. Class: Eumycetes

The Eumycetes are mostly terrestrial and propagate vegetatively by conidia and oidia. Their subdivision into two subclasses is based upon the nature of the meiosporangia. In the *Ascomycetes*, a sac-like sporangium, the ascus, is typical. Within the ascus, the ascospores, usually eight in number, are produced endogenously by free cell formation. The *Basidiomycetes* form their spores exogenously; they usually number four and are formed on a club-shaped basidium. In both the ascus and basidium, meiosis occurs immediately following karyogamy.

1. Subclass: Ascomycetes

The primitive representatives of this taxon lack a dikaryotic phase. The zygote is transferred directly into an ascus and no fruiting body is produced (Protoascomycetes). The second group, the Euascomycetes, form, after plasmogamy, dikaryotic hyphae, at the tips of which the asci are cut off. The latter are grouped in large numbers into a hymenium, which is formed within a fruiting body. A literature survey concerned especially with the morphology of the Ascomycetes was made by MARTENS (1946).

a) Protoascomycetes

The Protoascomycetes are noteworthy in that they include the genetically interesting family, Saccharomycetaceae (yeasts, order: Endomycetales) with the genera *Saccharomyces* and *Schizosaccharomyces*. The species of these genera usually do not produce true hyphae; their vegetative phase rather consists of spherical, ellipsoidal, or cylindrical, single cells, which reproduce vegetatively through budding (*Saccharomyces*) or fission (*Schizosaccharomyces*). Their sexual reproduction is characterized by somatogamy. (Literature surveys: STELLING-DEKKER, 1931; MRAK and PHAFF, 1948; ROMAN, 1957; WINDISCH and LASKOWSKI, 1960.)

Saccharomyces. S. cerevisiae HANSEN, commonly known as beer or baking yeast, is the most significant species of yeast in genetic research. This physiologically dioecious species, characterized by two mating types, is subject to far-reaching modification through genetic as well as non-genetic factors. Copulation usually occurs between germinating ascospores, while still in the ascus. The resulting cells are thus diploid. The four spores may be more or less linearly arranged (HAWTHORNE, 1955).

Literature: GORODKOWA (1908), GUILLIERMOND (1920, 1928, 1940), KATER (1927), WINGE (1935, 1939), WINGE and LAUTSEN (1937), HENRICI (1941), GRAHAM and HASTINGS (1941), ADAMS (1949), LINDEGREN (1949), FOWELL (1952), HARTELINS and DITLEVSEN (1953), PAZONYI (1954), TREMAINE and MILLER (1954).

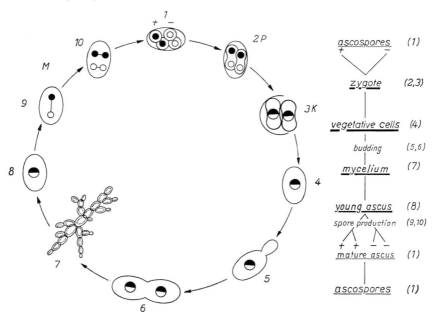

Fig. I-2. Life cycle of *Saccharomyces cerevisiae*

WINGE, especially, deserves credit for integrating the morphological work of the older authors into a unified, experimentally confirmed explanation of the life history of *S. cerevisiae* (see Fig. I-2).

The diploid vegetative yeast cells have an ellipsoidal form (4—8) which is able to reproduce itself by budding indefinitely until the exhaustion of nutrients (5—7). Under certain physiological conditions, e.g., on gypsum or cement blocks, or on nutrient-deficient agar or broth, the cells become transformed into asci. The diploid nucleus divides meiotically into four haploid nuclei, each of which is incorporated into a spore (9, 10, 1). Copulation, which ordinarily occurs between ascospores of different mating types (2, 3), can be prevented experimentally by isolating (with a micromanipulator) the haploid spores before their germination. By this technique, haploid clones can be established and used for crossing experiments; the cells are easily distinguished by their round form.

Schizosaccharomyces. The genus *Schizosaccharomyces* is in many characteristics identical with *Saccharomyces*. *Schizosaccharomyces pombe* has proven especially interesting for genetic experimentation. This yeast was originally described in 1893 by LINDNER, although it was not until more than a half a century later in the work of LEUPOLD (1950, 1958), that it became an object of genetic study. It is physiologically dioecious and possesses two morphologically identical mating types (*plus* and *minus*).

LEUPOLD utilized a strain that OSTERWALDER had originally isolated from grape juice and had first described as *S. liquefaciens* OSTERWALDER. Since the strain differed from *S. pombe* LINDNER only in its ability to liquefy gelatine, it was classified by STELLING-DEKKER as *S. pombe* LINDNER, race *liquefaciens* OSTERWALDER.

Literature: LINDNER (1893), BEYERINK (1894), GUILLIERMOND (1903, 1940), OSTERWALDER (1924), LEUPOLD (1950).

The vegetative phase is characterized by haploid cells, which are cylindrical on beerwort and ellipsoidal to round on malt agar. Vegetative reproduction occurs, in contrast to *Saccharomyces*, not through budding, but through division, i.e., septation, into daughter cells of equal size. Spore production is initiated by copulation in pairs of isogamous *plus* and *minus* cells. The diploid phase is limited to the zygote, which immediately undergoes reduction division to yield an ascus usually containing four haploid spores. Copulation and spore production are particularly easy to induce on malt agar. When the spores are transferred to fresh nutrient medium, they germinate to form haploid vegetative cells.

b) Euascomycetes

Except in some primitive forms, the development of the ascus involves a sequence of nuclear divisions and septations. Through a developmental process called hook formation, a paternal and a maternal nucleus become included in the ascus primordium (see Fig. I-3).

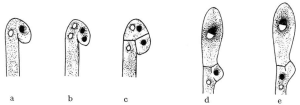

a   b   c   d   e

Fig. I-3 a—e. Hook formation in the Ascomycetes. The end of an ascogoneous hypha bends back upon itself to form a stalked hook (a). The pair of nuclei in this cell divides synchronously (b). One pair of nuclei remains in the tip; the two members of the other pair move into the hook and the stalk, respectively. Cross walls are then laid down across the stalk and hook to form a three-cell structure (c). The hook and the stalk cell thereafter fuse with one another; the stalk cell thus becomes dikaryotic again (d) and may serve as the origin of a new hook. (From GREIS, 1943)

After karyogamy and meiosis, a further division, in this case mitotic, generally occurs in the ascus. After the formation of spore walls around the eight nuclei, each pair of which represents a single meiotic product, further mitotic divisions occur during maturation of the spores, so that the mature spores are multinucleate. In many cases, the ascospores lie linearly arranged because of the typical orientation of the spindles during meiosis (see Fig. IV-4). In such cases, they can be isolated and analyzed as *ordered* tetrads (p. 148).

Among the Euascomycetes, which comprise nine orders according to the form of their fruiting bodies, representatives of four orders have significance for genetics. The *Plectascales* produce spherical, closed fruiting bodies (cleistothecia); the asci are released after the fruiting body breaks down. The *Pseudosphaeriales* and the *Sphaericales* produce flask-shaped perithecia usually a little larger than the cleistothecia. The spores are actively expelled from an opening at the apex of the perithecium. The *Pezizales* produce cup- or bowl-shaped apothecia, several millimeters in diameter, that bear hymenia on their upper surfaces. Upon maturation of the apothecium, the spores are discharged from the hymenial elements.

Plectascales

From the genetic point of view, the family[1] of interest in this order is the Aspergillaceae; it includes the genera *Aspergillus* and *Penicillium*, comprised mostly of imperfect species. The perfect forms of *Aspergillus* and *Penicillium* are frequently called *Eurotium* and *Talaromyces*, respectively. Representatives of both genera commonly live saprophytically as mildews on various organic substrates such as food, leather, fruit, etc. They propagate vegetatively by means of conidia formed exogenously in chains upon elaborate conidiophores. A few forms are important as producers of antibiotics, such as the members of the *Penicillium notatum-chrysogenum* group which yield penicillin.

Aspergillus. A. nidulans (EIDAM) WINT. is the species which has been most frequently used for genetic investigations; its sexual behavior is characterized by gametangiogamy. The cleistothecia, which contain asci with eight ascospores, are not suited for tetrad analysis because of their unusually small size.

Other important representatives of the Aspergillaceae are *A. niger* v. T. and *P. chrysogenum* THOM. Both are imperfects and can therefore be used only for analysis via mitotic recombination processes (see p. 91, 204 ff.).

The latter two species and three others (Table II-5) that have been less commonly used, show a vegetative development which does not deviate in any important features from that of *A. nidulans*. Only the life cycle of *A. nidulans* (Fig. I-4) is given here.

Literature: DE BARY and WORONIN (1870), BREFELD (1872), EIDAM (1883), DANGEARD (1907), OLIVE (1944), THOM and RAPER (1945), PONTE-CORVO et al. (1953), PONTECORVO (1954, 1956), PONTECORVO and SERMONTI (1954).

---

[1] Certain species of the genus *Ceratocystis* (syn. *Ophiostoma*), of the closely related family Ophiostomaceae, have been used to some extent for mutation experiments. The life cycle of *Ophiostoma* does not differ significantly from that of the perfect Aspergillaceae.

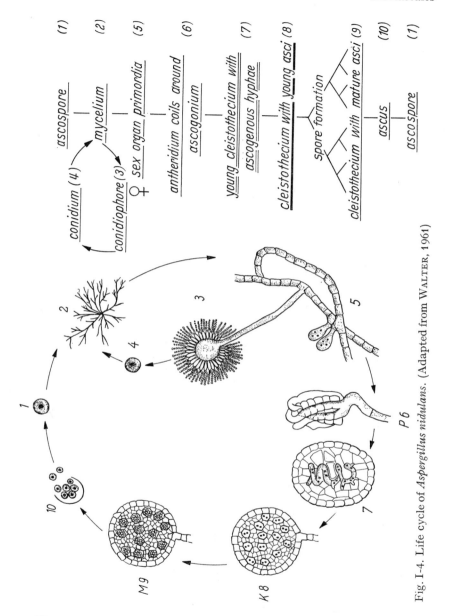

Fig. I-4. Life cycle of *Aspergillus nidulans*. (Adapted from WALTER, 1961)

The ascospore (1) produces a mycelium (2), that bears masses of conidia (4); the latter, which may be multinucleate, are pinched off from short sterigmata and adhere to form chains. These develop in a radial arrangement from spherically swollen ends of condiophores (3). (In *Penicillium* the conidiophores are branched.) The conidia are haploid and grow into mycelia.

15

Sexual reproduction is initiated with the development of two enlarged, club-shaped hyphae which grow spirally around one another (5). One becomes differentiated into an antheridium, the other into an ascogonium (6). Following plasmogamy, the ascogonium becomes septate and gives rise to many branching ascogenous hyphae (7). Hook formation occurs in this dikaryotic phase and is subsequently followed by karyogamy, meiosis, and the production of eight-spored asci (9), as the ascogonium becomes transformed into a cleistothecium.

The extremely small spores of these spherical asci (10) are not released until the outer layer (periderm) of the cleistothecium breaks down. The ascospores germinate to establish haploid mycelia.

Pseudosphaeriales

Venturia inaequalis (CKE.) WINT.[syn. *Endostigme inaequalis* (CKE.) SYD.] has been the main object of genetic investigation in this taxonomic group. The more recent developmental and genetic studies on this pathogenic fungus of higher plants (the cause of apple and pear scab) have been carried out by KEITT and coworkers.

Literature: KILLIAN (1917), KEITT and PALMITTER (1938), KEITT (1952, for survey of the literature), MÜLLER and v. ARX (1962).

The life cycle of *V. inaequalis* is essentially the same as that of *Neurospora crassa* (p. 7 and Fig. I-5). The ascospores are two celled. Each cell contains a single nucleus. The ascospores infect the young shoots of fruit trees in the spring. The mycelium is only able to penetrate between the cuticle and the epidermal cell wall in the leaves, but in fruits and shoots, it can enter the tissue as well. A few days after infection, the cuticle breaks open at the point of entrance. Conidiophores bearing countless uninucleate conidia arise; these produce new infections (asexual reproduction). The sex organs (ascogonia and antheridia) are not produced until autumn. Fertilization of the ascogonium occurs by means of a trichogyne. The asci do not develop until the following spring. The mycelium is able to overwinter in living shoots of the host and produces conidia as soon as spring arrives.

V. inaequalis is self-incompatible. The two mating types react according to the bipolar incompatibility system. This fungus can be cultured and fruited on a synthetic medium. The ascospores, however, require a dormant period of 2—3 months before germinating.

Sphaeriales

With the exception of *Glomerella*, all of the genera and species treated here belong to the family Sordariaceae. This taxon includes primarily saprophytic fungi the perithecia of which are borne free and are not surrounded by a plectenchymatous stroma. Except for *Neurospora*, they are found mostly on the feces of herbivores.

The different genera are separated mainly according to the structure of their asci and ascospores.

Neurospora. This genus was established by SHEAR and DODGE in 1927. It includes the pink bread molds, previously identified as imperfects under the genus *Monilia.* For genetic investigations, the species *N. crassa, N. sitophila,* and *N. tetrasperma* have been utilized primarily. *N. crassa* and *N. sitophila,* separated only on minor morphological differences such as size of conidia, perithecia, and ascospores, exhibit identical life cycles. They are self-incompatible species and possess two mating types, which have been widely designated in the literature as *A* and *a* instead of *plus* and *minus*. Plasmogamy between different mating types is oogamous, but somatogamy is also possible. The asci contain eight linearly arranged ascospores, which can be utilized in genetic studies as ordered tetrads. Vegetative reproduction is achieved by conidia, as is also the case in *N. tetrasperma*. In light of the great significance that *N. crassa* has attained as an object of genetic research, the life cycle of this fungus will be presented in detail (Fig. I-5); the features that are distinctive for *N. tetrasperma* will also be noted.

Literature: DODGE (1927, 1928, 1930, 1932, 1935, 1936a, 1946), WILCOX (1928), KÖHLER (1930), MORUZI (1932), ARONESCU (1933), SCHÖNFELDT (1935), WÜLKER (1935), MOREAU and MORUZI (1936), BACKUS (1939), LINDEGREN and LINDEGREN (1941a, 1941b), McCLINTOCK (1945), SANSOME (1946, 1947), SINGLETON (1953), SOMERS et al. (1960).

An excellent summary of the life history of *Neurospora* is given by BEADLE (1945).

Unfortunately many authors, who have dealt with the life history of *Neurospora*, have arrived at contradictory conclusions, a fact which must be considered in the following discussion.

A haploid *plus* or *minus* ascospore (1) germinates to form an abundantly branched, rapidly growing mycelium (2) with many aerial hyphae. Under optimal conditions, the radial growth rate reaches about 9 cm per day. After a few days, the tips of many of the aerial hyphae begin to pinch off macroconidia (10—12 microns in diameter) in chains. Most of these conidia possess more than one nucleus (3). In addition to the macroconidia, which are produced exogenously, uninucleate microconidia (1—2 microns in diameter) (4) arise endogenously in constricted cells. After release from the parent cell, they remain attached to it in grape-like clusters. Both types of conidia are capable of germinating into new mycelia without any resting stage. The vegetative reproduction of *N. crassa* can thus occur in three ways: by means of 1. hyphal or mycelial fragments, 2. macroconidia, 3. microconidia.

The female sex organs, the ascogonia, arise as coiled branches on 3 to 4 day old hyphae (5). After a few hours, they become surrounded by a thick web of "mantle" hyphae. At the same time, a receptive hypha, the trichogyne, grows out of the end of each cell destined to become an ascogonium. The mantle of hyphae, along with the female sex organ which it surrounds (now called the protoperithecium), possesses numerous radiating hyphae, to which many authors have ascribed a trichogynal function.

Microconidia, macroconidia, (6) or vegetative hyphae of the opposite mating type which fuse with the trichogynal tip (plasmogamy) contribute the male nucleus. The nuclei of the fertilizing element migrate through the trichogyne into the ascogonium to initiate the dikaryotic phase.

Protoperithecia of the *plus* mating type can only be fertilized by nuclei of the *minus* mating type, and conversely, minus protoperithecia by nuclei of the plus type. The paternal and maternal nuclei in the ascogonium multiply by many conjugate mitoses. It is still not clear if these divisions can be traced back to one or more than one pair of nuclei, since no data on the nuclei within the ascogonium are available. According to the genetic data of SANSOME (1947), however, they arise from more than a single pair of nuclei. At the ends of the ascogenous hyphae growing out of the ascogonium, the ascus-primordia develop successively through hook formation (7). Following karyogamy (8), meiosis (9, 10), and postmeiotic mitosis, four *plus* and four *minus* nuclei are present in the ascus. During ascus development, the hyphal mantle of the protoperithecium enlarges and takes on the typical form of the perithecium, which becomes black through the accumulation of melanin pigments. At maturity, the asci elongate and eject

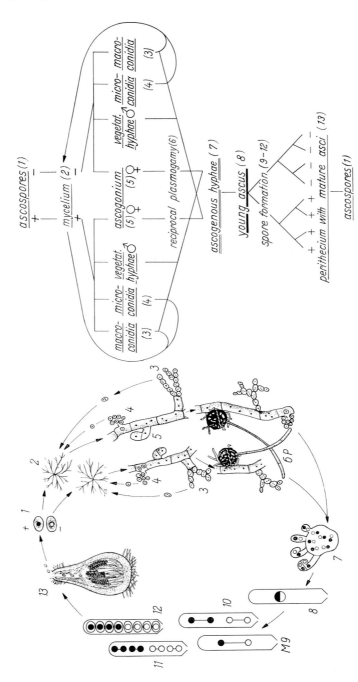

Fig. I-5. Life cycle of *Neurospora crassa*

their eight ascospores (12) through an opening at the apex of the peri-thecium (13).

The entire sexual cycle of *N. crassa* requires about 14 days under optimal cultural conditions. Since the ascospores require at least a fourteen day resting period before germination, practically speaking, the cycle has a duration of four weeks. After the resting period, almost 100 per cent spore germination can be obtained, if the spores are subjected to a 30—45 minute heat treatment at 60° C. Details concerning the relevant methodologies are available in Ryan (1950).

Neurospora tetrasperma is distinguished from the previously discussed representatives of this genus in that the ascus contains only four linearly arranged ascospores. Each spore receives two of the eight nuclei pro-duced by three successive nuclear divisions in the ascus (p. 152 ff.). If the two nuclei of a single spore carry different mating type alleles, a self-fertile, heterokaryotic mycelium (mictohaplont) results. This pheno-menon, called pseudocompatibility, will be discussed in greater detail in the following paragraph on *Podospora anserina* and in the chapter on *reproduction* (p. 66 ff.).

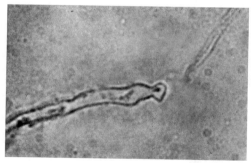

Fig. I-6. Tip of a trichogyne of a protoperithecium in *P. anserina* which has fused with a pear-shaped spermatium. The illustration shows the nucleus of the male gamete (recognizable by its nucleolus) migrating into the tricho-gyne. At the upper right is the hypha producing the male gametes. The photograph is of living material. × 2400. (From Esser, 1959)

Podospora. The synonyms, *Bombardia* Fr., *Pleurage* Fr., *Schizothe-cium* Corda, *Schizotheca* C.A. Mey and *Philocopra* Speg. have been used for this genus in the literature. A revision of the genus has recently been published by Cain (1962). The two species that are relevant here are *P. anserina* (Ces.) Rehm and *Bombardia lunata* Zckl. They differ from *Neurospora* in the following features of their life cycle:

1. Plasmogamy occurs through spermatization. Spermatia are produced in spermogonia. The spermatia correspond to the microconidia of *Neuro-spora* in appearance and manner of formation, but differ in function: the uninucleate, pear-shaped spermatia act as male gametes. They fuse with the tip of the trichogyne (Fig. I-6) and are only able to germinate to a very limited extent for asexual propagation under very special culture conditions (Beisson-Schecroun, 1963).

In contrast to *Neurospora*, in which the details of fertilization have not been entirely clarified, we know that the ascogonium of *P. anserina* initially contains only a single nucleus (RIZET and ENGELMANN, 1949). The tip of the trichogyne fuses with a single uninucleate spermatium, the nucleus of which then migrates rapidly into the ascogonium (ESSER, 1959). Fertilization in *Bombardia lunata* occurs in the same manner (ZICKLER, 1952).

Both species of *Podospora* exhibit a *morphologically distinct hermaphroditism*.

2. *The asexual reproductive phase is absent*. Macroconidia or similar cells, which ordinarily function for asexual reproduction, are absent. Vegetative reproduction in both species is possible only through mycelial transfer.

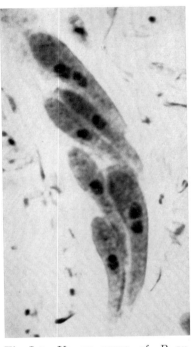

Fig. I-7. Young ascus of *P. anserina*. Each half of the ascus contains one binucleate, and two uninucleate spores. During maturation, the spores, originally club-shaped, become ellipsoidal. × 1000. (From FRANKE, 1958)

The differences between *B. lunata* and *P. anserina* are primarily the number and morphology of the ascospores and the duration of the life cycle (Table I-1).

B. lunata has eight, relatively small, linearly arranged, half-moon-shaped ascospores in the ascus. The life cycle requires from four to five weeks. *P. anserina*, on the other hand, possesses four, relatively large, ellipsoidal ascospores which, like those of *Neurospora tetrasperma*, initially have two nuclei (Fig. I-7 and I-8). The life cycle is completed in two weeks.

The reproductive behavior of both species is controlled by an incompatibility mechanism, comprised of two mating types (*plus* and *minus*). As a result of the binucleate nature of the spores and the high frequency of second division segregation at meiosis, most mycelia of *Podospora anserina* are heterokaryotic and pseudocompatible (see p. 67, 151 ff.). One to two per cent of the asci are atypical, in that each contains instead of one of the binucleate spores, two uninucleate spores (Fig. I-7, I-8). Since such uninucleate spores always give rise to homokaryotic mycelia, the use of these abnormal asci facilitates tetrad analysis (p. 152 ff.).

Literature: DOWDING (1931), AMES (1932, 1934), ZICKLER (1934, 1952), DODGE (1936b, 1936c), RIZET (1941a, b, c), RIZET and ENGELMANN (1949), MOREAU and MOREAU (1951), ESSER (1956, 1959), FRANKE (1957, 1962).

Gelasinospora. G. tetrasperma DOWDING, which is a species of some interest here, is similar to *P. anserina* in habit and physiological behavior. The asci contain four linearly arranged spores, most of which germinate to form pseudocompatible mictohaplonts.

The mycelia lack spermagonia and spermatia. Fertilization between different mating types, which arise, as in *P. anserina*, from small, mono-karyotic spores, is somatogamic. After the formation of anastomoses, *plus* and *minus* nuclei migrate into hyphae of the complementary mating type and multiply. In the resulting heterokaryons, *plus* and *minus* nuclei migrate in pairs into ascogonia. The development of the ascus and the perithecium is the same as in *Podospora* (DOWDING, 1933; DOWDING and BAKERSPIGEL, 1954, 1956).

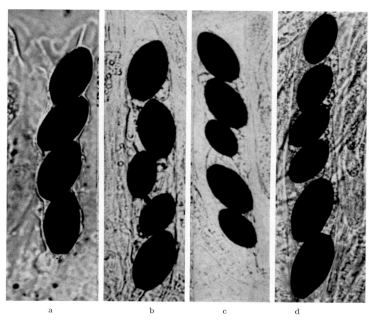

a   b   c   d

Fig. I-8a—d. Asci of *P. anserina*. (a) Normal ascus with 4 binucleate spores; (b, c) abnormal asci, each with a pair of uninucleate spores in place of one of the binucleate spores; (d) abnormal ascus with two pairs of uninucleate spores. ×460. (b and c from MARCOU, 1961)

Sordaria. Two species of this genus have been employed repeatedly in genetic experiments, namely, *S. fimicola* (ROB.) CES. et de NOT. and *S. macrospora* AUERSW. Both are self-fertile and possess eight linearly arranged ascospores, which can be utilized as ordered tetrads (see cover illustration and Fig. IV-3 for asci of *S. macrospora*). The two species differ primarily in spore size (Table I-1), but the method of fertilization has also been used as a criterion for differentiating the two species.

According to GREIS (1941), the trichogyne of the ascogonium in *S. fimi-cola* either fuses with a hypha-like antheridium or with a neighboring vegetative cell. In *S. macrospora* neither a trichogyne nor an antheridium is produced. According to the observations of DENGLER (1937), the asco-gonium fuses with a neighboring hypha.

In contradiction to these reports, recent work has shown that the development of the fruiting bodies is unquestionably apandrous and fertilization autogamous (CARR and OLIVE, 1958, for *S. fimicola*, and ESSER and STRAUB, 1958, for *S. macrospora*).

As noted earlier by RITCHIE (1937), trichogynes are not produced in *S. fimicola*; antheridia, or even hyphae functioning as such have not been demonstrated in either species. The hyphae to which an antheridial function has been ascribed are actually "mantle" hyphae that initiate the development of the perithecial wall.

The life cycle diagram that applies to both species is shown in Fig. II-1 (p. 45).

The ascospore gives rise to a mycelium (1) which produces ascogonia (2) as the only differentiated sex organs; male sex organs or conidia are not produced. The ascogenous cell becomes surrounded by "mantle" hyphae and is transformed into a protoperithecium. The nuclei of the ascogonium divide conjugately and migrate in pairs into the ascogenous hyphae, from which the asci develop in the same manner as in *Neurospora crassa*. The generation time is seven days. By means of genetic experiments, it was shown that the ascogenous cell in *S. macrospora* initially contains more than a single pair of nuclei (ESSER and STRAUB, 1958).

Crossing of *Sordaria* strains occurs, as in *Gelasinospora*, by means of heterokaryon formation in the zone of contact between two strains. The genetically different nuclei arrive in the ascogonial initial purely by chance. Because the species is self-fertile, perithecia arising from the selfing of each parent as well as from crosses, occur together (see illustration on dust cover and Fig. IV-3). Genes that determine spore color have been employed as markers to differentiate the two types of fruiting bodies (p. 148ff.).

Glomerella (Gnomoniaceae). The genus *Glomerella* includes those plant pathogenic fungi that cause bitter rots. The species that has been used most often for genetic research is *G. cingulata* (STONEMAN) SPAULD. & V. SCHR. It was isolated in 1912 by EDGERTON and was probably the first higher fungus to be used for genetic experiments (EDGERTON, 1914). The investigations begun by this author were later continued by WHEELER, CHILTON, MARKERT, etc. (see MARKERT, 1949; WHEELER, 1954, and p. 44). *G. cingulata* propagates vegetatively by conidia. Sexual reproduction of this self-fertile fungus is initiated by gametangiogamy. The asci contain eight ascospores, which are not linearly arranged. The duration of the life cycle is nine days.

Literature: LUCAS (1946), WHEELER et al. (1948), McGAHEN and WHEELER (1951), OLIVE (1951).

The sexual cycle begins with two uninucleate initials that grow out as short branches from neighboring hyphal septae. HÜTTIG (1935) designated these two cells as antheridium and archegonium on the basis of investigations on *G. lycopersici* KRÜGER. McGAHEN and WHEELER were able, in studies of living and stained material, to refute this interpretation. They found that later in the course of differentiation, the wall of the protoperithecium arises from one of the cells; this surrounds the cell-like ascogonium which develops from the other cell. Before the development of the perithecial wall is complete, the ascogonium fuses with a copulatory hypha. This arises from a hypha adjacent to that from which the protoperithecium is initiated. After plasmogamy, subsequent development to spore maturation corresponds to that described for *Neurospora crassa*.

In crosses between different self-fertile strains, the ascogonia may be fertilized by antheridial hyphae from the mating partner. Since self-fertilization may also occur, perithecia from selfing as well as from crossing are produced, just as in *Sordaria*.

Pezizales

In this order, the genus *Ascobolus*, a member of the family Pyronema-ceae, is of greatest interest. The two species, *A. stercorarius* (BULL.) SCHROET (Synonym *A. furfurcareus* PERS.) and *A. immersus* PERS. occur on the feces of herbivores and on decaying plant remains. The apothecia of these species are not surrounded by a stromatic tissue. The eight ellipsoidal ascospores are not linearly arranged; the spore walls are dark violet when mature. The essential difference between the two species is spore size (Table I-1). The spores of *A. immersus* (up to 60 μ) are the largest spores found in all genetically studied Ascomycetes and can be isolated with ease. Both *A. stercorarius* and *A. immersus* are incompatible and exhibit two mating types.

The *life cycle of A. stercorarius* essentially parallels that of *N. crassa*. Vegetative reproduction occurs by means of oidia. The sexual cycle is initiated by the dikaryotization of an ascogonium. The apex of the ascogonial trichogyne may fuse either with oidia or ordinary hyphal tips. Specific sexual substances are responsible for initiation and differentiation of gametangia and for plasmogamy. The ascogonium becomes surrounded by "mantle" hyphae, usually before plasmogamy, and these, in the course of subsequent development, form a typical apothecium. The dikaryotic phase and the development of the asci, as far as is known, is the same as in *N. crassa*.

The ontogeny of *A. immersus* PERS. has not been investigated in detail, but, according to RAMLOW (1915), this fungus exhibits an apandrous development. As in *Sordaria*, only ascogonia occur on the mycelium. Male gametangia and gametes are lacking. Fertilization between *plus* and *minus* mycelia occurs somatogamously.

Literature: JANCZEWSKI (1871), WELSFORD (1907), DOWDING (1931), SCHWEIZER (1932), RIZET (1939), BJÖRLING (1941), DODGE and SEAVER (1946), OLIVE (1956), BISTIS (1956, 1957).

2. Subclass: Basidiomycetes

The *basidium*, the characteristic structure of this subclass, is *unicellular* in the *Holobasidiomycetes* (Fig. I-9) and septate in the *Phragmobasidio-mycetes* (Fig. I-10).

a) Holobasidiomycetes

The larger fleshy fungi that grow on plant remains, wood, and dung belong in this taxon. The fruiting bodies of these forms may attain diameters of several decimeters and are commonly called "mushrooms". The Holobasidiomycetes are distinguished by the following characters:

1. Sexual differentiation is lacking; plasmogamy occurs through the fusion of vegetative hyphae (somatogamy).

2. Their sexual behavior is determined in most cases by an incompatibility system. As a result, at least two mating types exist in such species (p. 55).

In spite of the absence of sex organs, sexual reproduction in the Holobasidiomycetes falls into the category of incompatibility, because, in plasmogamy, there is a regular exchange of nuclei; this indicates a bisexual potency for each mating type. This problem will be discussed in greater detail in the chapter on *reproduction* (p. 51).

3. In the dikaryotic hyphae that result from plasmogamy, each cell contains a nucleus of each of the two mating types. This distribution of nuclei results, at each cellular division, from the formation of characteristic septal appendages or clamp connections (Fig. I-9). The formation of clamp connections involves essentially the same developmental processes as hook formation in the dikaryotic hyphae of the Ascomycetes (see Fig. I-5).

4. In contrast to the Ascomycetes, in which the jacket and the stroma of the fruiting bodies are composed of haploid hyphae, the entire fruiting bodies in the Holobasidiomycetes are composed of dikaryotic hyphae.

5. The basidia, which develop in typical fruiting bodies, are homologous as well as analogous to asci. Karyogamy, and subsequently, reduction division occur within the basidium. There is typically no postmeiotic mitosis, so that only four spores ordinarily result; these develop as outgrowths of the basidium rather than within it (Fig. I-9).

It was previously assumed that basidiospores were cut off exogenously from the basidium as in the formation of conidia. Recent investigations have shown, however, that the spores actually develop endogenously with the outgrowths of the basidium (WELLS, 1965). Each spore possesses a double membrane, namely, its own and that of the basidium and thus arises "exotopically".

The many hundreds of species of Holobasidiomycetes are classified as belonging to two orders, the Hymenomycetales and the Gastromycetales. In the former, the basidia form a layer or hymenium which often covers a part or the whole of the surface of the fruiting body; in the latter order, they are arranged as inner linings of the roughly spherical fruiting bodies. The forms of greatest genetic significance are the much studied genera *Schizophyllum* and *Coprinus*, which belong to the Agaricaceae (Hymenomycetales). Representatives of the Gastromycetales are only of slight genetic interest. The Agaricaceae typically have the hymenium on lamellae differentiated from the under surface of a stalked fruiting body (gill fungi). The genera *Collybia*, *Pleurotus*, and *Lentinus*, all briefly mentioned in this treatment (p. 65), belong to this same family.

Schizophyllum. The best known representative of this genus is *S. commune* FR., a species growing on wood. Its fruiting body, which is either sessile or pendant, normally reaches a size of about a centimeter. The gills are radially arranged (Fig. I-9); when mature, they split lengthwise, and the edges curl outward (herefore the name of the genus). The white basidiospores are very small as compared to most ascospores

(Table I-1) and are very difficult to isolate as tetrads (PAPAZIAN, 1950, has described a method). Asexual reproduction by means of oidia or conidia is lacking. The life cycle is illustrated in Fig. I-9.

Literature: WAKEFIELD (1909), LINDER (1933), SINGER (1936), BULLER (1941), WHITEHOUSE (1949), RAPER (1953), RAPER and MILES (1958), PRÉVOST (1962).

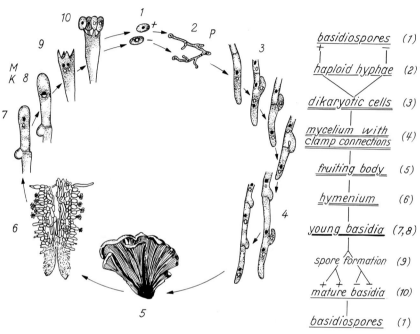

Fig. I-9. Life cycle of *Schizophyllum commune*. (Adapted from WALTER, 1961)

As apparent from Fig. I-9, a mycelium (2) composed of uninucleate cells arises from a haploid basidiospore (1). As soon as two hyphae of different mating types come into contact, plasmogamy and nuclear exchange occurs. As a probable result of partial dissolution of the septa of the hyphae, the nuclei are able, after plasmogamy, to migrate throughout the mycelium of the opposite mating type. At the same time, they multiply by mitosis. Either at the point of hyphal fusion or from cells which have become binucleate, a dikaryotic mycelium with clamp connections (3, 4) arises, and upon which, fruiting bodies (5) differentiate under suitable nutritional conditions. The gills and hymenia (6, 7) develop in the fruiting body.

After karyogamy (8) and the subsequent meiotic division in the basidium, basidiospores are produced. During maturation of the spore, its nucleus usually undergoes a mitotic division. The mature spores are thrown from their sterigmata by a specific discharge mechanism.

Coprinus. Species of the genus *Coprinus* occur on dung or soil rich in humus. The fruiting body, consisting of a cap and stipe, may reach a height of 10 cm. The spores are at first forcibly discharged, but later, they become entrapped in drops of liquid resulting from the autolysis of the hymenium as the cap matures. Species of *Coprinus* reproduce asexually by means of oidia. Sexual reproduction corresponds to that of *Schizophyllum commune*.

The different species are distinguished by the appearance of the fruiting body. The following species have been used for genetic investigations: *C. macrorhizus* Fr. ex Bolt (synonym: *C. stercorarius* ss. Rick, *C. cinereus* ss. Kohr), *C. radiatus* Fr. ex Bolt (synonym: *C. fimetarius* ss. Rick, *C. lagopus* ss. Lange), *C. lagopus* Fr., *C. sphaerosporus* Kuhn Joss (synonym: *C. funarium* Metrod).

Literature: Prévost (1962) has recently reviewed the published work on the morphology and genetics of the genus *Coprinus*.

b) Phragmobasidiomycetes

Only two of the four orders of this group, the *Uredinales* or rusts, and the *Ustilaginales* or smuts, deserve mention in this discussion. The species of these groups are plant parasites and can be cultured in the laboratory only with considerable difficulty. With the exception of studies on plant pathogenicity, the interest of geneticists in these species has been limited to the study of the incompatibility that they show in their reproductive behavior (see chapter on *reproduction*). These obligate plant pathogens are characterized by the failure in general to produce clamp connections or fruiting bodies.

Uredinales

The life history of the Uredinales is often correlated with an alternation of host. Plasmogamy involves somatogamy or dikaryon formation. Only the life cycle of *Puccinia graminis* Pers. (Pucciniaceae) is described here (Fig. I-10). Other species of this genus show deviations from this type in their development, but these variations are described in the references given below, especially in that of Gäumann (1964). *P. graminis* can only infect cereal grains in its dikaryotic phase, and this arises through plasmogamy of *plus* and *minus* types on the alternate host. Asexual reproduction occurs on the primary host by means of conidia (uredospores).

Literature: Craigie (1927, 1931, 1942), Hanna (1929), Allen (1932, 1934), Lamb (1935), Brown (1935), Mitter (1936), Savile (1939), Kulkarni (1956).

Plus and *minus* basidiospores (1) infect leaves of the alternate host (species of *Berberis*). The germ tubes (2) of the basidiospores grow through epidermal cells of the leaf and develop mycelia within the leaf. On the mycelium near the upper surface of the leaf, spermatogonia (also called pycnia) are produced (3). New infections may result from the spermatia (pycniospores) produced in the spermatogonia. At the same time, aecial primordia (5) develop on the lower surface of the leaves. Aeciospores are, however, only produced after dikaryotization (plasmogamy) has occurred. The latter may take place in the following ways: A spermatium (4) fuses with a receptive hypha of a spermatogonium of the opposite mating type (6)

on alternate host haploid stage   on cereal dicaryotic stage

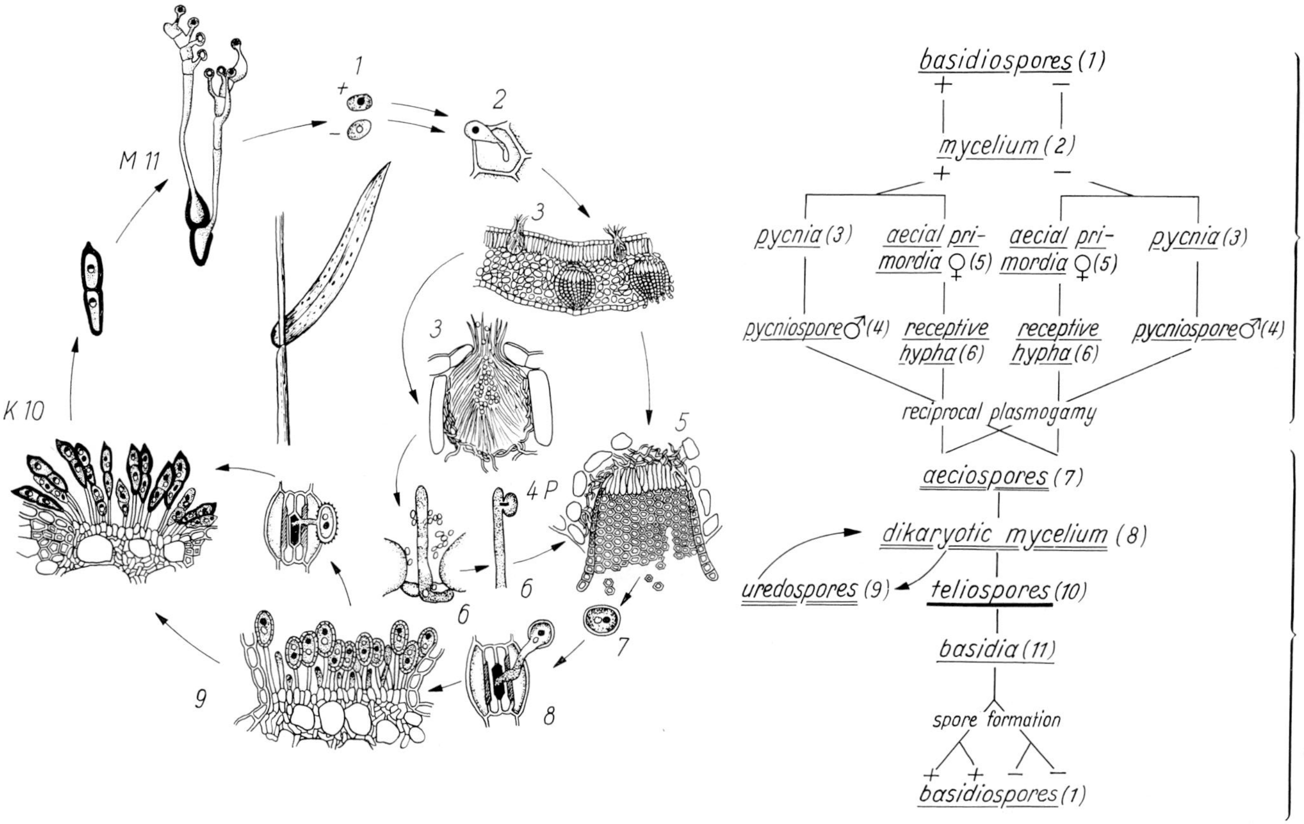

Fig. I-10. Life cycle of *Puccinia graminis*. In the center, a rust-infected cereal leaf. (Adapted from WALTER, 1961)

27

or two vegetative hyphae of different mating type may fuse. If a leaf is infected by spores of only one mating type, spermatia of the other mating type may be transferred to it from other leaves by insects. The spermatial nucleus migrates through the mycelium to the basal cell of the aecial primordium and forms a dikaryon. The spermatogonium can accordingly be considered as the male, the basal cell of the aecial primordium, the female sex organ. Plasmogamy can also occur through somatogamous fusion of *plus* and *minus* mycelia. The dikaryotic basal cell pinches off aeciospore mother cells in chain-like fashion. These divide into two cells. The apical cell becomes a dikaryotic aeciospore (7), which is disseminated (5) after the aecium opens, while the remaining cell of the pair degenerates. The aeciospore infects the cereal host by the entry of the germ tube through the stomate (8). In the new host, the mycelium produces dikaryotic uredospores, primarily on the lower surfaces of the leaves; these spread the infection on the same plant and to others (asexual cycle) (9). After a period of vegetative growth, two celled teliospores (10) are produced on the dikaryotic mycelium. The dikaryotic phase comes to an end with karyogamy in each of the two cells of the teliospore. These spores overwinter and subsequently each cell of the diploid teliospore germinates, undergoing a reduction division to form a septate basidium (11), from which arises four basidiospores (1).

Ustilaginales

In contrast to the rusts, the haploid phase in the smuts is greatly reduced or completely supressed. Only the dikaryotic phase is infectious. In rare instances, the dikaryotic mycelium will show clamp connections (SEYFERT, 1927). The basidium may either be septate or composed of a single cell (see Fig. I-11). Plasmogamy is somatogamous. Sex organs are absent. In plasmogamy, any vegetative cell may function either as a nuclear donor or receptor. There is typically an incompatibility system which may consist of either two or several mating types (p. 51, 60ff.).

The life cycle of *Ustilago tritici* (PERS.) JENS (loose smut of wheat) will serve as an example of the smuts. For the special features of basidiospore production, *Tilletia tritici* (BJERK.) WINT. (stinking smut of wheat) will be cited.

Literature: FISCHER's book (1951), which has a comprehensive bibliography, can be consulted for details. For specific references on *U. tritici*: RAWITSCHER (1912), WANG (1934), THREN (1937, 1941), WESTERN (1937); on *T. tritici*: RAWITSCHER (1922), SARTORIS (1924), WANG (1934), BECKER (1936).

As shown in Fig. I-11, haploid *plus* and *minus* basidiospores (1), also called sporidia, give rise to primary mycelia (2) which can grow saprophytically. The host plant is not infected until fusion of the plus and minus cells (3, 4) to form a dikaryon. The mycelium spreads through the plant (5) from the point of infection to the inflorescence. Here the mycelium fragments into dikaryotic smut spores (6) as in the formation of oidia. The fruiting head of the affected host plant becomes completely deformed by the mass of smut spores. Smut spores are probasidia and correspond to teliospores of the Uredinales. After a resting period, each germinates into a septate basidium, which bears four basidiospores (8). Karyogamy occurs prior to germination (7), and meiosis occurs during the development of the basidium. The basidia of *Tilletia* lack cross walls, the four or eight spores developing at the apex of the basidium (9). The basidiospores ordinarily fuse immediately to form dikaryotic mycelia (10, 11).

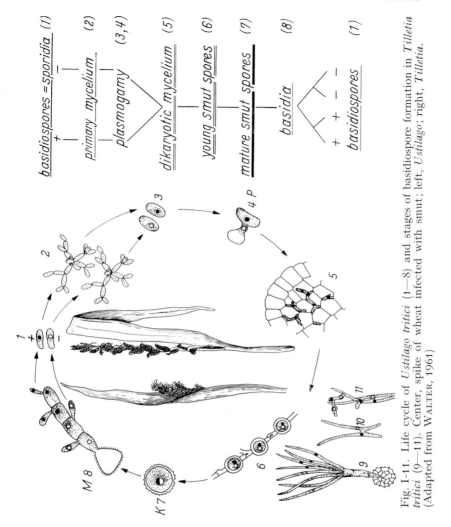

Fig. I-11. Life cycle of *Ustilago tritici* (1—8) and stages of basidiospore formation in *Tilletia tritici* (9—11). Center, spike of wheat infected with smut; left, *Ustilago*; right, *Tilletia*. (Adapted from WALTER, 1961)

Conclusions

In summary, we would like to emphasize that the foregoing treatment of the life histories of representative fungi is not intended to be exhaustive. The presentation has been oriented towards the needs of fungal genetics. The fungi that have served as objects of genetic research and their characteristics of particular interest to geneticists are shown in Table I-1. The tabulation reveals both the type of reproductive behavior as well as the suitability of a fungus for a particular kind of genetic experiment.

Table I-1. *Summary of important diagnostic characters of fungi that have been employed extensively in genetic studies*

The designation (+) under the heading "tetrad analysis" indicates that the analysis of either ordered or unordered tetrads is possible within limits. The data which do not come from original papers were obtained by personal communications with Drs. HAWTHORNE, KEEPING, LASKOWSKI, LEUPOLD, OLIVE, RAPER and WHEELER. The chromosome numbers were taken in part from the survey of DELAY (1953).

object	asexual reproduction	sexual behaviour	mode of syngamy	number of spores in sporangium	size of spores length	size of spores width	tetrad analysis possible ordered	tetrad analysis possible unordered	tetrad analysis impossible	generation time under optimal conditions (days)	haploid number of chromosomes
					averages in microns						
1. Phycomycetes *Phycomyces blakesleeanus*	sporangiospores		gametangiogamy	100—200	13	9			+	130—180	6
2. Eumycetes a) Ascomycetes *Saccharomyces cerevisiae*	vegetive cells	physiologically dioecious	somatogamy	4	3	3		+		12	15
Schizosaccharomyces pombe			gametangiogamy	4	3.5	3.5	(+)			8—14	≧2
Aspergillus nidulans	conidia	compatible		8	4.2	3.8		+		7—10	8
Neurospora crassa		incompatible	oogamy	8	28	14	+			28	7
Neurospora sitophila				8	24	14	+			28	7
Neurospora tetrasperma		pseudocompatible		4	31	15	+			28	7
Podospora anserina	—			4	37	19	+			14	7
Bombardia lunata	—	incompatible		8	20	8	+			28—35	7

Table I-1 (Continued)

object	asexual reproduction	sexual behaviour	mode of syngamy	number of spores in sporangium	size of spores length (averages in microns)	size of spores width (averages in microns)	tetrad analysis possible ordered	tetrad analysis possible unordered	tetrad analysis impossible	generation time under optimal conditions (days)	haploid number of chromosomes
Gelasinospora tetrasperma	—	pseudo-compatible	somatogamy	4	24	15	+			7	7
Sordaria fimicola	—	compatible	autogamy; in crosses somatogamy	8	17	10	+			7	8
Sordaria macrospora	—	compatible	somatogamy	8	28	18	+			7	7
Glomerella cingulata	conidia	compatible	gametangiogamy	8	24	5		+		9	4
Ascobolus immersus	—	incompatible	oogamy	8	60	32		+		10—13	16
Ascobolus stercorarius	oidia	incompatible	oogamy	8	25	13		+		8—10	16
b) Basidiomycetes Schizophyllum commune	—	incompatible	somatogamy	4	6	3		(+)	+	7	3
Coprinus radiatus	oidia	incompatible	somatogamy	4	10	7		+		12	8—9

31

Literature

ADAMS, A. M.: A convenient method of obtaining ascospores from bakers yeast. Canad. J. Res. **27**, 179—189 (1949).

ALEXOPOULOS, C. J.: Introductory mycology. New York: Wiley 1962.

ALLEN, R. F.: A cytological study of heterothallism in *Puccinia coronata*. J. agric. Res. **45**, 513—541 (1932).

— A cytological study of heterothallism in flax rust. J. agric. Res. **49**, 765—791 (1934).

AMES, L. M.: A hermaphrodite self-sterile but cross-fertile condition in *Pleurage anserina*. Bull. Torrey bot. Club **59**, 341—345 (1932).

— Hermaphroditism involving self-sterility and cross-fertility in the ascomycete *Pleurage anserina*. Mycologia (N.Y.) **26**, 392—414 (1934).

ARONESCU, A.: Further studies on *Neurospora sitophila*. Mycologia (N.Y.) **25**, 43—54 (1933).

BACKUS, M. P.: The mechanism of conidial fertilization in *Neurospora sitophila*. Bull. Torrey bot. Club **66**, 63—76 (1939).

BARY, A. DE, u. M. WORONIN: Beiträge zur Morphologie und Physiologie der Pilze. III. Abh. senckenberg. naturforsch. Ges. **7**, 95 S. (1870).

BEADLE, G. W.: Genetics and metabolism in *Neurospora*. Physiol. Rev. **25**, 643—663 (1945).

BECKER, T.: Untersuchungen über Sexualität bei *Tilletia tritici* (BJERK.) im Rahmen der Immunitätszüchtung. Phytopath. Z. **9**, 187—228 (1936).

BEISSON-SCHECROUN, J.: Incompatibilité cellulaire et interactions nucleocytoplasmiques dans les phénomènes de barrage chez *Podospora anserina*. Ann. Génét. **4**, 3—50 (1963).

BESSEY, E. A.: Morphology and taxonomy of the fungi. New York: McGraw-Hill Book Co. 1950.

BEYERINK, M. W.: *Schizosaccharomyces octosporus*, eine achtsporige Alkoholhefe. Zbl. Bakt., **16**, 49—58 (1894).

BISTIS, G.: Sexuality in *Ascobolus stercorarius*. I. Morphology of the ascogonium; plasmogamy; evidence for a sexual hormonal mechanism. Amer. J. Bot. **43**, 389—394 (1956).

— Sexuality in *Ascobolus stercorarius*. II. Preliminary experiments on various aspects of the sexual process. Amer. J. Bot. **44**, 436—443 (1957).

BJÖRLING, K.: Zur Kenntnis der Kernverhältnisse im Ascus von *Ascobolus stercorarius*. Förh. K. Fysiogr. Sällsk. Lund **11**, 42—62 (1941).

BLAKESLEE, A. F.: Sexual reproduction in the Mucorineae. Proc. Amer. Acad. Arts Sci. **40**, 205—319 (1904).

— I. Zygospore germinations in the Mucorineae. Ann. Mycol. **4**, 1—28 (1906a).

— II. Differentiation of sex in thallus, gametophyte and sporophyte. Bot. Gaz. **42**, 161—178 (1906b).

BREFELD, O.: Untersuchungen aus dem Gesamtgebiet der Mycologie. II. *Penicillium*. Leipzig u. Münster 1872.

BROWN, C. A.: Morphology and biology of some species of *Odontia*. Bot. Gaz. **96**, 640—675 (1935).

BULLER, A. H. R.: The diploid cell and the diploidization process in plants and animals with special reference to higher fungi. Bot. Rev. **7**, 335—331 (1941).

BURGEFF, H.: Untersuchungen über Variabilität, Sexualität und Erblichkeit bei *Phycomyces nitens* KUNZE. II. Flora (Jena) **108**, 353—448 (1915).

— Untersuchungen über Sexualität und Parasitismus bei Mucorineen. I. Bot. Abh., herausgeg. v. GOEBEL **4**, 135 S. (1924).

— Über Arten und Artkreuzung in der Gattung *Phycomyces* KUNZE. Flora (Jena) **18**, 40—46 (1925).

— Variabilität, Vererbung und Mutation bei *Phycomyces blakesleeanus* BGFF. Z. indukt. Abstamm.- u. Vererb.-L. **49**, 26—94 (1928).

— u. A. SEYBOLD: Zur Frage der biochemischen Unterscheidung der Geschlechter. Z. Bot. **19**, 497—537 (1927).

CAIN, R. F.: Studies on coprophilous ascomycetes. VIII. New species of *Podospora*. Canad. J. Bot. **40**, 447—490 (1962).

CANTINO, E. C.: Metabolism and morphogenetics in a new *Blastocladiella*. Antonie v. Leeuwenhoek **17**, 325—362 (1951).

CARR, A. J. H., and L. S. OLIVE: Genetics of *Sordaria fimicola*. II. Cytology. Amer. J. Bot. **45**, 142—150 (1958).

CRAIGIE, J. N.: Experiments on sex in rust fungi. Nature (Lond.) **120**, 116—117 (1927).

— An experimental investigation of sex in the rust fungi. Phytopathology **21**, 1001—1040 (1931).

— Heterothallism in the rust fungi and its significance. Trans. roy. Soc. Can., Sect. V **36**, 19—40 (1942).

DANGEARD, P. A.: Recherches sur le développement du périthèce des Ascomycètes. 2me partie. Botaniste **10**, 1—385 (1907).

DELAY, C.: Nombres chromosomiques chez le cryptogames. Rev. Cit. Biol. végét. **14**, 59—107 (1953).

DENGLER, I.: Die Entwicklungsgeschichte von *Sordaria macrospora, S. uvicola* und *S. Brefeldii*. Jb. wiss. Bot. **84**, 427—448 (1937).

DODGE, B. O.: Nuclear phenomena associated with heterothallism and homothallism in the Ascomycete *Neurospora*. J. agric. Res. **35**, 289—305 (1927).

— Unisexual conidia from bisexual mycelia. Mycologia (N.Y.) **20**, 226—234 (1928).

— Breeding albinistic strains of the *Monilia* bread mold. Mycologia (N.Y.) **22**, 9—38 (1930).

— The non-sexual and the sexual functions of microconidia of *Neurospora*. Bull. Torrey bot. Club **59**, 347—360 (1932).

— The mechanics of sexual reproduction in *Neurospora*. Mycologia (N.Y.) **27**, 418—438 (1935).

— Reproduction and inheritance in ascomycetes. Science **83**, 169—175 (1936a).

— Spermatia and nuclear migration in *Pleurage anserina*. Mycologia (N.Y.) **28**, 284—291 (1936b).

— Facultative and obligate heterothallism in ascomycetes. Mycologia (N.Y.) **28**, 399—409 (1936c).

— Self-sterility in "bisexual" heterokaryons of *Neurospora*. Bull. Torrey bot. Club **73**, 410—416 (1946).

—, and F. J. SEAVER: Species of ascobolus for genetic study. Mycologia (N.Y.) **38**, 639—651 (1946).

DOWDING, E. S.: The sexuality of *Ascobolus stercorarius* and the transportation of the oidia by mites and flies. Ann. Bot. **45**, 621—638 (1931).

— *Gelasinospora*, a new genus of pyrenomycetes with pitted spores. Canad. J. Res. **9**, 294—305 (1933).

—, and A. BAKERSPIGEL: The migrating nucleus. Canad. J. Microbiol. **1**, 68—78 (1954).

— — Poor fruiters and barrage mutants in *Gelasinospora*. Canad. J. Bot. **34**, 231—240 (1956).

EDGERTON, C. W.: Plus and minus strains in an ascomycete. Science **35**, 151 (1912).

— Plus and minus strains in the genus *Glomerella*. Amer. J. Bot. **1**, 244—254 (1914).

EIDAM, E.: Zur Kenntnis der Entwicklung bei den Ascomyceten. Cohns Beitr. Biol. Pflanzen **3**, 377—433 (1883).

ESSER, K.: Die Incompatibilitätsbeziehungen zwischen geographischen Rassen von *Podospora anserina* (CES.) REHM. I. Genetische Analyse der Semi-Incompatibilität. Z. indukt. Abstamm.- u. Vererb.-L. **87**, 595—624 (1956).

— Die Incompatibilitätsbeziehungen zwischen geographischen Rassen von *Podospora anserina* (CES.) REHM. II. Die Wirkungsweise der Semi-Incompatibilitäts-Gene. Z. Vererbungsl. **90**, 29—52 (1959).

—, u. J. STRAUB: Genetische Untersuchungen an *Sordaria macrospora* AUERSW., Kompensation und Induktion bei genbedingten Entwicklungsdefekten. Z. Vererbungsl. **89**, 729—746 (1958).

FISCHER, G. W.: The smut fungi. A guide to the literature with bibliography. New York 1951.

FOWELL, R. R.: Sodium acetate agar as a sporulation medium for yeast. Nature (Lond.) **170**, 578 (1952).

FRANKE, G.: Die Cytologie der Ascusentwicklung von *Podospora anserina*. Z. indukt. Abstamm.- u. Vererb.-L. **88**, 159—160 (1957).

— Versuche zur Genomverdoppelung des Ascomyceten *Podospora anserina* (CES.) REHM. Inaug.-Diss. der Math.-Naturwiss. Fakultät der Univ. Köln 1958.

— Versuche zur Genomverdoppelung des Ascomyceten *Podospora anserina*. Z. Vererbungsl. **93**, 109—117 (1962).

GÄUMANN, E.: Die Pilze. 2. Aufl. Basel 1964.

GIESY, R. M., and P. R. DAY: The septal pores of *Coprinus lagopus* in relation to nuclear migration. Amer. J. Bot. **52**, 287—293 (1965).

GIRBARDT, M.: Licht- und elektronenoptische Untersuchungen an *Polystictus versicolor* (L.) I. Der Wassergehalt des Hyaloplasmas vegetativer Zellen. Ber. dtsch. bot. Ges. **73**, 227—240 (1960).

GORODKOWA, A. A.: Über das Verfahren, rasch die Sporen von Hefepilzen zu gewinnen. Bull. jard. imp. bot. St. Petersburg **8**, 169—170 (1908).

GRAHAM, V. E., and E. G. HASTINGS: Studies on film-forming yeasts. I. Media and methods. Canad. J. Res., Sect. C **19**, 251—256 (1941).

GREIS, H.: Mutations- und Isolationsversuche zur Beeinflussung des Geschlechtes von *Sordaria fimicola* (ROB.). Z. Bot. **37**, 1—116 (1941).

— Bau, Entwicklung und Lebensweise der Pilze. In: Die natürlichen Pflanzenfamilien (H. HARMS u. J. MATTFELD, Hrsg.), Bd. 5a I. Leipzig 1943.

GUILLIERMOND, A.: Contributions à l'étude cytologique des Ascomycètes. C. R. Acad. Sci. (Paris) **137**, 938—939 (1903).

— The yeasts. New York 1920.

— Clef dichotomique pour la détermination des levures. Paris 1928.

— Sexuality, developmental cycle and phylogeny of yeasts. Bot. Rev. **6**, 1—24 (1940).

HANNA, W. F.: Studies in the physiology and cytology of *Ustilago zeae* and *Sorosporium reilanum*. Phytopathology **19**, 415—442 (1929).

HARDER, R., F. FIRBAS, W. SCHUMACHER u. D. v. DENFFER: Lehrbuch der Botanik für Hochschulen, 28. Aufl. Stuttgart 1962.

HARM, H.: *Phycomyces blakesleeanus*, its life history, cultural characteristics and some genetical considerations. (Unveröffentl.)

HARTELINS, V., and E. DITLEVSEN: Cement blocks, heat-stable blocks for ascospore-formation in yeast. C. R. Lab. Carlsberg **25**, 7, 213—239 (1953).

HAWTHORNE, D. C.: The use of linear asci for chromosome mapping in *Saccharomyces*. Genetics **40**, 511—518 (1955).

HENRICI, A. T.: The yeasts. Genetics, cytology, variation, classification and identification. Bact. Rev. **5**, 97—179 (1941).

HESSELTINE, C. W.: Genera of mucorales with notes on their synonymy. Mycologia (N.Y.) **47**, 344—363 (1955).

HÜTTIG, W.: Die Sexualität bei *Glomerella lycopersici* KRÜGER und ihre Vererbung. Biol. Zbl. **55**, 74—83 (1935).

INGOLD, C. T.: The biology of fungi. London 1961.

JANCZEWSKI, E. G.: Morphologische Untersuchungen über *Ascobolus furfuraceus*. Bot. Z. **29**, 271—279 (1871).

JOHANNES, H.: Ein sekundäres Geschlechtsmerkmal des isogamen *Phycomyces blakesleeanus* BURGEFF. Biol. Zbl. **69**, 463—468 (1950).

KATER, J. McA.: Cytology of *Saccharomyces cerevisiae* with special reference to nuclear division. Biol. Bull. Mar. biol. Labor. **52**, 436—448 (1927).

KEENE, M. L.: Studies of zygospore formation in *Phycomyces nitens* KUNZE. Trans. Wisconsin Acad. Sci. **19**, 1195—1220 (1919).

KEITT, G. W.: Inheritance of pathogenicity in *Venturia inaequalis* (CKE.) WINT. Amer. Naturalist **86**, 373—390 (1952).

—, and D. H. PALMITTER: Heterothallism and variability in *Venturia inaequalis*. Amer. J. Bot. **25**, 338—345 (1938).

KILLIAN, K.: Über die Sexualität von *Venturia inaequalis* (LOOKE) HEL. Z. Bot. **9** 353—398 (1917).

KNIEP, H.: Die Sexualität der niederen Pflanzen. Jena 1928.

KÖHLER, E.: Zur Kenntnis der vegetativen Anastomosen der Pilze. II. Ein Beitrag zur Frage der spezifischen Pilzwirkungen. Planta (Berl.) **10**, 495—522 (1930).

KULKARNI, V. K.: Initiation of the dicaryon in *Puccinia penniseti* ZIM. Trans. Brit. mycol. Soc. **39**, 48—50 (1956).

LAMB, I. M.: The initiation of the dicaryophase in *Puccinia phragmitis* (SCHUM.) KORN. Ann. Bot. **49**, 403—438 (1935).

LEUPOLD, U.: Die Vererbung von Homothallie und Heterothallie bei *Schizosaccharomyces pombe*. C. R. Lab. Carlsberg, Sér. Physiol. **24**, 381—480 (1950).

— Studies on recombination in *Schizosaccharomyces pombe*. Cold Spr. Harb. Symp. quant. Biol. **23**, 161—170 (1958).

LINDEGREN, C. C.: The yeast cell, its genetics and cytology. St. Louis (Mo.) USA 1949.

—, and G. LINDEGREN: X-ray and ultraviolet induced mutations in *Neurospora*. J. Hered. **32**, 404—414 (1941a).

— — X-ray and ultraviolet mutations in *Neurospora*. II. Ultraviolet mutations. J. Hered. **32**, 435—440 (1941b).

LINDER, D. H.: The genus Schizophyllum. I. Species of the western hemisphere. Amer. J. Bot. **20**, 552—564 (1933).

LINDNER, P.: *Schizosaccharomyces pombe n. sp.*, ein neuer Gärungserreger. Wschr. Brauerei **10**, 1298—1300 (1893).

LUCAS, G. B.: Genetics of *Glomerella*. IV. Nuclear phenomena in the ascus. Amer. J. Bot. **33**, 802—806 (1946).

MARCOU, D.: Notion de longévité et nature cytoplasmique de déterminant de la sénescence chez quelques champignons. Ann. Sci. Nat. Bot. **2**, 653—764 (1961).

MARKERT, C. L.: Sexuality in the fungus *Glomerella*. Amer. Naturalist **83**, 227—231 (1949).

MARTENS, P.: Cycle de développement et sexualité des ascomycètes. Cellule **50**, 125—310 (1946).

McCLINTOCK, B.: Cytogenetic studies of *Maize* and *Neurospora*. Dept. of Genetics, Carn. Inst. of Washington Year Book **44**, 108—112 (1945).

McGAHEN, J. W., and H. E. WHEELER: Genetics of *Glomerella*. IX. Perithecial development and plasmogamy. Amer. J. Bot. **38**, 610—617 (1951).

MITTER, J. H.: Some contributions to our knowledge of heterothallism in fungi. J. Indian bot. Soc. **15**, 183—192 (1936).

MOREAU, F., et Mme MOREAU: Observations cytologiques sur les Ascomycètes du genre *Pleurage* FR. Rev. Mycol. **16**, 198—207 (1951).

—, et C. MORUZI: Recherches sur la génétique des Ascomycètes du genre *Neurospora*. Rev. gén. Bot. **48**, 393 (1936).

MORUZI, C.: Recherches cytologiques et expérimentales sur la formation des périthèces chez les Ascomycètes. Thèse Fac. Sci. Paris 1932. Rev. gén. Bot. **44**, 217 (1932).

MRAK, E. M., and H. J. PHAFF: Yeasts. Ann. Rev. Microbiol. **2**, 1—46 (1948).

MÜLLER, E., u. J. A. v. ARX: Einige Aspekte zur Systematik pseudosphärialer Ascomycetes. Ber. schweiz. bot. Ges. **60**, 329—397 (1962).

OLIVE, L. S.: Development of the perithecium in *Aspergillus* FISCHERI WEHMER, with a description of crozier formation. Mycologia (N.Y.) **36**, 266—275 (1944).

— Homothallism and heterothallism in *Glomerella*. Trans. N.Y. Acad. Sci. **13**, 238—242 (1951).

— The structure and behavior of fungus nuclei. Bot. Rev. **19**, 439—586 (1953).

— Taxonomic differentiation between *Ascobolus stercorarius* and *A. furfuraceus*. Mycologia (N.Y.) **46**, 105—109 (1956).

3*

OORT, A. J. P.: The spiral-growth of *Phycomyces*. Proc. roy. Acad. Amsterd. **34**, 564—575 (1931).

ORBAN, G.: Untersuchungen über die Sexualität von *Phycomyces nitens*. Beih. bot. Zbl. I **36**, 1—59 (1919).

OSTERWALDER, A.: *Schizosaccharomyces liquefaciens n. sp.* eine gegen schweflige Säure widerstandsfähige Gärhefe. Mitt. Lebensmittelunters. **15**, 5 (1924).

PAPAZIAN, H. P.: A method of isolating the four spores from a single basidium in *Schizophyllum commune*. Bot. Gaz. **112**, 139—140 (1950).

PAZONYI, B.: Studies on sporulation in yeasts and some problems of improving yeast strains. A method of the submerged culture type for inducing mass ascospore formation in yeasts. Acta microbiol. Acad. Sci. hung. **1**, 49—70 (1954).

PONTECORVO, G.: Mitotic recombination in the genetic system of filamentous fungi. Caryologia, Suppl. **6**, 192—200 (1954).
— The parasexual cycle in fungi. Ann. Rev. Microbiol. **10**, 393—400 (1956).
—, J. A. ROPER, and E. FORBES: Genetic recombination without sexual reproduction in *Aspergillus nidulans*. J. gen. Microbiol. **8**, 198—210 (1953).
—, and G. SERMONTI: Parasexual recombination in *Penicillium chrysogenum*. J. gen. Microbiol. **11**, 94—104 (1954).

PRÉVOST, G.: Étude génétique d'un basidiomycète *Coprinus radiatus* FR. ex BOLT. Thèse Fac. Sci. Univ. Paris 1962.

RAMLOW: (1915) zit. nach GREIS 1943.

RAPER, J. R.: Sexuality in *Achlya ambisexualis*. Mycologia (N.Y.) **32**, 710—727 (1940).
— Tetrapolar sexuality. Quart. Rev. Biol. **28**, 233—259 (1953).
—, and K. ESSER: The Fungi. The Cell, Bd. VI, S. 139—244. New York 1964.
—, and P. G. MILES: The genetics of *Schizophyllum commune*. Genetics **43**, 530—546 (1958).

RAWITSCHER, F.: Beiträge zur Kenntnis der Ustilagineen. I. Z. Bot. **4**, 673—706 (1912).
— Beiträge zur Kenntnis der Ustilagineen. II. Z. Bot. **14**, 273—296 (1922).

RITCHIE, D.: The morphology of the perithecium of *Sordaria fimicola* (ROB.) CES. and DE NOT. J. Elisha Mitchell Sci. Soc. **53**, 334—342 (1937).

RIZET, G.: Sur les spores dimorphes et l'hérédité et leur caractère chez un nouvel *Ascobolus* hétérothallique. C. R. Acad. Sci. (Paris) **208**, 1669—1671 (1939).
— Sur l'analyse génétique des asques du *Podospora anserina*. C. R. Acad. Sci. (Paris) **212**, 59—61 (1941a).
— La ségrégation des sexes et de quelques caractères somatiques chez *Podospora anserina*. C. R. Acad. Sci. (Paris) **213**, 42—45 (1941b).
— La valeur génétique des périthèces sur des souches polycaryotiques chez *Podospora anserina*. Bull. Soc. bot. France **88**, 517—520 (1941c).
—, et C. ENGELMANN: Contribution à l'étude génétique d'un Ascomycète tétraspore: *Podospora anserina* (CES.) REHM. Rev. Cytol. Biol. végét. **11**, 202—304 (1949).

ROBBINS, W. J., and F. KAVANAGH: Hypoxanthine, a growth substance for Phycomyces. Proc. nat. Acad. Sci. (Wash.) **28**, 65—69 (1942).
— V. W. KAVANAGH, and F. KAVANAGH: Growth substances and dormancy of spores of *Phycomyces*. Bot. Gaz. **104**, 224—242 (1942/43).

ROMAN, W. (edit.): Yeasts. New York: Academic Press 1957.

RYAN, F. J.: Selected methods of *Neurospora* genetics. Meth. Med. Res. **3**, 51—75 (1950).

SANSOME, E. R.: Heterokaryosis, mating-type factors and sexual reproduction in *Neurospora*. Bull. Torrey bot. Club **73**, 339—396 (1946).
— The use of heterokaryons to determine the origin of the ascogeneous nuclei in *Neurospora crassa*. Genetica **24**, 59—64 (1947).

SARTORIS, G. B.: Studies in the life history and physiology of certain smuts. Amer. J. Bot. **11**, 617—647 (1924).

SAVILE, D. B. O.: Nuclear structure and behavior in species of Uredinales. Amer. J. Bot. **26**, 585—609 (1939).

SCHÖNFELDT, M.: Entwicklungsgeschichtliche Untersuchungen bei *Neurospora tetrasperma* und *N. sitophila*. Z. indukt. Abstamm.- u. Vererb.-L. **69**, 193—20 (1935).

SCHWEIZER, G.: Studien über die Kernverhältnisse im Archikarp von *Ascobolus furfuraceus*. Ber. dtsch. bot. Ges. **50**A, 14—23 (1932).

SEYFERT, R.: Über Schnallenbildung im Paarkernmyzel der Brandpilze. Z. Bot. **19**, 577—601 (1927).

SHATKIN, A. J., and E. L. TATUM: Electron microscopy of *Neurospora crassa* mycelia. J. biophys. biochem. Cytol. **6**, 423—426 (1959).

SHEAR, C. L., and B. O. DODGE: Life histories and heterothallism of the red bread mould fungi of the *Monilia sitophila* group. J. agric. Res. **34**, 1019—1042 (1927).

SINGER, R.: Studien zur Systematik der Basidiomyceten. I. Beih. bot. Zbl. B **56**, 137—156 (1936).

SINGLETON, J. R.: Chromosome morphology and the chromosome cycle in the ascus of *Neurospora crassa*. Amer. J. Bot. **40**, 124—144 (1953).

SJÖWALL, M.: Studien über Sexualität, Vererbung und Zytologie bei einigen diözischen Mucoraceen. Lund 1945.

SOMERS, C. E., R. P. WAGNER, and T. C. HSU: Mitosis in vegetative nuclei of *Neurospora crassa*. Genetics **45**, 801—810 (1960).

STELLING-DEKKER, N. M.: Die Hefesammlung des „Centraalbureau voor Schimmelcultures". I. Teil. Die sporogenen Hefen. Verh. kon. Akad. Wet. Amsterd., Afd. Natuurk., Sect. 2 **28**, 1—524 (1931).

STUMM, C.: Die Analyse von Genmutanten mit geänderten Fortpflanzungseigenschaften bei *Allomyces arbuscula* BUTL. Z. Vererbungsl. **89**, 521—539 (1958).

THOM, C., and K. B. RAPER: A manual of the *Aspergilli*. New York: Williams & Wilkins Co. 1945.

THREN, R.: Gewinnung und Kultur von monokaryotischem und dikaryotischem Myzel. Z. Bot. **31**, 337—391 (1937).

— Über Zustandekommen und Erhaltung der Dikaryophase von *Ustilago nuda* (JENSEN) KELLERM. et SW. und *Ustilago tritici* (PERSOON) JENSEN. Z. Bot. **36**, 449—498 (1941).

TREMAINE, J. H., and J. J. MILLER: Effect of six vitamins on ascospore formation by an isolate of bakers yeast. Bot. Gaz. **115**, 311—322 (1954).

TROLL, W.: Allgemeine Botanik, 3. Aufl. Stuttgart 1959.

WAKEFIELD, E. M.: Über die Bedingungen der Fruchtkörperbildung, sowie das Auftreten fertiler und steriler Stämme bei Hymenomyceten. Naturwiss. Z. Forst- u. Landwirtschaft **7**, 521—551 (1909).

WALTER, H.: Einführung in die Phytologie. II. Grundlagen des Pflanzensystems, 2. Aufl. Stuttgart 1961.

WANG, D. T.: Contribution à l'étude des Ustilaginées. (Cytologie du parasite et pathologie de la cellule hôte.) Botaniste **26**, 539—647 (1934).

WELLS, K.: Ultrastructural features of developing and mature basidia and basidiospores of *Schizophyllum commune*. Mycologia 57, 236—261 (1965).

WELSFORD, E. J.: Fertilization in *Ascobolus furfuraceus* PERS. New Phytologist **6**, 151—161 (1907).

WESENDONCK, J.: Über sekundäre Geschlechtsmerkmale bei *Phycomyces blakesleeanus* BGFF. Planta (Berl.) **10**, 456—494 (1930).

WESTERN, J. H.: Sexual fusion in *Ustilago avenae* under natural conditions. Phytopathology **27**, 547—553 (1937).

WHEELER, H. E.: Genetics and evolution of heterothallism in *Glomerella*. Phytopathology **44**, 342—345 (1954).

— L. S. OLIVE, C. T. ERNEST, and C. W. EDGERTON: Genetics of *Glomerella*. V. Crozier and ascus development. Amer. J. Bot. **35**, 722—729 (1948).

WHITEHOUSE, H. L. K.: Multiple allelomorph heterothallism in the fungi. New Phytologist **48**, 212—244 (1949).

Morphology

WILCOX, M. S.: The sexuality and the arrangement of the spores in the ascus of *Neurospora sitophila*. Mycologia (N.Y.) **20**, 3—17 (1928).
WINDISCH, S., u· W. LASKOWSKI: Die Hefen, Bd. I, S. 23—208. Nürnberg 1960.
WINGE, Ö.: On haplophase and diplophase in some Saccharomycetes. C. R. Lab. Carlsberg, Sér. Physiol. **21**, 77 (1935).
— *Saccharomyces Ludwigii* HANSEN, a balanced heterozygote. C. R. Lab. Carlsberg, Sér. Physiol. **22**, 357—370 (1939).
—, and O. LAUTSEN: On two types of spore germination and on genetic segregation in *Saccharomyces*, demonstrated through single-spore cultures. C. R. Lab. Carlsberg, Sér. Physiol. **22**, 99—116 (1937).
WOLF, F. A., and F.T.WOLF: The fungi, vols. I and II. New York 1947.
WÜLKER, H.: Untersuchungen über die Tetradenaufspaltung bei *Neurospora sitophila* SHEAR et DODGE. Z. indukt. Abstamm.- u. Vererb.-L. **69**, 210—248 (1935).
ZICKLER, H.: Genetische Untersuchungen an einem heterothallischen Ascomyceten *(Bombardia lunata nov. spec.)*. Planta (Berl.), **22** 573—613 (1934).
— Zur Entwicklungsgeschichte des Ascomyceten *Bombardia lunata* ZCKL. Arch. Protistenk. **98**, 1—70 (1952).

References

which have come to the authors' attention after conclusion of theGerman manuscript

A

JINKS, J. L., and G. SIMCHEN: A consistent nomenclature for the nuclear status of fungal cells. Nature (Lond.) **210**, 778—780 (1966).
OLIVE, L. S.: Nuclear behavior during meiosis. Fungi (N. Y.) **1**, 143—161 (1965).

B I

BARKSDALE, A. W.: *Achlya ambisexualis* and a new cross-conjugating species of *Achlya*. Mycologia (N.Y.) **57**, 493—501 (1965).
GALLE, H. K.: Untersuchungen über die Entwicklung von *Phycomyces blakesleeanus* unter Anwendung des Mikrozeitrafferfilmes. Protoplasma **59**, 423—471 (1964).
SANSOME, E.: Meiosis in diploid and polyploid sex organs of *Phytophthora* and *Achlya*. Cytologia **30**, 103—117 (1965).
WEIJER, J., and S. H. WEISBERG: Karyokinesis of the somatic nucleus of *Aspergillus nidulans*. I. The juvenile chromosome cycle (feulgen staining). Canad. J. Genet. Cytol. **8**, 361—374 (1966).

B II, 1a

HAEFNER, K.: A rapid method for obtaining zygotes and to determine the mating type in *Saccharomyces*. Z. allg. Mikrobiol. **5**, 77 (1965).
HWANG, Y. L., J. LINDEGREN, and C. C. LINDEGREN: The twelfth chromosome of *Saccharomyces*. Canad. J. Genet. Cytol. **6**, 373—380 (1964).
SCHEDA, R., and D. YARROW: The instability of physiological properties used as criteria in the taxonomy of yeasts. Arch. Mikrobiol. **55**, 209—225 (1966).

B II, 1b

BERG, C. M.: Biased distribution and polarized segregation in asci of *Sordaria brevicollis*. Genetics **53**, 117—129 (1966).
CHEN, K. C., and L. S. OLIVE: The genetics of *Sordaria brevicollis*. II. Biased segregation due to spindle overlap. Genetics **51**, 761—766 (1965).
DOWDING, E. S.: The chromosomes in *Neurospora* hyphae. Canad. J. Bot. **44**, 1121—1125 (1966).
ENGEL, H., u. H. P. KOOPS: Die Rhythmik des Sporenabschusses von *Sordaria macrospora* AUERSWALD. Ber. dtsch. bot. Ges. **79**, 92—100 (1966).

GAMUNDI, I. J., and M. E. RANALLI: Apothecial development in *Ascobolus stercorarius*. Trans. Brit. mycol. Soc. **46**, 393—400 (1965).
— — Estudio sistemático y biológico de las Ascoboláceas de Argentina II. Nova Hedwigia **10**, 339—366 (1966).
MITCHELL, M. B.: An extended model of periodic linkage in *Neurospora crassa*. Canad. J. Genet. Cytol. **7**, 563—570 (1965).
PATEMAN, J. A., and B. T. O. LEE: Segregation of polygenes in ordered tetrads. Heredity **15**, 351—361 (1960).
STRICKLAND, W. N., and D. THORPE: Sequential ascus collection in *Neurospora crassa*. J. gen. Microbiol. **33**, 409—412 (1963).
TURIAN, G.: Recherches sur la morphogenèse des *Neurospora*. Arch. Sci. (Genève) **18**, 371—381 (1965).
WEIJER, J., A. KOOPMANS, and D. L. WEIJER: Karyokinesis of somatic nuclei of *Neurospora crassa*: III. The juvenile and maturation cycles (feulgen and crystal violet staining). Canad. J. Genet. Cytol. **7**, 140—163 (1965).
ZUK, J., and Z. SWIETLINSKA: Cytological studies in *Ascobolus immersus*. Acta Soc. Bot. Polon. **34**, 171—179 (1965).

B II, 2a
LANGE, I.: Das Bewegungsverhalten der Kerne in fusionierten Zellen von *Polystictus versicolor* (L.). Flora (Jena) **156**, 487—497 (1966).
PARAG, Y.: Papillae secreting water droplets on aerial mycelia of *Schizophyllum commune*. Israel J. Bot. **14**, 192—195 (1966).

B II, 2b
KUKKONEN, I., and M. RAUDASKOSKI: Studies on the probable homothallism and pseudo-homothallism in the genus *Anthracoidea*. Ann. Bot. Fenn. **1**, 257—271 (1964).

Chapter II

Reproduction

Genetic control of reproduction consists of two phases: 1. *differentiation* during ontogeny which leads to the formation of asexual and sexual reproductive cells and 2. the *occurrence of karyogamy and meiosis*. The interaction of these regulatory functions throughout evolution is undoubtedly responsible for the great diversity which the fungi exhibit both in morphology and in sexual behavior.

Early genetic investigations of the fungi, carried out by BLAKESLEE, BURGEFF, KNIEP, and coworkers, were devoted almost exclusively to the study of the mechanism of sexual reproduction. Through the efforts of DODGE, LINDEGREN, BEADLE and TATUM, fungal genetics broadened to embrace control of morphogenetic processes and problems of formal and physiological genetics. Problems dealing with reproduction have received less attention during the last twenty-five years. Recently, however, a number of publications have been concerned with this area of research, in particular with the genetic analysis of sexual incompatibility. Nevertheless, our knowledge of the genetic control of morphogenesis involved in sexual and asexual reproduction remains incomplete.

Literature: KNIEP (1928, 1929a, b), BRIEGER (1930), CAYLEY (1931), VAN-DENDRIES (1938), BULLER (1941), CRAIGIE (1942), HARTMANN (1943 respectively 1956), MATHER (1942), LINDEGREN (1948), WHITEHOUSE (1949 a, b, 1951a, b), QUINTANILHA and PINTO-LOPES (1950), RAPER (1951, 1953, 1954, 1955a, b, 1959, 1960, 1963), LEWIS (1954, 1956), BURNETT (1956a), PAPA-ZIAN (1958), ESSER (1962, 1967), RAPER and ESSER (1964), JOLY (1964).

A. Morphogenesis

Morphogenesis in fungi involves three aspects of differentiation: that of (1) the mycelium, (2) the cells functioning in asexual reproduction and the structures which bear them, and (3) the sex organs and associated fruiting bodies. The expression of morphological characters is, of course, dependent not only upon the genotype, but also to a considerable extent upon external factors. In the present treatment we shall discuss only those results which deal with the *effect of hereditary factors on the morphogenetic process*. Reports on the role of environmental factors in fungal morphogenesis can be found in the books of LILLY and BARNETT (1951), HAWKER (1957) and COCHRANE (1958).

Genetic investigation of morphogenesis is primarily concerned with the problem of how the hereditary factors initiate and direct the physiological processes which are responsible for particular steps in differentiation. Therefore, it is first necessary to establish, with the help of deficiency mutants, how many genes take part in the expression of a specific morphological character. If more than a single gene is involved, the effect of each in the developmental sequence must be determined. Thus investigation of the biochemical effects of the individual genes becomes necessary.

In Table II-1 a number of fungi are listed in which gene-controlled morphological effects are known. This survey shows innumerable examples of *alteration of mycelial habit* through gene mutation.

Mutations may affect the following characters: differences in the growth of aerial and subterranean hyphae: the frequency and type of hyphal branching or abnormality; color of the mycelium, resulting either from the absence of a pigment or the production of a new pigment; the cyclic growth pattern, which, for example, in *Neurospora* manifests itself in a periodic production of aerial hyphae and macroconidia (STADLER, 1959).

In many of the organisms which have been investigated in detail (e.g. *Sordaria, Glomerella*) it has been shown that *morphological altera-*

tions in mycelial habit are frequently correlated with developmental changes in asexual and sexual reproductive structures (pleiotropy).

A great number of experiments with a variety of organisms (e.g. *Neurospora crassa:* MURRAY and SRB, 1962), in which attempts have been made to compensate for morphological defects by the addition of supplements of protein or nucleic acid components or vitamins, have proved fruitless. One exception is found in the work of FULLER and TATUM (1956).

A morphological mutant of *Neurospora crassa* was induced to grow *normally by the addition of inositol to the medium.* Since in normal mycelium 86% of the inositol is combined with phospholipid while in inositolless mutants only one-fifth of the normal phospholipid is present, FULLER and TATUM concluded that a relationship between differentiation of the wild type mycelium and phospholipid exists.

In none of the cases so far investigated has it been possible to relate a specific mycelial aberration to more than a single gene. Further, with the exception of the inositolless mutant, it is not known at what point in the metabolic pathway the genes responsible for mycelial morphology exercise their control. *Thus at the present time a satisfactory model of the physiological mechanism by which the genetic material controls mycelial habit is lacking.*

This is also true for differentiation involved in the production of asexual spores. Genes which affect the development of these structures are known in a number of organisms; in *Neurospora, Aspergillus,* and *Glomerella* several genes have been described each of which prevents conidial formation. Nevertheless, the relation of these genes to one another as well as to specific physiological reactions remains to be determined.

Investigations which have dealt with the *influence of the genotype on the production of sex organs and fruiting bodies* are somewhat more numerous. That a single gene mutation may interfere with the development of male or female sex organs (Table II-1) in bisexual Phycomycetes and Ascomycetes has long been known. On the basis of this observation it has been assumed that the processes necessary for differentiation of sex organs and gametes were determined by a single hereditary factor. The observations that *in certain fungi differentiation of the sex organs is induced by substances produced by the mating partner* appear to support this view.

The best known examples of the *induction of differentiation by substances from the mating partner* are the observations on the *aquatic fungus, Achlya ambisexualis and on the mold, Mucor mucedo,* which will be discussed in more detail later (p. 84 and 88ff.).

In the dioecious species, *A. ambisexualis,* the differentiation of oogonia and antheridia is induced by the interaction of manifold substances from the male and female mycelia (summarized by RAPER, 1947, 1951, 1952, 1957). Because the oospores of *Achlya* have not been observed to germinate (BARKSDALE, 1960; RAPER, personal communication), the genetic basis of these morphogenetic events is not known[1].

The differentiation and growth of progametangia toward each other in *Mucor mucedo* is similarly induced by an interaction of sexual substances

---

[1] Recently germination of *A. ambisexualis* oospores has been demonstrated (MULLINS and RAPER, 1965a, b).

Table II-1. *Gene-controlled morphological changes in the fungi (the list published by* MURRAY *and* SRB, 1962, *has been used for this compilation)*

effect of mutation	object	reference
Mycelium	*Aspergillus*	SCHWARTZ, 1928; RAPER et al., 1945; COY and TUVESON, 1964
	Colletotrichum	CHENA and HINGORANI 1950; HALDEMAN, 1950
	Coprinus	QUINTANILHA and BALLE, 1940; PRÉVOST, 1962
	Eurotium	BARNES, 1928
	Helminthosporium	STEVENS, 1922
	Mucor	BLAKESLEE, 1920
	Neurospora	LINDEGREN, 1933; WÜLKER, 1935; LINDEGREN and LINDEGREN, 1941 a, b; BARRATT and GARNJOBST, 1949; MITCHELL, 1954; FULLER and TATUM, 1956; SRB, 1957; MURRAY and SRB, 1962; SUSSMAN et al., 1964
	Penicillium	HAENICKS, 1916; SANSOME, 1947, 1949; RAPER and THOM, 1949
	Phoma	CHODAT, 1926
	Phycomyces	BURGEFF, 1914, 1915
	Phytophthora	LEONIAN, 1926; BUDDENHAGEN, 1958
	Podospora	RIZET and ENGELMANN, 1949; ESSER, 1956b
	Schizophyllum	PAPAZIAN, 1950; RAPER and SAN ANTONIO, 1954; DICK, 1960
Asexual spores and sporangia or sporophores	*Aspergillus*	SCHIEMANN, 1912; HAENICKS, 1916; CHODAT, 1926; RAPER et al., 1945; THOM and RAPER, 1945
	Glomerella	WHEELER and McGAHEN, 1952
	Neurospora	GRIGG, 1958, 1960
	Penicillium	STAUFER and BACKUS, 1956
	Phytophthora	LEONIAN, 1926
	Verticillium	PRESLEY, 1941
Sex organs and fruiting bodies	*Allomyces*	STUMM, 1958
	Alternaria	ROBERTS, 1924
	Aspergillus	MAHONY and WILKIE, 1962
	Bombardia	ZICKLER, 1934, 1937, 1952; LAIBACH et al., 1954, 1955
	Colletotrichum	BURGER, 1921; HALDEMAN, 1950
	Glomerella	WHEELER, 1954a
	Hypomyces	HANSEN and SNYDER, 1943, 1946; EL-ANI, 1954a, b
	Mucor	BLAKESLEE, 1920
	Neurospora	ARONESCU, 1933; WÜLKER, 1935; LINDEGREN and LINDEGREN, 1941a, b; DODGE, 1946; SRB, 1957; BARBESGAARD and WAGNER, 1959; HOROWITZ et al., 1960; MURRAY and SRB, 1961, 1962; FITZGERALD, 1963
	Phoma	CHODAT, 1926
	Podospora	ESSER, 1956b
	Schizophyllum	RAPER and KRONGELB, 1958; JÜRGENS, 1958
	Sordaria	ESSER and STRAUB, 1956, 1958; OLIVE, 1956; HESLOT, 1958; CARR and OLIVE, 1959
	Verticillium	TOMPKINS and ARK, 1941

43

from the two mating partners. The substances in question have been isolated. In this case it seems reasonably certain that the two mating strains differ in only a single allele (BURGEFF, 1924; PLEMPEL, 1957, 1960).

Similar problems have also been investigated in the Ascomycetes. DODGE (1920) found that *Ascobolus magnificus* only produced ascogonia or antheridia in the presence of the opposite mating type. BISTIS (1956, 1957) obtained similar results in his extensive studies on the closely related species, *Ascobolus stercorarius*. In this fungus differentiation of the ascogonial primordium can be induced by a substance from the oidia of the complementary mating type. In both cases the mating types differ in a single allele (p. 76).

Genetic and developmental studies on a number of other Ascomycetes have shown, on the contrary, that *not one, but a whole series of genes plays a role in differentiation of the sex organs and fruiting bodies*.

Species which illustrate this behavior are *Glomerella cingulata* (LUCAS et al., 1944; CHILTON et al., 1945; MARKERT, 1949, 1952; McGAHEN and WHEELER, 1951; WHEELER and McGAHEN, 1952; WHEELER, 1954a; WHEELER et al., 1959), *Sordaria fimicola* (OLIVE, 1954, 1956; CARR and OLIVE, 1959), *Sordaria macrospora* (ESSER and STRAUB, 1956, 1958; HESLOT, 1958) and *Podospora anserina* (ESSER, 1956b, 1966b).

If minor differences in the life cycles of these four species are disregarded, the investigations have led to results which are in complete agreement. These are presented in detail using *Sordaria macrospora* as an example.

In order for the entire developmental cycle of the monoecious, self-fertile *Sordaria macrospora* to run its course at least 15 non-allelic genes are necessary. Moreover, it is likely that with a more intensive search for morphological mutants an appreciably greater number of such morphologically important hereditary factors will be identified.

A mutation at a single locus blocks differentiation beyond the point in the developmental sequence where it produces its effect. If the block occurs at any time prior to spore formation, sterility results. With regard to the developmental cycle of *S. macrospora* (Fig. II-1), the morphological mutants may be arranged into a sequence on the basis of their effects.

Fig. II-1 shows that differentiation of female sex organs is controlled by five genes, two of which are responsible for the formation of the ascogonium, and three of which influence the subsequent development of the protoperithecium. Ten genes are involved in fruiting body formation: three in stages from the dikaryotic phase to ascus formation, three in the meiotic divisions and spore production, one in the linear arrangement of the ascospores, one in pigment development, and two in spore discharge. None of these genes is allelic. Multiple alleles which have been found are noted in the figure.

The gene *lu* (Fig. II-1, stage 5) suppresses the development of the black melanin pigments of the spores, but not the mechanism of their discharge. Such *lu* spores are yellow. The gene *ire* results in a non-linear spore arrangement and leads to abnormal development which may be significant in the taxonomy of the genus *Sordaria*. A gene with a similar effect has been found in *Neurospora crassa* by MURRAY and SRB (1962).

The fact that a number of non-allelic genes affect the same developmental stage does not mean that they have the same genetic effect. On the contrary, it has been shown that such genes have their effects at different points in the same developmental process.

For example, the mutants *min, pa, and s* all produce sterile fruiting bodies. Cytological investigation has shown that in the asci of *min* and *pa* strains meiosis does not occur, whereas the development of asci of the *s* strain is not blocked until after meiosis.

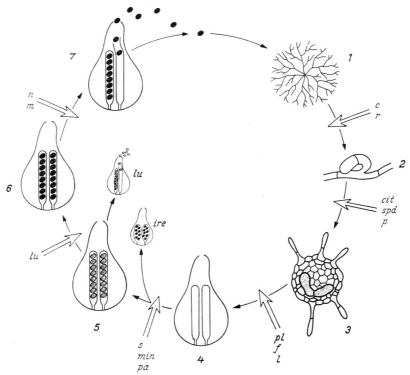

Fig. II-1. Life cycle of *Sordaria macrospora*. The significant developmental stages are: mycelium formation (1), differentiation of the ascogonium (2), transformation of the ascogonium into a protoperithecium (3), dikaryotic phase, karyogamy, and ascus formation (4), meiosis and spore formation (5), maturation of the ascospores (6), discharge of the ascospores (7). Male sex organs are not produced in this apandrous fungus (see chapter on *morphology*). The arrows with the gene symbols show at which stage of differentiation particular mutations block the developmental sequence. Further explanation in text. (Adapted from Esser and Straub, 1958)

General agreement among the observations in the genera *Glomerella, Sordaria,* and *Podospora* supports the *assumption that in fungi (at least in Ascomycetes) differentiation of sex organs and fruiting bodies is polygenically controlled.*

The work of Raper and Krongelb on *Schizophyllum commune* is pertinent in this connection. Fruiting body production in this Basidiomycete is a quantitative character varying in frequency and habit. These authors describe five different fruiting body anomalies which are all genetically controlled. The reaction norm of the individual types is dependent to a large extent upon external factors (light, nutrients, etc.). In no case has it been possible, however, by modifying the environmental factors of the mutants, to obtain a fruiting body showing the wild type character.

A clue to the *physiological effects of morphogenetic factors* is provided by complementation experiments with mutants of the Ascomycetes mentioned above. Heterokaryons of *S. macrospora* which carry nuclei with non-allelic mutations are phenotypically wild type because the individual defect in each is compensated for by the corresponding wild type allele in the other nuclear type. This becomes especially clear when mutants are used in which the sterility genes have a pleiotropic effect; i.e., in addition to blocking the developmental cycle they produce changes in mycelial morphology. The *complementation effect* here is identical to that of auxotrophic *Neurospora* mutants observed by BEADLE and COONRADT (1944) (p. 386).

The *"induction phenomenon"* in *S. macrospora* (ESSER and STRAUB, 1958) also holds promise for the physiological analysis of morphological mutants. By induction is meant the production not only of crossed perithecia but also selfed ones by a heterokaryon obtained from non-allelic *sterile* mutants.

While the asci of the hybrid perithecia contain both types of nuclei, those of selfed perithecia carry only nuclei of one or the other type. Induction is thus a type of complementation which extends beyond the individual cell. The assumption is therefore reasonable that the gene products responsible for induction (i.e. complementation), must also be present in a culture filtrate.

Despite numerous attempts in *Sordaria macrospora* it has not been possible to demonstrate that such substances diffuse into the culture medium. In this species induction appears to be an intracellular process which cannot be separated from the cytoplasm (ESSER and STRAUB, 1958).

DRIVER and WHEELER (1955) were able, by addition of a culture filtrate from a wild strain, to increase the perithecial production of a mutant of *Glomerella cingulata* which ordinarily produces very few fertile perithecia. Isolation of the substance responsible was not possible. Sterile mutants of *Glomerella* could not be induced to produce perithecia by addition of culture filtrates.

CARR and OLIVE (1959) have shown that the culture filtrate of a sterile mutant of *Sordaria fimicola* suppresses fruiting body production of the wild type. Whether or not this involves an inhibitory substance or a secondary effect such as, for example, a shift in pH, is not known. The fact that the effect of this mutant is not expressed in the heterokaryon argues against the existence of an inhibitory substance.

The essentially *negative results* of innumerable attempts *to isolate* from Ascomycetes *morphogenetically active gene products* lead to the conclusion that at least in those cases so far investigated, such substances are of high molecular weight and do not diffuse out of the hyphae. It is also possible that these compounds are very unstable and are inactivated upon extraction from the protoplast.

A number of investigators have attempted to relate specific enzymes to morphogenetic processes.

CANTINO and coworkers were able to demonstrate the relationships of several enzymes to specific steps in differentiation in *Blastocladiella emersonii* (McCURDY and CANTINO, 1960; LOVETT and CANTINO, 1960). In extensive biochemical and cytochemical investigations (TURIAN, 1960, 1961 a, b; TURIAN and SEYDOUX, 1962) discovered enzyme systems which are responsible for sexual differentiation in monoecious *Allomyces* strains (hybrids of *A. macrogynus × A. arbuscula*) and in *Neurospora crassa*. HIRSCH (1954) and WESTERGAARD and HIRSCH (1954) postulated that the phenoloxidase, tyrosinase (i.e. tyrosinase metabolism) is responsible for the production of perithecia. In both cases the experiments were performed with the wild type, and therefore the controlling genes are not known.

The studies of BARBESGAARD and WAGNER (1959) and HOROWITZ et al. (1960) can be considered extensions of HIRSCH's experiments. These authors showed that female sterile mutants of *N. crassa* had lost to a large extent their capacity to produce tyrosinase. It is doubtful, however, that tyrosinase itself is correlated with fruiting body formation, since enzyme production (p. 373) induced by the addition of aromatic amino acids does not alleviate the sterility.

A new contribution to the clarification of genetic control of the physiological basis of morphogenesis is the discovery in bacteria that the synthesis of enzymes may be controlled by cooperation between genetic factors (structural and regulator genes) and environmental conditions (presence or absence of inducers) (p. 407 ff.). Because the character of an organism is dependent to a large extent upon enzymatically controlled processes, one can assume that growth and differentiation are determined by a chain of enzymatic reactions, some of which function continuously while others are operative only under certain conditions of the cellular and extracellular environment. Whether or not this model is also applicable to the complex morphogenetic processes of multicellular organisms is yet to be seen. In any case, the model may lead to a more precise concept of morphogenesis.

Summary

1. Hereditary factors which control morphogenetic processes have been identified in a large number of fungi. Such genes often exhibit pleiotropic effects. On the one hand they influence the character of the mycelium and, on the other, they are responsible for the differentiation of the asexual and sexual reproductive structures.

2. Normal development in certain species of Ascomycetes has been shown to be determined by a series of genes. The developmental cycle is blocked at specific points by individual mutations. Non-allelic genes which produce defects complement each other in heterokaryons. The gene products which are responsible for the individual developmental steps have not as yet been identified.

3. Differentiation of sex organs in a number of Phycomycetes and Ascomycetes results from substances produced by the mating partners. With one exception (*Mucor mucedo*), it has not been possible to correlate these morphogenetic substances with specific genes.

4. A definitive model of the chain of biochemical and biophysical processes which connects the genetic information to cell, tissue, and organ differentiation is not yet available.

B. Systems of sexual reproduction

The fungi show great variability in the expression of sexual characteristics. Forms range from those showing no trace of morphological sexual
differentiation through intermediates to those which are strongly monoecious or dioecious. Therefore, it is extremely difficult to arrange the
fungi into a scheme of classification which is based on their reproductive
behavior. This difficulty is reflected in the numerous partly synonymous,
partly incorrect terms which have been applied to reproduction in the
fungi, and has led to confusion regarding the concepts of sexuality
and incompatibility which will be discussed in the following section.

As HARTMANN (1943) and WHITEHOUSE (1951 b) have already emphasized, the *essence of sexual reproduction is the alternation of meiosis with
karyogamy*. With regard to the *important consequence of these processes*,
namely, *genetic recombination*, it is of minor importance whether morphologically distinct sex organs or gametes which can be designated as
male or female are present. Thus, a clear distinction between sexual
reproduction and sexual differentiation must be made (DARLINGTON,
1937). *Sexual differentiation* involves the presence in an organism of
morphological structures which enable karyogamy to occur. This distinction is particularly significant among the fungi, since, in contrast to
other plants and animals, sexual reproduction does not necessarily involve
morphological differentiation.

Before we proceed to the genetic basis of sexual reproduction in the
fungi, it is appropriate to first analyze the concepts which have been
used in the literature in connection with sexual reproduction and to
attempt a classification of the reproductive systems of the fungi according to our current knowledge of the group. We clearly realize that such
a classification, like any taxonomic scheme, is only an approximation
of the situation as it actually exists.

I. Definitions

Among the higher plants the criterion for the two alternative reproductive mechanisms is the presence of the male and female sex organs
either in one individual (monoecism) or in two different individuals
(dioecism). A *monoecious, bisexual organism* is generally able to *fertilize
itself*, while in *dioecism* the *separation of the sexes* requires the presence
of *a male and a female plant for sexual reproduction*. The self-fertility
of monoecious species may, however, be limited by incompatibility. By
incompatibility is meant a *genotypically determined prevention of karyogamy* which is not the result of sterility defects in the gametes or their
nuclei.

When crossing relationships in a race or species are determined by
an incompatibility system, self-fertilization in the same individual as well
as cross-fertilization between different individuals may by prevented. In

the first instance we speak of self-incompatibility; in the latter, cross-incompatibility. Because both cases are generally an expression of the same mechanism, the term *incompatibility* applies to both situations.

In the literature, the term *self-sterility*, introduced by Jost (1907), is widely used instead of *incompatibility* (Stout, 1916). These terms are not equivalent, however, because self-sterility applies only to self-incompatibility. Accordingly, cross-incompatibility and cross-sterility should be the same. This is incorrect, however, since in contrast to true sterility, the term cross-incompatibility does not denote the failure of karyogamy resulting from defective zygotes or developmental defects in the sex organs or gametes. The terms *auto-inconceptibilité* (Sirks, 1917) and *Parasterilität* (Brieger, 1930), introduced to designate this phenomenon, have not come into general usage.

Since all flowering plants possess male and female sex organs whatever their arrangement, every higher plant can be classified as either monoecious or dioecious. When monoecious forms fail to produce seed, it is easy to determine, through appropriate crosses, whether incompatibility or sterility is involved. Among thallophytes, particularly the fungi, classification into a monoecious or a dioecious category is difficult because in many species the sex organs are absent or are not morphologically distinguishable as male and female. In order to circumvent this difficulty the concepts of homothallism and heterothallism have been employed (Blakeslee, 1904a, b). Blakeslee based these concepts on developmental relationships of the Mucoraceae, a group in which it is not possible to recognize sex differences among the gametangia.

Although differences between the gametangia of strains of the same species of mucors have been observed frequently (Burgeff, 1915; Orban, 1919; Wesendonck, 1930; Plempel and Braunitzer, 1958). Burgeff was able to show that this is not a matter of sexual differentiation, but rather, segregation of morphological characters among the progeny.

Blakeslee designated those species as *homothallic* which produced *zygospores* from a culture arising from *a single spore*. Species whose *life cycle* is completed only after *crossing two morphologically indistinguishable mycelia of different mating potential*, each having arisen from a single spore, were called *heterothallic* (for example, *Rhizopus nigricans*). The two types of a heterothallic species, distinguishable only through their crossing behavior, were designated as *plus* and *minus*. The introduction of this terminology for haplonts, in contrast to the use of monoecious and dioecious for flowering plants was based on the assumption by Blakeslee that the *plus* strain represented the female and the *minus* the male sex in a reduced form.

In order to test this hypothesis, Satina and Blakeslee (1928, 1929) undertook extensive tests between homothallic and heterothallic mucors. The self-fertile, homothallic species had micro- and macrogametangia and were therefore considered to be hermaphrodites with male and female sex organs. These workers found that the "microgametangia" of *Zygorhynchus moelleri*, *Absidia glauca*, and a species of *Dicranophora* reacted only with the *plus* mating type and correspondingly the "macrogametangia" only with the *minus* mating type. The conclusion based on these experiments that the *plus* type is female and the *minus* type is male did not prove correct, however, since *Zygorynchus heterogamus* behaved exactly opposite; the strains with microgametangia fused with the *minus* testers and the macrogametangial strains with the *plus* testers.

Because it has not yet been possible to refer the physiological differences between *plus* and *minus* strains of mucors to a sexual difference in the sense of dioecism, we do not speak of *plus* and *minus* sexes, but rather of *plus* and *minus mating types*.

Subsequent to BLAKESLEE's definition, all species of thallophytes which required two mating types for sexual reproduction were accordingly called heterothallic. This led eventually to the concept, adopted by some texts, that in general in the thallophytes the terms homothallic and heterothallic rather than monoecious and dioecious should be used (e.g. HARDER et al., 1962). In our opinion this is erroneous. Included in the concept of heterothallism are actually two different reproductive systems: 1. the truly dioecious type which occurs only occasionally among the fungi (p. 81 ff.) and the self-incompatible monoecious type which is relatively frequent in the group (p. 54 ff.) (see Fig. II-2). This inconsistency can be explained by the fact that at the time this classification was established the life histories of many fungi were not accurately known. Nevertheless, when a species showed two mating types, it was designated as heterothallic. A similar confusion existed for a long time with regard to homothallism. There is a series of apparently self-fertile Ascomycetes and Basidiomycetes (e.g. *Podospora anserina* and *Psalliota campestris* f. *bispora*) whose ascospores and basidiospores, respectively, possess two nuclei which represent different meiotic products (p. 152 ff.). When the two nuclei of a spore are heterogenic for mating type, a heterokaryotic mycelium is produced. The self-fertility of such a heterokaryon is only simulated, for in actuality the species in question are self-incompatible. Accordingly, they are often segregated from other homothallic fungi as so-called secondarily homothallic forms. There is no dearth in the literature of objections to the concepts of homo- and heterothallism and of suggestions for other classification criteria (CORRENS, 1913, 1928; HARTMANN, 1918, 1929, 1943, 1956; KNIEP, 1929a, b; WHITEHOUSE, 1949a; KORF, 1952; BURNETT, 1956a). WHITEHOUSE has suggested that heterothallism be subdivided into a physiological (incompatibility) and a morphological type (dioecism). Such a system does not, however, allow classification of the mucors (see Fig. II-2).

These Phycomycetes are not dioecious as far as those species with two mating types are concerned, since their gametangia are isomorphic. They cannot be designated as self-incompatible, however, because they lack differentiation of male and female structures.

KORF and HARTMANN recommend abandonment of the terms, homo- and heterothallism. HARTMANN emphasizes in particular the absurdity of applying these terms to animals among which parallel phenomena occur. BURNETT proposes a fully new, but extremely complex terminology.

Because of these difficulties in classifying the fungi according to their reproductive behavior, we have attempted (ESSER, 1959b) to utilize a *criterion for classification* which, analogous to that of the higher plants, employs *the concepts of monoecism and dioecism*. This demands, however, that the meaning of these terms be broadened (see also LEWIS, 1954) to take into account the incompleteness or absence of sexual differentiation which is widespread among the fungi. Since the essence of sexual reproduction involves alternation of meiosis and karyogamy, we feel justified in disregarding completely in the definition of monoecism and dioecism the presence or absence of sex organs and in defining both these terms on the basis of physiological rather than morphological criteria. As a *criterion for sex the capacity of an organism to contribute one or both nuclei to karyogamy is used.* By *monoecious* we mean an individual which can function both as a *donor* and a *recipient of a nucleus.* An individual which possesses *only one or the other potential,* we call *dioecious.*

For a botanist who is accustomed to associate monoecism and dioecism with the presence of male and female sex organs, the proposed generalization may not be acceptable. It is pertinent to emphasize, that between fungi which fail to produce sex organs and higher plants which possess sex organs, there is no fundamental difference; in both cases the sexual mechanism is the same. Moreover, we are reluctant to introduce new terms and thereby increase the confusion resulting from the already existing terminology. In this connection one might, for example, consider the introduction of the terms, "monogamous" and "digamous". With which category is each of these to be associated? From the words themselves, monogamous appears equivalent to monoecious, digamous to dioecious; from their meanings, the reverse might be considered correct.

By definition, then, the *monoecious* fungi include:

1. All species which produce *male and female sex organs on the same mycelium.* Such occur among the Euascales and the Uredinales.

2. All species *without sex organs* in which *each mycelium can function as a nuclear donor as well as a nuclear acceptor.* Representatives of this group are found in all fungal taxa.

Nuclear exchange between different mycelia was first shown clearly for the Holobasidiomycetes by BULLER (1931) and QUINTANILHA (1939); HOLTON (1942) discovered the phenomenon in the Ustilaginales.

When a *monoecious form* is unable to reproduce sexually without a partner, *incompatibility* is said to occur.

Two groups of fungi are designated as *dioecious:*

1. All species in which *mycelia bear either female or male sex organs (morphologically dioecious).* Only a few Phycomycetes and Ascomycetes belong to this group.

2. All *species with two mating types* and in which *zygote production* follows the fusion of *isogametes or isogametangia.* Innumerable Protoascomycetes and mucors belong to this category. The two mating types of such species (e.g. *Saccharomyces cerevisiae, Mucor mucedo*) cannot be distinguished morphologically and in the sexual process show no nuclear exchange, but simply a fusion of uninucleate or multinucleate cells. We are thus unable to designate them either as morphologically dioecious forms or as self-incompatible monoecious types. Since their sexual behavior is determined purely through the physiological difference of the two mating types, we are calling these *physiologically dioecious.*

This classification is to be regarded as an expedient which is comparable to the grouping together of all the fungi in which sexual reproduction is unknown into the Fungi Imperfecti. The artificial grouping of the physiologically dioecious forms serves to clarify the situation.

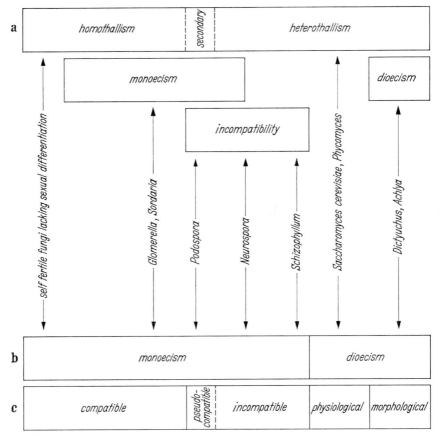

Fig. II-2a—c. Schematic representation of reproductive systems among the fungi. (a) Compilation of the classifications based on homothallism and heterothallism, which are currently in use. (b, c) Proposed classification based on a broadened concept of monoecism and dioecism. The pseudocompatible fungi belong basically to the incompatible forms; they are separated from the latter by a broken line. The extent of each rectangle is not correlated with the frequency of that particular pattern in nature. For a better understanding of the proposed classification a few well-known species have been included under their proper designation. For further information, see text

In Fig. II-2 the classification of reproductive systems of the fungi commonly used are compared with the system here proposed. As examples, we have included some of the best known objects of genetic research. One advantage of this system is the absence of overlap. Each fungus can be classified unequivocally according to whether it is monoecious (compatible or incompatible) or dioecious (morphologically or physiologically).

The following points are worth noting:

1. A difficulty of the old system of classification for understanding fungal reproductive systems has been the overlap of monoecism with homothallism-heterothallism on the one hand and with incompatibility on the other. This complication is absent in the proposed system.

2. By broadening the concepts of monoecism and dioecism the gap between them which existed in the old system is eliminated. Each species of fungus can be classified as either dioecious or monoecious without question.

3. By establishment of the category, "physiological dioecism", fungi such as the mucors may be classified without doubt. According to the old scheme, it is not possible to distinguish whether these heterothallic species are characterized by sexual incompatibility or by a reduced morphological dioecism (see Fig. II-2).

Summary

1. Classification of the reproductive systems of fungi as homothallic and heterothallic has led to manifold difficulties. For example, dioecious as well as incompatible species have been called heterothallic. In order to avoid this and other complications, a new basis for classification is proposed, which eliminates overlap and permits every fungus to be classified without contradiction.

2. As in the flowering plants, we are establishing for the fungi monoecious and dioecious categories. Each individual of a monoecious fungus can function both as a nuclear donor as well as recipient. Each individual of a dioecious form can perform only one or the other of the two functions. We believe that this extension of the concepts of monoecism and dioecism is meaningful, since the principle of sexual reproduction is based on regular alternation of meiosis and karyogamy and is not dependent upon the presence or absence of sex organs.

3. Monoecious fungi are either compatible or incompatible. This depends upon whether, for the completion of their normal life cycle, a single or two morphologically indistinguishable types of mycelia are required.

4. Apart from morphologically dioecious forms, physiologically dioecious types occur which fail to produce either male or female sex organs. All species whose zygotes arise without nuclear exchange and by fusion of isogametes or gametangia are included in this category.

II. Reproductive systems of monoecious forms

All classes of fungi include numerous monoecious forms which are self- and cross-compatible. Incompatible species are known with certainty only among the Ascomycetes and Basidiomycetes. Sexual incompatibility has been demonstrated in some 50 species of Ascomycetes and 380 species of Basidiomycetes.

Tables have been published in a number of cases in which compatible and incompatible species of Eumycetes have been classified together: WHITE-HOUSE, 1949a (Euascales), WHITEHOUSE, 1949b (Holobasidiomycetes), WHITEHOUSE, 1951b (Ustilaginales), QUINTANILHA and PINTO-LOPES, 1950 (Holobasidiomycetes), CRAIGIE, 1942 (Uredinales), NOBLES et al., 1957 (Holobasidiomycetes), ESSER, 1967 (Euascales, Holo- and Phragmobasidio-mycetes).

1. Compatible species

Compatible monoecious species which have been used for genetic experiments have been almost exclusively Ascomycetes. In these forms, when two self-fertile mycelia are mated, it is not possible, without further analysis, to distinguish, between fruiting bodies which have arisen from crossing and those which have resulted from selfing. This difficulty may be overcome in two ways:

1. By crossing more or less sterile mutants which form only a few fruiting bodies in the zone of contact.

2. By using spore color marker genes for the two mating partners. The asci resulting from crosses can be distinguished in the fruiting bodies by segregation of the spore color gene; in the asci from selfing the spores are all of one color.

The first method was used by LUCAS et al. (1944), EDGERTON et al. (1945) and later by WHEELER and his coworkers (see review of WHEELER, 1954a) in *Glomerella cingulata* and by GREIS (1941) in *Sordaria fimicola*.

Glomerella was the first self-fertile monoecious fungus to be used for genetic investigations (EDGERTON, 1912, 1914) as WHEELER (1954b) has pointed out in a brief survey of the literature.

Spore color genes and sterility factors were used by OLIVE (1956), CARR and OLIVE (1959), HESLOT (1958), ESSER and STRAUB (1958), ITO (1960) for their genetic experiments with *Sordaria fimicola* as well as *Sordaria macrospora* (see illustration on cover and Fig. IV-3).

These investigations have shown, as described in the section on *morphogenesis*, that a number of genes are responsible for differentiation of mycelium and sexual reproductive structures in the compatible monoecious fungi. No overall genetic control of this reproductive pattern is known.

2. Incompatible species

Two types of sexual incompatibility, homogenic and heterogenic, are known among the fungi (ESSER, 1959b, 1961, 1962, 1966a). The main difference between the two is found in the genetic mechanism responsible for the incompatibility. In *homogenic incompatibility* zygote formation is prevented when two *mating partners carry the same incompatibility allele*. When two incompatible mating partners carry *different alleles at all incompatibility loci*, the *incompatibility is heterogenic*. Table II-2 summarizes the distribution of sexual incompatibility among the Eumycetes; unequivocal evidence of incompatibility in the lower fungi is still lacking. The cases shown in the table are discussed in the following sections.

Table II-2. *Systems, mechanisms and distribution of sexual incompatibility in fungi*
The distribution of incompatibility by numbers of species is taken from a list to be published elsewhere. (ESSER, 1967)

System	homogenic						heterogenic		
Mechanism	bipolar			tetrapolar		unknown	semi-incompatibility	unknown	
Number of factors	1			2 in part genetically complex			2		
Occurrence	Euasco-mycetes	Holo-basidio-mycetes	Phragmo-basidio-mycetes	Holo-basidio-mycetes	Phragmo-basidio-mycetes	Basidio-mycetes	Euasco-mycetes	Euasco-mycetes	Basidio-mycetes
Number of species	46	93	42	172	6	71	1	2	4
Multiple alleles	unknown	21	4	35	2	—	unknown	—	—

a) Homogenic incompatibility

Homogenic incompatibility was discovered independently by BENSAUDE (1918) in *Coprinus fimetarius*, by KNIEP (1918, 1920) in *Schizophyllum commune* and by DODGE (1920) in *Ascobolus stercorarius*. In subsequent years KNIEP, his coworkers and other groups of investigators demonstrated that sexual incompatibility occurs widely among the Eumycetes; it is found in 90% of the Holobasidiomycetes, in numerous Phragmobasidiomycetes, and in the Ascomycetes (Table II-2 and the lists cited on p. 54). Insofar as genetic data are available, the incompatibility is almost entirely of the homogenic type.

Homogenic incompatibility is controlled by either *one or two genetic factors*. Since in the former case two, and, in the latter, four mating types are present, they are referred to as *bipolar* and *tetrapolar mechanisms*, respectively.

Bipolar mechanism

In the simplest case the locus controlling incompatibility exists in two allelic forms. These are generally designated as *plus* and *minus* and are responsible for the two mating types.

The two mating types in *Neurospora crassa* are usually designated as *A* and *a*, although LINDEGREN (1933) had originally introduced the

55

plus and *minus* designation. The use of *plus* and *minus* seems preferable, because A/a implies a dominance of A over a, which has never been demonstrated. The symbols a and α have been adopted in *Saccharomyces* genetics to designate the two mating types (p. 56).

By definition in homogenic incompatibility *mycelia of the same mating type are always self- and cross-incompatible* $(+ \times +; \; - \times -)$. *Karyogamy* occurs only *between two different mating types* $(+ \times -)$. Since each mating type may function as nuclear donor as well as acceptor, reciprocal crosses are possible between *plus* and *minus* mycelia (Fig. II-3).

Incompatible fungi may be fruited by genome doubling if the diploid nucleus is heterozygous for mating types (FRANKE, 1962 in *Podospora anserina*). The same result was obtained in *Neurospora crassa* through disomy (MARTIN, 1959).

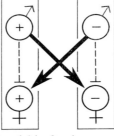

Fig. II-3. The genetic basis of bipolar incompatibility. The rectangles represent two homokaryotic individuals which are of different mating types. In each mating type a male and a female nucleus are indicated. The hermaphroditic character of these individuals is represented by sexually different nuclei, because this scheme applies to both the Ascomycetes (with sex organs) and the Basidiomycetes (partly without sex organs). The broken lines which are blocked indicate self-incompatibility (or cross-incompatibility between different individuals of the same mating type). Zygotes can only be derived from nuclei of different mating types (heavy arrows) and reciprocal crosses are possible $(\female + \times \male -; \; \female - \times \male +)$

As is apparent from Table II-2, homogenic incompatibility in the Euascomycetes is determined exclusively by the *bipolar mechanism*. All self-incompatible members of this group produce female sex organs. Depending upon the origin of the male nucleus, gametangiogamy, oogamy, or somatogamy are prevented through incompatibility, in most instances *prior to plasmogamy*.

Gametangiogamy. The ascogonium, which is usually provided with a trichogyne, fuses with a multinucleate male gametangium (plasmogamy). After transfer of the male nuclei to the multinucleate ascogonium, the male and female nuclei migrate in pairs into the ascogenous hyphae (dikaryotic phase) which develop from the ascogonium. Fusion of the nuclei occurs in the young ascus (karyogamy) (e.g. *Ascobolus magnificus*: DODGE, 1920).

Oogamy (spermatization). Fertilization of the ascogonium occurs with aid of the trichogyne by means of uninucleate gametes (= spermatia), which are produced in gametangia (= spermagonia). Following fusion of the trichogynal tip and spermatium (plasmogamy), the nucleus of the spermatium migrates through the trichogyne into the ascogonium which is generally uninucleate. Ascogenous hyphae develop from the ascogonium by conjugate nuclear divisions; their ends become transformed into asci (e.g. *Podospora anserina*: DOWDING, 1931; AMES, 1932; *Bombardia lunata*: ZICKLER, 1952).

Oogamy (dikaryotization): The trichogyne of an ascogonium may undergo plasmogamy with any viable mycelial structure that contains a nucleus (oidium, conidium, germ tube of an ascospore, hyphal tip, etc.). After penetration of the "male" nucleus the development is the same as that following spermatization (e.g. *Ascobolus stercorarius*: BISTIS, 1957; *Neurospora sitophilia*: BACKUS, 1939). According to BISTIS and RAPER (1963), plasmogamy may possibly occur between activated oidia and trichogynal tips of the same mating type of *Ascobolus stercorarius*: this does not lead to karyogamy, however.

Somatogamy: After fusion of *plus* and *minus* hyphae (plasmogamy) a nuclear exchange between mating partners takes place. Heterokaryotic hyphae, on which ascogonia develop, arise in the zone of contact; the ascogonia contain the *plus* and *minus* nuclei (dikaryotic phase). Subsequent development corresponds to that of the spermatization type. *Gelasinospora tetrasperma* is the only example of sexual incompatibility known for this type of fertilization (DOWDING, 1933). Hyphal fusion does not take place between imcompatible partners.

Among the Basidiomycetes bipolar incompatibility is frequently found in the Phragmobasidiomycetes (Table II-2). Since the Holobasidiomycetes do not possess sex organs, plasmogamy is somatogamous. Hyphal fusion between different mating types allows nuclear exchange. After lysis of cross walls and nuclear migration the monokaryotic cells become dikaryotic and form clamp connections (LANGE, 1966). A mycelium that exhibits clamp connections is generally considered dikaryotic. Clamp connection formation and conjugate division assure the equal distribution of the *plus* and *minus* nuclei until karyogamy. In contrast to the Ascomycetes, plasmogamy and nuclear exchange in the Holobasidiomycetes is possible between incompatible partners. In such a case either no clamp connections or incompletely developed ones form. Thus the presence of *clamp connections* serves as a *criterion for compatibility*.

A few cases are known in which the donor nucleus migrates through the acceptor mycelium following somatogamy and induces the formation of clamp connections at various sites (BULLER, 1931).

In the Ustilaginales, which also lack sex organs, fertilization occurs through copulation of *plus* and *minus* sporidia. Each sporidium may donate a nucleus to, as well as receive one from, its mating partner (HOLTON, 1942). Here the *criterion of compatibility is the capacity of the dikaryon to produce infection* after sporidial fusion.

The sexual behavior of the *Uredinales* is comparable to that of the Ascomycetes. Pycniospores (spermatia) are considered male gametes, and aecial primordia, female sex organs (LAMB, 1935; MITTER, 1936). Since the mycelium from a single aeciospore is always bisexual (ALLEN, 1932), reciprocal crosses between the *plus* and *minus* mating types are possible. The *production of aeciospores* serves as the *criterion for compatibility*.

Tetrapolar mechanism

Tetrapolar incompatibility has been demonstrated only in species of the Holobasidiomycetes and Ustilaginales (Table II-2). Two genes, that are generally unlinked and that are designated as *A* and *B*, are responsible for this mechanism. The individual alleles of these two loci are indicated by numerical subscripts. In the simplest case four mating types occur, $A_1B_1, A_2B_2, A_1B_2, A_2B_1$. The *relationships between the four self-incompatible mating types* are governed by the *rule* formulated by KNIEP (1920): *neither of the allelic pairs can unite to form a homozygote at karyogamy.*

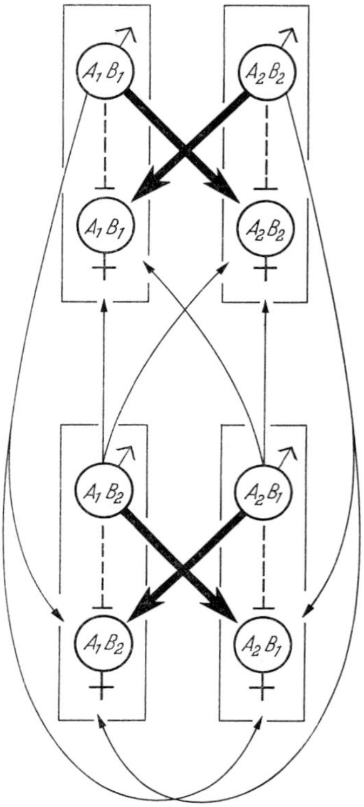

Let us first consider the Holobasidiomycetes. Of the six different crosses (see Fig. II-4) which are possible between the four mating types, only two are compatible, namely the reciprocal combinations, $A_1B_1 \times A_2B_2$ and $A_1B_2 \times A_2B_1$. No others produce zygotes. Detailed genetic and developmental studies on *Schizophyllum commune* (PAPAZIAN, 1950; RAPER and SAN ANTONIO, 1954; PARAG and RAPER, 1960) and *Coprinus lagopus* (SWIEZYNSKI and DAY, 1960a, b) have shown that the incompatible combinations do not behave uniformly. On the basis of these investigations the crossing relationships shown in Fig. II-4 fall into the following categories (RAPER, 1961a):

1. Compatible: Mycelia with different A and B factors fuse and exchange nuclei reciprocally. A dikaryon with clamp connections forms (the four combinations that are indicated by heavily drawn arrows).

Fig. II-4. The genetic basis of tetrapolar incompatibility in Holobasidiomycetes. Mode of presentation like Fig. II-3. Nuclei with different incompatibility factors are compatible (heavily drawn arrows) and form dikaryons. Nuclei with like A and B factors (lightly drawn arrows) are hemi-compatible and form heterokaryons. Nuclei with like incompatibility factors are incompatible

As in the Holobasidiomycetes which exhibit bipolar incompatibility, the donor nuclei are in a position to penetrate and migrate throughout the acceptor mycelium following dikaryon formation (BULLER, 1941; BRODIE, 1948; FULTON, 1950; KIMURA, 1954b; SNIDER and RAPER, 1958; TERRA, 1958; SWIEZYNSKI and DAY, 1960b; SNIDER, 1963a, b). The rate of migration depends on the genotype of the two mating partners and on the culture conditions (PRÉVOST, 1962).

It is not yet clear whether a cytoplasmic exchange accompanies nuclear exchange when a dikaryon is formed (p. 459). It is certain, however, that normal clamp connection formation occurs only in the presence of both nuclei. Dikaryons from which one of the nuclear components was removed (HARDER, 1927) lost the capacity to produce clamp connections after several dozen cell generations (p. 458).

2. Hemi-compatible-A: Nuclear exchange and migration occurs between mycelia identical in *A* but different in *B* factors. No clamp connections form, however. Fruiting bodies rarely develop on these sparsely growing heterokaryotic mycelia (four of the eight combinations that are indicated by lightly drawn arrows).

QUINTANILHA (1935) mentions the formation of fruiting bodies on a heterokaryon of *Coprinus fimetarius* with identical *A* factors; they resulted from a mutation of one of the *A* factors. SWIEZYNSKI and DAY (1960a) later confirmed these observations on the same material.

3. Hemi-compatible-B: Nuclear exchange but no migration takes place between mycelia with different *A* but identical *B* factors. The heterokaryotic mycelium is found in the zone of contact between the two mating partners and exhibits defective clamp connections. The hook does not fuse with the subterminal cell. Fruiting bodies do not generally form (four of the eight combinations that are shown by the lightly drawn arrows).

The zone of contact between the hemi-compatible *B* mycelia is called a barrage, because nuclear migration fails to occur and the heterokaryon that forms only in this zone exhibits a macroscopic linear barrier between the partners.

It has long been known that on heterokaryons with common *B* factors occasional fertile fruiting bodies eventually arise (BRUNSWICK, 1924; OORT, 1930; VANDENDRIES and BRODIE, 1933; QUINTANILHA, 1935; FULTON, 1950). QUINTANILHA first analyzed this phenomenon through tetrad analysis. He found that the number of such "illegitimately" formed fruiting bodies in *Coprinus fimetarius* depends upon the age of the monokaryons which are crossed. When he used freshly germinated mycelia, he always obtained heterokaryons with fruiting bodies. Since the distribution of both types of nuclei ($A_1 B_1$ and $A_2 B_1$) in the heterokaryons was irregular, different tetrad types were observed in each fruiting body: Most tetrads were biparental with a 1:1 segregation for the two parental types, but uniparental tetrads with four $A_1 B_1$ or four $A_2 B_1$ spores also occurred. Cytological analysis has shown that in the basidia of the biparental tetrads a normal meiosis follows karyogamy. In the formation of the uniparental tetrads, however, meiosis fails to occur and a haploid nucleus gives rise to the spores (apomictic spore development). In most cases the mononucleate basidia degenerate before or after the first nuclear division. This explains the low frequency of uniparental tetrads in the fruiting bodies.

Twenty-five years later the same problems were attacked by PARAG and RAPER (1960) in *Schizophyllum commune*. In this organism segregation of marker genes also revealed that meiosis precedes spore formation. In contrast to the findings of QUINTANILHA, however, the breakdown of incompatibility involved spontaneous mutation of one of the *B* factors with one exception. By mustard gas treatment of a "common *B* heterokaryon", PARAG (1962a) obtained two additional *B* factor mutants. Test crosses showed that the mutational site in both cases was within the genetic region of the *B* factor. Because the two mutants were compatible with 55 different *B* factors, PARAG concluded that the *B* function was suppressed through mutation, and thereby tetrapolar incompatibility was transformed into bipolar incompatibility. SWIEZYNSKI and DAY (1960a) have similarly explained fruiting body formation in hemi-compatible heterokaryons of *Coprinus lagopus*.

4. Incompatible: Plasmogamy and nuclear exchange occur in the zone of contact between mycelia with identical *A* and *B* factors, although only to a limited extent. Nuclear migration and clamp connection formation are absent, however. No fruiting bodies develop. These four combinations are indicated in Fig. II-4 by the blocked arrows between nuclei of individual mating types.

RAPER and RAPER (1964) recently showed that sexual incompatibility in *S. commune* is not determined exclusively by the two incompatibility factors *A* and *B*, but also by modifying factors. They found nine mutants, involving at least three loci (not linked with the *A* und *B* factors), which influence heterokaryosis between the particular combinations of the four mating types. All nine mutants increase the viability of the heterokaryons with common *A* factors and induce the production of false clamp connections. Eight of these affect the normal development of dikaryons arising from compatible combinations. False clamp connections also develop in these mycelia so that the mechanism of nuclear distribution is disturbed and formation of fruiting bodies is inhibited.

Table II-3. *Survey of phenocopies of dikaryons and heterokaryons observed in Schizophyllum and Coprinus which are controlled by gene mutation or disomy*

Mutations of the incompatibility factors are designated by "*x*"; mutations at other loci by "*M*". (From RAPER and ESSER, 1964, supplemented.)

type of mycelium	wild type genotype	genotype of the phenocopy altered by:			
		mutation	object	disomy	object
Dikaryon	$(A_1B_1 + A_2B_2)$	$A_1\text{-}MB_x$ or A_xB_x	(S. commune)	$A_1B_2B_1B_2$	(S. commune)
Hetero-karyon with common A factors	$(A_1B_1 + A_1B_2)$	A_1B_x	(S. commune)	$A_1B_1B_2$	(S. commune, C. fimetarius)
Hetero-karyon with common B factors	$(A_1B_1 + A_2B_1)$	A_xB_1 $A_1\text{-}MB_1$	(C. lagopus) (S. commune, C. lagopus)	$A_1B_2B_1$	(S. commune)

The *criterion for incompatibility in the Holobasidiomycetes* is the *failure of karyogamy to occur*. The regular occurrence of nuclear exchange and partial clamp connection formation in hemi-compatible combinations appears to be the same phenomenon observed by VANDENDRIES (1923) and BRUNSWICK (1924) and designated by the latter as "break-through copulation".

In exceptional cases homokaryons may also assume the habit of dikaryons or heterokaryons (Table II-3). Such phenocopies arise either as a result of suppressor-like mutations of the incompatibility factor or other genes, or of disomy (PRUD'HOMME and GANS, 1958; RAPER and OETTINGER, 1962).

A number of species of Ustilaginales (Table II-2) possess a tetrapolar mechanism similar to that of the Holobasidiomycetes. The work of BAUCH (1930, 1931, 1932a, b, 1934) provided the foundation for the investigation of ROWELL (1955), whose results led to the compatibility pattern shown in Table II-5. KNIEP's *rule*, namely, that only strains with *different incompatibility factors can produce a zygote*, applies here also. The difference from the Holobasidiomycetes is a functional one (compare Fig. II-5). Hemicompatibility exists only when the B factors are identical; however, heterokaryons of this type are not infectious. With

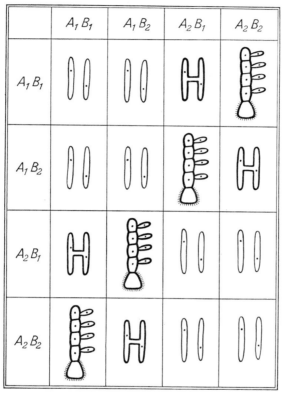

Fig. II-5. The genetic basis of tetrapolar incompatibility in Ustilaginales. In incompatible combinations there is no fusion of sporidia, which are shown in the figures as extended cells. Hemicompatibility is characterized by plasmogamy and compatibility by germinating smut spores

incompatibility (identity of A factors or A and B factors) *plasmogamy fails to occur.*

BAUCH (1934) used the genetic symbols A and B in the opposite sense; he designated as A those factors which blocked the developmental cycle after plasmogamy and as B those which prevented plasmogamy. This nomenclature was not adopted by other authors, however (see review of WHITEHOUSE, 1951 b).

Since the two incompatibility factors of the Ustilaginales, when homozygous, interfere with the life cycle at different stages (the A gene before and the B gene after plasmogamy), HARTMANN (1956) attempted to explain incompatibility in this group with a bipolar mechanism which is controlled only by the "copulation-blocking" A factor. The B gene, according to HARTMANN, is a lethal factor which merely interferes with development. This interpretation of the B factor does not recognize its true nature. The B factor inhibits development as all incompatibility factors do; whether or not this inhibition occurs before or after plasmogamy is relative and not a fundamental distinction. The same difference is also found between the incompatibility factors of the Ascomycetes and Holobasidiomycetes (see above). In general the former blocks before, the latter after plasmogamy. The B factor also can not be considered a lethal factor because its effect is not expressed in monokaryotic cells, but only in crosses involving specific alleles. In our opinion, therefore, both A and B are true incompatibility factors.

An interpretation similar to that of HARTMANN's was proposed by WHITEHOUSE (1951 b). In his view, the B factor is responsible not for incompatibility, but for pathogenicity. It can be argued, however, that pathogenicity is an essential aspect of the normal development of an obligate parasite. Any block of this function is at the same time an interruption of its life cycle. Genes which induce such a physiological block are by definition incompatibility genes.

We agree with WHITEHOUSE who, in his critical survey of the literature (1951 b), called attention to the obscure and inconclusive results of earlier investigations on "multipolar" incompatibility in rusts. Clarification of these disputed cases calls for a repetition of the experiments employing cytological methods, as ROWELL (1955) used, for example, to demonstrate tetrapolar incompatibility in *Ustilago zeae*.

Multiple alleles of incompatibility loci

KNIEP (1920, 1922) discovered *multiple alleles of incompatibility factors*. In time it became apparent that the phenomenon occurred widely *among bipolar as well as tetrapolar incompatible Holobasidiomycetes* (Table II-2). Multiple alleles are *also* known in two cases *for the B factor of the Ustilaginales* (*Ustilago longissima*, BAUCH, 1934; *U. zeae*, ROWELL, 1955). The existence of multiple alleles at the A locus of the rusts is disputed (WHITEHOUSE, 1951 b). No multiple alleles are known as yet for the incompatible species of the Euascales and Uredinales (WHITEHOUSE, 1949b; NELSON, 1957).

If multiple alleles occur in a species, more than two mating types exist. In such a case the designation a is used for the incompatibility locus in the *bipolar* mechanism, instead of *plus* and *minus*. The individual alleles are indicated by a subscript, e.g. a_1, a_2, a_3, ..., a_n. Correspondingly, for the *tetrapolar* mechanism the designations $A_1, A_2, A_3, ..., A_n$ and $B_1, B_2, B_3, ..., B_n$ are used.

The number of alleles known for each locus varies, and is, of course, dependent upon the degree to which a particular species has been analyzed. On the basis of the data available at that time, WHITEHOUSE (1949b) estimated the number of alleles possible for each locus of a tetrapolar Basidiomycete to be 100.

Roshal (1950) isolated 23 A und 21 B alleles from the progeny of 12 fruiting bodies of *Schizophyllum commune* collected in an area of about 28 hectares. Eggertson (1953) investigated 24 fruiting bodies of *Polyporus obtusus* collected in a limited area. He found a total of 39 alleles for the two factors.

The extensive researches of Raper et al. (1958b) with 114 homokaryotic strains of *Schizophyllum commune* obtained from different localities on five continents revealed 96 alleles at the A and 56 alleles at the B factor.

Assuming a random distribution of the two factors within the total population and an equal frequency of the alleles in both series, a total of 339 A alleles (limits of 217 and 562 with $P = 0.05$) and 64 B alleles (limits of 53 and 79 with $P = 0.05$) were estimated by statistical means.

Because investigations of this kind have not been carried out on Gastromycetes, we are dependent upon older data (Fries and Trolle, 1947; Fries, 1948) for an estimate of the number of incompatibility alleles in this group that is on the order of ten.

Among the spores of a fruiting body of a bipolar incompatible species there are only two different incompatibility alleles (e.g. a_2 and a_7) which with occasional exceptions segregate in a $1:1$ ratio. Correspondingly, in a tetrapolar species one finds only two alleles which occur in four different combinations, e. g. A_3B_9, A_1B_{10}, A_3B_{10}, A_1B_9. The *sexual compatibility of the innumerable mating types which exist as a result of multiple alleles* follows the basic rule of homogenic incompatibility: *two strains are compatible only if they are heterogenic for all incompatibility factors.*

Structure of the incompatibility loci

Kniep (1923) considered the problem of the origin of multiple alleles in incompatible fungi. His assumption that new incompatibility alleles arise through spontaneous mutation was reasonable, because new alleles were only rarely found among the progenies of crosses between incompatible strains. Papazian (1951) pointed out that spontaneous mutation was not the only explanation for the origin of new alleles. He showed through tetrad analysis in *Schizophyllum commune* that "new" alleles of the A locus may arise pairwise as a rare event during meiosis. Disregarding the B factors this can be diagrammed as follows:

$$A_1 \times A_2 \to \text{Meiosis} \to A_1, A_2, A_x, A_y.$$

These observations led to the conclusion that the A locus is complex, consisting of at least two genetic subunits. The two subunits act together to produce the A phenotype; they are very closely linked, because only a small percentage of recombination occurs between them. When a reciprocal recombination does occur, units with altered physiological properties arise, namely, the new factors A_x and A_y.

The extensive researches of Raper and his associates on the same organism (Raper, 1963; Vakili, 1953; Raper and Miles, 1958; Raper et al., 1958a, c, 1960; RAPER, 1961a) have not only confirmed the results of PAPAZIAN, but extended them and filled in details. On the basis of his experiments RAPER developed the following concept of the constitution and manner of action of the incompatibility factors:

Both the A and B factors of S. commune are composed of at least two subunits, called *alpha* and *beta;* both involve *mutational sites that may recombine through crossing over.*

Nine different A_α and twenty-six A_β factors have been identified. Crossing over values between A_α and A_β range from 1 to 23%. Three each of the B_α and B_β are known; these are two map units apart. No recombinants have been found within the *alpha* and *beta* subunits.

The physiological specificity of the A and B factors is determined by the particular combination of the alpha and beta subunits. The A or the B factors have the same physiological effect if they possess the same *alpha* and *beta* subunits. Their physiological specificity differs when either one or both of the subunits are different. For example, $A_{\alpha 1 \beta 1}$ reacts differently from $A_{\alpha 1 \beta 2}$. Thus two mating types are incompatible only if both A and both B factors consist of identical subunits. The simplest case of sexual compatibility occurs when each factor differs in an *alpha* or *beta* subunit (e.g. $A_{\alpha 1 \beta 1} B_{\alpha 3 \beta 3} \times A_{\alpha 1 \beta 2} B_{\alpha 2 \beta 3}$).

Because of these results RAPER has suggested that *A and B should be called factors and not genes.* This suggestion has been adopted in the foregoing discussion. By factor we mean here a complex consisting of two genes (*alpha* and *beta*). Both genes have a similar function. They provide the genetic information for the reaction norm of the incompatibility factors (e.g. A_1). RAPER speaks of a "physiological unit".

Because the use of this term may easily lead to confusion, it should be pointed out that "physiological unit" in this sense and "functional unit" are not equivalent. A single gene is generally considered to be the functional unit (p. 211ff.).

The proof that A and B are actually physiological units lies in the fact that the factor specificity is altered by a genetic constitution differing in only *one alpha* or *one beta.* There is no need to reinterpret the incompatibility mechanism (e.g. as an octopolar one!). We can continue to speak of *tetrapolar incompatibility* and to designate the *two physiological units, A and B, as determinants whose specificity is each controlled by two genes.*

This new interpretation of the structure of the A and B factors appears to be generally valid, since more detailed genetic analysis in a number of other forms has shown that either the A or the B, or both factors are composed of subunits (Table II-4).

For completeness we have included in Table II-4 the results of RAPER which have been discussed above. Since in *Collybia* and *Schizophyllum* the designation of the A and B factors is reversed, the distances between the subunits are in reverse relationship. While in *Collybia* the genes of factor B are relatively far apart, in *Schizophyllum* this is true for the genes of the A factor.

Table II-4. *Summary of the results of investigations on the complex structure of A and B factors in Basidiomycetes with tetrapolar incompatibility*
    Each factor consists of two subunits between which reciprocal recombination is possible (— = not yet analyzed).

object	percentage of recombination within factor		reference
	A	*B*	
Collybia velutipes	0.5—1.3	19.4	TAKEMARU, 1957a, b
Coprinus lagopus	0.068—0.88	—	DAY, 1960, 1963a, b
Coprinus sp.	—	7.4	TAKEMARU, 1961
Lentinus edodes	—	7.5	TAKEMARU, 1961
Pleurotus spodoleucus	—	8.3	TAKEMARU, 1961
Pleurotus ostreatus	—	—	TERAKAWA, 1957
Schizophyllum commune	0.9—22.8	2.0	RAPER et al., 1958a, 1960
Schizophyllum commune	18.3	1.9	TAKEMARU, 1961

Crosses between mono- and dikaryons (Buller phenomenon)

BULLER (1931, 1933, 1941) carried out extensive crosses between dikaryotic and monokaryotic mycelia. He found that nuclei migrated from dikaryon into monokaryon. This behavior was designated by QUINTANILHA (1937) as the "Buller phenomenon". The nomenclature suggested by PAPAZIAN (1950) for such "di-mon" crosses is useful here because of its clarity. The following combinations are possible in a species which exhibits tetrapolar incompatibility:

1. Compatible "di-mon" crosses are those in which the nucleus of the monokaryon is compatible with each nucleus of the dikaryon: $(A_1B_1 + A_2B_2) \times A_3B_3$.

A dikaryon is designated by the A and B factors for each of the two nuclear components; these are joined by a plus sign and bracketed. The specific genetic factors are represented by subscripts.

One might expect that in a series of different di-mon crosses the two nuclear components of the dikaryon would migrate into the monokaryon with equal frequency. However, this is not always the case. QUINTANILHA (1939) demonstrated in such crosses with *Coprinus fimetarius* deviations from a 1:1 ratio in the newly formed dikaryons. KIMURA (1954a, b, 1958), SWIEZYNSKI (1961), ELLINGBOE and RAPER (1962b), PRÉVOST (1962) and CROWE (1960, 1963) obtained similar results in their researches on *Collybia velutipes*, *Coprinus macrorhizus f. microsporus*, *Coprinus lagopus*, *Pleurotus ostreatus*, *Schizophyllum commune*, and *Coprinus radiatus*.

KIMURA concluded from his experiments that extrachromosomal determinants in addition to a number of loci which differ from the incompatibility factors, are responsible for dikaryotization (p. 460). PRÉVOST also adopted this view. In *Schizophyllum* extra-chromosomal factors do not come into question because ELLINGBOE and RAPER made extensive use of isogenic stocks. In the latter case incompatibility loci as well as other genes that affect sexual behavior play a role. In any case, since both types of nuclei migrate into the monokaryon, selection of the dikaryotizing nucleus must occur in the monokaryon. The recently published data of CROWE confirm this view.

2. Semi-compatible di-mon crosses are those in which the nucleus of the monokaryon is compatible with one and incompatible with the other nucleus of the dikaryon: e.g. $(A_1B_1 + A_2B_2) \times A_1B_1$.

3. Incompatible di-mon crosses are those in which the nucleus of the monokaryon is hemi-compatible with both nuclei of the dikaryon, i.e. it possesses one factor in common with each of the two nuclei: e.g. $(A_1B_1 + A_2B_2) \times A_1B_2$.

The fact that in compatible and semi-compatible crosses nuclei migrate from the dikaryon into the monokaryon is to be expected. In the former case the monokaryon may be dikaryotized by both nuclei, while in the latter, only by the compatible nucleus. However, one would not predict that *a dikaryon would arise in incompatible di-mon crosses in which neither of the two nuclear types of the dikaryon is compatible with the monokaryon*. This special case of the Buller phenomenon has been observed in various experimental subjects (CHOW, 1934; DICKSON, 1934, 1935, 1936; NOBLE, 1937; QUINTANILHA, 1937, 1938a, b, 1939; OIKAWA, 1939; PAPAZIAN, 1950, 1954; TERRA, 1953; KIMURA, 1954a, b, 1957, 1958; GANS and PRUD'HOMME, 1958; CROWE, 1960; TAKEMARU, 1961; PRUD'-HOMME, 1962; PRÉVOST, 1962; ELLINGBOE and RAPER, 1962b; SWIE-ZYNSKI, 1962, 1963; ELLINGBOE, 1963).

Dikaryotization in incompatible di-mon crosses may occur in two ways:

1. Both nuclei of the dikaryon migrate into the monokaryon, replacing the nuclei of the latter and producing a dikaryotic mycelium with clamp connections. This interpretation proposed by BULLER has been supported by CHOW, DICKSON, and OIKAWA. Experimental proof was first provided by QUINTANILHA (1939) and later by PAPAZIAN and CROWE.

2. A genetic exchange occurs between the two types of nuclei in the dikaryon leading to the formation of a nucleus which is compatible with that of the monokaryon. QUINTANILHA (1933) proposed this interpretation; he later confirmed it (1938b, 1939) through tetrad analysis in *Coprinus fimetarius*. Further evidence for this theory of recombination is found in the papers of PAPAZIAN, GANS and PRUD'HOMME, CROWE, PRUD'HOMME, KIMURA, SWIEZYNSKI, ELLINGBOE and RAPER. Dikaryons were used in which the *A* as well as *B* factors were marked with a closely linked gene. Genetic analysis of the spores obtained from the fruiting bodies of the original monokaryon showed unequivocally that gene exchange had taken place in the dikaryon.

The *mechanism* of exchange, which calls for a fusion of the two nuclei of the dikaryon, is not clear. CROWE's investigations have shown that the nuclei which migrate from the dikaryon are haploid. *Either a mitotic or meiotic recombination* is required for production of new types of haploid nuclei (see section on *alternatives to sexual reproduction*, p. 92ff.).

In addition to the two classical recombination mechanisms *another, completely unknown recombination process* occurs in dikaryons of *Schizophyllum commune* and leads to the formation of compatible nuclei.

ELLINGBOE (1963) found that in incompatible di-mon crosses in which the nuclear components were adequately marked, only the incompatibility factors, and not the markers in their vicinity, were exchanged. An explanation for this phenomenon is lacking.

Pseudocompatibility

There exist monoecious Ascomycetes and Basidiomycetes in which self-fertility does not depend upon self-compatibility (Table II-2, Fig. II-2). A number of terms have been used in the literature to describe such *pseudocompatibility:* secondary homothallism (DODGE, 1927), amphithallism (LANGE, 1952), pseudo-monothallism (AHMAD, 1954), homoheteromixis (BURNETT, 1956a). Pseudocompatibility is found among tetrasporic Ascomycetes and bisporic Basidiomycetes. A tetrasporic form has an ascus which contains four, generally binucleate, instead of eight uninucleate ascospores (e.g. *Podospora anserina, Neurospora tetrasperma, Gelasinospora tetrasperma*). The bisporic Basidiomycetes develop only two binucleate, instead of four uninucleate spores on the basidium. BAUCH (1926) has surveyed the distribution of bispory among the Hymenomycetes.

In speaking of uninucleate and binucleate spores we are referring to the nuclear condition at the time of spore wall formation. It is well known that during spore maturation the original nucleus or nuclei may divide further by mitosis (e.g. DOWDING and BAKERSPIGEL, 1954; FRANKE, 1962).

Pseudocompatibility has been investigated intensively genetically and cytologically in *Podospora anserina* (RIZET and ENGELMANN, 1949; FRANKE, 1957, 1962) (p. 151 ff.). As shown in the diagram of nuclear division and spore production in Fig. IV-6, each of the four ascospores of *Podospora anserina* receives two meiotic products. The allelic pair responsible for bipolar incompatibility exhibits a second division segregation frequency of about 97%. Thus spores, most of which are heterokaryotic, arise in the asci. Mycelia which develop from such spores are heterokaryons (miktohaplonts); they produce male and female sex organs which contain either a *plus* or a *minus* nucleus. Perithecia form only through a sexual reaction between *plus* and *minus* organs in accordance with the bipolar incompatibility mechanism. *Self-compatibility is simulated because of the heterokaryotic constitution of the mycelium.* The three percent of asci in which the $+/-$ allelic pair segregates at the first division (see Fig. IV-6) contain spores which develop into self-incompatible mycelia.

In 1—2% of all asci one or more pairs of uninucleate spores may develop instead of a binucleate spore or spores. The uninucleate spores are easily distinguished by their small size. A mycelium arising from a uninucleate spore is self-incompatible. It may be either of the *plus* or the *minus* mating type (p. 20, Fig. I-7 and Fig. I-8).

The situation is similar in *Neurospora tetrasperma* (DODGE, 1927, 1928) and *Gelasinospora tetrasperma* (DOWDING, 1933; DOWDING and BAKERSPIGEL, 1956).

The post reduction frequency of the $+/-$ allelic pair in *Neurospora tetrasperma* is very low so that a high percentage of self-incompatible mycelia are expected. However, this is not the case. According to DODGE (1927) and COLSON (1934) the increased frequency of heterokaryotic spores results from nuclear shifts that occur because of a positive affinity (p. 152ff.) between *plus* and *minus* nuclei. This observation deserves genetic and cytological confirmation. Data on the post reduction frequency of the *plus* and *minus* alleles are not available for *G. tetrasperma*.

5*

Pseudocompatibility in the bisporic Basidiomycetes also involves the formation of heterokaryotic spores in which the nuclei carry compatible combinations of incompatibility alleles. The production of such spores is believed to result, as in *N. tetrasperma*, from a positive affinity between genetically different nuclei so that they migrate into the same sterigma of the basidium together (SASS, 1929; OIKAWA, 1939; SKOLKO, 1944).

KUHNER (1954) has reported that only two thirds of the basidiospores of *Clitocybe lituus* germinate to produce heterokaryotic mycelia. The remaining spores produce monokaryons. Since most spores of the latter category contain only one nucleus, it is possible that they may in part arise from irregularities in nuclear division. The other monokaryons may be explained by assuming pre-reduction of the incompatibility factors. Because tetrad analysis has not been carried out in this species, it is impossible to distinguish between the two alternatives.

According to the investigations of LAMOURE (1957), pseudocompatibility is not restricted to the bisporic Basidiomycetes. It is said to also occur in the tetrasporic species, *Coprinus lagopus*. More detailed genetic and cytological investigations of this single example are lacking, however. Similarly, a genetic confirmation of the reports of pseudocompatibility in the rust, *Puccinia arenariae* (LINDFORS, 1924) and in the smut, *Cintracta montagnei* (RAWITSCHER, 1922) is needed.

Pseudocompatibility is not always obligatory for a species; it may be facultative, as shown by the work of BURNETT and BOULTER (1963) on the Gastromycete *Mycocalia denudata*.

The sexual behavior of this tetrasporic Basidiomycetes is determined through bipolar incompatibility. Normally the four spores of a basidium are uninucleate and show a 1 : 1 segregation for mating type alleles. When the dominant gene *P* is present, however, meiosis in the young basidium is followed by a mitosis. The eight nuclei which result become distributed at random among the four basidiospores. Depending upon the constitution of the binucleate spores with respect to mating types, either incompatible homokaryotic or pseudocompatible heterokaryotic mycelia are produced.

Summary

1. Incompatibility is homogenic if two incompatible mating partners carry the same incompatibility alleles. Homogenic incompatibility may be controlled by one or two factors. If one factor is involved, there are two alternative mating types and the mechanism is said to be bipolar; with two factors at least four mating types occur and the mechanism is termed tetrapolar.

2. The bipolar mechanism is responsible for the sexual behavior in all incompatible Ascomycetes and Uredinales. The incompatible Holobasidiomycetes possess predominantly the tetrapolar type of incompatibility. This type is also found in some members of the Ustilaginales.

3. In Ascomycetes plasmogamy is prevented by sexual incompatibility. In Basidiomycetes the incompatibility reaction generally does not occur until the interval between plasmogamy and karyogamy.

4. The incompatibility factors in Holobasidiomycetes often exist as multiple alleles. The number of alleles may exceed one hundred. Among tetrapolar Ustilaginales multiple alleles have been demonstrated in only two cases.

5. The incompatibility factors of some tetrapolar Holobasidiomycetes are complex loci. They consist of two linked genes which together form a functional unit.

6. Crosses between monokaryotic and dikaryotic mycelia that carry nuclei compatible with each other (Buller phenomenon) have revealed that mitotic and perhaps also meiotic recombination may occur between the nuclei of a dikaryon.

7. A number of fungi produce binucleate spores which are heterokaryotic for incompatibility factors because of the particular mechanics of nuclear division in the ascus and in the basidium. The miktohaplonts which develop from these spores are pseudocompatible.

b) Heterogenic incompatibility

Heterogenic incompatibility may be observed *when races of the same species* derived from different geographic sources *are intercrossed* and prove to be incompatible. As Table II-2 reveals, only a few cases of heterogenic incompatibility are known.

This may be a result of the fact that the geneticist tends to work with isogenic stocks of a single race rather than with different races. Moreover, until recently it was not possible to determine in crosses with self-fertile races whether fruiting bodies resulted from selfing or crossing. Only when suitable markers became available (e.g. spore color genes) did it become possible to determine whether or not two self-fertile races were cross-compatible (p. 54). In interracial crosses of homogenically incompatible fungi fruiting bodies frequently fail to form; this phenomenon has generally been called cross sterility and not investigated further.

Since heterogenic incompatibility was discovered in interracial crosses of the Ascomycete *Podospora anserina* and its genetic basis clarified (see the review of Esser, 1962, 1965), it has been shown that some instances of cross sterility previously reported in the literature are in fact heterogenic incompatibility.

This system of sexual incompatibility, like the homogenic one, is not restricted to the fungi. It may occur in the higher plants (*Oenothera:* Steiner, 1961). The sexual behavior of different geographic races of mosquitoes (*Culex pipiens:* Laven, 1957a, b) can also be explained by heterogenic incompatibility.

Heterogenic incompatibility occupies a unique position among the reproductive systems of fungi. It may *exist between races of a compatible species as well as between races of a homogenically incompatible species.* If we consider the investigations on geographic races of mosquitoes mentioned above, it is not unlikely that heterogenic incompatibility may determine the sexual behavior of ecotypic forms of dioecious fungi. This special role of heterogenic incompatibility has not been included in Fig. II-2, since the remaining systems of reproduction are valid for both intra- and interracial crosses. Heterogenic incompatibility applies only to interracial crosses and thus does not fit into this context.

Self-compatible species

In self-compatible species heterogenic incompatibility serves as the sole regulatory mechanism of sexual reproduction. Olive (1956) investigated the crossing relationship between nineteen different self-fertile races of *Sordaria fimicola*. By utilizing races with spore color markers he showed that not all races (e.g. races A_1 and C_1) can be crossed with one another. Both races can be crossed with a third, C_4. Thus one can conclude that the failure of perithecial formation between A_1 and C_1 is not the result of cross-sterility (chromosomal divergence etc.) but simply the expression of heterogenic incompatibility. Since the mycelium arising from a single spore in every race is self-fertile, the interracial incompatibility must be controlled by genic differences between the individual strains. Similar evidence of heterogenic incompatibility has been found in interracial crosses of *Sordaria macrospora* (ESSER, unpublished). The genetic mechanism which underlies both cases remains unknown.

Self-incompatible species

Sexual reproduction in self-incompatible species is controlled through a combination of heterogenic and homogenic incompatibility. While the former mechanism is responsible in crosses between individual races, the latter regulates sexual reproduction within each race. This is shown in the results of investigations on *Podospora anserina*. Each of the geographical races of this fungus, which stem from different localities in Germany and France, possesses two mating types (*plus* and *minus*) which react sexually according to the bipolar mechanism of homogenic incompatibility.

A mycelium arising from a homokaryotic spore produces male and female sex organs (ascogonia and spermatogonia). Since zygote formation occurs only in the combination $+ \times -$, a row of perithecia develops in the zone of contact between the two self-incompatible $+$ and $-$ strains. These represent a mixture of the reciprocal crosses $♀ + \times ♂ -$ and $♀ - \times ♂ +$ (left half of Fig. II-6).

In crosses *between eight different races of this fungus* the following *deviations from reciprocal compatibility between plus and minus mating types* are observed:

1. *Non-reciprocal incompatibility*, designated *semi-incompatibility*. The ascogonia of one mating partner may be fertilized by the male gametes of the other. The reciprocal cross is incompatible (Fig. II-6 top and bottom half, Fig. II-7a and b).

2. *Reciprocal incompatibility*. No fruiting bodies form between *plus* and *minus* strains of different races (Fig. II-6 right half, Fig. II-7c). In both cases plasmogamy is blocked.

An important aid in identifying semi-incompatibility is barrage formation. This phenomenon is characterized by a clearly visible, pigment-free zone between the two strains (Fig. II-6). In this zone the greenish-black melanin pigments fail to form in the hyphal tips of both partners. Barrage formation, which was first observed by RIZET (1952) (p. 447ff.), never occurs between different strains of the same race. It appears only in interracial crosses, and moreover, occurs independently of mating type, i.e. it is found also between two *plus* strains or two *minus* strains. Because no perithecia form within the barrage zone, the perithecia of reciprocal crosses between $+$ and $-$ types are separated. Ordinarily two rows of fruiting bodies develop, one along each side of the barrage. When only one row of perithecia forms, semi-incompatibility is indicated.

Non-reciprocal incompatibility may also be demonstrated in another way, namely by spermatization. The *plus* and *minus* strains which are to be tested are grown in separate Petri dishes and covered with a suspension of spermatia from their respective mating partners. If perithecia develop on only one of the two plates, semi-incompatibility is indicated.

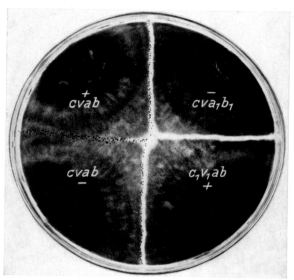

Fig. II-6. The genetic basis of heterogenic incompatibility in *Podospora anserina*, showing the crossing relationships between four different mycelia, the genotypes of which are indicated. For details see text. (From Esser, 1956c)

The occurrence of semi-incompatibility shows that prevention of fertilization between races of P. anserina results from sexual incompatibility and not cross-sterility. If cross-sterility were involved, it would be difficult to account for one cross being fertile and its reciprocal sterile. The phenomenon of semi-incompatibility provides a means of genetic analysis which is lacking in the case of reciprocal incompatibility. The results of such analysis (Rizet and Esser, 1953; Esser, 1954a, b, 1955, 1956c, 1958, 1959a) have revealed the genetic conditions for complete incompatibility. These are shown in Fig. II-6 and Fig. II-7. *Four loci (a, b, c, v), each with two allelic forms, are responsible for the incompatibility behavior between different races. The genes are not linked. Two genes control semi-compatibility,* which appears only in the crosses $ab \times a_1 b_1$ and $cv \times c_1 v_1$.

The sexual behavior of the recombinants between the two genic mechanisms ($ab_1 \times a_1 b$ and $cv_1 \times c_1 v$) cannot be tested because the genotypes $a_1 b$ and $c_1 v$ have a reduced viability and do not form ascogonia (p. 74). Semi-compatibility occurs only through the interaction of the a and b or the c and v genes. Combinations of a, b, c and v other than those indicated above do not interfere with reciprocal sexual compatibility.

Reciprocal incompatibility occurs only in the combination $abc_1v_1 \times a_1b_1cv$. As the illustration shows, it results *from the superposition of the two mechanisms responsible for semi-incompatibility*. The basis of reciprocal incompatibility is thus semi-incompatibility.

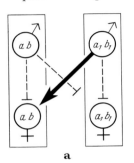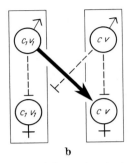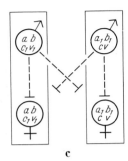

a   b   c

Fig. II-7a—c. Genetic basis of heterogenic incompatibility in *Podospora anserina*. Represented as in Fig. II-3. The mating partners of each of the three crosses have a different mating type. To avoid confusion the *plus* and *minus* symbols have been omitted. The arrows indicate the compatible and the blocked lines the incompatible combinations. (a) and (b) semi-incompatibility, (c) reciprocal incompatibility. For details see text

Thus heterogenic incompatibility is observed only between *plus* and *minus* strains of different races in which the requirements of the bipolar mechanism of homogenic incompatibility have been met. The two mating partners are heterogenic for the incompatibility genes of both systems.

Heterogenic incompatibility plays a role not only in the sexual, but also in the vegetative phase. The viability of nuclei in homo- and heterokaryons is influenced by specific combinations of incompatibility genes.

1. Homokaryotic mycelia of the genotype a_1b or c_1v that arise as recombinants among the progeny of semi-incompatible crosses grow very poorly. Most of the nuclei in the hyphae degenerate after a few hours, breaking into fragments. The mycelia soon die. No sex organs form (see also p. 74).

2. The genes a_1 and b (or c_1 and v) are not only incompatible with one another when they are present in the same nucleus, but also when they occur in two different nuclei of the same heterokaryon. Although such heterokaryons (e.g. $ab + a_1b_1$) do not exhibit any growth anomalies, they become transformed into homokaryons after a few days, because the capacity of the nuclei which carry the a_1 gene to divide is sharply reduced. The mycelium ultimately carries only ab nuclei. The corresponding phenomenon occurs also in the case of the c_1v incompatibility in the $(cv + c_1v_1)$ heterokaryon. Here only the cv nuclei survive.

In genetic investigation of another race of *Podospora anserina*, BERNET (1963 a, b) identified three pairs of alleles, C/c, D/d, and E/e which in specific combinations ($Ce \times cE$ and $Cd \times Cd$) reduce viability in homokaryons and heterokaryons and also produce semi-incompatibility. The gene action in this heterogenic situation corresponds in principle to that described above. The only difference is that the effect of gene D, which is comparable to that of *b* or *v*, is temperature-dependent. It is manifested only at 20° C and can not be detected at 32° C. It is still not clear whether the *C-D* genes are identical to those already known for heterogenic incompatibility or whether other loci are involved which produce the same phenomenon.

As mentioned previously (p. 69), there are reports of cross-sterility in the older literature which can be interpreted as clear-cut cases of heterogenic incompatibility.

BAUCH (1927) carried out extensive crosses with different races of *Ustilago violacea*. He found that within each race of this species, which shows bipolar incompatibility, both mating types are compatible with each other. On the other hand, in interracial crosses opposite mating types are either semi-incompatible or incompatible. BAUCH assumed that secondary sex characters of a physiological nature were involved here. Morphological differences between the individual races could not be established with certainty. In our opinion this is a clear-cut case of heterogenic incompatibility.

GRASSO (1955), in interracial crosses of *Ustilago avenae* and *U. levis* found that certain combinations were unsuccessful. These, which were designated "semi-lethal", represent heterogenic incompatibility. Both of these species show bipolar incompatibility. The races which were studied were isolated from two localities in Italy and one in the United States.

The phenomenon designated by VANDENDRIES (1927, 1929) as sterility in interracial crosses of *Coprinus micaceus*, a species with tetrapolar incompatibility, can likewise be interpreted as heterogenic incompatibility. Similar observations on other races of this species made by KUHNER et al. (1947) and on races of *Erysiphe cichoreacearum* by MORRISON (1960) have been described.

MOREAU and MORUZI (1933) found in interracial crosses of *Neurospora sitophila* (bipolar incompatibility) that two races compatible with each other were both incompatible with a third race. "Cross-sterility" has also been described in races of *N. crassa* by LINDEGREN (1934).

Unfortunately genetic analysis, which could have provided evidence of the genetic mechanism of heterogenic incompatibility, was not carried out.

Summary

1. Heterogenic incompatibility is present when both incompatible mating partners possess different alleles at all incompatibility loci. It can only occur between different races of a species and never within an ecotype. Heterogenic incompatibility is independent of other reproductive systems, since it can be effective within compatible as well as homogenically incompatible species.

2. Relatively few cases of heterogenic incompatibility are known. Its genetic basis has been analyzed in detail only in one case, namely in the ascomycete *Podospora anserina*.

3. At least four unlinked genes are responsible for heterogenic incompatibility between different races of *Podospora anserina*.

4. Sexual behavior as well as vegetative multiplication of nuclei in homokaryotic or heterokaryotic mycelia may be negatively influenced through heterogenic incompatibility.

c) Mutability and incompatibility

In genetic analysis of sexual incompatibility it has repeatedly been observed that in *certain combinations of incompatibility alleles specific mutations occur or the mutation frequency in a particular direction is significantly increased.*

Strains of *Podospora anserina*, which exhibit a lowered viability because of particular combinations of incompatibility alleles regularly undergo the same mutation. When in a mycelium of the genotype $a_1 b$ or $c_1 v$ (p. 72), the number of nuclei has reached about 10^7, mutations to ab or cv arise. Mutations at other loci have not been observed (ESSER, 1959a). These facts may be explained as follows: The gene b (or v) has a specific action which affects only the gene a_1 (or c_1), but not its allele a (or c). As a result the capacity of the $a_1 b$ or $c_1 v$ nuclei to multiply is very low. They become deformed and die prematurely. The mycelium grows slowly. Because of this anomaly theoretically every mutant which leads to a nucleus of normal viability has a selective advantage. Actually, however, this occurs only through a specific mutation, namely that of a_1 to a (or c_1 to c). Other mutations do not result in nuclei with normal viability. Thus through the action of b or v in the mycelium with reduced viability, a "selective sieve" exists which permits only a particular mutation to persist. This gene-controlled selective mechanism simulates "directed mutation".

Similar results with respect to the correlation between incompatibility factors and the production of mutations were obtained in the investigations of PAPAZIAN (1951), RAPER et al. (1958c), DICK (1960), DICK and RAPER (1961) on *Schizophyllum commune*. Heterokaryons of this fungus, in which the nuclei carry the same A factor, grow abnormally (p. 59). They fail to produce clamp connections and show only a sparse development. PAPAZIAN found that in such heterokaryons sectors with hyphae growing normally arise; the nuclei of these hyphae in many cases mutate in the same direction. RAPER and his coworkers confirmed these results. They found further that the spectrum of mutants arising from these heterokaryons is completely different from that which is obtained by irradiation of homokaryotic strains. DICK tested the hypothesis that the specific mutational spectrum is produced by the incompatibility factors. He demonstrated that in the homokaryotic mycelia which possess one of the two nuclear types of the heterokaryon, numerous morphological mutants also arise spontaneously. The mutation rate and spectrum of the heterokaryon correspond to those of the homokaryon. The difference between the two mycelial types is that the morphological mutants have a selective advantage only in the poorly growing heterokaryons and therefore are easily recognized. In the rapidly growing homokaryons this is not the case.

No correlation between mutability and incompatibility factors can be shown in either *Podospora anserina* or *Schizophyllum commune. In both cases the detection of spontaneous mutants is appreciably easier because they offer a selective advantage over certain mycelial types which grow abnormally as a result of the specific effects of incompatibility factors.*

Summary

In a number of cases specific mutations have arisen from certain combinations of incompatibility factors or the mutation frequency has increased in general. There is, however, no correlation between mutability and the effect of incompatibility factors. Rather, the incompatibility factors lower the viability of particular mycelial types and thereby provide a selective advantage for all spontaneous mutations which lead to more normal growth.

d) Physiology of incompatibility

At present we have no clear concept of the nature of the physiological processes that are controlled by the incompatibility genes. This can be attributed in part to technical difficulties. Before these are pointed out, it is well to recall the complexity of the normal sexual reaction. In the first place, it involves cell wall dissolution which is initiated by plasmogamy. This is followed by a unilateral nuclear migration (Euascomycetes, Uredinales) or a nuclear exchange (Hymenomycetes, Ustilaginales) and a more or less prolonged dikaryotic phase that eventually leads to karyogamy. Doubtless a large number of physiological reaction sequences in which macromolecular substances participate, are involved (41 ff.). The isolation of such substances from mycelial extracts or culture filtrates presupposes a suitable test for determining their effects. The test is whether or not the normal sexual reaction of incompatible strains can be initiated by treatment with extracts from a compatible type.

The sexual reaction in the Ascomycetes and Uredinales takes place between typical sex organs. Because of the minute size of these structures, it has not been possible to obtain sufficient quantities of extracts from them. In the Holobasidiomycetes all of the hyphae take part in the sexual reaction; preparing mycelial extracts presents no problem. Nevertheless, compatibility is correlated with the dikaryotic condition and the addition of extract can not substitute for it. Dikaryons from which one nuclear component has been surgically removed (HARDER, 1927) lose the capacity to produce clamp connections after a few nuclear generations. As far as we know, in the Ustilaginales the two nuclear components of the haploid gametes are also required for the production of a smut spore.

To arrive at a concept of incompatibility gene action, one must resort in the main to observations on development. Physiological and biochemical experiments on this problem are relatively few. These studies do allow the construction of models of gene action which remain highly speculative, however. We shall discuss two models which are based on extensive investigation of homogenic incompatibility in the flowering plants (see literature survey of LEWIS, 1954). These are the *complementary-stimulant* and the *oppositional-inhibitor models*, both of which are based on incompatibility resulting from genetically identical incompatibility factors. Neither of these, of course, applies to heterogenic incompatibility.

The complementary mechanism is effective in compatible combinations. The complementary effect of the gene products of heterogenic mating partners leads to zygote formation. Incompatibility stems from the failure of such an event to occur when the mating partners are homogenic and the same gene products do not complement one another.

In contrast, the oppositional mechanism is only effective in incompatible combinations. It involves an inhibition of sexual processes as a result of the reaction between identical gene products. Such an inhibition does not occur when the gene products are dissimilar (compatibility) (see also STRAUB, 1958).

We will attempt to apply these two concepts to the incompatible fungi as already done elsewhere in more detail (ESSER, 1966a). Since the criterion for incompatibility differs among individual groups of fungi, the physiological basis of incompatibility will be discussed separately for each group.

Euascomycetes, homogenic incompatibility

In the Euascomycetes homogenic incompatibility is determined exclusively by the bipolar mechanism; moreover, plasmogamy fails to occur in most cases (p. 56).

A number of authors have recognized in the growth of the tricho-gyne toward conidia or spermatia of opposite mating type a clue for the explanation of bipolar incompatibility. This phenomenon was first observed by DODGE (1912) in *Ascobolus carbonarius* and later analyzed in greater detail by ZICKLER (1952) *in Bombardia lunata.*

Bombardia lunata belongs to the group in which spermatization occurs (p. 56). ZICKLER demonstrated by means of ingenious physiological experiments that the *plus* and *minus* spermatia evidently release different substances which react only with the trichogynes of the complementary mating type. Although different gene products of the *plus* and *minus* nuclei may be involved here, the absence of attraction between the trichogyne and sperma-tium of the same mating type is not necessarily responsible for the incompatibility. As a matter of fact, if a trichogynal tip touches by chance or is artificially brought into contact with a genetically identical spermatium, plasmogamy does not occur. Similar investigations with *Ascobolus stercorarius* showed that the attraction of the trichogyne by the male gametes is non-specific (BISTIS and RAPER, 1963).

Bipolar incompatibility between sex organs of the same mating type does not depend upon the absence of an attraction, but on the incapacity of both organs to engage in plasmogamy.

FOX and GRAY (1950) have reported that in *Neurospora crassa* only one mating type (*A*) produces polyphenol oxidase (tyrosinase). Since these results involve an experimental error, as HOROWITZ and SHEN (1952) have demonstrated, the experiments may be disregarded.

KUWANA (1954, 1955) observed that in mixed cultures of the *A* and *a* mating types of *Neurospora crassa*, tyrosinase is produced in larger amounts than in a culture of either mating type alone. It was assumed at first that this phenomenon was related to the compatibility reaction between *A* and *a*. However, genetic analysis revealed that instead of the *A/a* locus, at least three loci not linked to it are responsible for the increased tyrosinase production. The alleles of the three loci act in a complementary manner in the heterokaryotic condition (KUWANA, 1956, 1958).

Clues to a mechanism which may explain the incompatibility reaction are found in the investigations of BISTIS (1956, 1957) on *Ascobolus sterco-rarius* and of ITO (1956) on *Neurospora crassa.*

In *Ascobolus stercorarius* (p. 57) oidia which act as male gametes are "activated" only by the hyphal tips of the complementary mating type. They induce the formation of ascogonial primordia. In *Neurospora crassa* protoperithecia of each mating type can be induced to develop into sterile fruiting bodies by the addition of a culture filtrate from the complementary mating type.

Since *Ascobolus stercorarius* belongs to the dikaryotization type, fertilization by oidia is only one of several possibilities. Therefore the specific genetic effect of the incompatibility factors cannot be directly related to sexual incompatibility. Rather, they play a role in directing morphogenetic processes prior to fertilization. On the other hand, the formation of fruiting body "shells" in *Neurospora crassa* stimulated by substances from the complementary mating type is directly related to the compatibility reaction. In the latter case stages of differentiation are involved which normally occur after the sexual reaction has taken place.

It is exceedingly difficult to arrive at a physiological basis of bipolar incompatibility in the context of these research results. One can only say that there is *no evidence against the complementary mechanism*, postulated by LEWIS (1954) for this group. This means that the *plus* and *minus* organs produce complementary gene products which complement each other according to a "lock and key" model, enabling the sexual reaction to occur. The results of ITO (1956) argue against an oppositional mechanism. It would be difficult to understand in this case why the addition of a culture filtrate from the complementary mating type should overcome the inhibition of development.

Euascomycetes, heterogenic incompatibility

Clues to the physiology of heterogenic incompatibility may be found in genetic and developmental studies of semi-incompatible crosses in *Podospora anserina* (ESSER, 1959a).

In the compatible combinations of a semi-incompatible cross (Fig. II-7) fertilization proceeds normally. The trichogyne grows toward the spermatia. As soon as the tip comes into contact with a spermatium, degeneration of the trichogynal nuclei occurs simultaneously with the fusion of the organs (plasmogamy, Fig. I-6). In the reciprocal combination, which is incompatible, the trichogyne is attracted but plasmogamy does not occur. The trichogynal nuclei do not degenerate and the trichogynal tip continues to grow and branch. Each branch grows toward a different spermatium without fusing with it.

The asymmetric fruiting body formation in semi-incompatible crosses (see Fig. II-6) may be explained as follows: semi-incompatibility (e.g. in the cross $+ab \times -a_1b_1$) results from an inhibition of plasmogamy by the gene b which is effective only when b is localized in the spermatial nucleus and a_1 is in trichogynal nucleus ($\female -a_1b_1 \times \male +ab$). In this combination the dissolution of the cell wall between the trichogyne and spermatium, which is ordinarily catalyzed by the *plus* and *minus* genes, is inhibited. This inhibition does not occur in the reciprocal cross $\female +ab \times \male -a_1b_1$; in this case the gene b is located in the trichogynal nuclei and the gene a_1 in the spermatial nucleus. The trichogynal nuclei degenerate with the initiation of wall dissolution and lose their capacity to react.

In the second semi-incompatibility mechanism, which is illustrated by the cross $cv \times c_1v_1$, plasmogamy is likewise prevented in the incompatible cross by the action of the factor v.

Both factors b and v express their effects in reciprocal incompatibility (Fig. II-6 and II-7). b blocks plasmogamy in the one, v in the other of the two reciprocal crosses between strains of the genotypes $a b c_1 v_1$ and $a_1 b_1 c v$.

With regard to the chemical substances which underlie heterogenic incompatibility, the gene b (or v) produces its effects only in the contact between trichogyne and spermatium or in the heterokaryon. The products of this gene do not diffuse into the culture medium. The effect of the gene can be recognized in mixed cultures of semi-incompatible mycelia in which the hyphae grow in close contact and form in part heterokaryons. As a matter of fact, b and v induce an inhibition of protein synthesis in these cultures which is expressed in a change in protein specificity, as shown by serological analysis (ESSER, 1959b).

Heterogenic incompatibility involves an incompatibility of two non-allelic genes, namely b and a_1 and v and c_1. Because of the characteristics of the fertilization events their reaction is expressed in only one of the reciprocal crosses between the bisexual semi-incompatible strains. The mechanism may be designated as heterogenic oppositional.

Holobasidiomycetes

The bipolar as well as the tetrapolar incompatibility mechanism of the Holobasidiomycetes prevents karyogamy. Plasmogamy, the absence of which indicates incompatibility in most incompatible Ascomycetes, occurs in all hemi-compatible and in many cases, in incompatible combinations of Basidiomycetes (p. 57ff. and Fig. II-4). Our knowledge of the physiological basis of incompatibility stems primarily from investigations on species with tetrapolar incompatibility. PAPAZIAN (1950, 1951, 1954) and RAPER and SAN ANTONIO (1954) found specific interaction between genetically different mycelial types in *Schizophyllum commune*, which permits an hypothesis concerning the mechanism of sexual incompatibility. These results were confirmed in studies of other tetrapolar incompatible species (ASCHAN, 1954, literature survey: PAPAZIAN, 1958). No experimental data are available for species with bipolar incompatibility so that conclusions about the physiology of incompatibility are not possible.

PAPAZIAN investigated heterokaryons which arise in hemi-compatible combinations. The abnormal habit, the slow growth, and the instability of the heterokaryons with common A or B factors suggest that each kind of nucleus has a negative influence on the other. The results of RAPER and SAN ANTONIO also support this view. This mutual antagonism influences not only the normal function of the nuclei but also the viability of the entire cell.

Such disturbances of cell function explain the fact that when a heterokaryon and homokaryon are crossed, the former can serve only as nuclear donor, the latter only as nuclear acceptor. Because the physiological characteristics described for the incompatibility factors have not been supplemented with biochemical data, every attempt to explain the physiological mechanism of tetrapolar incompatibility has been fraught with difficulties.

The existence of a *complementary* mechanism is unlikely because of the following:

1. The existence of multiple alleles of incompatibility factors.

For each of up to 100 incompatibility alleles a specific gene product is formed. A complementary mechanism requires that each specific product complement all others but not itself. This demands a structure of a complexity that hardly seems possible.

2. The nuclear antagonism in heterokaryons from hemi-compatible matings.

The absence of a complementary gene product cannot result in any inhibitory effects in crosses with the same A or B factors.

3. The breakdown of the incompatibility reaction through mutation.

Mutations within the B factor, which have been described by PARAG (1962a) (p. 59), cause breakdown of the incompatibility reaction. If this is interpreted according to a complementary mechanism, one would have to assume that the mutants produce a substance complementary for all products of the innumerable B factors. That such a "universal gene product" exists is very unlikely.

On the other hand, the hypothesis of an *oppositional* mechanism which is based on an inhibition resulting from an reaction of identical gene products provides the following explanation: Each of the two incompatibility factors produces a specific gene product. Identical gene products are incompatible and inhibit normal development prior to karyogamy. This is shown particularly in the incompatibility between mating partners with common A and B factors. In hemi-compatible crosses (common A or B factors) the antagonistic effect of the gene product of one factor is involved. Plasmogamy occurs, but the formation of clamp connections and karyogamy do not follow. In crosses between compatible mycelia (different A and B factors) no inhibitory effect occurs; the normal sexual cycle consisting of plasmogamy, nuclear exchange, the dikaryotic phase, karyogamy, and meiosis runs its course.

The disturbances of growth observed in heterokaryons from hemi-compatible mating partners may be considered as an oppositional effect of two processes: Physiological reactions essential for normal development are initiated as a result of the heterogeneity of one of the two loci (e.g. $B_1 + B_2$). Because of the homogeneity of the other locus ($A_1 + A_1$), an inhibitory mechanism is triggered that acts antagonistically to the former. In homokaryotic mycelia these opposing reactions do not occur; thus the mycelia show normal viability.

Recent investigations on *Schizophyllum* (RAPER and OETTINGER, 1962; MIDDLETON, 1964; SNIDER and RAPER, 1965) nevertheless contradict the oppositional hypothesis, since

1. Nuclei with identical A and B factors are compatible in heterokaryons under special circumstances.

Mycelia with identical A and B factors normally do not form hetero-karyons in the zone of contact (p. 59). However, MIDDLETON was able to show that non-allelic nutritional mutants produce prototrophic hetero-karyons on minimal medium relatively frequently, although these often show sectoring.

2. Additional identical factors do not effect the normal development of a dikaryon.

When disomy occurs (e.g. $A_1B_1A_2 + A_2B_2$), the extra chromosome with the A_2 factor in one of the nuclear components has no inhibitory effect on the other nuclear type with the same factor.

The present status of research on the *physiology of tetrapolar incom-patibility does not permit a choice between the two classical physiological mechanisms which explain sexual incompatibility.* The experimental data cannot be reconciled satisfactorily with either a complementary or an oppositional mechanism. It is questionable that even with genetic ana-lysis of the function of incompatibility factors further conclusions may be reached. Biochemical studies in which attempts are made to isolate and identify the active products of the incompatibility factors appear more promising.

The immunological studies on *Schizophyllum commune* (RAPER and ESSER, 1961) are a first step in this direction. These authors showed that a dikaryon and its two monokaryotic components possess different pro-tein spectra. Since the two homokaryotic strains from which the dikaryon was constructed were isogenic, the difference in proteins was attributed to differences in the action of the incompatibility factors in the homo- and dikaryon.

We must also point out that there is some evidence for a relationship between the cytoplasm and the incompatibility reaction (HARDER, 1927; FRIES and ASCHAN, 1952; KIMURA, 1954a, b; PAPAZIAN, 1958) (p. 458). These results are insufficient, however, to allow development of a hypothesis.

Phragmobasidiomycetes

No experimental data which give an insight into the physiology of sexual incompatibility are known for species of Uredinales which exhibit bipolar incompatibility.

With regard to the Ustilaginales, which exhibit tetrapolar incompati-bility, it is known that in hemicompatible combinations heterokaryons arise in which the physiological processes are disturbed in the same way as in the heterokaryons of the Holobasidiomycetes. However, to generalize from these data seems premature.

Summary

1. Hypotheses concerning the physiological genetics of sexual incom-patibility are based primarily on results of genetic and developmental investigations and only to a small degree on physiological and bioche-mical experiments.

2. Homogenic incompatibility appears to involve several different mechanisms. In one case (Euascomycetes) sexual compatibility is believed to be determined by gene products produced through the reciprocal complementation of *plus* and *minus* mating types. Such a complemen-tation is not possible between the products of two identical mating types

(incompatibility). In another case (Holobasidiomycetes) complementation may be ruled out. On the other hand, an oppositional mechanism based on antagonism of identical gene products fails to explain the experimental data satisfactorily. At the present time an acceptable hypothesis of incompatibility factor function has not been forthcoming.

3. Heterogenic incompatibility involves the incompatibility of two different genes, which express themselves both in the sexual process as well as in vegetative hyphae. The physiological effect of these genes appears to be one in which the product of one gene inhibits the activity of the other.

III. Reproductive systems in dioecious forms

1. Morphological dioecism

Only a few species of fungi are known which exhibit sexual differentiation into morphologically distinct male and female plants. *Morphological dioecism is known only in certain species of Phycomycetes and Ascomycetes* (Table II-5). Such a breeding system is unknown in the Myxomycetes and Basidiomycetes.

The dioecism postulated by GREIS (1942) for the Basidiomycete *Solenia anomala* has been shown, through analysis of GREIS' experimental data (RAPER, 1959), to be a case of incompatibility.

As shown in Table II-5, the existence of dioecism has been known for a long time. Genetic analysis of this phenomenon has been possible only in isolated cases, because crossing and culture of the progenies have often been hampered by technical difficulties (Blastocladiales and Chytridiales). In many cases the zygotes have failed to develop (*Achlya*, p. 43).

One of the few species for which genetic data are available is *Stromatinia narcissi*. Here the F$_1$ shows a 1:1 segregation for males and females. *Whether this inheritance is determined by a pair of alleles at a single locus or by a sex chromosome as in higher organisms is unknown.* The prospect of distinguishing between these alternatives by cytological means is poor, because chromosomes of fungi are notably small and difficult to distinguish from one another. According to COUCH (1926) more than a single hereditary factor is responsible for sexual dimorphism in the Oomycete *Dictyuchus monosporus*. Nothing is known about the action of these factors, however.

The observation that older cultures of dioecious strains of *Achlya ambisexualis* eventually give rise to bisexual sectors has added to the confusion regarding the basis of sexual differentiation (RAPER, 1947). This fact may be explained by assuming that single nuclei in the coenocytic hyphae of *Achlya ambisexualis* mutate. Since monoecious as well as male and female strains are known in this species the question arises whether or not sexually differentiated mycelia represent defect mutants which occasionally undergo reversion. Mutants which simulate dioecism are known in *Hypomyces solani*. Thus it seems worthwhile to discuss the work with *Hypomyces solani* in greater detail.

Reproduction

Table II-5. *Species of Phycomycetes and Ascomycetes which are morphologically dioecious.* (Taken in part from the list of WHITEHOUSE, 1949a)

Systematic taxa		Reference
Phycomycetes		
Entomophthorales	*Ancylistes closterii* PFIT.	PFITZNER, 1872
	Zoophagus insidians SOM.	ARNAUDOW, 1925
Peronosporales	*Phytophthora palmivora* BUTL.	ASHBY, 1922
	(syn. *Ph. faberi* MAUBL.)	
	Phytophthora omnivora DE BARY	LEONIAN, 1931
	Peronospora parasitica	BRUYN, 1935, 1936
	(PERS. ex FR.) TUL.	
Lagenidiales	*Lagenidium rabenhorstii* ZOPF	ZOPF, 1884; VAN-
	Lagena radicola VANT & LED	TERPOOL and LE-
		DINGHAM, 1930
Leptomitales	*Sapromyces reinschii* (SCHROET)	WESTON, 1938
	FRITSCH	BISHOP, 1940
Saproleginales	*Dictyuchus monosporus* LEITG.	COUCH, 1926
	Achlya bisexualis COKER	COKER, 1927;
		RAPER, 1936
	Achlya regularis COKER & LEITN.	COKER and
		LEITNER, 1938
	Achlya ambisexualis RAPER	RAPER, 1939a, 1940b
Blastocladiales	*Blastocladiella variabilis*	HARDER and
	HARDER & SÖRGEL	SÖRGEL, 1938
Chytridiales	*Zygorhizidium willei* LÖWENTH.	LÖWENTHAL, 1905
	Olpidiopsis saprolegniae CORNU	BARRET, 1912
	Dangeardia mammillata SCHRÖD.	CANTER, 1946
	Rhizophydium columaris CANTER	CANTER, 1947
	Polyphagus euglenae NOW.	NOWAKOWSKI, 1876;
		KNIEP, 1928
Ascomycetes		
Taphrinales	*Ascosphaera apis* MAASSEN	CLAUSSEN, 1921;
	(syn. *Pericystis apis* MAASSEN)	SPILTOIR, 1955
Helotiales	*Stromatinia narcissi*	DRAYTON and
	DRAYTON & GROVES	GROVES, 1952
Laboulbeniales	*Amorphomyces falagriae*	THAXTER, 1896
	THAXTER	
	Laboulbenium formicarum	BENJAMIN and
	THAXTER	SHANOR, 1950

The Ascomycete *H. solani* f. *cucurbitae* is monoecious. Its crossing behavior is determined by bipolar incompatibility (DIMOCK, 1937). Male and female strains arose occasionally in the cultures of HANSEN and SNYDER (1943, 1946). The male mycelia failed to produce perithecia and the female lacked conidia. A cross between the two sexes gave progeny which included among other types, neutral mycelia. These showed the habit of the male strain, but could not be crossed. HIRSCH (1949) believed that the different sexual potencies of *Hypomyces* could be correlated with differences in chromosome number. The monoecious wild type of *H. solani* possessed a haploid complement of four chromosomes, one large, one small, and two which were intermediate in size. The female mycelia were characterized by the absence of one of the chromosomes of intermediate size and the male by the absence of the other. The neutral mycelium lacked both intermediate sized chromosomes. EL-ANI (1954a, b) demonstrated with genetic and cytological studies that the observations of HIRSCH, which represented the first example of sex chromosomes in fungi, were erroneous. All four types have the same chromosome number, namely $n = 4$. The male and female strains are single gene mutants. These two genes, which are responsible for the defects in normal differentiation, appear to be unlinked. The sexually neutral mycelium is the double mutant.

Thus it is evident that at present there is *no satisfactory general explanation for the genetic basis of dioecism in the fungi*. This is regretable

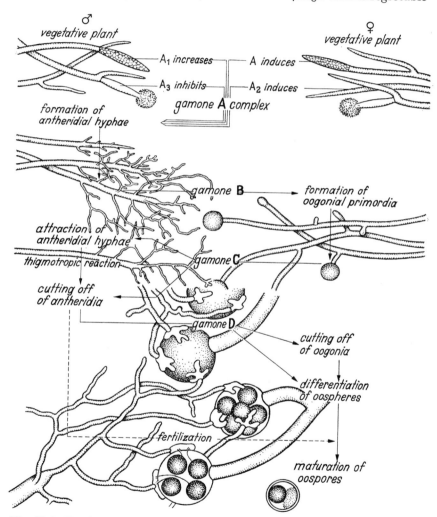

Fig. II-8. Partially diagrammatic representation of the sexual stages controlled by gamones in the male and female strains of *Achlya ambisexualis*. Each line which is designated by a letter shows the origin and effect of a particular gamone complex. The broken line (fertilization) is based on data still in question. (After RAPER, 1955b)

6*

because extensive data on the physiology of sex are available for the aquatic fungus, *Achlya ambisexualis*, which exhibits this kind of breeding system (RAPER, 1939a, b, 1940a, b, 1942a, b, 1950a, b; RAPER and HAAGEN-SMIT, 1942, review: RAPER, 1951).

RAPER showed that in dioecious strains of this Phycomycete *morphogenesis of the sex organs and the sexual reaction are determined by a series of sex hormones (gamones) which are correlated with one another*. Four groups of gamones (A-D) have been identified. The order in which they are produced depends on reciprocal interaction between male and female strains. The developmental stages and physiological steps which are controlled by the four gamone complexes are shown in Fig. II-8.

Since only recently it has been possible to germinate the oospores of *Achlya ambisexualis* to a very limited extent (MULLINS and RAPER, 1965a, b), one may hope that genetic studies of this fungus will be undertaken. It should then be possible to relate the outstanding discoveries described above to the genetics of morphogenesis and the hereditary basis of dioecism.

Summary

1. Fungi which are morphologically dioecious are relatively rare (18 species of Phycomycetes and 4 species of Ascomycetes).

2. It is not known whether dioecism is determined genically or by sex chromosomes.

3. Both the production of sex organs and the sexual reaction in the aquatic fungus, *Achlya ambisexualis*, are controlled by a series of gamones which are correlated with one another. The genetic basis of these morphological and physiological processes is unknown.

2. Physiological dioecism

Species with two physiologically distinct mating types occur among the Myxomycetes, Phycomycetes, and Ascomycetes. In the Myxomycetes these are mostly non-cellular species such as *Didymium nigripes*, *D. difforme*, *D. iridis*, and *Physarum polycephalum* [SKUPIENSKI, 1918, 1926; COLLINS, 1961 (literature review), 1963; COLLINS and LING, 1964; DEE, 1960]. The Mucoraceae among the Phycomycetes (see list in KNIEP, 1928) and the yeasts among the Ascomycetes include numerous physiologically dioecious species. Isolated cases of physiological dioecism have been described in the Chytridiales and the Blastocladiales (Phycomycetes) (see WHITEHOUSE, 1949a, for literature).

a) Genetic basis

Insofar as the reproductive behavior of physiologically dioecious forms has been investigated, only two complementary mating types have been found. These are generally designated as *plus* and *minus*, or in the yeasts, as *a* and *alpha* (p. 56). In all cases which have been studied genetically *the two mating types are determined by two alleles of a single locus* and not by sex chromosomes. Although no genetic studies of physiologically dioecious Myxomycetes that go beyond the identification of two mating types are known, work has been done on the Mucoraceae and extensively on the yeasts.

Mucoraceae

Genetic experiments with the Mucoraceae require a great deal of time because the dormancy of zygospores may extend as long as 6 months. Moreover it is not known for all species of these molds whether the 500—2000 spores of the germ sporangium arise from one or more meiotic divisions (p. 10). BURGEFF (1928) investigated this question genetically years ago in *Phycomyces blakesleeanus*. He carried out a three factor cross. In the F_1 eight different genotypes are expected, if the genes are unlinked and the products of more than a single meiosis are analyzed. BURGEFF found, however, only four genotypes among the progeny of a germ sporangium; this led to the conclusion that *only one nucleus divides meiotically in a zygosporangium*. The nuclei of the individual spores arise by mitotic divisions from the four products of meiosis. HARM (unpublished) confirms these results. The cytological investigations of SJÖWALL (1945) suggest the same conclusions. He found that diploid as well as haploid nuclei occur in the germ sporangium, but the former do not divide. The mechanisms by which their division is prevented is unknown.

Another problem in the Mucoraceae concerns the number of mating types. The work of BURGEFF (1912, 1914, 1915), BLAKESLEE et al. (1927) and KNIEP (1929a, b) *clearly demonstrated that a large number of species possesses only two mating types.*

BLAKESLEE and his coworkers made extensive crosses testing about 2000 mucoraceous strains from 34 species and 12 genera. Races of the same species were tested in about 10,000 different combinations. In no case were more than two mating types found.

Nevertheless, BLAKESLEE (1906a, b) found in his first studies on *Phycomyces nitens* among the *plus* and *minus* mycelia from a germ sporangium, a third mating type which exhibited a neutral reaction, i.e. it crossed with both *plus* and *minus* strains. This phenomenon was subsequently observed frequently and interpreted erroneously by many workers as proof of the existence of more than two mating types.

We know now that these *neutral mycelia are heterokaryons*. They arise at spore formation at a specific frequency through the inclusion of both a *plus* and *minus* nucleus in the spore. The possibility that diploid nuclei which undergo meiosis during their later development are responsible for the origin of the neutral mycelium cannot be excluded, however. Strains with sufficient markers to test this alternative are not available.

Two types may be distinguished among the neutral mycelia in *Phycomyces blakesleeanus*:

1. Strains with a *reduced viability* (BLAKESLEE, 1906a; ORBAN, 1919). These strains seldom produce vegetative sporangia but develop a large number of pseudophores which are similar in size and form to the suspensors of a zygospore. On occasion the pseudophores become transformed into

zygotes ("Binnenzygoten", ORBAN, 1919). More commonly small sporangio-spores grow out of them. In addition, sterile zygotes which fail to germinate are produced on this type of heterokaryon. Nothing is known about the nuclear and meiotic events involved in the formation of zygotes and sporangia. Generally a sexual reaction fails to occur between the heterokaryon and the homokaryotic *plus* and *minus* strains.

2. Strains with *normal viability* (HARM, unpublished). These heterokaryons form pseudophores in the region of mating with the two homokaryotic mating types. In many cases sporangia grow out of the suspensors which are produced by the heterokaryon and which lead to aborted zygotes. Zygotes from selfing also are formed on the heterokaryon. The degree of fertility of this type of heterokaryon when mated with the *plus* and *minus* strains is variable; it probably depends upon the distribution of nuclei in the heterokaryon.

The crossing behavior of the heterokaryons indicates that haploid *plus* and *minus* nuclei are present in these mycelia. This view was confirmed by experimentally producing heterokaryons. BURGEFF (1915) and ORBAN (1919) injected cytoplasm which contained nuclei from one homokaryon into mycelia of the complementary mating type. Using this technique, heterokaryons of type 1 were obtained.

The growth and sexual anomalies of the heterokaryons suggest a certain incompatibility of the plus and minus nuclei within the vegetative hyphae. There are no experimental data which indicate the nature of this incompatibility and its significance for physiological dioecism.

BLAKESLEE (1904a), in his first study on the Mucoraceae, observed that the sexual reaction, i.e. the formation of zygophores which come into contact or twist around one another, occurred between complementary mating types of different species and even different genera. However, such crosses do not yield germinable zygospores. Imperfect hybridization was extensively studied by BURGEFF (1924) and CALLEN (1940). By using this reaction BLAKESLEE and CARTLEDGE (1927) were able to classify the mating types of numerous species and genera using tester strains of *Mucor hiemalis* as standards. Monoecious Mucoraceae could also be classified in part in this scheme if they reacted with either the *plus* or *minus* tester. These investigations support the view *that the substances responsible for the sexual reaction are not species or genus specific*. This belief has been confirmed by the physiological experiments of PLEMPEL (p. 90).

Saccharomycetaceae

The sexual behavior of the physiologically dioecious yeasts is controlled by two mating types. This was first shown by GUILLIERMOND (1936, 1940) and later confirmed by WINGE and LAUTSEN (1939a, b). LINDEGREN and LINDEGREN (1943) established through crosses and analyses of asci that an allelic pair (*a/alpha*) determined the two mating types.

In certain *yeasts*, however, this *breeding system* may be *modified* by a series of different factors. RAPER (1960) identifies five different types of modification in his review:

1. Genetic lability of the mating-type locus. In contrast to the extreme stability of the *plus* and *minus* determinants of bipolar homogenic incompatibility, the *a* and *alpha* alleles mutate relatively frequently to either the complementary mating type or to a sterile type. This phenomenon has been described by various workers for different species (LINDEGREN and LINDEGREN, 1943, 1944; LEUPOLD, 1950; AHMAD, 1952, 1953).

The change of mating type in *Saccharomyces cerevisiae* may result not only from gene, but also chromosomal mutation. HAWTHORNE (1963) found that a change of mating type from *a* to *alpha* occurred as a result of a deletion which extended from the mating type locus for a distance of over 30 map units.

2. Complex structure of the mating-type locus. In *Schizosaccharomyces pombe* a multiple allelic series has been demonstrated for the mating type locus (LEUPOLD, 1950) which may produce self-fertility, complete sterility, or physiological dioecism. The individual alleles seem to be arranged linearly within the genetic mating-type region, since self-fertile strains have appeared among the progeny of a cross between different mating types at a frequency of 0.3% (LEUPOLD, 1958a, b). This is interpreted as intragenic recombination (p. 220).

3. Modifying factors which influence sexual behavior. WINGE and ROBERTS (1949) found a dominant gene, *D*, coming from the self-fertile *Saccharomyces chevalieri* in hybrids between this species and the dioecious *S. cerevisiae*. This gene induces a diploidization of haploid strains which occurs independently of mating type. Such diplonts do not react with either the *a* or the *alpha* type. Only haplonts with the recessive gene *d* pair with the complementary mating type. TAKAHASHI et al. (1958) found a similar situation in crosses between self-fertile and dioecious strains among races of *S. cerevisiae*. The difference between the results of WINGE and ROBERTS and those of TAKAHASHI is that the self-fertile strains used by the latter possess a positive pairing affinity for the complementary mating partner with respect to *a* or *alpha*. In a later study TAKAHASHI was able to extend these observations and show that the diploidization gene is composed of a series of complementary subunits.

4. Polyploidy. Polyploidy has been frequently observed in yeasts. Apart from certain unexplained mating reactions of the polyploids, most diploid yeasts behave according to the pattern of physiological dioecism, i.e. homozygous diplonts (*a/a* and *alpha/alpha*) are self-sterile and react only with strains of the opposite mating types. The heterozygotes (*a/alpha*) are self-fertile and do not react sexually with other strains (LINDEGREN and LINDEGREN, 1951; ROMAN et al., 1951, 1955; SUBRAMANIAM, 1951; POMPER et al., 1954; LEUPOLD, 1956). (See also the section on *polyploidy* p. 313 ff.)

5. Abnormal segregations. Irregularities in nuclear division and in the distribution of meiotic products may lead to abnormal segregations. Ratios may deviate from expectation through conversion (p. 223 ff.) or through postmeiotic mitoses (WINGE, 1951; WINGE and ROBERTS, 1954). In the latter case asci with more than four nuclei arise so that some of the four spores may be heterokaryotic (p. 223).

In spite of these complications yeasts have become valuable objects for genetic investigation.

Reproduction

b) Physiology

Mucoraceae

The Mucoraceae are not only the first fungi to be studied genetically, but also the first to be used for investigations of the physiology of the gene. Innumerable studies have been undertaken to determine the physiological differences between morphologically indistinguishable mating types and to relate them to their genetic determinants, the *plus* and *minus* genes.

Only a few examples of the physiological characters which have been investigated are mentioned below: Content of reducing sugars, lipids and oxidizing enzymes, Manilov reaction (SATINA and BLAKESLEE, 1925, 1926, 1927, 1928, 1929; BURGEFF and SEYBOLD, 1927; YOUNG, 1930, 1931); uptake of nutrients (SCHOPFER, 1928); free amino acids (RITTER, 1955); oxygen consumption of homokaryons in comparison with mixed cultures of *plus* and *minus* types (BURNETT, 1953a, b). There are differences of opinion regarding the role of carotenoids in the sexual reaction. A host of investigations reviewed by HESSELTINE (1961) have dealt with this problem. BURGEFF (1924), SCHOPFER (1928, 1930), CHODAT and SCHOPFER (1927) and KÖHLER (1935) have noted that differences in the carotene content of individual mating types have not been established with certainty. HESSELTINE (1961) showed that the carotene content of *plus/minus* mixed cultures was distinctly greater than in pure cultures or in mixed cultures of *plus* and *minus* mycelia which had lost their sexual activity. This suggests that increased carotene production begins only following the interaction of the two mating types. Thus it is not difficult to understand that individual cultures do not show differences. However, the results of BURNETT (1956b) do not agree with this interpretation. BURNETT demonstrated that in *Phycomyces blakesleeanus* a correlation between the sexual reaction and the formation of beta carotene, as postulated by others (GARTON et al., 1951; GOODWIN and LIJINSKY, 1951; GOODWIN, 1952; GOODWIN and WILLMER, 1952), does not exist.

In most of these investigations *physiological differences between mating types were actually found,* but in many cases only single strains were tested. *Moreover, genetic studies,* which show that such differences can be attributed to the mating type locus, *are generally lacking.* It is possible that other genes with purely physiological effects could be shown to segregate in a careful genetic analysis, e.g. for gametangial size, in many Mucoraceae (p. 49). BURNETT (1956b) has expressed similar views. In his critical review of the pertinent literature he found through analysis of earlier data that significant physiological differences between *plus* and *minus* strains did not exist or existed in the reverse relationship. Conclusive evidence demands (1) the testing of more than a single strain of the same species and (2) the use of isogenic stocks.

The following example illustrates the confusion that may result from experiments carried out without regard to the above criteria: UTIGER (1953) reported that *minus* mycelia crushed in 30% NaOH gave an intense red color, while the *plus* mycelia did not give such a reaction. A more detailed testing of this technique on the same material by BRUCKER (1954, 1955) led to different results. BRUCKER showed that the pigment was also produced in *plus* strains from other sources and is not correlated with mating type. He identified phenol derivatives of gallic acid as the precursors of the red pigment; these become oxidized to red hallochromes in alkaline medium.

The investigations to date do not indicate a correlation between physio-logical differences in the plus and minus strains and the alleles which are responsible for mating determination. In this respect a *working hypo-thesis of* BURGEFF *has been fruitful.* This was the idea of isolating the substances responsible for each of the steps in the differentiation of the sexual reaction. BURGEFF already postulated the existence of such sub-stances in 1924. He observed that *plus* and *minus* strains of *Mucor mucedo* which were separated by a collodion membrane begin to develop zygophores in 2—3 days on each side of the membrane. The zygo-phores of both strains bend in the direction of the other mating type, i.e. toward the membrane, as in the typical sexual reaction. Since the hyphae are unable to penetrate the membrane, the reaction must be produced by diffusible substances. KÖHLER (1935) made similar obser-vations on the same species and KRAFCZYK (1935) on *Pilobolus crystal-linus.*

After unsuccessful attempts by other workers (VERKAIK, 1930; RONS-DORF, 1931; KÖHLER, 1935; KEHL, 1937; BANBURY, 1954a, b), PLEMPEL, using ingenious techniques, succeeded in *isolating the sex hormones of Mucor mucedo* (BURGEFF and PLEMPEL, 1956; PLEMPEL, 1957; PLEMPEL and BRAUNITZER, 1958; PLEMPEL, 1960, 1963a, b; PLEMPEL and DAWID, 1961). The following scheme represents the course of the sexual reaction:

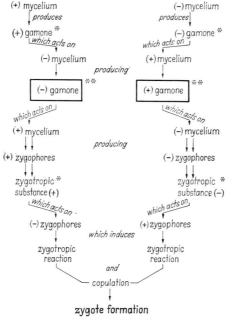

Fig. II-9. Mechanism of sexual reaction in *Mucor mucedo.* * demonstrated by biological assay, ** isolated in crystalline form. (Adapted from PLEMPEL, 1960)

As shown in Fig. II-9 a number of steps in differentiation are involved. The sexual reaction is initiated with formation of zygophores and concludes with fusion of gametangia (plasmogamy). *The complete process is controlled by three different sex hormones: progamones, gamones, and zygotropic substances.*

Each mating type produces in an individual culture a specific progamone that is released into the substrate and induces in the opposite mating type the formation of gamones. The gamones stimulate the production of typical sex organs (zygophores) in the complementary mating type. The specific *plus* and *minus* gamones have been isolated and prepared in crystalline form. They are composed of unsaturated aliphatic polyhydroxycarbonyl linkages. Elemental analysis and molecular weight determinations give the formula: $C_{20}H_{25}O_5$. The last step in the sexual reaction, the rapprochment of zygophores of opposite mating type, is controlled by specific gaseous substances released from the gametangia. These hormones have two properties: 1. they function as growth hormones and are responsible for the zygotropic reaction of the zygophores; 2. they determine the capacity of the zygotes to function; without the effect of these substances the zygophores become transformed into vegetative sporangiophores. The chemical nature of these gaseous hormones is not known. No experiments on the physiological processes related to the sexual reaction, such as plasmogamy and the germination of the zygotes, have been carried out.

The participation of gaseous hormones in the sexual reaction has also been demonstrated for *Rhizopus sexualis* by HEPDEN and HAWKER (1961). Preliminary studies indicate that these ephemeral substances are basic in nature. The existence of volatile sex hormones was assumed by BLAKESLEE (1904b), BURGEFF (1924) and BANBURY (1954a, b). The assumption came into question, however, because in spite of numerous attempts such substances could not be demonstrated. However, the work of PLEMPEL and of HEPDEN and HAWKER can be considered conclusive proof for the existence of gaseous hormones.

PLEMPEL has also demonstrated progamones and gamones in *Phycomyces blakesleeanus* and *Rhizopus nigricans*. *Neither sex hormone has a species specific effect.* The substances isolated from *Mucor* also induce the typical sexual reaction in other genera and species of the Mucoraceae and Choanophoraceae (PLEMPEL, 1957). Thus a hypothesis proposed earlier by a number of workers has been confirmed (p. 86).

In conclusion we would like to point out a parallel between the physiological mechanisms that determine the sexual reaction in *Achlya* and in the Mucoraceae. The sexual reaction is correlated with morphogenetic processes in both instances. Differentiation results from reciprocally induced hormones. While in *Achlya* the genetic basis of the sexual reaction is unknown, in the Mucoraceae it seems reasonably certain that an allelic pair is responsible for the *plus* and *minus* types.

The strains used by PLEMPEL were obtained from the BAARN collection and had been propagated vegetatively for many years. It would be desirable to analyze these strains genetically and prepare isogenic stocks from them. Conclusive demonstration that the *plus/minus* locus determines the sex hormones should present no difficulties.

Saccharomycetaceae

Physiological investigations of the development of the sexual reaction between two mating types in yeast were carried out by LEVI (1956) and BROCK (1958a, b, 1959). LEVI tested the copulatory reaction of three races of *Saccharomyces cerevisiae*. He found that only the *a* cells form a copulation process in all possible combinations between the *a* and *alpha* mating types of the three races. The differentiation is determined by a diffusible substance that is secreted by the *alpha* partner, since cells separated by a membrane from the *alpha* cells showed the same reaction. The *a* cells also form a copulation process if they are transferred to an agar medium on which *alpha* cells have previously been growing. Interpretation of these observations as gamone activity corresponding to that in *Achlya* or in the Mucoraceae is inevitable. More detailed investigations are necessary to show to what extent the parallel holds. No generalization is warranted as yet, since BROCK arrived at an entirely different conception of sexuality in the physiologically dioecious yeasts in his studies on *Hansenula wingei*. He viewed the cellular fusion of different haploid mating types as an antigen-antibody reaction: a protein (antibody) that occurs on the surface of the cells of one mating type is believed to react with a polysaccharide (antigen) that is localized on the cell surface of the opposite mating type: as a result agglutination of cells of opposite mating types occurs. The relation of such cell agglutination to the processes of plasmogamy and karyogamy remains undetermined, since no differentiation such as germ tube formation (*S. cerevisiae*, LEVI) or development of sex organs (*Achlya* and Mucoraceae) is correlated with aggregation.

Summary

1. Species with two physiologically different mating types occur among the Myxomycetes, Phycomycetes, and Ascomycetes. Extensive genetic and physiological investigations have been carried out on the Mucoraceae and yeasts.

2. The two mating types of a physiologically dioecious species are determined by two alleles of a single locus. Multiple alleles for mating type are known only in yeasts.

3. Only a single meiosis occurs in the multinucleate germ sporangium of the Mucoraceae. The nuclei of the individual spores arise by mitosis from the four products of this division. The spores may be heterokaryotic.

4. The breeding system of the physiologically dioecious yeasts may be influenced by a large number of modifying factors.

5. The innumerable experiments on the Mucoraceae that attempted to elucidate the physiological effect of the mating type gene have succeeded in only one instance. The substances responsible for the sexual reaction in *Mucor mucedo* have been identified and in part isolated. The effect of these sex hormones is not genus or species specific.

6. Physiological genetic studies on yeasts have not led to a unified hypothesis of the mechanism of the sexual reaction.

C. Alternatives to sexual reproduction

For a long time sexual reproduction was thought to be the only mechanism through which new combinations of genetic material could arise. In light of this monopoly of sexually reproducing organisms on recombination the evolution of many microorganisms (e.g. Bacteria and Fungi Imperfecti) which do not undergo karyogamy and meiosis during their life cycle was difficult to understand. The idea that mutation can provide the variability for evolution in such species in the absence of any genetic recombination was not very convincing (MULLER, 1947). It is to the credit of the microbial geneticists that an alternative explanation has been provided. A number of mechanisms are now known, which, although differing in their genetic bases, nevertheless provide for recombination of genetic information.

Most of these mechanisms have been found in bacteria: transformation (GRIFFITH, 1928), transduction (ZINDER and LEDERBERG, 1952), bacterial conjugation (TATUM and LEDERBERG, 1947), lysogeny (LWOFF, 1953; JACOB and WOLLMAN, 1957), and sexduction (JACOB and ADELBERG, 1959). The mechanism of each of these systems may be found in JACOB and WOLLMAN (1961).

The only known alternative to sexual reproduction in fungi is somatic recombination. This refers to genetic recombination which occurs in vegetative cells. Somatic recombination was demonstrated in Ascomycetes and related Fungi Imperfecti and in Basidiomycetes (Table II-6 and review by BRADLEY, 1962).

Only a single case is known in which, in spite of intensive analysis, somatic recombination could not be demonstrated. DUTTA and GARBER (1960) tested approximately one million conidia from heterokaryotic strains of the imperfect fungus *Colletrichum lagenarium* without recording a diploid heterozygous nucleus.

The *best known genetic mechanism* underlying somatic exchange is *mitotic recombination* (p. 204 ff.). This was first discovered by STERN (1963) in heterozygous tissue of *Drosophila* and was considered to represent the result of crossing over between homologous chromosomes during mitosis.

The prerequisites for mitotic recombination, namely diploidy and heterozygosis, may also be realized in the hyphae of haploid fungi in rare cases. ROPER (1952) found occasional diploid nuclei exhibiting such recombination in heterokaryons of the Ascomycete *Aspergillus nidulans*. Since these processes lead to the same result as meiotic recombination during sexual reproduction, PONTECORVO (1954) suggested the term parasexuality to describe the phenomenon.

The *parasexual cycle* consists basically of the following stages (for details see PONTECORVO et al., 1953b; PONTECORVO, 1956a, b, 1958):

1. Heterokaryon formation. Two haploid mycelia fuse to form a heterokaryon.

2. Nuclear fusion. Nuclear fusions occur on rare occasions in heterokaryons (at a frequency of approximately 10^{-6}), forming diploid heterozygous nuclei. The haploid and diploid nuclei in the heterokaryon multiply by mitosis. Eventually, however, a segregation occurs so that haploid heterokaryotic as well as diploid heterozygous sectors of the mycelium develop.

3. Mitotic crossing over. Mitotic crossing over takes place in the diploid nuclei at a frequency of about 10^{-2} per nuclear division.

4. Haploidization. The diploid nuclei are reduced to the haploid condition at a relatively constant frequency of 10^{-3} per nuclear division. During the process they may go through different stages or aneuploidy. The change in ploidy level does not result from a meiotic division (KÄFER, 1961).

Mitotic recombination and haploidization take place concurrently. The dissemination of haploid nuclei that are genetic recombinants may occur through segregation in the heterokaryon (sectoring) or by means of uninucleate conidia.

Genetically identical nuclei are also assumed to fuse and undergo mitotic recombination and reduction to the haploid condition in hetero- and homokaryons. This does not have a genetic effect, of course. Heterokaryosis is essential for the genetic effectiveness of the parasexual cycle. In this respect heterokaryosis corresponds to heterozygosis in sexual reproduction.

Parasexuality has been demonstrated with certainty in *Aspergillus niger* and *Penicillium chrysogenum*. Since both species are imperfect, parasexuality is the only mechanism by which they undergo genetic recombination. It is very likely that in other imperfect Ascomycetes listed in Table II-6 mitotic recombination is correlated with a parasexual cycle.

The practical significance of the discovery of parasexuality is that genetic investigations may also be carried out on the imperfect fungi.

It has not been possible as yet to relate the somatic exchanges observed in Basidiomycetes to a parasexual cycle. While unequivocal proof of mitotic crossing over has been shown in *Ustilago maydis*, its occurrence in rusts is doubtful. In the latter case the exchange may be of a meiotic type (ELLINGBOE, 1961).

Somatic recombinations occur regularly in the Holobasidiomycetes (*Schizophyllum* and *Coprinus*) in crosses between dikaryons and homokaryons, and, in fact, in incompatible combinations (see Buller phenomenon, p. 65). The genetic mechanisms that are involved are mitotic (CROWE, 1960) or meiosis-like recombination (ELLINGBOE and RAPER, 1962a; ELLINGBOE, 1963)[1].

Not all somatic recombination in dikaryon-monokaryon crosses can be explained by these two mechanisms, as the results of ELLINGBOE and RAPER have shown. As yet no hypothesis has been formulated to explain this type of recombinant.

Somatic exchange of genetic material in *Schizophyllum* does not occur only in connection with the Buller phenomenon; recombinants for incompatibility factors arise from dikaryons at a frequency of approximately 1% (RAPER, 1961b; PARAG, 1962b).

The prerequisite for elucidating this unknown mechanism of somatic recombination is the development of suitable methods to permit selection of rare recombinants. These processes of genetic recombination may very likely hold a greater significance for evolution than was previously believed, particularly in the imperfect fungi.

---

[1] Acknowledgement in proof: Recent investigations of ELLINGBOE (1964) support the view that meiotic recombination actually occurs.

Table II-6. *Examples of somatic recombination in fungi*

In some species somatic recombination may result from several different genetic mechanisms.

object	sexual phase known	mechanism of recombination			reference
		mitotic crossing over	meiotic crossing over	other	
Protoascomycetes					
Saccharomyces cerevisiae	+	?	?	?	WILKIE and LEWIS, 1963
Euascomycetes					
Aspergillus fumigatus		+			STRØMNAES and GARBER, 1963
nidulans	+	+			ROPER, 1952; PONTECORVO et al., 1953b, 1954; PONTECORVO and ROPER, 1952
niger		+			KÄFER, 1958, 1961; PONTECORVO et al., 1953a
oryzae		+			ISHITANI, 1956;
sojae		+			ISHITANI et al., 1956; IKEDA et al., 1957
Cephalosporium mycophyllum		+		+	TUVESON and COY, 1961
Cochliobolus sativus	+	+			TINLINE, 1962
Penicillium chrysogenum		+			PONTECORVRO and SERMONTI, 1954; SERMONTI, 1957
expansum		+			BARRON, 1962; BARRON and MACNELL, 1962
italicum		+			STRØMNAES et al., 1964
Fusarium oxysporum f. pisi		+			BUXTON, 1956; TUVESON and GARBER, 1959
oxysporum f. cubense		+			BUXTON, 1962
Verticillium albo-atrum		+			HASTIE, 1962
Holobasidiomycetes					
Coprinus fimetarius	+	+		+	GANS and PRUD'HOMME, 1958; PRUD'HOMME, 1962, 1963
lagopus	+	?		?	SWIEZYNSKI, 1962
macrorhizus	+	?		?	KIMURA, 1954a, b

Table II-6 (Continued)

object	sexual phase known	mechanism of recombination			reference
		mitotic crossing over	meiotic crossing over	other	
Schizophyllum commune	+	?	?	?	CROWE, 1960; RAPER, 1961a; ELLINGBOE and RAPER, 1962a; PARAG, 1962b; ELLINGBOE, 1963, 1964
Uredinales					
Puccinia recondita f. tritici	+			+	VAKILI and CALDWELL, 1957
graminis f. tritici	+		?	?	BRIDGEMON, 1959; WATSON, 1957, 1958; ELLINGBOE, 1961; WATSON and LUIG, 1962
Ustilaginales					
Ustilago maydis	+	+			HOLLIDAY, 1961a, b

Summary

1. In bacteria and fungi exchange of genetic material is not restricted to sexual reproduction. A series of mechanisms are known that allow new combinations of genetic material in somatic nuclei without karyogamy and meiosis; thus, true alternatives to sexual reproduction have been discovered.

2. The best known and best analyzed of these mechanisms in fungi is mitotic crossing over. The discovery of the parasexual cycle in the Aspergillaceae has made possible the genetic analysis of imperfect fungi.

3. Investigations on Basidiomycetes have led to the conclusion that meiosis-like processes are responsible for somatic recombination.

D. Relative sexuality

The principle that the strength of sex determination is relative is one of the three concepts upon which *Hartmann's theory of sexuality* is based (HARTMANN, 1956, p. 408). According to this principle, *differentiation of gametes into male or female is relative*, not absolute. Thus in an extreme case a normally female gamete A may react as a female with male gamete B, but behave as a male with a more strongly female gamete A_1 (HARTMANN, 1956, p. 121).

HARTMANN (1909) called this phenomenon relative sexuality. The extensive investigations of HARTMANN and his numerous students, which were carried out over many years, as well as those of other workers on sexual behavior in plants and animals (presented in detail and interpreted in the first edition of his book "Sexualität", 1943), led to the conclusion that relative sexuality is closely correlated with sexual differentiation.

The theory of relative sexuality is based primarily on work with algae and other protista. A variety of experimental results with *fungi* may also be explained by HARTMANN's theory. These will be discussed in some detail, because in the light of recent work *certain older data are no longer to be interpreted as relative sexuality*.

The investigations of GREIS (1941) on *Sordaria fimicola* were used by HARTMANN as unequivocal evidence for the occurrence of relative sexuality. GREIS irradiated this monoecious Ascomycete with X rays and obtained a large number of sterile variants whose development was blocked at different stages in the normal cycle. Mycelia which did not produce ascogonia were called "males", and those which produced ascogonia but failed to form mature fruiting bodies were called "females". "Males" and "females" could be crossed. In some cases males, and also females, could be crossed among themselves. GREIS attributed to these "males" and "females" different degrees of sexual potency and thus explained his results on the basis of relative sexuality. This interpretation can no longer be supported because of investigations on *Sordaria macrospora* as well as information obtained from the application of tetrad analysis to GREIS' data by ESSER and STRAUB (1958). As pointed out previously (p. 44 f.), GREIS' variants, which had different male and female "potencies", were actually *mutants with developmental defects. When such defects are determined by non-allelic genes in different mutants they may be crossed because of the complementary effect of the corresponding wild type alleles. Mutants with defects determined by alleles do not show complementation and cannot be crossed.* The same conclusions were reached by HESLOT (1958) and CARR and OLIVE (1959) in their investigations on *Sordaria macrospora* and *S. fimicola* respectively.

In another monoecious self-incompatible Ascomycete, *Glomerella lycopersici*, HÜTTIG (1935) likewise attributed the crossing ability of sterile strains with fertile variants to relative sexuality. That such a view is untenable has already been pointed out by WHEELER and his coworkers (summary in WHEELER, 1954a). They demonstrated that in *Glomerella cingulata* sterile mycelia, whether they produce male or female sex organs, can only be crossed with one another if their sexual defects result from non-allelic genes. Whether the defect involves the differentiation of the male or the female organs is irrelevant.

Therefore, the crossing ability of such morphological mutants in *Glomerella* has the same basis as in *Sordaria*, namely the complementation of non-allelic defects. Complementation in both instances occurs by means of the heterokaryon which always forms in the zone of contact between two mating partners. The heterokaryotic hyphae are capable of undergoing normal development characteristic of the wild type strain.

This can best be demonstrated by transferring small pieces of heterokaryotic mycelium from the contact zone between sterile partners to fresh medium. The mycelium which develops from such transfers cannot be distinguished from the corresponding homokaryotic wild type strain.

Apart from the investigations on *Sordaria* and *Glomerella*, which were cited as unequivocal proof of relative sexuality by HARTMANN, other data from fungi were interpreted as supporting this theory (HARTMANN, 1956, p. 124ff.), although, it was admitted, not entirely without question. For the sake of completeness, these cases will be discussed briefly.

COUCH (1926) observed in the dioecious Phycomycete *Dictyuchus monosporus* that a female strain (*N*) when crossed with another female strain (*A*) produces antheridia as well as oogonia, the former growing alongside the oogonia of strain *A*. A similar shift from dioecism to monoecism occurs in another Phycomycete, *Achlya ambisexualis* (RAPER, 1947 and pp. 81, 83), and can be correlated with environmental conditions. However, it is not clear whether either case represents true dioecism, since the genetic bases for these sexual differences are not clear.

BLAKESLEE et al. (1927), BLAKESLEE and CARTLEDGE (1927), as mentioned previously (p. 49), carried out extensive racial, specific and generic crosses with numerous Mucoraceae which served to establish and standardize mating types. In certain combinations they observed differences in the strength of reaction between *plus* and *minus* mating types which were expressed in the number of zygospores produced. Similar differences also occurred in matings between monoecious Mucoraceae.

In the first place, the fact that no sexual differentiation can be demonstrated in mating types of the Mucoraceae argues against relative sexuality. Further, it is well known that in *Achlya* and probably *Dictyuchus* the differentiation of sex organs is the result of the action of sex hormones of each partner on the other. That different races, species, and genera possess different genes for the production of sex hormones which permit the sexual reaction in only particular combinations is to be expected. As long as the exact genetic basis of the expression of different mating types remains unknown, such results cannot be attributed to expression of relative sexuality.

GREIS (1942) described morphological mutants in the Hymenomycete *Solenia anomala*. The male mycelia possess very slender hyphae and grow slowly. The female mycelia exhibit the habit and growth characteristics of the wild type. In crosses between the two types the "male" mycelia function exclusively as nuclear donors and the "female" as nuclear recipients. This donor-recipient relationship is relative, however, since crosses within the "male" and "female" strains are possible. RAPER and MILES (1958c) showed by genetic analysis in *Schizophyllum commune*, a relative of *Solenia*, that such unilateral crossing relationships occur frequently and are controlled by a gene which is closely linked to the mating type locus. Thus this type of sexual differentiation is illusory. The crossability of the different *Solenia* mutants can also be explained through complementation in the heterokaryon, as in the Ascomycetes discussed above.

The significance of the investigations of BAUCH (1923, 1934) on Ustilaginaceae and of QUINTANILHA (1935) on Hymenomycetes (Buller phenomenon) with regard to relative sexuality was based on the idea that in these Basidiomycetes with tetrapolar incompatibility one of the two factors is a sex factor and the other a sterility factor. We now know that this is not the case and that both are incompatibility factors (see also p. 61f.), making further discussion superfluous.

Summary

From the standpoint of genetic analysis the existence of relative sexuality in fungi has not been demonstrated.

E. Vegetative incompatibility

A number of authors (DODGE, 1935; BEADLE and COONRADT, 1944; SANSOME, 1945, 1946) have observed that the formation of heterokaryons in *Neurospora sitophila* and *N. crassa* is only possible between incompatible strains (identical mating types). SANSOME was able to demonstrate the formation of a heterokaryon between sexually compatible *plus* and *minus* strains only once in numerous attempts. These results have led to the view that heterokaryosis is inhibited by the action of incompatibility factors. SANSOME did point out, however, the difficulty of reconciling this hypothesis with the fact that in fruiting body formation in the ascogenous hyphae the heterokaryotic condition between *plus* and *minus* nuclei is not disturbed. Therefore, she assumed that *heterokaryon formation may be inhibited, not because of the incompatibility factors, but through other genes.* This view was confirmed by the experiments of GARNJOBST and coworkers on *Neurospora crassa* (GARNJOBST, 1953, 1955; GARNJOBST and WILSON, 1956, 1957; WILSON et al., 1961).

Using nutritional mutants GARNJOBST proved that *two unlinked allelic pairs (C/c and D/d) were responsible for heterokaryon formation.* A heterokaryon will result only if the partners carry like alleles at both loci. If one or both factors are heterogenic, no heterokaryon will form. *The effect of the "CD" genes is in no way correlated with the effect of the incompatibility genes.* On the one hand, whether or not the partners are sexually compatible in the same *"CD"* configuration is irrelevant for formation of a heterokaryon; on the other hand, in dissimilar *"CD"* configurations which result in the failure of heterokaryon formation sexual compatibility of the *plus* and *minus* strains is not affected, a fact which remains unexplained.

Microscopic observations showed that hyphal fusions occur normally in the heterokaryon-negative combinations. Immediately following protoplasmic transfer, a denaturation of the cytoplasm begins (vacuole formation, "balling up" of the cytoplasm) which may extend beyond the adjoining septa and in a few hours leads to the death of the hyphal region. The hyphal segments which lie more distant from the point of contact and have not been involved in the cytoplasmic mixing cut themselves off from the zone of degeneration either by forming a plug in the perforation of the crosswall or through the formation of a distinct plasma membrane. In heterokaryon-positive combinations there is no indication of such "protoplasmic incompatibility" (GARNJOBST). Following anastomosis and protoplasmic exchange which simultaneously involves nuclear exchange, heterokaryon formation results either through nuclear migration or production of new hyphae which are heterokaryotic.

In our opinion the choice of the expression "protoplasmic incompatibility" was unfortunate. Considered superficially, it leads readily to the conception that this kind of incompatibility is produced exclusively by factors in the plasmone. Moreover, the word "incompatibility" needs a clear distinction from sexual incompatibility for which the term has generally been used. To circumvent these objections we suggest that "vegetative incompatibility" be used instead of "protoplasmic incompatibility".

The use of a micro-injection technique (WILSON, 1961) has allowed a more detailed investigation of vegetative incompatibility. The injection of a cell-free compatible mycelial extract produces no hyphal damage. Following injection of an extract from an incompatible partner, the same protoplasmic denaturation occurs at the point of injection as in the case of anastomosis of heterokaryon-negative partners. Cell extracts which had been treated previously in a variety of chemical and physical ways (centrifugation, heating, ammonium sulfate precipitation) were tested with this method. The experiments led to the conclusion that the active principle for vegetative incompatibility is part of the soluble protein fraction.

The investigations of GROSS (1952), HOLLOWAY (1955) and DE SERRES (1962) confirmed the view that in *Neurospora crassa* failure of heterokaryon formation is controlled by a number of genes which are not identical to the factors determining homogenic incompatibility. Whether or not these are the same loci analyzed by GARNJOBST was not determined.

PITTENGER and BRAWNER (1961) have described another example of vegetative incompatibility. These authors found an allelic pair (I/i) which controls the capacity of nuclei in the heterokaryon to divide. When the proportion of i nuclei in an $(I+i)$-heterokaryon is less than 70%, the capacity of the i nuclei to divide is decreased to such an extent that the number of I nuclei increase and the heterokaryon becomes an I homokaryon. By making corresponding tests with homokaryons it was shown that cytoplasmic differences or a differential division frequency of the two kinds of nuclei could not account for the incompatibility between the I and i nuclei[1]. By using different marker genes in the nuclear components of the heterokaryons the I/i incompatibility was shown to be generally independent of the genetic background. Sexual incompatibility (as in the case of the CD genes) is not influenced by the I/i alleles. Tests in which strains carrying the I locus were compared with the C and D loci of GARNJOBST indicate that different genes are involved. This is confirmed by the fact that phenotypic expression of vegetative incompatibility differs in the two cases.

Vegetative incompatibility is not limited to *Neurospora*. GRINDLE (1963), JINKS and GRINDLE (1963) found that isolates of *Aspergillus nidulans* from the wild can be placed into different groups according to their ability to produce heterokaryons. As genetic analysis of the separated components of "heterokaryons" has shown, chromosomal factors are responsible for vegetative incompatibility, but their exact nature remains to be determined.

---

[1] Acknowledgement in proof: PITTENGER (1964) recently established that after prolonged vegetative multiplication the incompatibility between the two nuclear components of a heterokaryon no longer remains evident because of the genetic modification of the incompatibility locus. It is not clear whether this change results from mutation or somatic recombination.

The researches of BEISSON-SCHECROUN (1962) and BERNET (1963 a, b) have shown that *the barrage phenomenon of Podospora* (RIZET, 1952) *is a macroscopic expression of vegetative incompatibility*. The formation of a barrage (p. 71) depends in general upon the heterogenic constitution of the two mycelia. In the example analysed in detail by RIZET and BEISSON-SCHECROUN *not only are genic differences responsible, but also differences in the plasmone*. For this reason this type of vegetative incompatibility will be discussed in greater detail in the chapter on *extra-chromosomal inheritance* (p. 447 ff).

Another case of vegetative incompatibility which differs in its genetic basis from previously discussed examples was observed by CHEVAUGEON and VAN HUONG (1961) in *Pestalozzia annulata*. In this fungus anastomoses can occur only between hyphae of the same age in a mycelium germinating from a spore. Young hyphae do not fuse with hyphae which are more than 90 hours old. Since this phenomenon occurs in homokaryotic mycelia, it is simply a matter of hyphal differentiation and is not genetically determined. A comparison with the data obtained from *Neurospora crassa* is therefore not possible.

Vegetative incompatibility shows several parallels with heterogenic incompatibility.

1. The existence of both systems remained unrecognized for a long time.

The reasons for this are the following: In genetic work with fungi one is primarily interested in having two strains reproduce sexually. Whether or not they produce a heterokaryon in the vegetative phase is of secondary importance. Likewise, that races of the same species can or cannot be crossed is not of major significance, since stocks of a race which are as isogenic as possible are usually desired.

2. Incompatibility occurs when two partners carry different incompatibility genes in both systems.

Thus we should speak of heterogenic vegetative incompatibility. However, since "heterogenic incompatibility" is cumbersome enough as is, to add "vegetative" has no real purpose as long as only a single genetic mechanism is known for vegetative incompatibility and no alternatives need be differentiated.

3. In both systems incompatibility of the two partners leads to complications in heterokaryon formation.

The essential difference between the two systems is that vegetative incompatibility is expressed only in the vegetative phase, while heterogenic incompatibility also produces an effect in sexual reproduction.

Summary

1. The failure of particular pairs of mycelia to form heterokaryons has often been observed in *Neurospora crassa*. This phenomenon has been designated vegetative incompatibility.

2. Various workers have identified a number of genes which are responsible for vegetative incompatibility. The genetic mechanism of vegetative incompatibility is heterogenic, i.e. two mycelia form a heterokaryon only if they carry different alleles of the incompatibility factors involved.

3. Vegetative incompatibility, in contrast to heterogenic incompatibility, does not influence sexual behavior.

F. Evolution

Evolution is a process which constantly alters form as well as all aspects of living things. Its essential components are mutation and selection. Spontaneous mutations become distributed in a population and become finally established as a constant feature of the population only if they possess a selective advantage. Environmental factors thus play a decisive role. Sexual reproduction is a primary condition for the distribution of changes in the genetic material. Genetic information is constantly recombined through the regular alternation of karyogamy and meiosis. To a limited extent this function may be assumed by somatic recombination (p. 204ff.).

The fungi represent the product of a long evolution. They differ from most other phylogenetic lines of the plant kingdom in exhibiting many kinds of sexual differentiation and numerous reproductive systems. The path of evolution from the ancestral types to the modern forms is difficult to trace. Fossil representatives are almost completely lacking (which in other groups, e.g. Pteridophyta, allow a relatively accurate determination of phylogenetic relationships). Clues to the evolution of fungi must be found in studies of comparative anatomy, morphology, and genetics. Unfortunately conclusions from such investigations remain to a large extent speculative. Unequivocal evidence is meagre.

Our purpose is not to develop a general concept of the evolution of the fungi. Rather we wish, with a few examples and a minimum of speculation, to evaluate the evolutionary significance of certain genetic data.

I. Reduction of sex organs

A major evolutionary trend in the fungi is the reduction of the sex organs. In almost all the larger taxa evolutionary lines can be identified in which all gradations between individuals with distinct male and female sex organs and those in which sexual differentiation is completely absent occur. Most texts present the idea that *the production of sex organs is a primitive character and that all reductions are to be considered derived.* Genetic data which are relevant to this question support such a view. No mutation has led to more highly differentiated sex organs; only reductions have occurred. As discussed in the section on *morphology* (p. 41 ff.), certain cases are known in the Ascomycetes in which sex organ formation is inhibited by single gene mutations.

The *consequences of such mutations with regard to evolution* may vary. If a fungus which loses the capacity to produce either male or female sex organs is hermaphroditic (e.g. *Bombardia lunata*, ZICKLER, 1934) unisexual strains arise which may be the basis for the origin of a dioecious species. In species with apandric development, which do not produce male sex organs (e.g. *Sordaria macrospora*, ESSER and STRAUB, 1958), sexual differentiation may be inhibited by a single mutational step. For all practical purposes such a mutant is an *imperfect fungus*.

101

It should be emphasized that the Fungi Imperfecti are not "dead ends" of evolution, since in a number of these forms genetic exchange occurs through somatic recombination.

The two evolutionary trends toward dioecism and an imperfect condition are possible only if the mutants possess a selective advantage. Such a mutant may succeed through better adaptation to natural nutritional conditions or through a forced ecological isolation.

An example of the former instance is a type of female-sterile mutant in the Ascomycete, *Podospora anserina*. Such *"incoloris"* mutants arise spontaneously after extended vegetative propagation. They can be recognized as colorless sectors in older *Podospora* cultures. Since the growth rates of the mutants are greater than that of the wild type, every wild type strain sooner or later loses the capacity to serve as a female. A number of loci, some of which are unlinked, are responsible for the *"incoloris"* character (ESSER, 1966b).

Reproductive isolation may occur through genetic mechanisms (e.g. through vegetative incompatibility, p. 98 ff.). Sterile strains (*Sordaria*, Fig. II-1, p. 45 ff.) normally produce heterokaryons as long as non-allelic defects are complemented. Fruiting body formation is thus not a problem. This is impossible, however, if heterokaryosis is prevented by vegetative incompatibility. The same effect may result from heterogenic incompatibility. More work is needed on the isolating effect of this type of incompatibility system.

An interesting example of a naturally occurring isolation based on an adaptation is described by GORDON (1954). He investigated the sexual relationships between different strains of *Gibberella cyanogena* collected in Canada, England, New Zealand, and Tasmania. *G. cyanogena* shows bipolar incompatibility and is the perfect form of *Fusarium sambucinum*. GORDON found that the two mating types of this species, with one exception, never occur together in nature. Apparently this phytopathogenic fungus exists in optimal environmental situations which do not necessitate sexual reproduction for the maintenance of the species.

Genetic investigations on sterile deficiency mutants have implications for another problem, namely, *the evolution of breeding systems*. The problem as to *which breeding system is primitive — monoecism with self-compatibility, monoecism with homogenic incompatibility, or dioecism —* has been discussed frequently (WHITEHOUSE, 1949a, 1951a; BURNETT, 1956a; OLIVE, 1958). OLIVE concerned himself in his review particularly with the alternatives of compatibility and incompatibility. He concluded that incompatibility was a derived condition. He has also recently presented experimental evidence for this view (EL-ANI and OLIVE, 1962).

Investigations were carried out using the self-compatible monoecious Ascomycete, *Sordaria fimicola*. Many mutants are known in *Sordaria* which block normal development at different stages (Fig. II-1 and p. 43f.). Sterile mutants will cross only if the defects result from non-allelic factors. An intensive analysis of mutants induced by X-rays led to the discovery of two closely-linked sterile mutants which nevertheless crossed with one another. The two mutants, which were characterized by aborted asci and ascospores produced perithecia with normal spores in the cross. 504 asci were analyzed genetically. Recombinations between the two mutants did not occur. These experiments support the view that the two markers represent two mutational sites within the same gene.

The results are considered by the authors as a model for the origin of bipolar incompatibility. Two self-incompatible mating types have arisen from a self-compatible fungus by independent mutational events.

Following this step from self-fertility to sterility, the individual mutants must have lost the capacity to form sterile fruiting bodies in the course of further evolution.

Generalizing from this example leads to the view that the different incompatibility mechanisms found among the fungi have arisen independently of one another. The fact that only relatively few incompatibility mechanisms are known and that these are widely distributed throughout particular groups argues against this idea. Since these mechanisms possess the same characteristics (e.g. plasmogamy never occurs in Ascomycetes which possess bipolar incompatibility) they appear to be phylogenetically related if not of the same origin.

One can arrive at the opposite view (OLIVE's), if the data of RAPER (Table II-3) on mutation of incompatibility factors are considered. RAPER found in *Schizophyllum commune* that the *A* as well as the *B* factor could lose its incompatibility action through mutation. Tetrapolar incompatibility was transformed into the bipolar type through two independently occurring mutations. Through appropriate crosses these mutant *A* and *B* types can easily give rise to a self-compatible strain.

Thus, in light of contradictory evidence from RAPER and OLIVE, a *decision as to whether the ancestral breeding system in fungi was characterized by compatibility or incompatibility* is not possible at present.

II. Incompatibility and dioecism

The effectiveness of sexual reproduction as a means of genetic recombination increases when inbreeding, i.e. fertilization involving genetically identical gametes, is reduced or prevented. This is particularly true for dioecious forms. The same end is attained in many monoecious species through homogenic incompatibility. Incompatibility systems found in both fungi and flowering plants have one characteristic in common in spite of numerous genetic and physiological differences, namely, the incompatibility of genetically identical gametes. *Homogenic incompatibility limits inbreeding and favors outbreeding.* Therefore, wild strains of homogenically incompatible species are to a large extent heterogenic, or in the case of diplonts, heterozygous. This heterogeneity is absent in races of pseudocompatible species the spores of which include almost always two nuclei with different incompatibility genes (p. 67f.). The outbreeding effect of homogenic incompatibility is alleviated because of the self-fertility of mycelia arising from the binucleate spores. This explains the fact that all wild strains of *Podospora anserina* which have been isolated are homogenic (with the exception of the *plus/minus* gene) (ESSER, 1959a; BERNET et al., 1960).

A further reduction in outbreeding in homogenically incompatible species results from *heterogenic incompatibility* (p. 69ff.), since this involves incompatibility of genetically different gametes. Genetic exchange between races of a species is reduced and *the isolation of individual races is enforced through inbreeding.* Thus, in *Podospora anserina* the inbreeding effect produced by pseudocompatibility is further strengthened by heterogenic incompatibility, which determines the crossing relationship of the wild strains.

Reciprocal heterogenic incompatibility leads to complete isolation of *Podospora* races. Semi-incompatibility also favors isolation of individual races by preventing the formation of particular recombinant types through the specific effects of their genes (p. 72).

Heterogenic incompatibility acts as an isolating factor even when it does not occur in combination with homogenic incompatibility, as, for example, in the self compatible species, *Sordaria fimicola* (p. 70).

The influence of the various genetic components of a particular breeding system on evolution leads to the following generalization: A continuous recombination of hereditary material occurs in species in which the sexual behavior is determined by dioecism or homogenic incompatibility. Spontaneous mutations are quickly distributed throughout the species which therefore takes part in evolution as a whole. Such an outbreeding effect may be inhibited either through pseudocompatibility, heterogenic incompatibility or both. In such cases, mutative changes become distributed to other races to only a slight degree or not at all. Thus the race and not the species becomes the unit of evolution.

Summary

1. The results of genetic investigation confirm the generally held view that the formation of sex organs is a primitive character. All reductions are considered to be derived.

2. A decision as to whether the ancestral breeding system of monoecious fungi was characterized by compatibility or homogenic incompatibility is not possible. The existing data are compatible with both interpretations.

3. Homogenic incompatibility and dioecism have the same evolutionary effect. Both systems reduce inbreeding and encourage outbreeding.

4. On the other hand, heterogenic incompatibility is an isolating mechanism. It favors inbreeding and restricts outbreeding. Each race and not the species, becomes unit of evolution. Pseudocompatibility has a similar effect.

Literature

AHMAD, M.: Single-spore cultures of heterothallic *Saccharomyces cerevisiae* which mate with both tester strains. Nature (Lond.) **170**, 546—547 (1952).
— The mating system in *Saccharomyces*. Ann. Bot., N.S. **17**, 329—342 (1953).
— A consideration of the terms and mechanisms of heterothallism. Pak. J. Sci. **5**, 59 (1954).
ALLEN, C. E.: Sex-inheritance and sex-determination. Amer. Naturalist **66**, 97—107 (1932).
AMES, L. M.: A hermaphrodite self-sterile but cross-fertile condition in *Pleurage anserina*. Bull. Torrey bot. Club **59**, 341—345 (1932).
ARNAUDOW, N.: Untersuchungen über den tierefangenden Pilz *Zoophagus insidians* SOM. Flora (Jena) **18/19**, 1—16 (1925).

ARONESCU, A.: Further studies on *Neurospora sitophila* Mycologia (N.Y.) **25**, 43—54 (1933).

ASCHAN, K.: Some facts concerning the incompatibility groups, the dicaryotization and the fruitbody production in *Collybia velutipes*. Svensk bot. T. **48**, 603—625 (1954).

ASHBY, S. F.: Oospores in cultures of *Phytophtora faberi*. Kew Bull. 257—262 (1922).

BACKUS, M. P.: The mechanism of conidial fertilization in *Neurospora sitophila*. Bull. Torrey bot. Club **66**, 63—76 (1939).

BANBURY, G. H.: Processes controlling zygophore formation and zygotropism in *Mucor mucedo* BREFELD. Nature (Lond.) **173**, 499—500 (1954a).

— Physiological studies in the Mucorales. III. J. exp. Bot. **6**, 235—274 (1954b).

BARBESGAARD, P. O., and S. WAGNER: Further studies on the biochemical basis of protoperithecia formation in *Neurospora crassa*. Hereditas (Lund) **45**, 564—572 (1959).

BARKSDALE, A. W.: Inter-thallic sexual reactions in *Achlya*, a genus of the aquatic fungi. Amer. J. Bot. **47**, 14—23 (1960).

BARNES, B.: Variations in *Eurotium herbariorum* (WIGG.) LINK induced by the action of high temperatures. Ann. Bot. **42**, 783—812 (1928).

BARRATT, R. W., and L. GARNJOBST: Genetics of a colonial microconidiating mutant strain of *Neurospora crassa*. Genetics **34**, 351—369 (1949).

BARRET, J. D.: Development and sexuality of some species of *Olpidiopsis* (CORNU) FISCHER. Ann. Bot. **26**, 209—238 (1912).

BARRON, G. L.: The parasexual cycle and linkage relationships in the storage root fungus *Penicillium expansum*. Canad. J. Bot. **40**, 1603—1614 (1962).

—, and B. H. MACNEILL: A simplified procedure for demonstrating the parasexual cycle in *Aspergillus*. Canad. J. Bot. **40**, 1321—1327 (1962).

BAUCH, R.: Über *Ustilago longissima* und ihre Varietät *macrospora*. Z. Bot. **15**, 241—279 (1923).

— Untersuchungen an zweisporigen Hymenomyceten. I. Haploide Par thenogenesis bei *Camarophyllus virgineus*. Z. Bot. **18**, 337—387 (1926).

— Rassenunterschiede und sekundäre Geschlechtsmerkmale beim Antherenbrand. Biol. Zbl. **47**, 370—383 (1927).

— Über multipolare Sexualität bei *Ustilago longissima*. Arch. Protistenk. **70**, 417—466 (1930).

— Geographische Verteilung und funktionelle Differenzierung der Erbfaktoren bei der multipolaren Sexualität von *Ustilago longissima*. Arch. Protistenk. **75**, 101—132 (1931).

— *Sphacelotheca Schweinfurthiana*, ein neuer multipolar sexueller Brandpilz. Ber. dtsch. bot. Ges. **50**, 17—24 (1932a).

— Die Sexualität von *Ustilago Scorzonerae* und *Ustilago Zeae*. Phytopath. Z. **5**, 315—321 (1932b).

— Über Kreuzungen zwischen bipolar und multipolar sexuellen Brandpilzarten. Z. indukt. Abstamm.- u. Vererb.-L. **67**, 242—245 (1934).

BEADLE, G. W., and V. L. COONRADT: Heterocaryosis in *Neurospora crassa*. Genetics **29**, 291—308 (1944).

BEISSON-SCHECROUN, J.: Incompatibilité cellulaire et interactions nucléocytoplasmiques dans les phénomènes de "Barrage" chez le *Podospora anserina*. Ann. Génét. **4**, 4—50 (1962).

BENJAMIN, R. K., and L. SHANOR: The development of male and female individuals in the dioecious species *Laboulbenia formicarum*. Amer. J. Bot. **37**, 471—476 (1950).

BENSAUDE, M.: Recherches sur le cycle évolutif et la sexualité chez les Basidiomycètes. Thèse Némours (Paris) 1918.

BERNET, J.: Sur les modalités d'expression de gènes pouvant conduire à une incompatibilité cytoplasmique chez le champignon *Podospora anserina*. C. R. Acad. Sci. (Paris) **256**, 771—773 (1963a).

BERNET, J.: Action de la temperature sur les modifications de l'incompatibilité cytoplasmique et les modalités de la compatibilité sexuelle entrecertaines souches de *Podospora anserina*. Ann. Sci. nat. Bot., Sér. XII, **4**, 205—233 (1963b).
— K. ESSER, D. MARCOU et J. SCHECROUN: Sur la structure génétique de l'espèce *Podospora anserina* et sur l'interêt de cette structure pour certaines recherches de génétique. C. R. Acad. Sci. (Paris) **250**, 2053—2055 (1960).
BISHOP, H.: A study of sexuality in *Sapromyces Reinschii*. Mycologia (N.Y.) **32**, 505—529 (1940).
BISTIS, G.: Sexuality in *Ascobolus stercorarius*. I. Morphology of the ascogonium; plasmogamy; evidence for a sexual hormonal mechanism. Amer. J. Bot. **43**, 389—394 (1956).
— Preliminary experiments on various aspects of the sexual process. Amer. J. Bot. **44**, 436—443 (1957).
BISTIS, G. N., and J. R. RAPER: Heterothallism and sexuality in *Ascobolus stercorarius*. Amer. J. Bot. **50**, 880—891 (1963).
BLAKESLEE, A. F.: Zygospore formation, a sexual process. Science **19**, 864—866 (1904a).
— Sexual reproduction in the Mucorineae. Proc. Amer. Acad. Arts Sci. **40**, 205—319 (1904b).
— Zygospore germinations in the Mucorineae. Ann. Mycol. **4**, 1—28 (1906a).
— II. Differentiation of sex in thallus, gametophyte and sporophyte. Bot. Gaz. **42**, 161—178 (1906b).
— Mutations in Mucors. J. Heredity **11**, 278—284 (1920).
—, and J. L. CARTLEDGE: Sexual dimorphism in mucorales. II. Interspecific reactions. Bot. Gaz. **84**, 51—58 (1927).
— D. S. WELCH, and A. D. BERGNER: Sexual dimorphism in mucorales. II. Intraspecific reactions. Bot. Gaz. **84**, 27—50 (1927).
BRADLEY, S. G.: Parasexual phenomena in Microorganisms. Ann. Rev. Microbiol. **16**, 35—52 (1962).
BRIDGEMON, G. H.: Production of new races of *Puccinia graminis var. tritici* by vegetative fusion. Phytopathology **49**, 386—388 (1959).
BRIEGER, F.: Selbststerilität und Kreuzungssterilität im Pflanzenreich und Tierreich. Berlin 1930.
BROCK, T. D.: Mating reaction in the yeast *Hansenula wingei*. J. Bact. **75**, 697—701 (1958a).
— Protein as a specific cell surface component in the mating reaction of *Hansenula wingei*. J. Bact. **76**, 334—335 (1958b).
— Biochemical basis for mating in yeast. Science **129**, 960—961 (1959).
BRODIE, H. J.: Tetrapolarity and unilateral diploidization in the bird's nest fungus, *Cyathus stercoreus*. Amer. J. Bot. **35**, 312—320 (1948).
BRUCKER, W.: Über *Utigers* Reaktion zur Geschlechtsbestimmung von *Phycomyces blakesleeanus*. Naturwissenschaften **41**, 309 (1954).
— Über *Utigers* Reaktion zur geschlechtlichen Differenzierung von Phycomyces-Stämmen. Arch. Protistenk. **100**, 339—350 (1955).
BRUNSWICK, H.: Untersuchungen über die Geschlechts- und Kernverhältnisse bei der Hymenomycetengattung *Coprinus*. Bot. Abh. **5**, 152 (1924).
BRUYN, H. G. L.: Heterothallism in *Peronospora parasitica*. Phytopathology **25**, 8 (1935).
— Heterothallism in *Peronospora parasitica*. Genetica **19**, 553—608 (1936).
BUDDENHAGEN, I. W.: Induced mutations and variability in *Phytophtora cactorum*. Amer. J. Bot. **45**, 355—365 (1958).
BULLER, A. H. R.: Researches on fungi., vol. 4. London 1931.
— Researches on fungi, vol. 5. London 1933.
— The diploid cell and the diploidization process in plants and animals with special reference to higher fungi. Bot. Rev. **7**, 335—431 (1941).

Burgeff, H.: Über Sexualität, Variabilität und Vererbung bei *Phycomyces nitens* Kunze. Ber. dtsch. bot. Ges. **30**, 679—685 (1912).
— Untersuchungen über Variabilität, Sexualität und Erblichkeit bei *Phycomyces nitens* Kunze. I. Flora, N.F. **107**, 259—316 (1914).
— Untersuchungen über Variabilität, Sexualität und Erblichkeit bei *Phycomyces nitens* Kunze. II. Flora, N.F. **108**, 353—448 (1915).
— Untersuchungen über Sexualität und Parasitismus bei Mucorineen. I. Bot. Abh. **4**, 135 (1924).
— Variabilität, Vererbung und Mutation bei *Phycomyces blakesleeanus*. Z. indukt. Abstamm.- u. Vererb.-L. **48**, 26—94 (1928).
—, u. M. Plempel: Zur Kenntnis der Sexualstoffe bei Mucorineen. Naturwissenschaften **43**, 473—474 (1956).
—, u. A. Seybold: Zur Frage der biochemischen Unterscheidung der Geschlechter. Z. Bot. **19**, 497—537 (1927).
Burger, O. F.: Variations in *Colletotrichum gloeosporioides*. J. Agr. **20**, 723—736 (1921).
Burnett, J. H.: Oxygen consumption during sexual reproduction of some Mucoraceae. New Phytologist **52**, 58—64 (1953a).
— Oxygen consumption of mixtures of heterothallic and homothallic species in relation to "imperfect hybridization" in the Mucoraceae. New Phytologist **52**, 86—88 (1953b).
— The mating systems of fungi. I. New Phytologist **55**, 50—90 (1956a).
— Carotene and sexuality in Mucoraceae, especially *Phycomyces blakesleeanus*. New Phytologist **55**, 45—49 (1956b).
—, and M. E. Boulter: The mating system of fungi. II. Mating systems of the Gasteromycetes *Mycocalia denudata* and *M. duriaeana*. New Phytologist **62**, 217—236 (1963).
Buxton, E. W.: Heterokaryosis and parasexual recombination in pathogenic strains of *Fusarium oxysporum*. J. gen. Microbiol. **15**, 133—139 (1956).
— Parasexual recombination in the banana-wilt *Fusarium*. Trans. Brit. mycol. Soc. **45**, 274—279 (1962).
Callen, O. E.: The morphology, cytology and sexuality of the homothallic *Rhizopus sexualis* (Smith) Callen. Ann. Bot. (Lond.) N.S. **4**, 791—818 (1940).
Canter, H. M.: Studies on British Chytrids. I. *Dangeardia mammillata* Schröder. Trans. Brit. mycol. Soc. **29**, 128—134 (1946).
— Studies on British Chytrids. III. *Zygorhizidium Willei* Löwenthal and *Rhizopodium columnaris*, n.sp. Trans. Brit. mycol. Soc. **31** 128—135 (1947).
Carr, A. J. H., and L. S. Olive: Genetics of *Sordaria fimicola*. III. Cross-compatibility among self-fertile and self-sterile cultures. Amer. J. Bot. **46**, 81—91 (1959).
Cayley, D. M.: The inheritance of the capacity for showing mutual aversion between monospore-mycelia of *Diaporthe perniciosa* Marchal. J. Genet. **24**, 1—63 (1931).
Chena, B. L., and M. K. Hingorani: Mutation in *Colletotrichum falacatum* Went. The casual organism of sugarcane red rot. Phytopathology **40**, 221—227 (1950).
Chevaugeon, J., et N. van Huong: L'auto-incompatibilité. Conséquence régulière de la différenciation chez le *Pestalozzia annulata*. C.R. Acad. Sci. (Paris) **252**, 4183—4185 (1961).
Chilton, S. J. P., G. B. Lucas, and C. W. Edgerton: Genetics of *Glomerella*. III. Crosses with a conidial strain. Amer. J. Bot. **32**, 549—554 (1945).
Chodat, F.: Mutations chez les champignons. Bull. Soc. bot. Genève **18**, 41—144 (1926).
—, et W. H. Schopfer: Carotène et sexualité. C. R. Soc. Phys. Hist. Nat. Genève **44**, 176—179 (1927).
Chow, C. H.: Contribution à l'étude de développement des coprins. Botaniste **26**, 89—232 (1934).

Reproduction

CLAUSSEN, P.: Entwicklungsgeschichtliche Untersuchungen über den Erreger der als „Kalkbrut" bezeichneten Krankheit der Bienen. Arb. biol. Abt. (Reichsanst.) Berl. **10**, 467—521 (1921).

COCHRANE, V. W.: Physiology of fungi. New York and London 1958.

COKER, W. C.: Other water molds from the soil. J. Elisha Mitchell Sci. Soc. **42**, 207—226 (1927).

—, and J. LEITNER: New species of *Achlya* and *Apodachlya*. J. Elisha Mitchell Sci. Soc. **54**, 311—318 (1938).

COLLINS, O. R.: Heterothallism and homothallism in two Myxomycetes. Amer. J. Bot. **48**, 674—683 (1961).

— Multiple alleles at the incompatibility locus in the myxomycete *Didymium iridis*. Amer. J. Bot. **50**, 477—480 (1963).

—, and H. LING: Further studies in multiple allelomorph heterothallism in the myxomycete *Didymium iridis*. Amer. J. Bot. **51**, 315—317 (1964).

COLSON, B.: The cytology and morphology of *Neurospora tetrasperma*. Ann. Bot. **48**, 211—224 (1934).

CORRENS, C.: Geschlechtsverteilung und Geschlechtsbestimmung bei Pflanzen. In: Handwörterbuch der Naturwissenschaften, Bd. 4, S. 975. Jena 1913.

— Bestimmung, Vererbung und Verteilung des Geschlechts bei den höheren Pflanzen. In: Handbuch der Vererbungswissenschaften, II C, S. 138. Berlin 1928.

COUCH, J. N.: Heterothallism in *Dictyuchus*, a genus of the water molds. Ann. Bot. **40**, 848—881 (1926).

COY, D. O., and R. W. TUVESON: Genetic control of conidiation in *Aspergillus rugulosus*. Amer. J. Bot. **51**, 290—293 (1964).

CRAIGIE, J. N.: Heterothallism in the rust fungi and its significance. Trans. roy. Soc. Can., Sect. V **36**, 19—40 (1942).

CROWE, L. K.: The exchange of genes between nuclei of a dikaryon. Heredity **15**, 397—405 (1960).

— Competition between compatible nuclei in the establishment of a dikaryon in *Schizophyllum commune*. Heredity **18**, 525—533 (1963).

DARLINGTON, C. D.: What is hybrid? J. Hered. **28**, 308 (1937).

DAY, P. R.: The structure of the *A* mating type locus in *Coprinus lagopus*. Genetics **45**, 641—650 (1960).

— Mutants of the *A* mating type factor in *Coprinus lagopus*. Genet. Res. Camb. **4**, 55—64 (1963a).

— The structure of the *A* mating type factor in *Coprinus lagopus*: Wild alleles. Genet. Res. Camb. **4**, 323—325 (1963b).

DEE, J.: A mating type system in an acellular slime mould. Nature (Lond.) **185**, 780—781 (1960).

DICK, S.: The origin of expressed mutations in *Schizophyllum commune*. Thesis Havard Univ. (Cambridge, Mass., USA) 1960.

— and J. R. RAPER: Origin of expressed mutations in *Schizophyllum commune*. Nature (Lond.) **189**, 81—82 (1961).

DICKSON, H.: Studies in *Coprinus sphaerosporus*. I. The pairing behavior and the characteristics of various haploid and diploid strains. Ann. Bot. **48**, 527—547 (1934).

— Studies on *Coprinus sphaerosporus*. II. Ann. Bot. **49**, 181—204 (1935).

— Observations on inheritance in *Coprinus macrorhizus* (PERS.) REA. Ann. Bot. **50**, 719—734 (1936).

DIMOCK, A. W.: Observations on sexual relations in *Hypomyces ipomeae*. Mycologia (N.Y.) **29**, 116—127 (1937).

DODGE, B. O.: Methods of culture and the morphology of the archicarp of certain species of the Ascobolaceae. Bull. Torrey bot. Club **39**, 139—197 (1912).

— The life history of *Ascobolus magnificus*. Mycologia (N.Y.) **12**, 115—134 (1920).

— Nuclear phenomena associated with heterothallism and homothallism in the Ascomycete *Neurospora*. J. agric. Res. **35**, 289—305 (1927).

DODGE, B. O.: Unisexual conidia from bisexual mycelia. Mycologia (N.Y.) **20**, 226—234 (1928).
— The mechanism of sexual reproduction in *Neurospora*. Mycologia (N.Y.) **27**, 418—438 (1935).
— Self-sterility in "bisexual" heterocaryons of *Neurospora*. Bull. Torrey bot. Club **73**, 410—416 (1946).
DOWDING, E. S.: The sexuality of the normal, giant and dwarf spores of *Pleurage anserina* (CES.) KUNTZE. Ann. Bot. **45**, 1—14 (1931).
— *Gelasinospora*, a new genus of pyrenomycetes with pitted spores. Canad. J. Res. **9**, 294—304 (1933).
—, and A. BAKERSPIGEL: The migrating nucleus. Canad. J. Microbiol. **1**, 68—78 (1954).
— — Poor fruiters and barrage mutants in *Gelasinospora*. Canad. J. Bot. **34**, 231—240 (1956).
DRAYTON, F. D., and J. W. GROVES: *Stromatinia narcissi*, a new dimorphic discomycete. Mycologia (N.Y.) **44**, 119—140 (1952).
DRIVER, C. H., and H. E. WHEELER: A sexual hormone in *Glomerella*. Mycologia (N.Y.) **47**, 311—316 (1955).
DUTTA, S. K., and E. D. GARBER: Genetics of phytopathogenic fungi. III. An attempt to demonstrate the parasexual cycle in *Colletotrichum lagenarium*. Bot. Gaz. **122**, 118—121 (1960).
EDGERTON, C. W.: Plus and minus strains in an ascomycete. Science **35**, 151 (1912).
— Plus and minus strains in the genus *Glomerella*. Amer. J. Bot. **1**, 244—254 (1914).
— S. J. P. CHILTON, and G. B. LUCAS: Genetics of *Glomerella*. II. Fertilization between strains. Amer. J. Bot. **32**, 115—118 (1945).
EGGERTSON, E.: An estimate of the number of alleles at the loci for hetero-thallism in a local concentration of *Polyporus obtusus* BERK. Canad. J. Bot. **31**, 750—759 (1953).
EL-ANI, A. S.: The genetics of sex in *Hypomyces solani f. cucurbitae*. Amer. J. Bot. **41**, 110—113 (1954a).
— Chromosomes of *Hypomyces solani f. cucurbitae*. Science **120**, 323—324 (1954b).
—, and L. S. OLIVE: The induction of balanced heterothallism in *Sordaria fimicola*. Proc. nat. Acad. Sci. (Wash.) **48**, 17—19 (1962).
ELLINGBOE, A. H.: Somatic recombination in *Puccinia graminis tritici*. Phytopathology **51**, 13—15 (1961).
— Illegitimacy and specific factor transfer in *Schizophyllum commune*. Proc. nat. Acad. Sci. (Wash.) **49**, 286—292 (1963).
— Somatic recombination in dikaryon *K* of *Schizophyllum commune*. Genetics **49**, 247—251 (1964).
—, and J. R. RAPER: Somatic recombination in *Schizophyllum commune*. Genetics **47**, 85—98 (1962a).
— — The Buller phenomenon in *Schizophyllum commune*. Nuclear selection in fully compatible dikaryotic-homokaryotic matings. J. Bot. **49**, 454—459 (1962b).
ESSER, K.: Sur le déterminisme génétique d'une nouveau type d'incompatibilité chez *Podospora*. C. R. Acad. Sci. (Paris) **238**, 1731—1733 (1954a).
— Genetische Analyse eines neuen Incompatibilitätstypes bei dem Ascomyceten *Podospora anserina*. Compt. rend. 8, Congr. intern. Bot. (Paris) Sect. **10**, 72—77 (1954b).
— Genetische Untersuchungen an *Podospora anserina*. Ber. dtsch. bot. Ges. **68**, 143—144 (1955).
— Wachstum, Fruchtkörper- und Pigmentbildung von *Podospora anserina* in synthetischen Nährmedien. C. R. Lab. Carlsberg, Sér. Physiol. **26**, 103—116 (1956a).
— Gen-Mutanten von *Podospora anserina* (CES.) REHM mit männlichem Verhalten. Naturwissenschaften **43**, 284 (1956b).

Reproduction

Esser, K.: Die Incompatibilitätsbeziehungen zwischen geographischen Rassen von *Podospora anserina* (Ces.) Rehm. I. Genetische Analyse der Semi-Incompatibilität. Z. Indukt. Abstamm.- u. Verb.-L. **87**, 595—624 (1956c).
— The significance of semi-incompatibility in the evolution of geographic races in *Podospora anserina*. Proc. X. intern. Congr. of Genetics **2**, 76—77 (1958).
— Die Incompatibilitätsbeziehungen zwischen geographischen Rassen von *Podospora anserina* (Ces.) Rehm. II. Die Wirkungsweise der Semi-Incompatibilitäts-Gene. Z. Vererbungsl. **90**, 29—52 (1959a).
— Die Incompatibilitätsbeziehungen zwischen geographischen Rassen von *Podospora anserina* (Ces.) Rehm. III. Untersuchungen zur Genphysiologie der Barragebildung und der Semi-Incompatibilität. Z. Vererbungsl. **90**, 445—456 (1959b).
— Incompatibilität bei Pilzen. Ber. dtsch. bot. Ges. **74**, 324—325 (1961).
— Die Genetik der sexuellen Fortpflanzung bei den Pilzen. Biol. Zbl. **81**, 161—172 (1962).
— Heterogenic incompatibility. In: K. Esser and J. R. Raper (edts.), Incompatibility in fungi, p. 6—13. Berlin-Heidelberg-NewYork: Springer 1965.
— Incompatibility. In: G. C. Ainsworth and A. S. Sussman (edts.), The fungi, vol. II, p. 661—676. New York 1966a.
— Die Phenoloxydasen des Ascomyceten *Podospora anserina*. III. Quantitative und qualitative Enzymunterschiede nach Mutation an nicht gekoppelten Loci. Z. Vererbungsl. **97**, 327—344 (1966b).
— Die Verbreitung der Incompatibilität bei Tallophyten. In: Handbuch der Pflanzenphysiologie, **18**, 321—343. Berlin-Göttingen-Heidelberg 1967.
—, u. J. Straub: Fertilität im Heterocaryon aus zwei sterilen Mutanten von *Sordaria macrospora* Auersw. Z. indukt. Abstamm.- u. Vererb.-L. **87**, 625—626 (1956).
— — Genetische Untersuchungen an *Sordaria macrospora* Auersw. Kompensation und Induktion bei genbedingten Entwicklungsdefekten. Z. Vererbungsl. **89**, 729—746 (1958).
Fitzgerald, P. H.: Genetic and epigenetic factors controlling female sterility in *Neurospora crassa*. Heredity **18**, 47—62 (1963).
Fox, A. S., and W. D. Gray: Immunogenetic and biochemical studies of *Neurospora crassa:* differences in tyrosinase activity between mating types of strain. 15,300 (albino-2). Proc. nat. Acad. Sci. (Wash.) **36**, 538 (1950).
Franke, G.: Die Cytologie der Ascusentwicklung von *Podospora anserina*. Z. indukt. Abstamm.- u. Vererb.-L. **88**, 159—160 (1957).
— Versuche zur Genomverdoppelung des Ascomyceten *Podospora anserina*. Z. Vererbungsl. **93**, 109—117 (1962).
Fries, N.: Heterothallism in some Gasteromycetes and Hymenomycetes. Svensk bot. T. **42**, 158—168 (1948).
—, and U. Trolle: Combination experiments with mutant strains of *Ophiostoma multiannulatum*. Hereditas (Lund) **33**, 377—384 (1947).
—, and K., Aschan: The physiological heterogeneity of the dikaryotic mycelium of *Polyporus abietinus* investigated with the aid of micrurgical technique. Svensk bot. T. **46**, 429—445 (1952).
Fuller, R. C., and E. L. Tatum: Inositol phospholipoid in *Neurospora* and its relationship to morphology. Amer. J. Bot. **43**, 361—365 (1956).
Fulton, I.: Unilateral nuclear migration and the interactions of haploid mycelia in the fungus *Cyathus stercoreus*. Proc. nat. Acad. Sci. (Wash.) **36**, 306—312 (1950).
Gans, M., et N. Prud'Homme: Echanges nucléaires chez le basidiomycète *Coprinus fimetarius* (Fr.). C. R. Acad. Sci. (Paris) **247**, 1895—1897 (1958).
Garnjobst, L.: Genetic control of heterokaryosis in *Neurospora crassa*. Amer. J. Bot. **40**, 607—614 (1953).
— Further analysis of the genetic control of heterokaryosis in *Neurospora crassa*. Amer. J. Bot. **42**, 444—448 (1955).

GARNJOBST, L., and J. F. WILSON: Heterokaryosis and protoplasmic incompatibility in *Neurospora crassa*. Proc. nat. Acad. Sci. (Wash.) **42**, 613—618 (1956).

— — Heterokaryosis and protoplasmic incompatibility in *Neurospora crassa*. Proc. intern. Growth Symp. (Suppl. Vol. Cytologia) 539—542 (1957).

GARTON, J. A., T. W. GOODWIN, and W. LIJINSKY: Studies in carotogenesis. I. General conditions governing carotene synthesis by the fungus *Phycomyces blakesleeanus* BURGEFF. Biochem. J. **48**, 154—163 (1951).

GOODWIN, T. W.: Studies in carotenogenesis. III. Identification of minor polyene components of the fungus *Phycomyces blakesleeanus* and a study of their synthesis under various cultural conditions. Biochem. J. **50**, 550—558 (1952).

—, and W. LIJINSKY: Studies in Carotenogenesis. II. Carotene production by *Phycomyces blakesleeanus*: the effect of different aminoacids when used in media containing low concentrations of glucose. Biochem. J. **50**, 268—273 (1951).

—, and J. S. WILLMER: Studies in Carotogenesis. IV. Nitrogen metabolism and carotine synthesis in *Phycomyces blakesleeanus*. Biochem. J. **51**, 213—217 (1952).

GORDON, W. L.: Geographical distribution of mating types in *Gibberella cyanogena* (DESM.) SACC. Nature (Lond.) **173**, 505—506 (1954).

GRASSO, V.: Studies sulla genetica dei carboni dell'avena: *Ustilago avenae* e *U. levis*. Boll. staz. Pat. veget. **12**, 115—126 (1955).

GREIS, H.: Mutations- und Isolationsversuche zur Beeinflussung des Geschlechts von *Sordaria fimicola* (ROB.). Z. Bot. **37**, 1—116 (1941).

— Relative Sexualität und Sterilitätsfaktoren bei dem Hymenomyceten *Solenia*. Biol. Zbl. **62**, 46—92 (1942).

GRIFFITH, F.: The significance of pneumococcal types. J. Hyg. (Camb.) **27**, 113—156 (1928).

GRIGG, G. W.: The genetic control of conidiation in a heterokaryon of *Neurospora crassa*. J. gen. Microbiol. **19**, 15—22 (1958).

— Temperature-sensitive genes affecting conidiation in *Neurospora*. J. gen. Microbiol. **22**, 667—670 (1960).

GRINDLE, M.: Heterokaryon compatibility of closely related wild isolates *Aspergillus nidulans*. Heredity **18**, 397—405 (1963).

GROSS, S. R.: Heterokaryosis between opposite mating types in *Neurospora crassa*. Amer. J. Bot. **39**, 574—577 (1952).

GUILLIERMOND, A.: Nouvelles observations sur la sexualité des levures et quelques considerations sur la phylogénie de ces champignons. Rev. gén. Bot. **48**, 403—426 (1936).

— Sexuality, developmental cycle and phylogeny of yeasts. Bot. Rev. **6**, 1—24 (1940).

HAENICKS, A.: Vererbungsphysiologische Untersuchungen an Arten von *Penicillium* und *Aspergillus*. Z. Bot. **8**, 225—343 (1916).

HALDEMAN, Q. L.: Some falcate-spored *Colletotrichums* on legumes. Phytopathology **40**, 12 (1950).

HANSEN, H. N., and W. C. SNYDER: The dual phenomenon and sex in *Hypomyces solani f. cucurbitae*. Amer. J. Bot. **30**, 419—422 (1943).

— — Inheritance of sex in fungi. Proc. nat. Acad. Sci. (Wash.) **32**, 272—273 (1946).

HARDER, R.: Zur Frage nach der Rolle von Kern und Protoplasma im Zellgeschehen und bei Übertragung von Eigenschaften. Z. Bot. **19**, 337—407 (1927).

— F. FIRBAS, W. SCHUMACHER u. D. V. DENFFER: Lehrbuch der Botanik für Hochschulen, 28. Aufl. Stuttgart 1962.

—, u. G. SÖRGEL: Über einen neuen plano-isogamen Phycomyceten mit Generationswechsel und seine phytogenetische Bedeutung. Nachr. Ges. Wiss. Göttingen, math.-physik. Kl. N.F. VI (Biol.) **3**, 119—127 (1938).

HARTMANN, M.: Autogamie bei Protisten und ihre Bedeutung für das Befruchtungsproblem. Verh. Dtsch. Zool. Ges. 24. Jahresverslg Freiburg, 1909, S. 15.

HARTMANN, M.: Theoretische Bedeutung und Terminologie der Vererbungs-erscheinungen bei haploiden Organismen. *(Chlamydomonas, Phycomyces, Honigbiene.)* Z. indukt. Abstamm.- u. Vererb.-L. **20**, 1—26 (1918).
— Fortpflanzung und Befruchtung als Grundlage der Vererbung. In: Handbuch der Vererbungswissenschaften, Bd. 1 A, S. 1—103. Berlin 1929.
— Die Sexualität, 1. Aufl. Jena 1943; 2. Aufl. Stuttgart 1956.
HASTIE, A. C.: Genetic recombination in the hop-wilt fungus *Verticillium albo-atrum*. J. gen. Microbiol. **27**, 373—382 (1962).
HAWKER, L. E.: The physiology of reproduction in fungi. Cambridge 1957.
HAWTHORNE, D. C.: A deletion in yeast and its bearing on the structure of the mating type locus. Genetics **48**, 1727—1729 (1963).
HEPDEN, P. M., and L. E. HAWKER: A volatile substance controlling early stages of zygospore formation in *Rhizopus sexualis*. J. gen. Microbiol. **24**, 155—164 (1961).
HESLOT, H.: Contribution a l'étude cytogénétique et génétique des Sor-dariacies. Rev. Cytol. Biol. végét. **19** (Suppl. 2), 1—209 (1958).
HESSELTINE, C. W.: Carotinoids in the fungi Mucorales. Techn. Bull. No 1245, US Dept. of Agriculture. 33 P. Washington (D.C.) 1961.
HIRSCH, H. E.: The cytogenetics of sex in *Hypomyces solani f. cucurbitatae*. Amer. J. Bot. **36**, 113—121 (1949).
HIRSCH, H. M.: Environmental factors influencing the differentiation of protoperithecia and their relation to tyrosinase and melanin formation in *Neurospora crassa*. Physiol. Plantarum (Copenh.) **7**, 72—97 (1954).
HOLLIDAY, R.: The genetics of *Ustilago maydis*. Genet. Res. **2**, 204—230 (1961a).
— Induced mitotic crossing-over in *Ustilago maydis*. Genet. Res. **2**, 231—248 (1961b).
HOLLOWAY, B. W.: Genetic control of heterocaryosis in *Neurospora crassa*. Genetics **40**, 117—129 (1955).
HOLTON, C. S.: Extent of pathogenicity of hybrids of *Tilletia tritici* and *T. levis*. J. agric. Res. **65**, 555—563 (1942).
HOROWITZ, N. H., and S. C. SHEN: *Neurospora* tyrosinase. J. biol. Chem. **197**, 513—520 (1952).
— M. FLING, H. L. MACLEOD, and N. SUEOKA: Genetic determination and enzymatic induction of tyrosinase in *Neurospora*. J. molec. Biol. **2**, 96—104 (1960).
HÜTTIG, W.: Sexualität bei *Glomerella lycopersici* KRÜGER und ihre Ver-erbung. Biol. Zbl. **55**, 74—83 (1935).
IKEDA, Y., C. ISHITANI, and K. NAKAMURA: A high frequency of hetero-zygous diploids and somatic recombination induced in imperfect fungi by ultraviolet light. J. gen. appl. Microbiol. **3**, 1—11 (1957).
ISHITANI, C.: A high frequency of heterozygous diploids and somatic recom-bination produced by ultra-violet light in imperfect fungi. Nature (Lond.) **178**, 706 (1956).
—, Y. IKEDA, and K. SAKAGUCHI: Hereditary variation and genetic recombi-nation in Koji-molds *(Aspergillus oryzae* and *Asp. sojae)*. VI. Genetic recombination in heterozygous diploid. J. gen. appl. Microbiol. (Japan) **2**, 401—430 (1956).
ITO, T.: Fruit body formation in red bread mould, *Neurospora crassa*. I. Effect of culture filtrate on perithecial formation. Bot. Mag. (Tokyo) **69**, 369—372 (1956).
— Genetic study on the expression of the color factor of the ascospores in *Sordaria fimicola*. I. Segregation of the dark- and lightcolored ascospores. Res. Bull. Obihiro Zootechn. Univ., Ser. I **3**, 223—230 (1960).
JACOB, F., et E. A. ADELBERG: Transfer de caractères génétique par incorpo-ration au facteur sexuel d'*Escherichia coli*. C. R. Acad. Sci. (Paris) **249**, 189—191 (1959).
—, and E. L. WOLLMAN: Genetic aspects of lysogeny. In: McELROY and B. GLASS (edits.), The chemical basis of heredity, pp. 468—499. Baltimore 1957.
— — Sexuality and the genetics of bacteria. New York 1961.

Jinks, J. L., and M. Grindle: The genetical basis of heterokaryon incompatibility in *Aspergillus nidulans*. Heredity **18**, 407—411 (1963).

Joly, P.: Données récentes sur la génétique des champignons supérieurs (Ascomycètes et Basidiomycètes). Rev. Mycol. (Paris) **29**, 115—186 (1964).

Jost, L.: Über die Selbststerilität einiger Blüten. Bot. Ztg **65**, Abt. 1, 77—117 (1907).

Jürgens, C.: Physiologische und genetische Untersuchungen über die Fruchtkörperbildung bei *Schizophyllum commune*. Arch. Mikrobiol. **31**, 388—421 (1958).

Käfer, E.: An 8-chromosome map of *Aspergillus nidulans*. Advanc. Genet. **9**, 105—145 (1958).

— The processes of spontaneous recombination in vegetative nuclei of *Aspergillus nidulans*. Genetics **46**, 1581—1609 (1961).

Kehl, H.: Ein Beitrag zur Morphologie und Physiologie der Zygosporen von *Mucor mucedo*. Arch. Mikrobiol. **8**, 379—406 (1937).

Kimura, K.: On the diploidisation by the double compatible diploid mycelium in the Hymenomycetes. Bot. Mag. (Tokyo) **67**, 238—242 (1954a).

— Diploidisation in the Hymenomycetes. Biol. J. Okayama Univ. **1**, 226—233 (1954b).

— Nuclear conjugation in diploidisation by the double compatible diploid mycelium in the Hymenomycetes. Bot. Mag. (Tokyo) **70**, 391—395 (1957).

— Diploidisation in the Hymenomycetes. II. Nuclear behavior in the Buller phenomenon. Biol. J. Okayama Univ. **4**, 1—59 (1958).

Kniep, H.: Über die Bedingungen der Schnallenbildung bei den Basidiomyceten. Flora (Jena) **111**, 380—395 (1918).

— Über morphologische und physiologische Geschlechtsdifferenzierung. Verh. phys.-med. Ges., N.F. **46**, 1—18 (1920).

— Über Geschlechtsbestimmung und Reduktionsteilung. Verh. phys.-med. Ges., N.F. **47**, 1—28 (1922).

— Über erbliche Änderungen von Geschlechtsfaktoren bei Pilzen. Z. indukt. Abstamm.- u. Vererb.-L. **31**, 170—183 (1923).

— Die Sexualität der niederen Pflanzen. Jena 1928.

— Vererbungserscheinungen bei Pilzen. Bibl. genet. **5**, 371—475 (1929a).

— Geschlechtsverteilung bei den Pflanzen. Tabul. biol. ('s-Grav.) **5**, 115—171 (1929b).

Köhler, F.: Beitrag zur Kenntnis der Sexualreaktion von *Mucor mucedo* Bref. Planta (Berl.) **23**, 358—375 (1935).

Korf, R. P.: The terms homothallism and heterothallism. Nature (Lond.) **170**, 534—535 (1952).

Krafczyk, H.: Die Bildung und Keimung der Zygosporen von *Pilobolus crystallinus* und sein heterokaryotisches Mycel. Beitr. Biol. Pflanz. **23**, 349—396 (1935).

Kuhner, R.: L'amphithallie et ses causes dans la forme bisporique tétrapolaire de *Clitocybe lituus* Fr. Bull. Soc. mycol. France **69**, 307—325 (1954).

— H. Romagnesi et H. C. Yen: Différences morphologiques entre plusieurs souches de *coprins* de la section *micacei* et confrontation de leur haplontes. Bull. Soc. mycol. France **63**, 169—186 (1947).

Kuwana, H.: Melanine formation in the sexual generation in *Neurospora*. J. Genet. **29**, 163 (1954).

— Tyrosinase appearing in the sexual generation in *Neurospora*. Medicine and Biol. **36**, 187—191 (1955).

— Mating type alleles and tyrosinase activity in relation to genetic background in *Neurospora crassa*. Ann. Rep. Sci. Works Fac. Sci. **4**, 117—131 (1956).

— Melanization in the mycelium due to the interaction of two strains of *Neurospora crassa*. Bot. Mag. (Tokyo) **71**, 841—842 (1958).

Laibach, F., F. J. Kribben u. F. Heilinger: Die Sexualvorgänge bei *Bombardia lunata* Zckl. I. Beitr. Biol. Pflanz. **30**, 239—248 (1954).

— — — Die Sexualvorgänge bei *Bombardia lunata* Zckl. II. Beitr. Biol. Pflanz. **31**, 137—152 (1955).

Reproduction

Lamb, I. M.: The initiation of the dicaryphase in *Puccinia phragmitis* (Schum.) Korn. Ann. Bot. **49**, 403—438 (1935).
Lamoure, D.: Hétérocaryose chez les Basidiomycètes amphitalles. C. R. Acad. Sci. (Paris) **244**, 2841—2843 (1957).
Lange, I.: Das Bewegungsverhalten der Kerne in fusionierten Zellen von *Polystictus versicolor* (L.). Flora, Abt. A **156**, 487—497 (1966).
Lange, M.: Species in the genus *Coprinus*. Dansk. bot. Ark. **14**, 6 (1952).
Laven, H.: Vererbung durch Kerngene und das Problem der außerkaryotischen Vererbung bei *Culex pipiens*. I. Kernvererbung. Z. indukt. Abstamm.- u. Vererb.-L. **88**, 443—477 (1957a).
— Vererbung durch Kerngene und das Problem der außerkaryotischen Vererbung bei *Culex pipiens*. II. Außerkaryotische Vererbung. Z. indukt. Abstamm.- u. Vererb.-L. **88**, 478—516 (1957b).
Leonian, L. H.: The morphology and the pathogenicity of some *Phytophthora* mutations. Phytopathology **16**, 723—730 (1926).
— Heterothallism in *Phytophthora*. Phytopathology **21**, 941—955 (1931).
Leupold, U.: Die Vererbung von Homothallie und Heterothallie bei *Schizosaccharomyces pombe*. C. R. Lab. Carlsberg, Ser. Physiol. **23**, 349—39 (1950).
— Some data on polyploid inheritance in *Schizosaccharomyces pombe*. C. R. Lab. Carlsberg, Ser. Physiol. **26**, 221—251 (1956).
— Studies on allelism in *Schizosaccharomyces pombe*. Proc. X. Int. Congr. Gen. **2**, 165—166 (1958a).
— Studies on recombination in *Schizosaccharomyces pombe*. Cold Spr. Harb. Symp. quant. Biol. **23**, 161—170 (1958b).
Levi, J. D.: Mating reaction in yeast. Nature (Lond.) **177**, 753—754 (1956).
Lewis, D.: Comparative incompatibility in angiosperms and fungi. Advanc. Genet. **6**, 235—285 (1954).
— Incompatibility and plant breeding. Brookhaven Symp. in Biol. **9**, 89—100 (1956).
Lilly, V. G., and H. L. Barnett: Physiology of fungi. New York 1951.
Lindegren, C. C.: The genetics of *Neurospora*. III. Pure breed stocks and crossing-over in *N. crassa*. Bull. Torrey bot. Club **60**, 133—154 (1933).
— The genetics of *Neurospora*. VI. Bisexual and akaryotic ascospores from *N. crassa*. Genetica **16**, 315—320 (1934).
— Genetics of the fungi. Ann. Rev. Microbiol. **2**, 47—70 (1948).
—, and G. Lindegren: X-ray and ultraviolet induced mutations in *Neurospora*. J. Hered. **32**, 404—414 (1941a).
— — X-ray and ultraviolet induced mutations in *Neurospora*. II. Ultraviolet mutations. J. Hered. **32**, 435—440 (1941b).
— — Segregation, mutation and copulation in *Saccharomyces cerevisae*. Ann. Missouri bot. Gard. **30**, 453—468 (1943).
— — Instability of the mating type alleles in *Saccharomyces*. Ann. Missouri bot. Gard. **31**, 203—218 (1944).
— — Tetraploid *Saccharomyces*. J. gen. Microbiol. **5**, 885—893 (1951).
Lindfors, T.: Studien über den Entwicklungsverlauf bei einigen Rostpilzen aus zytologischen und taxonomischen Gesichtspunkten. Svensk bot. T. **18**, 34—37 (1924).
Löwenthal, W.: Weitere Untersuchungen an Chytridiaceen. Arch. Protistenk. **5**, 221—239 (1905).
Lovett, J. S., and E. C. Cantino: The relation between biochemical and morphological differentiation in *Blastocladiella emersonii*. II. Nitrogen metabolism in synchronous cultures. Amer. J. Bot. **47**, 550—560 (1960).
Lucas, G. B., S. J. P. Chilton, and C. W. Edgerton: Genetics of *Glomerella*. I. Studies on the behavior of certain strains. Amer. J. Bot. **31**, 233—239 (1944).
Lwoff, A.: Lysogeny. Bact. Rev. **17**, 269—337 (1953).
Mahony, M., and D. Wilkie: Nucleo-cytoplasmic control of perithecial formation in *Aspergillus nidulans*. Proc. roy. Soc. B **156**, 524—532 (1962).
Markert, C. L.: Sexuality in the fungus *Glomerella*. Amer. Naturalist **83**, 227—231 (1949).

MARKERT, C. L.: Radiation induced nutritional and morphological mutants of *Glomerella*. Genetics **37**, 339—352 (1952).

MARTIN, P. G.: Apparent self-fertility in *Neurospora crassa*. J. gen. Microbiol. **20**, 213—222 (1959).

MATHER, K.: Heterothally as an outbreeding mechanism in fungi. Nature (Lond.) **149**, 54—56 (1942).

McCURDY jr., H. D., and E. C. CANTINO: Isocitritase, glycine-alanine transaminase and development in *Blastocladiella emersonii*. Plant Physiol. **35**, 463—476 (1960).

McGAHEN, J. W., and H. E. WHEELER: Genetics of *Glomerella*. IX. Perithecial development and plasmogamy. Amer. J. Bot. **38**, 610—617 (1951).

MIDDLETON, R. B.: Evidences of common-*A B* heterokaryosis in *Schizophyllum commune*. Amer. J. Bot. **51**, 379—387 (1964).

MITCHELL, M. B.: A partial map of linkage group D. in *Neurospora crassa*. Proc. nat. Acad. Sci. (Wash.) **40**, 436—440 (1954).

MITTER, J. H.: Some contributions to our knowledge of heterothallism in fungi. J. Indian bot. Soc. **15**, 183—192 (1936).

MOREAU, F., et C. MORUZI: Sur de nouvelles irrégularités de la bipolarité sexuelle chez les ascomycètes du genre *Neurospora*. Bull. Soc. bot. France **80**, 574—576 (1933).

MORRISON, R. M.: Compatibility of several clonal lines of *Erysiphe cichoracearum*. Mycologia (N.Y.) **52**, 786—794 (1960).

MULLER, H. J.: Genetic fundamentals: the dance of the genes. In: Genetics, Medicine and Man, pp. 35—65. Cornell Univ. Press, Ithaka (N.Y., USA). 1947.

MULLINS, I. T., and J. R. RAPER: The genetical basis of heterothallism in biflagellatae aquatic fungi. Amer. J. Bot. **52**, 634 (1965a).
— — Heterothallism in biflagellatae aquatic fungi: preliminary genetic analysis. Science **150**, 1174—1175 (1965b).

MURRAY, J. C., and A. M. SRB: A mutant locus determining abnormal morphology and ascospore lethality in *Neurospos*. J. Hered. **4**, 149—153 (1961).
— — The morphology and genetics of wild-type and seven morphological mutant strains of *Neurospora crassa*. Canad. J. Bot. **40**, 337—349 (1962).

NELSON, R. R.: A major locus for compatibility in *Cochliobolus heterostrophus*. Phytopathology **47**, 742—743 (1957).

NOBLE, M.: The morphology and cytology of *Typhula trifolii* ROSTR. Ann. Bot., N.S. **1**, 67—98 (1937).

NOBLES, M. K., R. MACRAE, and B. P. TOMLIN: Results of interfertility tests on some species of Hymenomycetes. Canad. J. Bot. **35**, 377—387 (1957).

NOWAKOWSKI, L.: Beitrag zur Kenntnis der Chytridiaceen. II. *Polyphagus Euglenae*. Beitr. Biol. Pflanz. **2**, 201—219 (1876).

OIKAWA, K.: Diploidisation and fruit-body formation in the *Hymenomycetes*. Sci. Rep. Tohoku Univ. **14**, 245—260 (1939).

OLIVE, L. S.: Cross-karyogamy and segregation in a homothallic fungus. Bull. Torrey bot. Club **81**, 95—97 (1954).
— On the evolution of heterothallism in fungi. Amer. Naturalist **92**, 233—251 (1958).
— Genetics of *Sordaria fimicola*. I. Ascospore color mutants. Amer. J. Bot. **43**, 97—107 (1956).

OORT, A. J. P.: Die Sexualität von *Coprinus fimetarius*. Rec. Trav. bot. néerl. **27**, 85—148 (1930).

ORBAN, G.: Untersuchungen über die Sexualität von *Phykomyces nitens*. Beih. bot. Zbl. I. **36**, 1—59 (1919).

PAPAZIAN, H. P.: Physiology and incompatibility factors in *Schizophyllum commune*. Bot. Gaz. **112**, 143—163 (1950).
— The incompatibility factors and a related gene in *Schizophyllum commune*. Genetics **36**, 441—459 (1951).
— Exchange of incompatibility factors between the nuclei of a dikaryon. Science **119**, 691—693 (1954).
— The Genetics of Basidiomycetes. Advanc. Genet. **9**, 41—69 (1958).

8*

Reproduction

PARAG, Y.: Mutation in the *B* incompatibility factor of *Schizophyllum commune*. Proc. nat. Acad. Sci. (Wash.) **48**, 743—750 (1962a).
— Studies on somatic recombination in dikaryons of *Schizophyllum commune* Heredity **17**, 305—318 (1962b).
—, and J. R. RAPER: Genetic recombination in a common-B cross of *Schizophyllum commune*. Nature (Lond.) **188**, 765—766 (1960).
PFITZNER, E.: *Ancylistes Closterii*, ein neuer Algen-Parasit aus der Ordnung der Phycomyceten. Mber. Akad. Wiss. Berl. 379—398 (1872).
PITTENGER, T. H.: Spontaneous alterations of heterokaryon compatibility factors in *Neurospora*. Genetics **50**, 471—484 (1964).
—, and T. G. BRAWNER: Genetic control of nuclear selection in *Neurospora* heterokaryons. Genetics **46**, 1645—1663 (1961).
PLEMPEL, M.: Die Sexualstoffe der Mucoraceae, ihre Abtrennung und die Erklärung ihrer Funktion. Arch. Mikrobiol. **26**, 151—174 (1957).
— Die zygotropische Reaktion bei Mucorineen. I. Planta (Berl.) **55**, 254—258 (1960).
— Die Mucorineen-Gamone. Naturwissenschaften **50**, 226 (1963a).
— Die chemischen Grundlagen der Sexualreaktion bei Zygomyceten. Planta (Berl.) **59**, 492—508 (1963b).
—, u. G. BRAUNITZER: Die Isolierung der Mucorineen-Sexualstoffe. I. Z. Naturforsch. **13**, 302—305 (1958).
—, u. W. DAWID: Die zygotropische Reaktion bei Mucorineen. II. Planta (Berl.) **56**, 438—446 (1961).
POMPER, S., K. M. DANIELS, and D. W. McKEE: Genetic analysis of polyploid yeast. Genetics **39**, 343—355 (1954).
PONTECORVO, G.: Mitotic recombination in the genetic system of filamentous fungi. Caryologia, Suppl. **6**, 192—200 (1954).
— Allelism. Cold Spr. Harb. Sym. quant. Biol. **21**, 171—174 (1956a).
— The parasexual cycle in fungi. Ann. Rev. Microbiol. **1**, 393—400 (1956b).
— Trends in genetic analysis. New York 1958.
—, and J. A. ROPER: Genetic analysis without sexual reproduction by means of polyploidy in *Aspergillus nidulans*. J. gen. Microbiol. **6** VII (Abstr.) (1952).
—, and G. SERMONTI: Parasexual recombination in *Penicillium chrysogenum*. J. gen. Microbiol. **11**, 94—104 (1954).
— — and E. FORBES: Genetic recombination without sexual reproduction in *Aspergillus niger*. J. gen. Microbiol. **8**, 198—210 (1953a).
— J. A. ROPER, L. M. HEMMONS, K. D. MacDONALD, and A. W. J. BUFTON: The genetics of *Aspergillus nidulans*. Adv. Genet. **5**, 141—238 (1953b).
PRESLEY, J. T.: Saltants from a monosporic culture of *Verticillium albo-atrum*. Phytopathology **31**, 1135—1139 (1941).
PRÉVOST, G.: Etude génétique d'un Basidiomycète *Coprinus radiatus*. FR. et BOLT. Thèse Fac. Sci. Paris 1962.
PRUD'HOMME, N.: Echanges nucléaires chez *Coprinus fimetarius* au cours du phénomène de Buller compatible. C. R. Acad. Sci. (Paris) **253**, 3044—3046 (1962).
— Recombinaisons chromosomiques extra-basidiales chez un basidiomycète *"Coprinus radiatus"*. Ann. Génét. **4**, 63—66 (1963).
—, et M. GANS: Formation de noyaux partiellement diploides au cours du phénomène de Buller. C. R. Acad. Sci. **247**, 2419—2421 (1958).
QUINTANILHA, A.: Le problème de la sexualité chez les hyménomycètes. Bull. Soc. Brot. **8**, 1—99 (1933). — Cytologie et génétique de la sexualité chez les Hyménomycètes. Bol. Soc. Brot. **10**, 289—332 (1935).
— Contribution à l'étude génétique du phénomène de Buller. C. R. Acad. Sci. (Paris) **205**, 745 (1937).
— Deuxième contribution à l'étude génétique du phénomène de Buller. C.R. Soc. Biol. (Paris) **127**, 1245 (1938a).
— Troisième contribution à l'étude génétique du phénomène du Buller. C. R. Soc. Biol. (Paris) **129**, 730—734 (1938b).

QUINTANILHA, A.: Etude génétique du phénomène de Buller. Bol. Soc. Brot. 13, 425—486 (1939).

—, and S. BALLE: Etude génétique des phénomènes de nanisme chez les Hyménomycètes. Bol. Soc. Brot. 14, 17—46 (1940).

—, et J. PINTO-LOPES: Apercu sur l'état actuel de nos connaissances concernent la "conduite sexuelle" des espèces d' Hyménomycètes. I. Bol. Soc. Brot. 24, 115—290 (1950).

RAPER, C. A., and J. R. RAPER: Mutations affecting heterokaryosis in Schizophyllum commune. Amer. J. Bot. 51, 503—512 (1964).

RAPER, J. R.: Heterothallism and sterility in Achlya and observations on the cytology of Achlya bisexualis. J. Elisha Mitchell Sci. Soc. 52, 274—289 (1936).

— Sexual hormons in Achlya. I. Indicative evidence for a hormonal coordinating mechanism. Amer. J. Bot. 26, 639—650 (1939a).

— Role of hormones in the sexual reaction of heterothallic Achlyas. Science 89, 321—322 (1939b).

— Sexual hormones in Achlya. II. Distance reactions, conclusive evidence for a hormonal coordinating mechanism. Amer. J. Bot. 27, 162—173 (1940a).

— Sexuality in Achlya ambisexualis. Mycologia (N.Y.) 32, 710—727 (1940b).

— Sexual hormones in Achlya. III. Hormone A and the initial male reaction. Amer. J. Bot. 29, 159—166 (1942a).

— Sexual hormones in Achlya. V. Hormone A 1, a male-secreted augmenter or activator of hormone A. Proc. nat. Acad. Sci. (Wash.) 28, 509—516 (1942b).

— On the distribution and sexuality of Achlya ambisexualis. Amer. J. Bot. 34, 31a (1947).

— VI. The hormones of the A-complex. Proc. nat. Acad. Sci. (Wash.) 36, 524—533 (1950a).

— VII. The hormonal mechanism in homothallic species. Bot. Gaz. 112, 1—24 (1950b).

— Sexual hormones in Achlya. Amer. Sci. 39, 110—121 (1951).

— Chemical regulation of sexual processes in the Thallophytes. Bot. Rev. 18, 447—545 (1952).

— Tetrapolar sexuality. Quart. Rev. Biol. 28, 233—259 (1953).

— Life cycles, sexuality and sexual mechanism in the fungi. In: D. H. WENRICH et al. (edits.), Sex in microorganism, pp. 42—81. Washington (D.C.) 1954.

— Heterokaryosis and sexuality in fungi. Trans. N.Y. Acad. Sci., Ser. II 17, 627—635 (1955a).

— Some problems of specifity in the sexuality of plants. In: E. G. BUTLER (edit.), Biological specificity and growth, pp. 119—140. Princeton (New Jersey) 1955b.

— Hormones and sexuality in lower plants. Symp. Soc. exp. Biol. 11, 143—165 (1957).

— Sexual versatility and evolutionary processes in fungi. Mycologia (N.Y.) 51, 107—125 (1959).

— The control of sex in fungi. Amer. J. Bot. 47, 794—808 (1960).

— Incompatibilität bei dem Basidiomyceten Schizophyllum commune. Ber. dtsch. bot. Ges. 74, 326—328 (1961a).

— Parasexual phenomena in Basidiomycetes. In: Recent Advances in Botany. Toronto (Canad.): Univ. of Toronto Press. 1961b.

— Patterns of sexuality in fungi. Mycologia (N.Y.) 55, 79—92 (1963).

— M. G. BAXTER, and A. H. ELLINGBOE: The genetic structure of the incompatibility factors of Schizophyllum commune. The A factor. Proc. nat. Acad. Sci. (Wash.) 46, 833—842 (1960).

— — and R. B. MIDDLETON: The genetic structure of the incompatibility factors in Schizophyllum commune. Proc. nat. Acad. Sci. (Wash.) 44, 889—900 (1958a).

Reproduction

RAPER, J. R., and K. ESSER: Antigenic differences due to the incompatibility factors in *Schizophyllum commune*. Z. Vererbungsl. **92**, 439—444 (1961).
— — The fungi. The Cell, vol. VI, pp. 139—244. New York 1964.
—, and A. J. HAAGEN-SMIT: Sexual hormones in *Achlya*. IV. Properties of hormone *A* of *A. bisexualis*. J. biol. Chem. **143**, 311—320 (1942).
—, and G. S. KRONGELB: Genetic and enviromental aspects of fruiting in *Schizophyllum commune*. Fr. Mycologia (N.Y.) **50**, 707—740 (1958).
— — — and M. G. BAXTER: The number and distribution of incompatibility factors in *Schizophyllum*. Amer. Naturalist **92**, 221—232 (1958b).
—, and P. G. MILES: The genetics of *Schizophyllum commune*. Genetics **43**, 530—546 (1958).
—, and M. T. OETTINGER: Anomalous segregation of incompatibility factors in *Schizophyllum commune*. Rev. Biol. (Lisboa) **3**, 205—221 (1962).
—, and J. P. SAN ANTONIO: Heterokaryotic mutagenesis in Hymenomycetes. I. Heterokaryosis in *Schizophyllum commune*. Amer. J. Bot. **41**, 69—86 (1954).
— — and P. G. MILES: The expression of mutations in common *A* heterokaryons of *Schizophyllum commune*. Z. Vererbungsl. **89**, 540—558 (1958c).
RAPER, K. B., and C. THOM: Manual of *Penicillia*. Baltimore 1949.
— R. D. COGHILL, and A. HOLLAENDER: The production and characterization of U.V. induced mutations in *Aspergillus terreus*. II. Cultural and morphological characteristics of the mutations. Amer. J. Bot. **32**, 165—176 (1945).
RAWITSCHER, F.: Beiträge zur Kenntnis der Ustilaginaceen. Z. Bot. **14**, 273—296 (1922).
RITTER, R.: Physiologische Untersuchungen an Zygomyceten. Arch. Mikrobiol. **22**, 248—284 (1955).
RIZET, G.: Les phénomènes de barrage chez *Podospora anserina*. I. Analyse génétique des barrages entre souches *S*. et *s*. Rev. Cytol. Biol. végét. **13**, 51—92 (1952).
—, et C. ENGELMANN: Contribution a la étude génétique d'un ascomycète tétrasporé: *Podospora anserina*. (CES.) REHM. Rev. Cytol. Biol. végét. **11**, 202—304 (1949).
—, et K. ESSER: Sur des phénomènes d'incompatibilité entre souches d'origines différentes chez *Podospora anserina*. C. R. Acad. Sci. (Paris) **237**, 760—761 (1953).
ROBERTS, J. W.: Morphological characters of *Alternaria mali* ROBERTS. J. agric. Res. **27**, 699—708 (1924).
ROMAN, H., D. C. HAWTHORNE, and H. C. DOUGLAS: Polyploid in yeasts and its bearing on the occurrence of irregular genetic ratios. Proc. nat. Acad. Sci. (Wash.) **37**, 79—84 (1951).
— M. M. PHILLIPS, and S. M. SANDS: Studies of polyploid *Saccharomyces*. I. Tetraploid segregation. Genetics **40**, 546—561 (1955).
RONSDORF, L.: Über die chemischen Bedingungen von Wachstum und Zygotenbildung bei *Phycomyces blakesleeanus*. Planta (Berl.) **14**, 482—514 (1931).
ROPER, J. A.: Production of heterozygous diploids in filamentous fungi. Experientia (Basel) **8**, 14—15 (1952).
ROSHAL, J. Y.: Incompatibility factors in a population of *Schizophyllum commune*. Thesis Univ. of Chicago (Ill., USA) 1950.
ROWELL, J. B.: Functional role of compatibility factors and an in vitro test for sexual compatibility with haploid lines of *Ustilago zea*. Phytopathology **45**, 370—374 (1955).
SANSOME, E.: Heterokaryosis and the mating type factors in *Neurospora*. Nature (Lond.) **156**, 47 (1945).
— Heterokaryosis, mating type factors and sexual reproduction in *Neurospora*. Bull. Torrey bot. Club **73**, 397—409 (1946).
— Spontaneos variation in *Penicillium notatum* strain. N.R.R.L. 1249b 21. Trans. brit. mycol. Soc. **31**, 66—79 (1947).

SANSOME, E.: Spontangous mutation in standard and "gigas" forms of *Penicillium notatum* strains. Trans. Brit. mycol. Soc. **32**, 305—314 (1949).

SASS, J. E.: The cytological basis for homothallism and heterothallism in the Agaricaceae. Amer. J. Bot. **16**, 663—701 (1929).

SATINA, S., and A. F. BLAKESLEE: Studies on biochemical differences between *(+)* and *(—)* sexes in Mocurs. I. Tellurium salts as indicator of the reduction reaction. Proc. nat. Acad. Sci. (Wash.) **11**, 528—534 (1925).

— — II. A Preliminary report on the MANILOV reaction and other tests. Proc. nat. Acad. Sci. (Wash.) **12**, 191—196 (1926).

— — Further studies on biochemical differences between sexes in plants. Proc. nat. Acat. Sci. (Wash.) **13**, 115—122 (1927).

— — Studies on biochemical differences between sexes in Mucors. V. Quantitative determinations of sugars in *(+)* and *(—)* races. Proc. nat. Acad. Sci. (Wash.) **14**, 308—316 (1928).

— — Criteria of male and female in bread moulds (Mucors). Proc. nat. Acad. Sci. (Wash.) **15**, 735—740 (1929).

SCHIEMANN, E.: Mutationen bei *Aspergillus niger* VAN TIEGHEM. Z. indukt. Abstamm.- u. Vererb.-L. **8**, 1—35 (1912).

SCHOPFER, W. H.: Recherches sur le dimorphisme sexuel biochemique. C. R. Soc. Phys. Hist. Nat. Genève **45**, 14—18 (1928).

— Recherches physiologiques sur la sexualité d'un champignon *(Phycomyces)*. C. R. Soc. Phys. Hist. Nat. Genève **47**, 101—105 (1930).

SCHWARTZ, W.: Entwicklungsphysiologische Untersuchungen über die Gattung *Aspergillus* und *Penicillium*. Flora (Jena) **123**, 386—440 (1928).

SERMONTI, G.: Analysis of vegetative segregation and recombination in *Penicillium chrysogenum*. Genetics **42**, 433—443 (1957).

SERRES, F. J. DE: Heterokaryon-incompatibility factor interaction in tests between *Neurospora* mutants. Science **138**, 1342—1343 (1962).

SIRKS, M. J.: Stérilité, auto-inconceptabilité et différentiation sexuelle physiologique. Arch. néerl. Sci. exact. nat., Sér. B **3**, 205—235 (1917).

SJÖWALL, M.: Studien über Sexualität, Vererbung und Zytologie bei einigen diözischen Mucoraceen. Lund 1945.

SKOLKO, A. J.: A cultural and cytological investigation of a two-spored Basidiomycete. *Aleurodiscus canadiensis* n. sp. Canad. J. Res. **22**, 251—271 (1944).

SKUPIENSKI, F.-X.: Sur la sexualité chez les champignons Myxomycètes. C. R. Acad. Sci. (Paris) **167**, 31—33 (1918).

— Sur le cycle évolutif chez une espèce de Myxomycète endosporée *Didymium difforme* (DUBY). C. R. Acad. Sci. (Paris) **182**, 150—152 (1926).

SNIDER, P. J.: Estimation of nuclear ratios directly from heterokaryotic mycelia in *Schizophyllum*. Amer. J. Bot. **50**, 255—262 (1963a).

— Genetic evidence for nuclear migration in Basidiomycetes. Genetics **48**, 47—55 (1963 b).

—, and J. R. RAPER: Nuclear migration in the Basidiomycete *Schizophyllum commune*. Amer. J. Bot. **45**, 538—546 (1958).

— — Nuclear ratios and genetic complementation in common-*A* heterokaryons. Amer. J. Bot. **52**, 547—552 (1965).

SPILTOIR, C. F.: Life cycle of *Ascosphaera apis (Pericystis apis)*. Amer. J. Bot. **42**, 501—508 (1955).

SRB, A. M.: Exotic growth forms in *Neurospora*. J. Hered. **48**, 146—153 (1957).

STADLER, D. R.: Genetic control of a cyclic growth pattern in *Neurospora*. Nature (Lond.) **184**, 170 (1959).

STAUFER, J. R., and M. P. BACKUS: Spontaneous and induced variation in selected strains of *Penicillium chrysogenum* series. Ann. N.Y. Acad. Sci. **60**, 35—49 (1956).

STEINER, E.: Incompatibilität bei den Komplex-Heterozygoten von *Oenothera*. Ber. dtsch. bot. Ges. **74**, 379—381 (1961).

STERN, C.: Somatic crossing over and segregation in *Drosophila melanogaster*. Genetics **21**, 625—730 (1936).

Reproduction

STEVENS, F. L.: The *Helminthosporium* foot-rot of wheat with observation on the morphology of *Helminthosporium* and on the occurence of saltation in the genus. III. Nat. Hist. Surv. Bull. **14**, 78—185 (1922).

STOUT, A. B.: Self- and cross-pollinations in *Cichorium intybus* with reference to sterility. Mem. N.Y. bot. Gard. **6**, 333—354 (1916).

STRAUB, J.: Das Überwinden der Selbststerilität. Z. Bot. **46**, 98—110 (1958).

STRØMNAES, Ø., and E. D. GARBER: Heterocaryosis and the parasexual cycle in *Aspergillus fumigatus*. Genetics **48**, 653—662 (1963).

— — and L. BERAHA: Genetics of phytopathogenic fungi. IX. Heterocaryosis and the parasexual cycle in *Penicillium italicum* and *Penicillium digitatum*. Canad. J. Bot. **42**, 423—427 (1964).

STUMM, C.: Die Analyse von Genmutanten mit geänderten Fortpflanzungs-eigenschaften bei *Allomyces arbuscula*. BUTL. Z. Vererbungsl. **89**, 521—539 (1958).

SUBRAMANIAM, M. K.: Tetraploidy in yeasts. Cellule **54**, 143—148 (1951).

SUSSMAN, A. S., R. J. LOWRY, and T. DURKEE: Morphology and genetics of a periodic colonial mutant of *Neurospora crassa*. Amer. J. Bot. **51**, 243—252 (1964).

SWIEZYNSKI, K. M.: Exchange of nuclei between dikaryons in *Coprinus lagopus*. Acta Soc. Bot. Pol. **30**, 535—552 (1961).

— Analysis of an incompatible di-mon mating in *Coprinus lagopus*. Acta Soc. Bot. Pol. **31**, 169—184 (1962).

— Somatic recombination of two linkage groups in *Coprinus lagopus*. Genetica Pol. **4**, 21—36 (1963).

—, and P. R. DAY: Heterokaryon formation in *Coprinus lagopus*. Genet. Res. **1**, 114—128 (1960a).

— — Migration of nuclei in *Coprinus lagopus*. Genet. Res. **1**, 129—139 (1960b).

TAKAHASHI, T.: Complementary genes controlling homothallism in *Saccharomyces*. Genetics **43**, 705—715 (1958).

— T. H. SAITO, and Y. IKEDA: Heterothallic behavior of a homothallic strain in *Saccharomyces* yeast. Genetics **43**, 249—260 (1958).

TAKEMARU, T.: Genetics of *Collybia velutipes*. IV. "Interpolarity" occuring in the strain NL. 55. Bot. Mag. (Tokyo) **70**, 238—243 (1957a).

— V. Mating patterns between F_1 mycelia of legitimate and illegitimate origins in the strain NL. 55. Bot. Mag. (Tokyo) **70**, 244—249 (1957b).

— Genetical studies on fungi. X. The Mating in hymenomycetes and its genetical mechanism. Biol. J. Okayama Univ. **7**, 133—211 (1961).

TATUM, E. L., and J. LEDERBERG: Gene recombination in the bacterium *Escherichia coli*. J. Bact. **53**, 673—684 (1947).

TERAKAWA, H.: The nuclear behavior and the morphogenesis in *Pleurotus ostreatus*. Sci. Papers of Gen. Educ. Univ. Tokyo **7**, 61—68 (1957).

TERRA, P.: Détermination de la polarité sexuelle de trente espèces de Basidiomycètes saprophytes. C. R. Acad. Sci. (Paris) **236**, 115—117 (1953).

— Recherches expérimentales sur l'hétérothallie et l'amphithallie des Basidiomycètes. Thèse Lyon 1958.

THAXTER, R.: Contribution towards a monograph of the Laboulbeniaceae. Mem. Amer. Acad. Arts Sci. **12**, 187—429 (1896).

THOM, C., and K. B. RAPER: Manual of the *Aspergilli*. Baltimore 1945.

TINLINE, R. D.: *Cochliobolus sativus*. V. Heterokaryosis and parasexuality. Canad. J. Bot. **40**, 425—437 (1962).

TOMPKINS, G. M., and P. A. ARK: *Verticillium* wilt of strawflower. Phytopathology **31**, 1130—1134 (1941).

TURIAN, G.: Déficiences du métabolisme oxydatif et la différenciation sexuelle chez *Allomyces* et *Neurospora*, activité d'une DPN-deshydrogénase lactique chez *Allomyces*. Path. et Microbiol. (Basel) **23**, 687—699 (1960).

— Cycle glyoxylique, transaminase-alanine-glyoxalate et différenciation sexuelle chez *Allomyces* et *Neurospora*. Path. et Microbiol. (Basel) **24**, 819—839 (1961a).

Turian, G.: Nucleic acids and sexual differentiation in *Allomyces*. Nature (Lond.) **190**, 825 (1961b).

—, et J. Seydoux: Déficience d'activité de la deshydrogénase succinique dans les mitochondries isolées du *Neurospora* en condition d'induction isocitratasique par culture sur acétate. C. R. Acad. Sci. (Paris) **255**, 755—757 (1962).

Tuveson, R. W., and D. O. Coy: Heterocaryosis and somatic recombination in *Cephalosporium mycophyllum*. Mycologia (N.Y.) **53**, 244—253 (1961).

—, and E. D. Garber: Genetics of phytopathogenic fungi. II. The parasexual cycle in *Fusarium oxysporum f. pisi*. Bot. Gaz. **121**, 69—74 (1959).

Utiger, H.: Eine neue chemische Reaktion zur geschlechtlichen Differenzierung von *Phycomyces blakesleeanus* + und —. Naturwissenschaften **40**, 292 (1953).

Vakili, N. G.: On the genetics of the *A* factor in *Schizophyllum commune*. Thesis University of Chicago (Ill., USA) 1953.

—, and R. M. Caldwell: Recombination of spore color and pathogenicity between uredial clones of *Puccinia recondita f. sp. tritici*. Phytopathology **47**, 536 (1957).

Vandendries, R.: Nouvelles recherches sur la sexualité des Basidiomycètes. Bull. Soc. bot. Belg. **56**, 73—97 (1923).

— Les mutations sexuelles, l'héterothallisme et la sterilité entre races geographiques de *Coprinus micaceus*. Mem. Acad. roy. Belg., Cl. Sci. **9**, 1—50 (1927).

—- Les relations entre souches étrangères expliquées par les aptitudes sexuelles de individus parthénogéniques chez *Coprinus micaceus*. Bull. Soc. mycel. France **45**, 216—248 (1929).

— Les multiples aspects de la sexualité dans le monde des champignons. Bull. Acad. roy. Belg., Cl. Sci. **24**, 842—856 (1938).

—, et H. J. Brodie: Nouvelles investigations dans le domaine de la sexualité des Basidiomycètes et étude expérimentale des barrages sexuels. Cellule **42**, 163—210 (1933).

Vanterpool, T. C., and G. A. Ledingham: Studies on browning root rot of cereals. I. The association of *Lagena radicicola n. gen. n. sp.* with root injury of wheat. Canad. J. Res. **2**, 171—194 (1930).

Verkaik, C.: Über das Entstehen von Zygophoren von *Mucor mucedo (+)* unter Beeinflussung eines von *Mucor mucedo (—)* abgeschiedenen Stoffes. Proc. roy. Acad. (Amsterd.) **33**, 656—658 (1930).

Watson, I. A.: Further studies on the productions of new races of *Puccinia graminis var tritici* on wheat seedlings. Phytopath. **47**, 510—512 (1957).

— Somatic hybridization in *Puccinia graminis var. tritici*. Proc. Linnéan Soc. N. S. Wales **83**, 190—195 (1958).

—, and N. H. Luig: Asexual intercrosses between somatic recombinants of *Puccinia graminis*. Proc. Linnean Soc. N. S. Wales **87**, 99—104 (1962).

Wesendonck, J.: Über sekundäre Geschlechtsmerkmale bei *Phykomyces blakesleeanus* Bgff. Planta (Berl.) **10**, 456—494 (1930).

Westergaard, M., and H. Hirsch: Enviromental and genetic control of differentiation in *Neurospora*. Proc. VII. Symp. Colston Res. Soc. 171—183 (1954).

Weston, W. H.: Heterothallism in *Sapromyces reinschii*. Mycologica (N.Y.) **30**, 245—253 (1938).

Wheeler, H. E.: Genetics and evolution of heterothallism in *Glomerella*. Phytopathology **44**, 342—345 (1954a).

— Genetics of homothallic fungi. Science **120**, 718—719 (1954b).

—, C. H. Driver, and C. Campa: Cross- and self-fertilization in *Glomerella*. Amer. J. Bot. **46**, 361—365 (1959).

—, and J. W. McGahen: Genetics of *Glomerella*. X. Genes affecting sexual reproduction. Amer. J. Bot. **39**, 110—119 (1952).

Whitehouse, H. L. K.: Heterothallism and sex in the fungi. Biol. Rev. **24**, 411—447 (1949a).

Reproduction

WHITEHOUSE, H. L. K.: Multiple allelomorph heterothallism in the fungi. New Phytologist **48**, 212—244 (1949b).
— The significance of some sexual phenomena in the fungi. Indian Phytopath. **4**, 91—105 (1951a).
— A survey of heterothallism in the Ustilaginales. Trans. Brit. mycol. Soc. **34**, 340—355 (1951b).
WILKIE, D., and D. LEWIS: The effect of ultraviolet light on recombination in yeast. Genetics **48**, 1701—1716 (1963).
WILSON, J. F.: Micrurgical techniques for *Neurospora*. Amer. J. Bot. **48**, 46—51 (1961).
— L. GARNJOBST, and E. L. TATUM: Heterocaryon incompatibilities in *Neurospora crassa*, microinjection studies. Amer. J. Bot. **48**, 299—305 (1961).
WINGE, Ö.: The relation between yeast cytology and genetics. A critique. C. R. Lab. Carlsberg, Sér. Physiol. **25**, 85—99 (1951).
—, and O. LAUTSEN: On 14 new yeasts types, produced by hybridization. C. R. Trav. Lab. Carlsberg, Sér. Physiol. **22**, 337—355 (1939a).
— *Saccharomyces Ludwigii*, a balanced heterozygote. C. R. rend. Trav. Lab. Carlsberg, Sér. Physiol. **22**, 357—370 (1939b).
—, and C. ROBERTS: A gene for diploidization in yeast. C. R. Lab. Carlsberg, Sér. Physiol. **24**, 341—346 (1949).
— Causes of deviations from 2:2 segregations in the tetradie of monohybrid yeasts. C. R. Lab. Carlsberg, Sér. Physiol. **25**, 285—329 (1954).
WÜLKER, H.: Untersuchungen über die Tetradenaufspaltung bei *Neurospora sitophila* SHEAR et DODGE. Z. indukt. Abstamm.- u. Vererb.-L. **69**, 210—248 (1935).
YOUNG, L.: Etude biologique des phénomènes de la sexualité chez les Mucorinées. Rev. gén. Bot. **42**, 144—158, 205—218, 283—296, 348—365, 409—428, 491—504, 535—552, 618—639, 681—704, 722—752 (1930). **43**, 30—43 (1931).
ZICKLER, H.: Genetische Untersuchungen an einem heterothallischen Askomyceten (*Bombardia lunata* nov. spec.) Planta (Berl.) **22**, 573—613 (1934).
— Die Vererbung des Geschlechts bei dem Askomyceten *Bombardia lunata* ZCKL. Z. indukt. Abstamm.- u. Vererb.-L. **73**, 403—418 (1937).
— Zur Entwicklungsgeschichte des Askomyceten *Bombardia lunata*. Arch. Protistenk. **98**, 1—70 (1952).
ZINDER, N. D., and J. LEDERBERG: Genetic exchange in *Salmonella*. J. Bact. **64**, 679—699 (1952).
ZOPF, W.: Zur Kenntnis der Phycomyceten. I. Zur Morphology und Biologie der Ancylisteen und Chytridiaceen. Nova Acta Leopold. Carol. **47**, 143—236 (1884).

References

which have come to the authors' attention after conclusion of the German manuscript

It should be noted, that a symposium held at the 10th International Congress of Botany at Edinburgh August 1964 has been edited by K. ESSER and J. R. RAPER under the title "Incompatibility in fungi". Berlin-Heidelberg-New York: Springer 1965. The book of J. R. RAPER (Genetics of sexuality in higher fungi. New York 1966) exhibits a very detailed treatment of incompatibility in basidiomycetes.

A

BERLINER, M. D., and P. W. NEURATH: The rhythms of three clock mutants of *Ascobolus immersus*. Mycologia (N.Y.) **57**, 809—817 (1965).
— — The band forming rhythm of *Neuropora* mutants. J. cell. comp. Physiol. **65**, 183—194 (1965).
BRODY, S., and E. L. TATUM: The primary biochemical effect of a morphological mutation in *Neurospora crassa*. Proc. nat. Acad. Sci. (Wash.) **56**, 1290—1297 (1966).

Durkee, T. L., A. S. Sussman, and R. J. Lowry: Genetic localization of the clock mutant and a gene modifying its band-size in *Neurospora*. Genetics **53**, 1167—1175 (1966).

Galzy, P.: Étude génétique et physiologique du métabolisme de l'acide lactique chez *Saccharomyces cerevisiae* Hansen. Ann. Technol. agric. **13**, 109—259 (1964).

—, et C. Bizeau: Étude du contrôle génétique du caractère «colonie lisse» chez. *Saccharomyces cerevisiae* Hansen. Heredity **20**, 31—36 (1956).

— — Sur le controle génétique de la mutation « colonie lisse » chez *Saccharomyces cerevisiae*. Étude de mutants sélectionnés par culture sur éthanol. C. R. Acad. Sci. (Paris) **261**, 3490—3493 (1965).

Gerisch, G.: Eine Mutante von *Dictyostelium minutum* mit blockierter Zentrengründung. Z. Naturforsch. **20**b, 298—301 (1965).

Gregg, J. H.: Developmental processes in cellular slime molds. Physiol. Rev. **44**, 631—656 (1964).

Mishra, N. C., and S. F. H. Threlkeld: Variation in the expression of the ragged mutant in *Neurospora*. Genetics **55**, 113—121 (1967).

Nelson, R. R.: Genetic inhibition of perithecial formation in *Cochliobolus carbonum*. Phytopathology **54**, 876—877 (1964).

— The genetic control of conidial morphology and arrangement in *Cochliobolus carbonum*. Mycologia (N.Y.) **58**, 208—214 (1966).

Pandhi, P. N., and E. C. Cantino: Differentiation of glucose-6-phosphate dehydrogenase isozymes and morphogenesis in *Blastocladiella emersonii*. Arch. Mikrobiol. **55**, 226—244 (1966).

Raper, C. A., and J. R. Raper: Mutations modifying sexual morphogenesis in *Schizophyllum*. Genetics **54**, 1151—1168 (1966).

Solomon, E. P., E. M. Johnson, and J. H. Gregg: Multiple forms of enzymes in a cellular slime mold during morphogenesis. Develop. Biol. **9**, 314—326 (1964).

Sussman, A. S., T. L. Durkee, and R. J. Lowry: A model for rhythmic and temperature-independent growth in "clock" mutants of *Neurospora*. Mycopathologia (Den Haag) **25**, 381—396 (1965).

Sussman, M., and M. J. Osborn: UDP-Galactose polysaccharide transferase in the cellular slime mold, *Dictyostelium discoideum:* Appearance and disappearance of activity during cell differentiation. Proc. Nat. Acad. Sci. (Wash.) **52**, 81—87 (1964).

Viswanathan, M. A., et G. Turian: Teneur des gamètes mâles et femelles d'*Allomyces* en acides nucléiques et composition comparée d'une fraction saline de l'acide ribonucléique des gamétanges. Experientia (Basel) **22**, 377 (1966).

Wessels, J. G. H.: Control of cell-wall glucan degradation during development in *Schizophyllum commune*. Antonie v. Leeuwenhoek **32**, 341—355 (1966).

Yu-Sun, C. C. C.: Biochemical and morphological mutants of *Ascobolus immersus*. Genetics **50**, 987—998 (1964).

B II, 2a

Boidin, J.: Valeur des caractères culturaux et cytologiques pour la taxinomie des *Thelephoraceae* résupinés et étalés-réfléchis (Basidiomycètes). Bull. Soc. Bot. Fr. **111**, 309—315 (1964).

—, et P. Lanquetin: Hétérobasidiomycètes saprophytes et Homobasidiomycétes résupinés. X. — Nouvelles données sur la polarité dite sexuelle. Rev. Mycologie **30**, 3—16 (1965).

Burnett, J. H., and E. J. Evans: Genetical homogeneity and the stability of the mating-type factors of "fairy rings" of *Marasmius oreades*. Nature (Lond.) **210**, 1368—1369 (1966).

Casselton, L. A.: The production and behavior of diploids of *Coprinus lagopus*. Genet. Res. Camb. **6**, 190—208 (1965).

—, and D. Lewis: Compatibility and stability of diploids in *Coprinus lagopus*. Genet. Res. Camb. **8**, 61—72 (1966).

Reproduction

COLLINS, O. R.: Plasmodial compatibility in heterothallic and homothallic isolates of *Didymium iridis*. Mycologia (N.Y.) **58**, 362—372 (1966).

CROFT, J. H., and G. SIMCHEN: Natural variation among monokaryons of *Collybia velutipes*. Amer. Naturalist **99**, 451—462 (1965).

GARIBOVA, L. V.: The race of the cultivated true mushroom *Agaricus bisporus* TRESCHOW and its possible way of origin. Otd. Biol. **69**, 118—125 (1964).

HALISKY, P. M.: Physiologic specialization and genetics of the smut fungi III. Bot. Rev. (N.Y.) **31**, 114—150 (1965).

HOWE, H. B.: Determining mating type in *Neurospora* without crossing tests. Nature (Lond.) **190**, 1036 (1961).

— Sources of error in genetic analysis in *Neurospora tetrasperma*. Genetics **50**, 181—189 (1964).

— Vegetative traits associated with mating type in *Neurospora tetrasperma*. Mycologia (N.Y.) **56**, 519—525 (1964).

KOLTIN, Y., and J. R. RAPER: *Schizophyllum commune:* New mutations in the *B* incompatibility factor. Science **154**, 510—511 (1966).

— J. R. RAPER, and G. SIMCHEN: The genetic structure of the incompatibility factors of *Schizophyllum commune:* The *B* factor. Proc. nat. Acad. Sci. (Wash.) **57**, 55—62 (1967).

KWON, K. J., and K. B. RAPER: Sexuality and cultural characteristics of *Aspergillus heterothallicus*. Amer. J. Bot. **54**, 36—48 (1967).

— — Heterokaryon formation and genetic analyses of color mutants in *Aspergillus heterothallicus*. Amer. J. Bot. **54**, 49—60 (1967).

MIDDLETON, R. B.: Sexual and somatic recombination in common-*AB* heterokaryons of *Schizophyllum commune*. Genetics **50**, 701—710 (1964).

MUKHERJEE, K. L., and G. G. ZABKA: Studies of multiple allelism in the myxomycete *Didymium iridis*. Canad. J. Bot. **42**, 1459—1466 (1964).

PARAG, Y.: Common-B heterokaryosis and fruiting in *Schizophyllum commune*. Mycologia (N.Y.) **57**, 543—561 (1965).

—, and B. NACHMAN: Diploidy in the tetrapolar heterothallic Basidiomycete *Schizophyllum commune*. Heredity **21**, 151—154 (1966).

RAPER, J. R., D. H. BOYD, and C. A. RAPER: Primary and secondary mutations at the incompatibility loci in *Schizophyllum*. Proc. nat. Acad. Sci. (Wash.) **53**, 1324—1332 (1965).

MILES, P. G., T. TAKEMARU, and K. KIMURA: Incompatibility factors in the natural population of *Schizophyllum commune*. I. Analysis of the incompatibility factors present in fruit bodies collected within a small area. Bot. Mag. **79**, 693—705 (1966).

RAVISE, A.: Observations sur la reproduction sexuée de souches du *Phytophthora palmivora* (BUTL.) Butl. parasite de cultures tropicales. Cah. Orstom Biol. **2**, 91—101 (1966).

SIMCHEN, G.: Variation in a dikaryotic population of *Collybia velutipes*. Genetics **51**, 709—721 (1965).

— Monokaryotic variation and haploid selection in *Schizophyllum commune*. Heredity **21**, 241—263 (1966a).

— Fruiting and growth rate among dikaryotic progeny of single wild isolates of *Schizophyllum commune*. Genetics **53**, 1151—1165 (1966b).

— Genetic control of recombination and the incompatibility system in *Schizophyllum commune*. Genet. Res. Camb. **9**, 195—210 (1967).

—, and J. L. JINKS: The determination of dikaryotic growth rate in the basidiomycete *Schizophyllum commune:* A biometrical analysis. Heredity **19**, 629—649 (1964).

B II, 2b

BERNET, J.: Mode d'action des gènes de « barrage » et relation entre l'incompatibilité cellulaire et l'incompatibilité sexuelle chez *Podospora anserina*. Ann. Sci. natur. Bot. (Paris) **6**, 611—768 (1965).

GRUN, P., and M. AUBERTIN: The inheritance and expression of unilateral incompatibility in *Solanum*. Heredity **21**, 131—138 (1966).

NELSON, R. R.: Interspecific hybridization in the fungi. Ann. Rev. Microbiol. **17**, 31—48 (1963).
— Bridging interspecific incompatibility in the ascomycete genus *Cochliobolus*. Evolution **18**, 700—704 (1965).
— Assessing biological relationships in the fungi. Phytopathology **55**, 823—826 (1965).
—, and D. M. KLINE: Gene systems for pathogenicity and pathogenic potentials. I. Interspecific hybrids of *Cochliobolus carbonum* × *Cochliobolus victoriae*. Phytopath. **53**, 101—105 (1963).
— — Evolution of sexuality and pathogenicity. III. Effects of geographic origin and host association on cross-fertility between isolates of *Helminthosporium* with similar conidial morphology. Phytopathology **54**, 963—967 (1964).

B II, 2c

MILES, P. G.: Possible role of dikaryons in selection of spontaneous mutants in *Schizophyllum commune*. Bot. Gaz. **125**, 301—306 (1964).

B II, 2d

BISTIS, G., and M. ANCHEL: Evidence for genetic control of polyacetylene production in a Basidiomycete. Mycologia (N.Y.) **58**, 270—274 (1966).
CASSELTON, L. A., and D. LEWIS: Dilution of gene products in the cytoplasm of heterokaryons in *Coprinus lagopus*. Genet. Res. Camb. **9**, 63—71 (1967).
WANG, C. S., and P. G. MILES: The physiological characterization of dikaryotic mycelia of *Schizophyllum commune*. Physiol. Plant. **17**, 573—588 (1964).

B III, 1

BARKSDALE, A. W., M. J. CARLILE, and L. MACHLIS: A comparative study of hormone *A* and sirenin. Mycologia (N.Y.) **57**, 138—140 (1965).
SHERWOOD, W. A.: Evidence for a sexual hormone in the water mold *Dictyuchus*. Mycologia (N.Y.) **58**, 215—220 (1966).

B III, 2a

OESER, H.: Genetische Untersuchungen über das Paarungstypverhalten bei Saccharomyces und die Maltose-Gene bei *Saccharomyces* und die Maltose-Gene einiger untergäriger Bierhefen. Arch. Mikrobiol. **44**, 47—74 (1962).
SHERMAN, F., and H. ROMAN: Evidence for two types of allelic recombination in yeast. Genetics **48**, 255—261 (1963).

C

CLUTTERBUCK, A. J., and J. A. ROPER: A direct determination of nuclear distribution in heterokaryons of *Aspergillus nidulans*. Genet. Res. Camb. **7**, 185—194 (1966).
COWAN, J. W., and D. LEWIS: Somatic recombination in the dikaryon of *Coprinus lagopus*. Genet. Res. Camb. **7**, 235—244 (1966).
GARBER, E. D., and L. BERAHA: Genetics of phytopathogenic fungi. XVI. The parasexual cycle in *Penicillium expansum*. Genetics **52**, 487—492 (1965).
HASTIE, A. C.: The parasexual cycle in *Verticillium albo-atrum*. Genet. Res. Camb. **5**, 305—315 (1964).
HOFFMANN, G. M.: Heterokaryose und parasexuelle Vorgänge bei Fusarium oxysporum. Naturwissenschaften **53**, 45—46 (1966a).
— Untersuchungen über die Heterokaryosebildung und den Parasexualcyclus bei *Fusarium oxysporum*. I. Anastomosenbildung im Mycel und Kernverhältnisse bei der Conidienentwicklung. Arch. Mikrobiol. **53**, 336—347 (1966b).

Reproduction

HOPFMANN, G. M.: Untersuchungen über die Heterokaryosebildung und den Parasexualcyclus bei *Fusarium oxysporum*. II. Gewinnung und Identifierung auxotropher Mutanten. Arch. Mikrobiol. **53**, 348—357 (1966c).
— Untersuchungen über die Heterokaryosebildung und den Parasexualcyclus bei *Fusarium oxysporum*. III. Paarungsversuche mit auxotrophen Mutanten von *Fusarium oxysporum* f. *callistephi*. Arch. Mikrobiol. **56**, 40—59 (1967).
JINKS, J. L., C. E. CATEN, G. SIMCHEN, and J. H. CROFT: Heterokaryon incompatibility and variation in wild populations of *Aspergillus nidulans*. Heredity **21**, 227—239 (1966).
MALUZYNSKI, M.: Recombination in crosses between biochemical mutants of *Coprinus lagopus*. Acta Soc. bot. Pol. **35**, 191—199 (1966).
SANDERSON, K. E., and A. M. SRB: Heterokaryosis and parasexuality in the fungus *Ascochyta imperfecta*. Amer. J. Bot. **52**, 72—81 (1965).

E

CATEN, C. E., and J. L. JINKS: Heterokaryosis: Its significance in wild homothallic Ascomycetes and Fungi Imperfecti. Trans. Brit. mycol. Soc. **49**, 81—93 (1966).
JONES, D. A.: Heterokaryon compatibility in the *Aspergillus glaucus* link group. Heredity **20**, 49—56 (1965).
STEPHAN, B. R.: Heterokaryose bei *Colletotrichum gloeosporioides* PENZIG. Naturwisschenschaften **53**, 532—533 (1966).
— Untersuchung über die Variabilität bei *Colletotrichum gloeosporioides* PENZIG in Verbindung mit heterokaryose. I. Morphologische Variabilität bei *C. gloeosporioides* PENZ. II. Cytologische Grundlagen der *Heterokaryose*. III. Versuche zum Nachweis der *Heterokaryose*. Zbl. Bakt. **121**, 41—83 (1967).
WILSON, J. F., and L. GARNJOBST: A new incompatibility locus in *Neurospora crassa*. Genetics **53**, 621—631 (1966).

F I

EL-ANI, A. S.: Self-sterile auxotrophs and their relation to heterothallism in *Sordaria fimicola*. Science **145**, 1067—1068 (1964).

F II

PAPA, K. E., A. M. SRB, and W. T. FEDERER: Selection for increased growth rate in inter- and intrastrain crosses of *Neurospora*. Heredity **21**, 595—613 (1967).
WEIJER, J., and J. CHYONG-EN YANG: Third division segregation for bisexuality at the mating-type locus of *Neurospora* and its genetic implications. Canad. J. Genet. Cytol. **8**, 807—817 (1966).

Chapter III

Replication

The transmission of genetic information from cell generation to cell generation necessitates a duplication of the hereditary material prior to each cell division. Such self-duplication, which occurs according to a predetermined pattern, consists of the copying of the hereditary determiners of the cell and is called replication.

Since deoxyribosenucleic acid (DNA) was identified as the carrier of genetic information, innumerable experiments have been undertaken to clarify its structure and mechanism of replication (p. 129 ff. and p. 131 ff.). The definitive investigations have been carried out on bacteria and viruses. In contrast, fungi have not proved to be suitable experimental material. Thus no data from fungi relate to the problem of replication. Nevertheless, as background for the research results discussed in subsequent chapters a brief textbook discussion of replication is presented here.

Literature of a general nature: In addition to textbooks of biochemistry, the following reviews are useful: ZAMENHOF (1959), SAGER and RYAN (1961), KORNBERG (1962), SINSHEIMER (1962), PERUTZ (1962), CAIRNS (1963), BRESCH (1964), HAYES (1964), STAHL (1964).

A series of other studies are found in the publications edited by McELROY and GLASS (1957), GIERER (1961), KASHA and PULLMAN (1962), ALLEN (1962) and TAYLOR (1963) as well as in volumes **23** (1958) and **28** (1963) of the *Cold Spring Harbor Symposia of Quantitative Biology.*

A. DNA as the carrier of genetic information

DNA was assumed to be directly involved in the transmission of the hereditary determiners even at a time when little was known about its structure and replication. The first experimental proof that the genetic information resided in the DNA molecule was provided in 1944 by the investigations of AVERY and his coworkers. The significant lines of evidence underlying the concept that DNA is the genetic material are briefly summarized below:

1. Chromosomes as DNA-containing carriers of the hereditary determiners. The complete parallel between the distribution of genes and chromosomes suggests that the chromosome carry the genetic information. The results of experiments with chromosomal mutations also support this view. Chemical analysis of chromosomes indicates that they consist in large part of DNA.

2. Constancy of DNA content of the cell. Diploid somatic nuclei from different cells of the same organism possess the same quantity of DNA. Gametes or other haploid cells have half the amount (p. 314f.). This result conforms to the theoretical expectation.

3. Transformation. This refers to the incorporation of foreign DNA into the genome of a cell (known to occur in bacteria; its occurrence is disputed in fungi, p. 291). Transformation involves the transfer of DNA from a donor to a recipient cell, the latter thereby acquiring and retaining in a stable manner particular hereditary characters of the donor. Proof that the transforming substance is actually DNA was first presented by AVERY and coworkers (1944).

4. *Phage infection.* During infection by phage the DNA contents of the phage particle are injected into the bacterial host cell while the protein coat of the phage remains on the surface of the bacterium. HERSHEY and CHASE (1952) demonstrated this by labelling T 2 phage DNA with radioactive phosphorus and phage protein with radioactive sulfur.

5. *Stability of DNA during metabolism.* In contrast to other cellular constituents DNA remains stable and intact as a large molecule during the course of metabolism. This stability is related to the constancy with which genetic information is transmitted from cell generation to cell generation.

6. *Induction of mutations by UV radiation.* The strongest mutagenic activity of UV occurs for the wave lengths between 250 and 270 mμ. These are the same wave lengths for which DNA shows an absorption maximum (p. 285). This parallel suggests that UV induced mutations are changes in DNA.

7. *Production of mutations by chemical agents.* The mutagenic effect of base analogues, diethyl sulfate, nitrous acid, and other substances — as discussed in detail in the chapter on mutation (p. 291 ff.) — can be correlated with changes in the base sequence of DNA.

All the above evidence indicates the primary role of DNA as the hereditary material. For the sake of completeness we should point out that RNA may also perform this function as, for example, in tobacco mosaic virus (GIERER and SCHRAMM, 1956).

B. Structure of DNA

Polymeric structure of DNA. Nucleic acids are high molecular weight polynucleotides; in a DNA molecule the number of *nucleotides* may reach 10^6. Each of these building blocks consists of three components, 1. a nitrogenous *base* (purine or pyrimidine), 2. a *sugar* (pentose), and 3. *phosphoric acid*. These are arranged, as shown in Fig. III-1, in the order phosphate-pentose-phosphate-pentose-etc., with the bases existing as side chains of the sugar molecules.

DNA differs from RNA in the sugar component and in the base composition. While the nucleotides of DNA contain the pentose 2'-deoxyribose and the bases, *adenine, guanine, cytosine* and *thymine* (Fig. III-1 and III-2), RNA contains ribose and instead of thymine, the base, *uracil*. Other bases which are found rarely are disregarded here (5-methyl cytosine, 5-hydroxymethylcytosine). The combination of base and sugar is called a nucleoside; this is a nucleotide minus the phosphate group.

Double helical structure of DNA. WATSON and CRICK in 1953 developed a structural model of DNA on the basis of X-ray diffraction and chemical data; later investigations have shown this structure to be correct in its essential features (WATSON and CRICK, 1953a, b, c; WILKINS and RANDALL, 1953; WILKINS et al., 1953; CHARGAFF, 1950). This model, named after its discoverers, has been generally acclaimed as valid and provides the hypothetical basis for the molecular mechanism of recombination (p. 243 ff. and Fig. IV-24), mutation (p. 291 ff. and Fig. V-6), and function (p. 342 ff. and Fig. VI-1). Fig. III-2 reveals the following:

1. A DNA molecule consists of *two unbranched polynucleotide strands* which have a common axis in the *right-handed double helix*. The two strands have *opposite polarity*.

Polarity or direction can be assigned to each sugar-phosphate chain because each phosphate group is bound on one side the 5' carbon atom and on the other side to the 3' carbon atom of the sugar.

2. The two strands of the double helix are held together by the *specific pairing of bases which lie opposite each other*. Their linkage consists of hydrogen bonds between neighboring amino and keto groups of the corresponding bases (Fig. III-1). Adenine is always paired with thymine and guanine with cytosine. Therefore, *the two strands are not identical but complementary*.

The double helical structure of DNA was worked out from crystallographic investigations (review by LUZZATI, 1963). One turn of the helix encompasses ten nucleotides per strand. The complementary nature of the purine and pyrimidine bases was established through quantitative biochemical analyses (CHARGAFF, 1950, in fungi: STORCK, 1966). The relative amounts of adenine and thymine on the one hand, and guanine and cytosine on the other, are always equal, while the molar ratio of thymine to cytosine and adenine to guanine in different organisms varies widely.

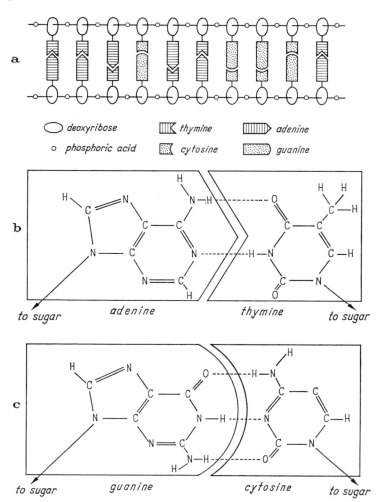

Fig. III-1 a—c. Structure of DNA. (a) Portion of the double strand of DNA with the spirals not represented (compare Fig. III-2). The two DNA strands are held together by specific pairing between complementary bases. Adenine pairs with thymine and guanine with cytosine. Note the reciprocal polarity. (b) and (c) Section of the figure in (a) showing the structural formulae: pairing between the purine base adenine and the pyrimidine base thymine (b) and between guanine and cytosine (c). In the former case the bases are joined by two hydrogen bonds, in the latter, three hydrogen bonds are present. See further in text

3. *The sequence of the different base pairs is aperiodic,* i.e. the sequence of the individual building blocks (nucleotide pairs) is not regular. This unidimensional, non-periodic arrangement of four elements provides a code of genetic information for organisms (see genetic code, p. 344 ff.).

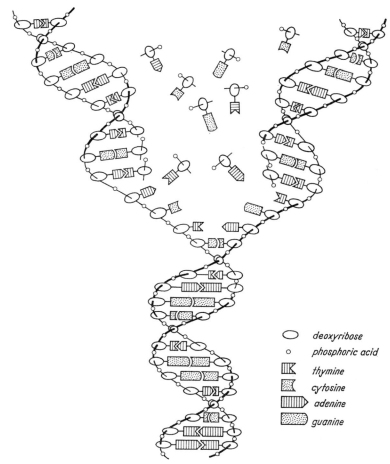

⬭	deoxyribose
∘	phosphoric acid
𝕂	thymine
𝕂	cytosine
⬜	adenine
⬜	guanine

Fig. III-2. Structure and replication of the DNA double helix (Watson-Crick model). The single strands of the righthanded double spiral separate when the hydrogen bonds break (see Fig. III-1). Each strand serves as a template to which complementary nucleotides attach (center). In this way two new double strands arise (above) each of which is composed of one of the strands of the original double helix and one which has been synthesized anew (semiconservative replication). See further in text

C. Biosynthesis of DNA

On the basis of the Watson-Crick model and the results of both in vivo and in vitro investigations in various laboratories, the essential features of DNA replication have been elucidated.

In vivo: 1. MESELSON and STAHL (1958) showed by means of tracer studies in *E. coli* that the replication of a DNA double helix leads to two double helices each of which possesses an "old" and a newly synthesized strand. This manner of replication is called "semi-conservative". Results with the phage λ and with *Chlamydomonas* support the idea of a semi-conservative DNA replication (MESELSON and WEIGLE, 1961; ARBER and DUSSOIX, 1962; SUEOKA, 1960). These experiments show that replication does not involve the synthesis of an entirely new double stranded molecule (conservative replication).

2. Further work with *E. coli* has shown that replication of a DNA molecule proceeds from a single locus (CAIRNS, 1963).

In vitro: Investigations on cell-free systems carried out in particular by KORNBERG and his coworkers have contributed significantly to the understanding of enzyme action during replication of DNA (KORNBERG et al., 1959; KORNBERG, 1960, 1961; see BOLLUM, 1963, for other references). The following points are noteworthy:

1. DNA of exact nucleotide sequences can be replicated *in vitro* with the aid of specific polymerases.

2. The polymerases require the triphosphates of all four nucleotides as substrates: deoxyadenosine, deoxyguanosine, deoxycytidine and deoxythymidine triphosphate. In addition DNA is necessary for the synthesis. This is called "primer" DNA and acts as a "starter" for the reaction. The DNA synthesized *in vitro* and the DNA primer correspond in base sequence. Primer DNA thus acts as a template which is copied with the help of the polymerases.

3. The enzyme responsible for DNA synthesis is not species specific. Polymerases extracted from *E. coli* can replicate DNA of different origins (e.g. calf thymus, wheat germ).

The mechanism of DNA replication, shown diagrammatically in Fig. III-2, has been derived on the basis of these facts. Although some details still remain hypothetical, the following steps in synthesis have been identified:

1. The two DNA strands separate like two parts of a zipper through the breaking of the hydrogen bonds between base pairs. It is assumed that separation proceeds from a specific locus.

2. The exposed bases of a single strand combine with complementary bases. In this way a sequence of new nucleotides arises which is identical to that of the "old" complementary strand.

3. By action of the DNA polymerases the new nucleotides which have become attached to the single strands are tied together in chains. Thus each of the old strands serves as a template on which a new strand with a complementary nucleotide sequence is built up (*semi-conservative replication*).

As a result, two complete DNA molecules are formed. *The original double helix has duplicated itself precisely.* The opposite polarity of the two strands is retained in the process.

A separation of the two sub-units occurs not only in the individual DNA molecules, but also in the duplication of entire chromosomes. This has been shown by experiments with labelled thymidine (*Vicia faba:* TAYLOR, 1957, 1958; *E. coli:* CAIRNS, 1963). This suggests that the two chromatids may be identified with the single strands of the DNA double helix (p. 229f.).

Summary

1. Deoxyribose nucleic acid is the carrier of genetic information. The discovery that genetically marked DNA from one cell may become integrated into the genome of another cell (transformation) is primary evidence in support of this concept.

2. A DNA molecule consists of two unbranched polynucleotide strands which are in the form of right-handed spirals and have a common axis (Watson-Crick model). Each strand is composed of many nucleotides joined together to form a chain. Pentoses and phosphate groups alternate with one another in this chain; the four bases of the nucleotides are attached as side chains of the sugar molecule. The two DNA strands of the double helix are held together by hydrogen bonds between the complementary bases which lie opposite one another (adenine with thymine, guanine with cytosine).

3. The synthesis of DNA always takes place according to a prescribed plan. After separation of two polynucleotide strands, the complementary bases become attached to the exposed bases of each strand. These are joined together by enzymatic action to form new strands of exactly complementary nucleotide sequence. The final result of the synthesis is the production of two complete DNA double helices, each composed of one "new" and one "old" strand (semi-conservative replication).

Literature

ALLEN, J. M. (edit.): The molecular control of cellular activity. New York-Toronto-London 1962.

ARBER, W., and D. DUSSOIX: Host specificity of DNA produced by *Escherichia coli*. I. Host controlled modification of bacteriophage λ. J. molec. Biol. **5**, 18—36 (1962).

AVERY, O. T., C. M. MACLEOD, and M. MCCARTY: Studies on the chemical nature of the substance inducing transformation of pneumococcal types Induction of transformation by a desoxyribonucleic acid fraction isolated from *pneumococcus* type III. J. exp. Med. **79**, 137—158 (1944).

BOLLUM, F. J.: "Primer" in DNA polymerase reactions. Progr. in Nucl. Ac. Res. **1**, 1—26 (1963).

BRESCH, C.: Klassische und molekulare Genetik. Berlin-Göttingen-Heidelberg: Springer 1964.

CAIRNS, J.: The bacterial chromosome and its manner of replication as seen by autoradiography. J. molec. Biol. **6**, 208—213 (1963).

CHARGAFF, E.: Chemical specificity of nucleic acids and mechanism of their enzymatic degradation. Experientia (Basel) **6**, 201—240 (1950).

GIERER, A.: Molekulare Grundlagen der Vererbung. Naturwissenschaften **48**, 283—289 (1961).

—, and G. SCHRAMM: Infectivity of ribonucleic acid from tobacco mosaic virus. Nature (Lond.) **177**, 702—703 (1956).

HAYES, W.: The genetics of bacteria and their viruses. Studies in basic genetics and molecular biology. Oxford 1964.

HERSHEY, A. D., and M. CHASE: Independent functions of viral protein and nucleic acid in growth of bacteriophage. J. gen. Physiol. **36**, 39—56 (1952).

KASHA, M., and B. PULLMAN (edits.): Horizons in biochemistry. New York and London 1962

KORNBERG, A.: Biologic synthesis of deoxyribonucleic acid. Science **131**, 1503—1508 (1960).
— Enzymatic synthesis of DNA. New York 1961.
— Biologic synthesis of deoxyribonucleic acid. In: J. M. ALLEN, The molecular control of cellular activity, p. 245—257. New York-Toronto-London 1962.
— S. B. ZIMMERMAN, S. A. KORNBERG, and J. JOSSE: Enzymatic synthesis of deoxyribonucleic acid. VI. Influence of bacteriophage *T2* on the synthetic pathway in host cells. Proc. nat. Acad. Sci. (Wash.) **45**, 772—785 (1959).
LUZZATI, V.: The structure of DNA as determined by x-ray scattering techniques. Progr. in Nucl. Ac. Res. **1**, 347—368 (1963).
McELROY, W. D., and B. GLASS (edits.): The chemical basis of heredity. Baltimore 1957.
MESELSON, M., and F. W. STAHL: The replication of DNA in *Escherichia coli*. Proc. nat. Acad. Sci. (Wash.) **44**, 671—682 (1958).
—, and J. J. WEIGLE: Chromosome breakage accompanying genetic recombination in bacteriophage. Proc. nat. Acad. Sci. (Wash.) **47**, 857—868 (1961).
PERUTZ, M. F.: Proteins and nucleic acids. Structure and function. Amsterdam-London-New York 1962.
SAGER, R., and F. J. RYAN: Cell heredity. An analysis of the mechanism of heredity at the cellular level. New York and London 1961.
SINSHEIMER, R.: The structure of DNA and RNA. In: J. M. ALLEN, The molecular control of cellular activity, p. 221—243. New York-Toronto-London 1962.
STAHL, F. W.: The mechanics of inheritance. Englewood Cliffs, New Jersey: Prentice-Hall, Inc. 1964.
STORCK, R.: Nucleotide composition of nucleic acids of fungi. II. Deoxyribonucleic acids. J. Bacteriol. **91**, 227—230 (1966).
SUEOKA, N.: Mitotic replication of deoxyribonucleic acid in *Chlamydomonas reinhardi*. Proc. nat. Acad. Sci. (Wash.) **46**, 83—91 (1960).
TAYLOR, J. H.: The time and mode of duplication of chromosomes. Amer. Naturalist **91**, 209—221 (1957).
— Sister chromatid exchanges in tritium-labeled chromosomes. Genetics **43**, 515—529 (1958).
— (edit.): Molecular genetics, part I. New York and London 1963.
WATSON, J. D., and F. H. C. CRICK: A structure for deoxyribose nucleic acid. Nature (Lond.) **171**, 737—738 (1953a).
— — Genetical implications of the structure of deoxyribonucleic acid. Nature (Lond.) **171**, 964—967 (1953b).
— — The structure of DNA. Cold Spr. Harb. Symp. quant. Biol. **18**, 123—131 (1953c).
WILKINS, M. H. F., and J. T. RANDALL: Crystallinity in sperm heads: Molecular structure of nucleoprotein in vivo. Biochim. biophys. Acta (Amst.) **10**, 192 (1953).
— A. R. STOKES, and H. R. WILSON: Molecular structure of deoxypentose nucleic acids. Nature (Lond.) **171**, 738—740 (1953).
ZAMENHOF, S.: The chemistry of heredity. Springfield 1959.

Chapter IV

Recombination

Genetic material possesses the capacity for self-duplication. The total amount of genetic information as well as its order within the genome remains the same from nuclear generation to nuclear generation. Each daughter genome arising from a replication generally consists of an exact copy of the original genetic information. With each mitotic division such copies are transmitted to the daughter cells. The identity of the genetic information is thus assured for each cell of a multicellular organism. Nevertheless, such an inflexible transmission of hereditary material would prevent evolution. This problem is overcome by two fundamental properties of the genetic material, namely recombination and mutation.

Recombination is the reassortment of the genome during nuclear division. It occurs regularly in meiocytes during reduction division (*meiotic recombination*). New combinations also occur in vegetative cells in rare cases; this is called *somatic recombination* (p. 149f. and p. 204ff.). It takes place most commonly as a recombination in diploid somatic cells during mitosis (*mitotic recombination*). In some instances somatic recombination may involve *meiosis-like processes* (p. 204).

In both meiotic and mitotic recombination entire chromosomes as well as parts of chromosomes may recombine. In the former case we speak of interchromosomal, in the latter, of intrachromosomal recombination. The genetic material remains unchanged in its information content, arrangement, and quantity; this is in contrast to mutation, which will be discussed in the following chapter.

In order to detect the recombination of genetic material one must have genetic markers of which the transmission can be followed easily through successive generations. Genes are generally used as markers, but the centromere may also serve this purpose. For convenience, all such loci are referred to as markers. In this chapter the genes will be considered only as markers of the genetic material and not from the standpoint of their functional characteristics.

Literature of a general nature: In addition to pertinent textbooks (e.g. BRESCH, 1964) the following reviews should be mentioned: LUDWIG (1938), PONTECORVO (1958), SHULT and LINDEGREN (1959), KAPLAN (1960), DEMEREC (1962), EMERSON (1963), PRITCHARD (1963), ROMAN (1963), STENT (1963), HAYES (1964), STAHL (1964), WESTERGAARD (1964), BERNSTEIN (1964), JOLY (1964).

Other treatments of the subject may be found in the *Cold Spring Harbor Symposia of Quantitative Biology*, volumes **21** (1956) and **23** (1958); The *Chemical Basis of Heredity* 1957 (edited by McELROY and GLASS).

A. Interchromosomal recombination

The prerequisites for the mechanism of interchromosomal recombination are: 1. union of two different haploid genomes through karyogamy and 2. the reduction of the diploid to the haploid chromosome complement in meiosis or in a series of irregular mitotic divisions.

I. Meiosis

Cytogenetic investigations of meiosis in fungi have revealed that the process is basically the same as that in higher organisms (e.g. *N. crassa*, McCLINTOCK, 1945; SINGLETON, 1953). Our understanding of meiosis is not based on studies with the fungi because the small size of their chromosomes makes them poorly suited for microscopic investigation. The following account of meiosis will consider only those stages which are pertinent to an understanding of the recombination process (Fig. IV-1).

Meiosis I. Prophase I, Leptotene: At this stage each chromosome has already divided longitudinally into a pair of sister chromatids. Self-duplication of the genetic material has therefore already taken place. The exact time of duplication remains in doubt. *Zygotene:* This stage begins with the pairing of homologous chromosomes (synapsis), a requirement for the intrachromosomal recombination process (p. 232). As yet there is no satisfactory explanation for the attraction of homologous chromosomes and the precision of their pairing in order to form bivalents. *Pachytene:* Pairing is complete. The chromosomes begin to spiralize and as a result appear thicker and shorter. *Diplotene:* The four chromatids of a bivalent are now clearly visible as a tetrad. The two sister chromatids remain associated with one another because of the undivided centromere. When pairs of chromatids separate, a crossing of non-sister chromatids, called a chiasma, may frequently be observed. *Diakinesis:* The coiling and resultant shortening of the chromosomes reaches its maximum at this time. *Metaphase I:* The nuclear membrane disappears and a spindle is formed of which the fibers, in the fungi (in contrast to most other plants) may develop from previously divided centrosomes (HESLOT, 1958). The bivalents become arranged in the equatorial plane. The spindle fibers attach to the centromeres (spindle fiber attachments). The two centromeres of one bivalent are joined to fibers coming from different poles. *Anaphase I*: In this stage the homologous centromeres, which as yet are undivided, move to opposite poles of the cell each carrying with it one of the two chromatid pairs which composed the bivalent (e.g. to the upper and lower part of the young ascus). The chiasmata are terminalized and eliminated as the homologues separate to the poles. *Telophase I:* Separation is complete. A daughter nucleus forms at each pole of the spindle; the chromosomes in these nuclei show a loosening of the spirals. *Interphase:* The chromosomes are strongly despiralized and appear as thin threads.

Meiosis II. Prophase II: The chromosomes undergo spiralization again, becoming shorter. The centromeres divide longitudinally. *Metaphase II:* The chromosomes are sharply contracted. A spindle develops from the centrosomes. Its axis is generally at a fixed angle to the direction of the spindle of metaphase I. *Anaphase II:* The sister chromatids separate from one another and move to opposite poles. Each centromere is now associated with only one chromatid. *Telophase II:* At the end of the division four nuclei have arisen from the four chromosome complements originally present in the tetrad. These often assume a definite arrangement with respect to one another (p. 148). Each of the four nuclei either becomes enclosed immediately in a spore (e.g. *Saccharomyces cerevisiae* Fig. I-2; *Schizophyllum commune*, Fig. I-9) or divides further mitotically. In the latter case eight nuclei are formed (e.g. *Neurospora crassa*, Fig. I-5) or frequently there are multiple duplications of four nuclei (e.g. *Phycomyces blakesleeanus*, Fig. I-1) which lead to many monokaryotic or polykaryotic spores.

II. Mechanism of chromosomal distribution

The segregation of the chromosomes during meiosis is such that each of the two daughter nuclei receives one chromosome of each bivalent (Fig. IV-1). In this way *reciprocal genomes* arise because two homologues are prevented from going to the same pole. How the spindle apparatus controls the mechanism of segregation remains unknown.

In general, the *assortment of different chromosomes to the poles of the spindle proceeds according to chance* (exceptions p. 141 ff. and p. 144 ff.). Therefore, parental combinations of chromosomes occur only rarely. A combination of maternal and paternal chromosomes is the most common result. Because each chromosome of a bivalent ordinarily has the same chance of going to one or the other of the poles, a recombination for every two non-homologous chromosomes occurs in half the cases (Fig. IV-1). If each chromosome carries a marker gene, in a cross involving two markers on different chromosomes, the recombination frequency is 50% (p. 159 ff.).

III. Prereduction and postreduction

At which of the two maturation divisions the maternal and paternal chromosomes separate from one another remained a question into the 1930's (Literature in GOLDSCHMIDT, 1932, and BRIEGER, 1933). Genetic

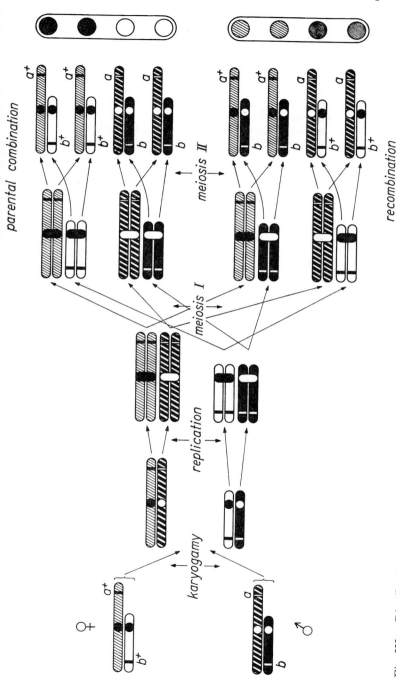

Fig. IV-1. Distribution of two pairs of chromosomes at meiosis. Homologous chromosomes are indicated by units of equal size. Chromosomes of the male parent are black, or white with heavy lines, those of the female parent white, or white with thin lines. Centromeres are represented by black or white circles or ellipsoids while the marker genes are shown by narrow, vertical black or white bands. The segregation of the chromosomes and marker genes in the four-spored ascus is shown at the right: black = parental combination of the female parent: a^+b^+; white = parental combination of the male parent: ab; lined = recombinant: a^+b; stippled = recombinant: ab^+

139

investigations revealed that this question could not be summarily answered either for all the chromosomes or for all the genes of single chromosome. From the arrangement of genetically marked spores in the asci of certain fungi, it was concluded that one and the same allelic pair may be reduced at certain times in the first meiotic division and at others in meiosis II. In the first case the gene is "*pre*reduced", and in the latter, "*post*reduced". In addition to asci containing only parental-type spores or only recombinant spores, some containing both types may be found. Such tetrads are called tetratypes (p. 153) because each of the four meiotic products exhibits a different genetic constitution. Tetratypes are observed not only in cases in which the genes are on different chromosomes, but also in those in which they are on the same chromosome. Although such complicated observations were originally made forty years ago, it remained for LINDEGREN (1933) to explain them on the basis of his experiments with *Neurospora crassa*.

The original hypothesis that reduction to the haploid condition occurs in meiosis I was challenged on the basis of cytological investigations, tetrad analysis (p. 148ff.) in lower plants, and indirect tetrad analysis in animals which could not be explained through exclusive prereduction of homologous chromosomes and the genes located thereon. Since both pre- and postreduction can be observed for the same allelic pair, it was argued that the entire chromosome followed the same type of reduction as the alleles which were located on it. This idea became untenable, however, after tetra-types involving two genes on the same chromosome were demonstrated. BRUNSWICK (1926) and BOHN (1933) discussed this discovery and arrived at the same explanation as LINDEGREN (1933).

The novel feature of LINDEGREN's interpretation was the clear distinction that he made between the segregation of entire homologous chromosomes and the segregation of the genes which are located on them. LINDEGREN postulated: 1. *Homologous centromeres are always segregated at the first meiotic division*, i.e. they are always prereduced. The adjacent chromosomal segments up to the first points of exchange are prereduced with the centromeres. 2. *Postreduction of an allelic pair results from an exchange (crossover, see p. 148) between the centromere and a particular genetic locus.* 3. *Crossing over occurs in the four-strand stage;* only two of the four strands take part in a single crossover. These considerations led to the mechanism of segregation represented in Fig. IV-2.

In order to prove his hypothesis LINDEGREN (1933) began with the assumption that the frequency of recombination between two linked genes is a measure of the distance between them (p. 195f.). He further reasoned that the postreduction frequency of a gene must be a measure of the distance between the genetic locus and the centromere. One would then expect the distance between the two genes to be either equal to the sum or to the difference of the distances that each gene is removed from the centromere, depending upon whether the genes lie on the same or on different sides of the centromere (p. 161f.). LINDEGREN (1933, 1936a) supported these ideas with the results of his experiments on *Neurospora crassa*.

That LINDEGREN's theory is correct has been confirmed not only in numerous experiments with *Neurospora crassa*, but also in all other organisms which have been studied genetically and cytologically. Nevertheless, reservations regarding these concepts continue to be put forward, particularly with regard to the universality of prereduction of the centromere.

For example, map distances are not additive if genes used as markers are far apart on the chromosome. LUDWIG (1937a) and RYAN (1943) showed that such discrepancies (e.g. in WÜLKER, 1935) could result from multiple exchanges (p. 196). Further, the high postreduction frequency of certain

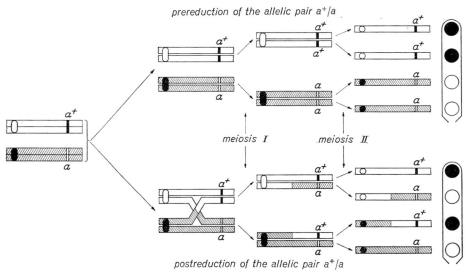

prereduction of the allelic pair a⁺/a

meiosis I   *meiosis II*

postreduction of the allelic pair a⁺/a

Fig. IV-2. Segregation of the allelic pair a^+/a in pre- and post-reduction, i.e. the separation of alleles in the first or in the second meiotic division. *Above:* Segregation of the four chromatids without a crossover having occurred between the centromere and the genetic marker (compare Fig. IV-1). *Below:* Segregation of the four chromatids two of which are new combinations resulting from a crossover between the centromere and marker gene. Unlined and lined units represent homologous chromosomes of the two parents. Centromeres and marker genes are shown as in Fig. IV-1. Open and solid circles (right) represent spores with the a and a^+ markers respectively, in a four-spored ascus.

genes (e.g. in *Podospora anserina*, Fig. IV-15) cannot be explained through a high frequency of postreduction of the centromere as LEDERBERG proposed (see PERKINS, 1955), because such an interpretation does not account for the high prereduction frequencies of other genetic markers.

IV. Polarized segregation of homologous chromosomes

Each chromosome and its centromere generally become oriented toward the "upper" or toward the "lower" pole of the spindle with equal probability (e.g. toward the upper or lower end of the ascus). Such a random arrangement is expected both in meiosis I and meiosis II. A deviation from the norm, i.e. *preferential orientation of a particular centromere toward a particular pole is called non-random or polarized segregation.* Spore color mutants of certain Euascomycetes are particularly suited for studies of this phenomenon (p. 148f.). The pre- or postreduction of the color markers can be observed directly from the arrangement of differently pigmented spores in the ascus (p. 140, Fig. IV-2 and Fig. IV-3).

The ratio of the two pre-reductional types (see Fig. IV-4) gives information regarding the *segregation of homologous centromeres at meiosis I.* On the basis of a random distribution the two types are expected in equal frequency (Type 1 : Type 2 = 1 : 1, Table IV-1). *Segregation of the centromere in meiosis II* affects only the postreduction, and not the prereduction frequency. Moreover, the chromatid pair which takes part in crossing over also determines the ratio of the two types. If both events proceed according to chance, the four postreductional types (Fig. IV-4) segregate as type 3 : type 4 : type 5 : type 6 = 1 : 1 : 1 : 1 (Table IV-1). The same relationship holds if the segregation is random for only one of the two processes.

Since one generally proceeds from the unproven assumption that homologous chromosomes segregate at random in the two meiotic divisions, deviations from such a segregation have rarely been noted. Thus few data exist which permit an analysis of the distributional mechanism at meiosis. In general the results are compatible with random distributions at meiosis I and II. Nevertheless, certain data show a significant deviation from the random expectation (Table IV-1).

PRAKASH (1963a) found deviations from the expected random segregation in *N. crassa* if he used different wild type strains for his crosses (see also p. 168ff.). In *Ascobolus stercorarius* BISTIS (1956) not only obtained an excess of asymmetric postreduction types (Table IV-1) but also higher postreduction values at lower temperatures (Fig. IV-10). The author interpreted both of these phenomena to be a result of overlapping spindles in the second meiotic division. This explanation is probably correct considering the fact that *A. stercorarius* produces asci in which the spores are only partly linearly oriented. SHAW (1962) explained data from *Sordaria brevicollis* in the same way. Like BISTIS, he found a significant excess of postreduction types showing asymmetric distribution. THRELKELD (1962a) similarly demonstrated an increased occurrence of asymmetric types in *N. crassa*, if he grew the strains on a medium containing 5-bromouracil. WHITEHOUSE and HALDANE (1946) attributed such deviations from expectation in *Neurospora sitophila* and in *Bombardia lunata* to an abnormal distribution of the centromeres at meiosis II.

In contrast to BISTIS, SHAW, and THRELKELD, MATHIESON (1956) obtained results with *Bombardia lunata* similar to those of LINDEGREN and STADLER using *N. crassa* (compare Table IV-1). The symmetrical ascus type was more frequent than the asymmetrical in the postreduction of spore color markers. Deviations from the expected 1 : 1 ratio could also be demonstrated for two prereduction types. The author concluded on the basis of cytological observations that chromosomal differences between the mating partners are not responsible for polarized segregation. He assumed that polarization occurs after the meiotic division and involves the effect of physiological factors which relate to the viability of the mutant spores.

The significance of these results is difficult to determine because secondary phenomena such as spindle overlap or passing of nuclei may also lead to abnormal segregations. *The value of these investigations lies*

Table IV-1. *Examples of segregation of markers during meiosis in ordered tetrads* (p. 147 ff.)

The tetrad types 1—6 correspond to ascus types 1—6 in Fig. IV-4. Individual crosses with *Sordaria macrospora* (KUENEN, unpublished) and *Ascobolus stercorarius* (BISTIS, 1956) were carried out at different temperatures. Those with *Neurospora crassa* (LINDEGREN, 1932b; STADLER, 1956a) employed strains with different markers. The values which are framed deviate from the expected random distribution. See explanation in text.

object and markers	cross no.	prereduction types		type 1 / type 2	postreduction types				type (3+4) / (5+6)
					asymmetric		symmetric		
		type 1	type 2		type 3	type 4	type 5	type 6	
		a^+a^+aa	aaa^+a^+		a^+aa^+a	aa^+aa^+	aa^+a^+a	a^+aaa^+	
Sordaria macrospora marker: *al*	1	150	164	0.9	200	174	171	153	1.2
	2	208	184	1.1	176	184	184	169	1.0
	3	131	192	0.7	191	185	185	157	1.1
	4	161	171	0.9	179	188	165	147	1.1
	5	173	194	0.9	192	185	172	174	1.1
total of 10 crosses		1,660	1,748	0.9	1,772	1,807	1,706	1,635	1.1
Neurospora crassa marker: *a*	1	105	129	0.8	9	5	16	10	0.5
marker: *asco*	1	5,020			280		276		1.0
	2	4,157			300		297		1.0
	3	793			86		106		0.8
	4	553			53		88		0.6
	5	481			62		90		0.7
total of 16 crosses		18,375			2,202		2,287		0.9
Ascobolus stercorarius marker: *l*	1	1,071			422		311		1.4
	2	1,964			388		382		1.0
marker: *t*	1	160			174		185		0.9
	2	266			260		232		1.1
	3	942			1,030		1,000		1.0

largely in calling attention to the fact that a random distribution of pre- and postreduction types does not always occur. The results of MITCHELL (1959, 1960a, b, 1964) in *Neurospora crassa* further show that the segregation of two linked markers does not always occur independently in the asci of the same perithecium. MITCHELL frequently observed that asci isolated successively showed the same distribution of markers if the asci came from the same fruiting body and were at the same stage of maturity.

The mechanism which underlies the non-random distribution of homologous chromosomes remains unknown. Such irregularities may be the result of affinities between non-homologous chromosomes, a phenomenon which is discussed in the following section.

V. Polarized segregation of non-homologous chromosomes (affinity)

It has often been noted that unlinked genes do not always assort independently at meiosis in *Saccharomyces cerevisiae, Neurospora crassa*, and *Ascobolus immersus* (*Saccharomyces:* SHULT and LINDEGREN, 1956b, 1959; SHULT and DESBOROUGH, 1960; HAWTHORNE and MORTIMER, 1960; SHULT et al., 1962; LINDEGREN et al., 1962; *Neurospora:* PRA-KASH, 1963a, b; *Ascobolus:* SURZYCKI and PASZEWSKI, 1964). Instead of the recombination value of 50% expected on the basis of the random assortment of two genes, appreciably higher or lower values were observed. These ranged from 20—35% or 58—73% for individual gene pairs (Table IV-2; for criteria of linkage or non-linkage see Table IV-6). Since in the first case more parental than recombinant types occur, a linkage of genes which are located on different chromosomes is simulated (*quasi-linkage*, compare Table IV-6). In the second case the recombination types exceed the parentals. The term *"reverse linkage"* (see Table IV-6) was introduced to describe this phenomenon. A non-random distribution of chromosomes at meiosis is apparently responsible for both phenomena.

Before such irregularities were observed in fungi, they were noted in the investigations of MICHIE and WALLACE on the house mouse (MICHIE, 1953, 1955; WALLACE, 1953, 1958a, b, 1959, 1961). A similar phenomenon was demonstrated in tomatoes and in cotton by WALLACE (1960a, b).

MICHIE (1953) and WALLACE (1953) suggested the following hypothesis to explain these observations: Certain loci on the chromosomes have an affinity for similar loci on non-homologous chromosomes. There are two different types of affinity loci (e.g. α and β) in every organism. When two non-homologous chromosomes possess affinity loci of the same type (α-α or β-β) they tend to segregate toward the same pole in meiosis I (quasi-linkage). If the affinity types are different (α-β) segregation at random or to the same pole is impeded (reverse linkage).

Some hypothetical examples will serve to clarify this mechanism of distribution. Only two pairs of homologous chromosomes will be considered. Chromosomes I and II are derived from one parent and chromosomes I' and II' from the other.

Case 1: I (α) II (α) $\times$ I' (β) II' (β). Since chromosomes I and II on the one hand and I' and II' on the other, have the same affinity loci (respectively α and β), the parental chromosome combinations are favored (quasi-linkage).

Case 2: I (α) II (β) $\times$ I' (β) II' (α). The chromosomes I and II' and II and I' respectively possess the same affinities, i.e. they move preferentially to the poles in the non-parental combinations (reverse linkage).

Case 3: I (α) II (β) $\times$ I' (α) II' (β). A random distribution of the chromosomes occurs, since the non-homologous chromosomes (e.g. I—II and I—II') possess different affinity loci.

Case 4: I (α) II (α) $\times$ I' (β) II' (α). In this case the chromosomes are also distributed at random since no preferential affinity exists between I or I' to II or II' respectively.

According to this hypothesis quasi- and reverse linkage are produced by the effect of the same affinity loci, the only difference being that the parents possess the same affinity type in the first case, and different affinity types in the second. Through an exchange of affinity loci by inter- or intrachromosomal recombination quasi-linkage is converted into reverse linkage and vice versa. The various linkage relationships of the same allelic pairs in different races of *Saccharomyces cerevisiae* (Table IV-2) is evidence in support of this hypothetical scheme.

Table IV-2. *Examples of quasi-linkage and reverse linkage of markers which are located on non-homologous chromosomes (Saccharomyces cerevisiae)*

For criteria for quasi- and reverse linkage see Table IV-6. The tetrad types *P*, *R*, and *T* (unordered tetrads) are shown in Fig. IV-5 and Fig. IV-8 and explained in the accompanying text. *p*, *r*, and *t* represent the frequencies of the corresponding tetrad types (see also p. 159). (From data of DES-BOROUGH and LINDEGREN, 1959.)

marker combination	"family"	tetrad distribution			percentage of tetratypes	recombination value (%)
		P	*R*	*T*		
Quasi-linkage: p > r and t < 2/3						
th—ur	108	64	3	69	50.7	27.6
ch—th	108	82	0	62	43.1	21.5
cu—ch	98	15	4	33	63.5	39.4
	99	16	2	11	37.9	25.9
	111	38	1	37	48.7	25.7
	118	13	1	13	48.1	27.7
cu—ur	118	12	4	14	46.7	36.7
cu—th	98	23	2	31	55.4	31.3
	99	8	1	10	52.6	31.6
	111	30	2	34	51.5	28.8
α—ch	107	13	3	17	51.5	34.8
ga—ur	97	7	1	12	60.0	35.0
α—ga	108	52	27	67	45.9	41.4
ad—ga	97	11	0	7	38.9	19.4
Reverse linkage: p < r and t < 2/3						
cu—th	108	10	37	40	46.0	73.1
cu—ch	108	6	48	56	50.9	69.1
cu—ur	108	12	35	88	65.2	58.5
α—ur	118	4	13	11	39.3	66.1
ad—ur	91	10	26	7	16.3	68.6
α—ga	85	3	12	16	51.6	64.5
	107	7	22	18	38.3	66.0
ad—ga	86	9	21	4	11.8	67.6

Analysis of the cross *UR-ch CU × ur-CH cu* in "family" 108 (as shown in Table IV-2) indicated a reverse linkage between *CU/cu* and *UR/ur-CH/ch*. One would therefore expect the affinity loci, α and β, to be distributed as follows: *UR-ch* (α) *CU* (β) × *ur-CH* (β) *cu* (α). In a second cross strains with the same genes but marked in a new combination were selected as parents ("family" 118): *UR-ch cu × ur-CH CU*. Quasi-linkage is observed between *CU/cu* and the other genes instead of reverse linkage. Apparently the affinity loci in "family" 118 were linked in the same manner with "their" markers as in "family" 108: *UR-ch* (α) *cu* (α) × *ur-CH* (β) *CU* (β). Such linkage reversals can be produced exeprimentally in considerable frequency in *Saccharomyces* (see also SHULT and LINDEGREN, 1959).

MICHIE and WALLACE assumed that the affinity locus and centromere were identical. The results with *Saccharomyces* argue in favor of this notion; the affinity locus was found to be tightly linked to the centromere (DESBOROUGH and LINDEGREN, 1959; LINDEGREN et al., 1962). However, this hypothesis could not be confirmed with results from *Ascobolus immersus*. In this species the affinity locus was separable from the centromere (SURZYCKI and PASZEWSKI, 1964).

It remains to be seen if and to what extent the results of these investigations can be generalized. At the moment there is still considerable disagreement regarding the chromosome number in *S. cerevisiae* (see Table I-1). Furthermore, no correlation between chromosomes and particular linkage groups has been established in this species. Thus one cannot exclude the possibility that translocations or other chromosomal mutations may account for quasi- or reverse linkage, particularly because LINDEGREN and his coworkers used different races in their work.

VI. Mitotic recombination

In addition to the reassortment of entire chromosomes, which regularly occurs at meiosis, an *interchromosomal recombination in somatic cells* may take place. *This involves an occasional irregular chromosome distribution in diploid mitoses which, in a few cases, may lead to balanced haploid nuclei* (mitotic recombination: PONTECORVO, 1953 a, b). These may carry combinations of genes which were previously located in different nuclei.

Mitotic recombinations have been described primarily in the genera *Aspergillus* and *Penicillium* (ROPER, 1952; PONTECORVO, 1953 a, b, 1954; PONTECORVO and ROPER, 1952; KÄFER, 1958, 1961; and PONTECORVO and SERMONTI, 1954; SERMONTI, 1957). The phenomenon is also known in certain Basidiomycetes, namely *Coprinus radiatus* (PRUD'HOMME, 1963) and *C. lagopus* (SWIEZYNSKI, 1963). Detailed investigations have been carried out by KÄFER (1961) on *Aspergillus nidulans*. She believes that the recombination of chromosomes occurs in two steps:

1. Heterozygous diploid nuclei arise through rare fusions of haploid nuclei within a heterokaryon during the parasexual cycle. Because of an abnormality in the mitotic division (e.g. non-disjunction) these lead in some instances (1—2%) to two reciprocally aneuploid nuclei one of which is trisomic and the other monosomic.

In the case of trisomy, not only may one chromosome be represented three times ($2n + 1$), but also two non-homologous chromosomes may each be present in triple dose ($2n + 2$). The configurations of the corresponding monosomic nucleus are then $2n - 1$ and $2n - 2$, respectively. KÄFER found that the majority of the aneuploids analyzed were of the $2n + 1$ or the $2n + 2$ types (56 out of 78). The corresponding monosomic configurations were not observed, however.

2. The irregular chromosome distribution may be repeated as the aneuploid nuclei divide; as a result, viable diploid, aneuploid, or haploid nuclei may arise. Reduction to the haploid condition takes place (at a

frequency of 0.02%) by a stepwise loss of chromosomes and generally results in a new combination of genetic information.

Interchromosomal recombinations produced through non-disjunction are rare. MORPURGO (1962a, b) found in *Aspergillus nidulans* that a mitotic recombination is much more often a result of mitotic crossing over (p. 204ff.) than of failure of homologous chromosomes to separate and their subsequent reassortment. According to KÄFER (1961) the frequencies of inter- and intrachromosomal recombination in somatic cells are approximately equal.

The significance of mitotic recombination for genetic analysis of the Fungi Imperfecti has already been pointed out in an earlier chapter (p. 93).

Summary

1. The assortment of whole chromosomes during meiosis is controlled by a spindle mechanism which is not yet understood. The recombinations which result are always reciprocal.

2. Reduction from the diploid to the haploid chromosomal complement occurs in the first meiotic division. However, all gene pairs whose linkage with the centromere is broken by a crossover are not separated until the second meiotic division, i.e. they are post-reduced. On the other hand, the centromere and the chromosomal segment between the centromere and the nearest crossover is always prereduced.

3. The separation of the chromosomes to the poles of the spindle is generally at random. However, on occasion centromeres and their chromosomes have been observed to move preferentially to a particular pole (polarized segregation). In these cases unlinked genes, instead of assorting independently as expected give recombination values which are significantly higher or lower than 50% (affinity). Both phenomena apparently involve a non-random distribution of chromosomes at meiosis. The mechanism responsible for these deviant distributions is unknown.

4. A reassortment of entire chromosomes during mitosis has also been observed. Mitotic recombination results from aneuploid nuclei which arise occasionally in the division of diploid somatic cells. These are reduced stepwise to normal haploid nuclei.

B. Intrachromosomal intergenic recombination

In interchromosomal recombination all the genes on a chromosome recombine as a block with those of other chromosomes; recombination within a chromosome involves the breakage of linkage, i.e. the rearrangement of alleles of genes in the same linkage group (p. 158ff.). The molecular aspects of the mechanism of intrachromosomal recombination remain unclear in spite of many experimental approaches undertaken primarily with phages and bacteria (see hypotheses of intrachromosomal recombination, p. 231ff.).

10*

According to the classical conception recombination occurs as a result of breakage at homologous loci in two of the four chromatids of a tetrad, followed by reunion of the broken ends (breakage-fusion hypothesis, p. 234 ff.). This mechanism, called crossing over, explains the reciprocal nature of recombination; it was not seriously questioned for some fifty years until the discovery of rare, non-reciprocal recombinants. In the following discussion the expression *"crossing over"* is not meant to imply a particular molecular basis of recombination but refers simply to the *cause of reciprocal recombination between different markers*. The illustrations likewise represent reciprocal recombination.

I. Methods of analysis

Certain fungi have an advantage over the classical materials of genetics (e.g. *Drosophila*, maize) in that the four products of each meiosis remain together as a tetrad of asco- or basidiospores and can be analyzed directly. Further, in some instances particular exchanges and the distribution of chromatids can be inferred from the order of the spores in the sporangium. This is possible when the spindles in both meiotic divisions and the post-meiotic mitosis are so oriented that *the nuclei, and thus the spores, assume a specific and predictable order in the sporangium.* Such groups of nuclei or spores are called *ordered* tetrads. If the spindles overlap or if the nuclei slip by one another without any regularity, *the spore arrangements in the sporangia cannot provide any information about the events occurring at meiosis.* In such a case the tetrads are said to be *unordered*.

1. Ordered tetrads

The total information which can be obtained through the analysis of ordered tetrads is greater than that obtained by analysis of unordered tetrads (Table IV-5). The additional information obtained in the former case results primarily from being able to distinguish between the pre- and postreduction of an allelic pair. Since postreduction is a consequence of crossing over between a gene and its centromere (p. 140), *the centromere serves as an additional marker in the analysis of ordered tetrads.* This is important in dealing with certain corollary problems (see Table IV-5).

Representatives of the genera *Neurospora*, *Sordaria*, and *Podospora* (Sordariaceae, Table I-1) are suitable for the analysis of ordered tetrads. The mutants of these species which differ from the wild type in spore characters, e.g. color and size, are especially useful.

In a cross between such a mutant and the wild type the segregation of the allelic pair can be determined directly by viewing the spores in the asci (Fig. IV-3 and illustration on cover).

Spore color mutants have been observed in the following: *Neurospora crassa* (STADLER, 1956a, b; NAKAMURA, 1961; THRELKELD, 1965), *Sordaria fimicola* (BISTIS and OLIVE, 1954; OLIVE, 1956; ITO, 1960), *Sordaria macrospora* (HESLOT, 1958; ESSER and STRAUB, 1958), *Podospora anserina* (KUENEN, 1962b; MARCOU, personal communication), *Ascobolus immersus* (RIZET et al.,

1960a, b), and *Aspergillus nidulans* (PONTECORVO and KÄFER, 1958; APIRION, 1963). The latter two Ascomycetes do not exhibit ordered tetrads. Mutants with differences in spore size have been found, for example, in *Podospora anserina* and *Glomerella cingulata* (RIZET and ENGELMANN, 1949; WHEELER and DRIVER, 1953).

Fig. IV-4 shows how the eight genetically marked spores come to be arranged linearly in the ascus using *Neurospora crassa* as an example. The following are the determining factors: 1. orientation of the spindles, 2. distribution of the nuclei, and 3. time of reduction of the marker genes.

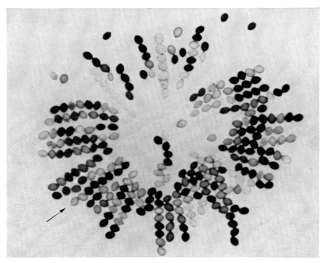

Fig. IV-3. Contents of a perithecium of *Sordaria macrospora*. The asci were derived from a cross of a black-spored (g^+) with gray-spored (g) strain. With one exception all asci show a segregation of 4 g^+:4 g. The exceptional ascus (arrow) exhibits six black and two gray spores (6 g^+:2 g). (The explanation of this abnormal segregation is found on page 222ff.)

The spindles in meiosis I and II are oriented parallel to the longitudinal axis of the ascus (Fig. IV-4). The nuclei formed in the first meiotic division move far enough apart so that the daughter nuclei rarely slip by one another in anaphase II. The spindles of the postmeiotic division lie obliquely to the longitudinal axis without overlapping one another (Fig. IV-4). As a result of this mechanism of division each nucleus and each spore assumes a particular position in the ascus (McCLINTOCK, 1945; SINGLETON, 1953).

A *single factor cross* such as $a^+ \times a$ in *Neurospora crassa* and certain other Euascomycetes (p. 148) results in *six distinguishable tetrad types* (Fig. IV-4, compare also with Table IV-1 and IV-4).

Two types (1, 2) result from the separation of the two alleles at meiosis I (prereduction type). The remaining four types (3—6) are the result of second division segregation (postreduction type). Types 3 and 4 represent asymmetric, types 5 and 6 symmetric segregations. The prereduction types occur in a 1:1 ratio and the postreduction types in a 1:1:1:1 ratio with a random distribution of the centromeres in the two maturation divisions (for exceptions see p. 141 ff.).

Recombination

In a *two-factor cross*, e.g. $a^+b^+ \times ab$ *seven tetrad types* can be differentiated depending upon the time of reduction and the combination of alleles of the two different genes (Fig. IV-5). If, in addition, the different division patterns for pre- and postreduction are considered (1—6 in Fig. IV-4), 36 tetrad types are possible (compare Table IV-4).

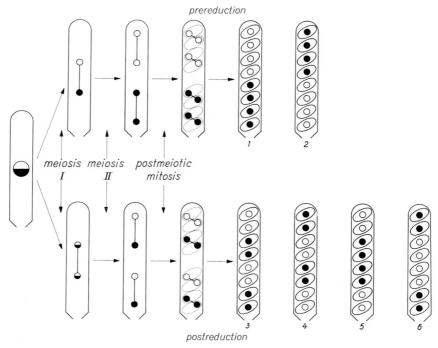

Fig. IV-4. Development of the linear spore arrangement in an ascus of an eight-spored Ascomycete (e.g. *Neurospora crassa*). The asci are derived from a unifactorial cross. Two prereduction types (1, 2) and four postreduction types (3—6) may be distinguished (compare Table IV-1 and IV-4). Open circle and semi-circle = allele of one parent; solid circle and solid semi-circle = allele of the other parent (see explanation in text)

In Fig. IV-5 the situation in *Neurospora crassa* before the postmeiotic mitosis is shown, since the latter division leads to a pair of spores the members of which are genetically identical; it does not involve any shifting of nuclei. As the figure reveals, the four groups, A, B, C, and D, which are determined by the time of reduction of the two genes can be subdivided in part. The following types can be distinguished: 1. tetrads with exclusively parental combinations (A_1 and D_1) so called parental ditypes (P); 2. tetrads with exclusively non-parental combinations (A_2 and D_2) so called non-parental or recombination ditypes (R); 3. tetrads whose four nuclei are all genetically different (A_3, B and C) so called tetratypes (T).

If the individuals crossed differ in more than two factors the number of tetrad types which can be distinguished increases exponentially (compare Table IV-4). Ordered tetrads have the advantage of allowing a larger number of types to be identified than unordered tetrads or single spores selected at random.

The tetrad types described for uni- or multifactorial crosses are characteristic of fungi which exhibit the same division mechanism in the ascus as *Neurospora crassa*. These include the eight-spored Euascomycetes, *N. sitophila*, *Sordaria macrospora*, *S. fimicola* and *Bombardia lunata* (Table I-1).

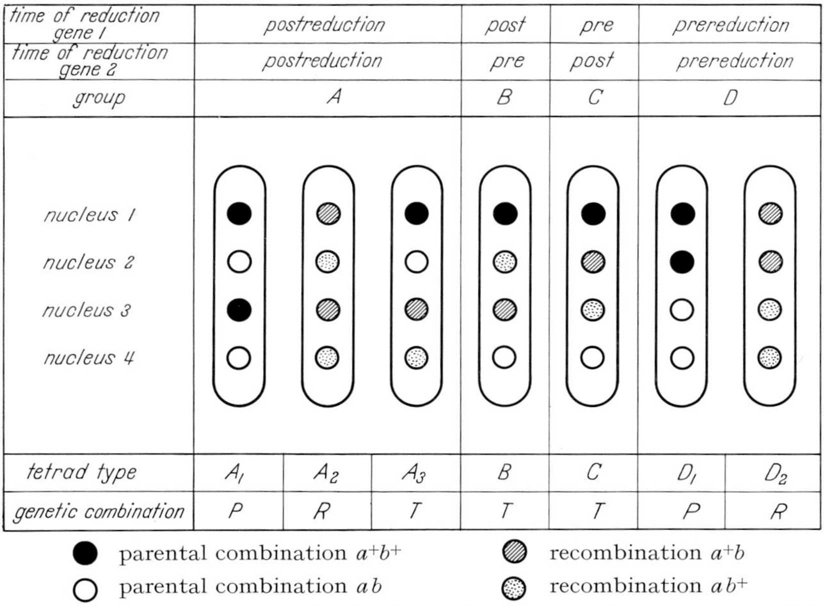

time of reduction gene 1	postreduction			post	pre	prereduction	
time of reduction gene 2	postreduction			pre	post	prereduction	
group	A			B	C	D	
nucleus 1 nucleus 2 nucleus 3 nucleus 4							
tetrad type	A_1	A_2	A_3	B	C	D_1	D_2
genetic combination	P	R	T	T	T	P	R

● parental combination a^+b^+ ◑ recombination a^+b
○ parental combination ab ◉ recombination ab^+

Fig. IV-5. The seven types of ordered tetrads which are theoretically possible in the two-factor cross $a^+b^+ \times ab$ (compare with Table IV-4). The nuclei, which arise through the two divisions of meiosis, are designated by circles. In eight-spored Ascomycetes each of the meiotic products divides by mitosis into two identical daughter nuclei. Explanation in text. (From KUENEN, 1962a)

The formation of spores occurus in a similar way in *Podospora anserina* (Fig. IV-6), *Neurospora tetrasperma* and *Gelasinospora tetrasperma*. The asci of these fungi contain only four spores. However, each spore is dikaryotic (p. 19 and p. 20f.). Analysis of ordered tetrads is possible.

The distribution of nuclei in *Podospora anserina* is controlled by the position of the spindles as in *Neurospora crassa*. The spindles are oriented longitudinally in the two meiotic divisions, but in the postmeiotic mitosis they lie obliquely to the long axis of the ascus (Fig. IV-6). Because of the arrangement of nuclei which results from these spindle orientations a pair of non-sister nuclei from the postmeiotic mitosis is included in each spore. Such a mechanism of nuclear distribution was postulated by RIZET and ENGELMANN (1949) on the basis of genetic investigations and was later confirmed cytologically by FRANKE (1957, 1962) (Fig. I-7). FRANKE also observed that the spindles are frequently obliquely oriented in meiosis II and longitudinally positioned in the postmeiotic division. The effect of this latter orientation on nuclear distribution is the same as the more common one. With prereduction of an allelic pair the two spores in an ascus half are homokaryotic for one allele; the two in the other half are homokaryotic for the other allele (Types 1 and 2 in Fig. IV-6). With postreduction all four spores are heterokaryotic. In contrast to the situation in *Neurospora crassa*, distinguishing the four postreductional types (Types 3—6 in Fig. IV-4)

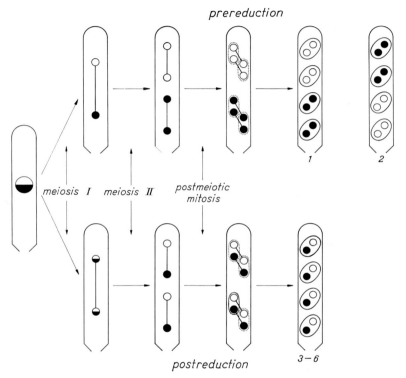

Fig. IV-6. Ascospore formation in *Podospora anserina* showing the segregation of a single pair of alleles. Symbols are the same as in Fig. IV-4. Further explanation in the text

is not possible. In two-factor crosses only five, instead of seven (compare with Fig. IV-5), tetrad types can be recognized directly. Nevertheless, it may be possible to detect the missing tetrad types through further analysis of dikaryotic spores of ascus type *A* (RIZET and ENGELMANN, 1949; ESSER, 1956, 1959a, b; KUENEN, 1962b). An additional possibility lies in the use of abnormal asci for tetrad analysis (p. 20 and Fig. I-8).

The zygotic nucleus of *Neurospora tetrasperma* also divides three times before the spores are formed. The daughter nuclei move far enough apart in meiosis II to allow the two inner nuclei to slip by one another so that non-sister nuclei lie in each ascus half (DODGE, 1928; LINDEGREN, 1932b; COL-SON, 1934). After the postmeiotic division daughter nuclei arising from different nuclei are included in each spore. With this type of distributional mechanism half of all postreduction produces the same allelic combinations as prereduction. As a matter of fact, asci with four genetically identical spores occur in both cases (Type I; corresponding to the postreduction type in *Podospora anserina* in Fig. IV-6). The remaining half of the postreductions lead to asci with a 2:2 segregation for the corresponding allelic pair (Type II; corresponding to the prereduction types in *Podospora anserina* in Fig. IV-6). When the two postreduction types occur in equal frequency, the post-reduction frequency of an allelic pair is equal to twice the frequency of type II asci (e.g. HOWE, 1963, 1964). A tetrad analysis is also possible here as in *P. anserina* by using asci which possess more than four spores. Such asci arise at a low frequency by the formation of two spores each with a single nucleus instead of a binucleate spore (HOWE, 1964.)

The cytological events leading to spore formation in *Gelasinospora tetrasperma* have been worked out (DOWDING, 1933; DOWDING and BAKER-SPIGEL, 1956) Determination of the postreduction frequency for particular markers is also possible for this species.

2. Unordered tetrads

These differ from ordered tetrads in that the spore arrangement does not allow any conclusions regarding the time of reduction of marker genes (p. 148). Thus the centromere cannot be used as a marker for genetic studies.

Unordered tetrads occur in certain eight-spored Euascomycetes, in the yeasts, which are four-spored, as well as in all Basidiomycetes (Table I-1). Obviously, ordered tetrads may be treated as unordered tetrads, if the linear order of the spores in the ascus is disregarded. Special methods have been devised for isolating unordered tetrads quickly, e.g. in *Schizophyllum commune* (PAPAZIAN, 1950b), *Neurospora crassa* (STRICKLAND, 1960), *Ascobolus immersus* (LISSOUBA and RIZET, 1960).

Single factor crosses $(a^+ \times a)$ reveal only whether segregation is reciprocal $(4\,a^+:4\,a)$ or non-reciprocal (e.g., $6\,a^+:2\,a$; Fig. IV-3, compare with Table IV-5 and p. 222ff.). Pre- and postreduction types can not be distinguished. Three tetrad types occur in *two-factor crosses*, namely the *parental ditype (P), and the non-parental or recombination ditype (R)* and the *tetratype (T)* (Fig. IV-5, bottom row, and Fig. IV-8).

If parents differ at three loci, the number of tetrad types is 12 (Table IV-4), provided complete analysis is possible. If only the tetrad types for each two loci, and not the genotypes of the single spores, are considered (incomplete analysis) only 11 types are distinguishable. In the latter case the TTT-types fall into a single group (Table IV-3 and IV-4).

The triple combinations TTT, which are brought together in two parts of Table IV-3 (last row, next to last column and next to last row, last column) can be related to two genotypically different tetrad types. With regard to the example given in the legend of Table IV-3, the following two tetrad types are designated by the symbol TTT: $a^+b^+c^+, a^+bc, abc^+, ab^+c$ (type 1) and $a^+b^+c, a^+bc^+, ab^+c^+, abc$ (type 2).

The number of distinguishable tetrad types increases exponentially with an increase in the number of allelic differences (Table IV-4).

Table IV-3. *The distributions of three marker genes which are theoretically possible in unordered tetrads*

The distribution of markers is shown for the three-factor cross $a^+b^+c^+ \times a\,b\,c$. With the order $a^+a^+a\,a$ constant (line 2, column 2) there are six sequences possible for b^+/b (column 2) and for c^+/c (line 2), giving a total of 6×6 or 36 combinations in all. These can be reduced to 11 or 12 different types (see text). In the triple combinations of P, R, and T, the first letter designates the tetrad type for a^+/a and b^+/b (column 1), the second letter the type for b^+/b and c^+/c, and the third the tetrad type for a^+/a and c^+/c (line 1). For example, the triple combination RRP (line 4, column 3) indicates the non-parental ditype (R) for a^+/a and b^+/b, i.e. a^+b, a^+b, ab^+, ab^+; the non-parental ditype (R) for b^+/b and c^+/c i.e., bc^+, bc^+, b^+c, b^+c; and the parental ditype (P) for a^+/a and c^+/c, i.e. a^+c^+. a^+c^+, ac, ac. The four products of meiosis represented by this tetrad therefore possess the genotypes, a^+bc^+, a^+bc^+, ab^+c, ab^+c. (Adapted from WHITEHOUSE, 1942)

tetrad types for a^+/a and b^+/b	tetrad types for a^+/a and c^+/c	P	R	T			
	a^+ a^+ a a	c^+ c^+ c c	c c c^+ c^+	c^+ c c^+ c	c c^+ c c^+	c^+ c c c^+	c c^+ c^+ c
P	b^+ b^+ b b	PPP	PRR	PTT			
R	b b b^+ b^+	RRP	RPR	RTT			
T	b^+ b b^+ b b b^+ b b^+	TTP	TTR	TPT TRT TRT TPT		TTT	
	b^+ b b b^+ b b^+ b^+ b			TTT		TPT TRT TRT TPT	

Because pre- and postreduction types cannot be distinguished directly in unordered tetrads (p. 153) indirect methods have been developed to determine the postreductional frequency and thus the *position of the centromere*.

One of these methods involves the determination of the frequencies of the three tetratypes from the results of a three-factor cross and the calculation of the postreductional frequencies of the three genes in question (WHITEHOUSE, 1949, 1950, 1957b; PERKINS, 1949; PAPAZIAN, 1951, 1952). The following assumptions are necessary: 1. The three genes are located on at least two different chromosomes. 2. The segregation of non-homologous centromeres at meiosis is random (p. 141f.). 3. Chromosome (p. 187ff.) and chromatid interference (p. 179ff.) are absent.

In another method centromere markers are utilized; these are genes which lie close to the centromere and are therefore almost always prereduced. The postreductional frequencies of the centromere markers must, however, have been previously determined in this case.

To investigate most genetic problems at least three markers are required (compare Table IV-5). Limited information can be obtained from the distribution of the tetrad types, P, R, and T in utilizing only two genes (SHULT and LINDEGREN, 1956b, 1957, 1959; DESBOROUGH and LINDEGREN, 1959). Five types of tetrad distribution are most significant; these are briefly described below (compare with Table IV-6):

The frequencies of parental ditypes (P), non-parental ditypes (R) and tetratypes (T) which can be tabulated from a two-factor cross, are designated as p, r, and t respectively.
Interpretation:

1. *N-distribution* (*no* linkage): $p:r:t=1:1:4$; $p=r$; $t=2/3$.
Interpredation: a) Two genes are on the same chromosome and recombine freely with one another. This can only occur if the markers are far apart, i.e. when the frequency of crossing over between them is high.

b) Two genes are on different chromosomes and either or both recombine freely with their respective centromeres. At least one of the markers is far from its centromere (see also WHITEHOUSE, 1949).

2. *F-distribution* (excess of *first* division segregation): $p=r$; $t<2/3$.
Interpretation: Two genes on different chromosomes recombine only rarely with their respective centromeres. The centromeres are distributed at random in meiosis I. The result is a low postreductional frequency for both genes (centromere-markers).

3. *L-distribution* (direct *linkage*): $p>r$; $t<2/3$.
Interpretation: a) Two genes lie close together on the same chromosome; the recombination frequency is low. Parental types predominate.

b) Two genes on different chromosomes lie close to their respective centromeres; the centromeres separate preferentially in parental combinations at meiosis I (quasi-linkage, p. 144 ff. and Table IV-6).

4. *R-distribution* (*reverse* linkage): $p<r$; $t<2/3$.
Interpretation: a) More frequent 4- than 2-strand double crossing over occurs between two genes located on the same chromosome (positive chromatid interference). The recombination value is greater than 50%.

b) Two genes on different chromosomes lie close to their respective centromeres; the centromeres segregate preferentially in non-parental combinations at meiosis I (reverse linkage; p. 144 ff. and Table IV-6). For example, this may occur through synapsis of non-homologous centromeres in the first meiotic division (LINDEGREN and SHULT, 1956; SHULT and LINDEGREN, 1957).

5. *T-distribution* (excess of *tetratype*): $p=r$; $t>2/3$.
Interpretation: a) Two genes are on the same chromosome or on different chromosomes; a high frequency of single crossovers occurs between the two genes or between at least one gene and its centromere (strong positive chromosome interference).

b) Two genes are on the same chromosome with a higher than expected frequency of three-strand doubles between them (chromatid interference).

Since two possible interpretations of a particular distribution are given as a rule, one must be excluded by obtaining additional data. This can be done in different ways, as DESBOROUGH and LINDEGREN (1959) have shown, e.g. through unequivocal proof of linkage or non-linkage or by determining the map distance between two genes or between a gene and its centromere.

Table IV-4. *The number of tetrad types or genotypes which are theoretically possible with* 1, 2, 3, 4, 5, ... *n markers.* (From FISHER, 1950; PAPAZIAN, 1952, and BENNETT, 1956)

type of analysis	number of possible combinations					
	1	2	3	4	5	n factors
Ordered tetrads (complete analysis)	6	36	216	1296	7776	6^n
Ordered tetrads (incomplete analysis)	2	7	32	172	4860	$\dfrac{6^n+5\cdot 2^n}{8}$
Unordered tetrads (complete analysis)	1	3	12	60	336	$\dfrac{6^{n-1}+3\cdot 2^{n-1}}{4}$
Unordered tetrads (incomplete analysis)	1	3	11	48	236	$\dfrac{6^n+3\cdot 4^n+15\cdot 2^n}{48}$
Single strands	2	4	8	16	32	2^n

Table IV-5. *Experimental data required for obtaining evidence of the various genetic events discussed in the current chapter*

event	minimum number of markers re-quired for the investigation	tetrads and
		ordered tetrads (o)
1. Prereduction of centromere	2 (o)	tetrad types $(A_1—D_2)$
2. Recombination in the four-strand stage	2 (o, u)	tetratypes (A_3, B, C)
3. Distribution of homologous chromosomes in		
a) meiosis I	1 (o)	prereduction types
b) meiosis II	1 (o)	postreduction types
4. Assortment of non-homolo-gous chromosomes in meiosis	2 (o, u)	tetrad types (A_1--D_2)
5. Reciprocal recombination	1 (o, u)	pre- or postreduction types
6. Linkage		
a) between gene markers	2 (o, u, s)	tetrad types $(A_1—D_2)$
b) between gene marker and centromere	1 (o)	pre- and postreduction types
7. Chromatid interference		
a) across the centromere	2 (o)	tetrad types (A_1, A_2, A_3)
	3 (u)	
b) within an arm	2 (o)	tetrad types (A_3, B)
	3 (u)	
8. Chromosome interference		
a) across the centromere	2 (o)	tetrad types $(A_1—D_2)$
	3 (u, s)	
b) within an arm	2 (o)	tetrad types $(A_1—D_2)$
	3 (u, s)	
9. Localization and mapping of		
a) gene markers	2 (o, u, s)	tetratypes (A_3, B, C)
b) centromeres	1 (o)	postreduction types
10. Intragenic recombination	2 (o, u, s)	recombination ditypes and tetratypes (A_2, A_3, B, C, D_2) and aberrant tetrad types
11. Negative interference	3 (o, u, s) (mostly 4)	tetrad types $(A_1—D_2)$
12. Non-reciprocal recombination	1 (o, u) (mostly 2)	abberrant pre- or post-reduction types

The abbreviations used in the second column have the following meanings: o = ordered tetrads, u = unordered tetrads, s = single strands. The explanation of the symbols used in columns 3—5 appears earlier in the text.

genotypes necessary for the investigation		page reference
unordered tetrads (u)	single strands (s)	
		140
tetratypes (T)		140
		141 ff.
tetrad types (P, R, T)		144 ff.
tetrad types		147 ff.
parental and non-parental ditypes (P, R)	parental and recombination types (P_1, P_2, R_1, R_2)	158 ff.
tetrad types (TPT, TTT, TRT), using a centromere marker		179 ff.
tetrad types (TPT, TTT, TRT)		
tetrad types (P, R, T), using a centromere marker	parental and recombination types (P_1, P_2, R_1, R_2), using a centromere marker	187 ff.
tetrad types (P, R, T)	parental and recombination types (P_1, P_2, R_1, R_2)	
tetratypes (T)	recombination types (R_1, R_2)	195 ff.
non-parental ditypes and tetratypes (R, T) and aberrant tetrad types	recombination types (R_1, R_2)	210 ff.
tetrad types (P, R, T)	parental and recombination types (P_1, P_2, R_1, R_2)	216 ff.
aberrant tetrad types		222 ff.

3. Single strands

When tetrad analysis can not be performed, the isolation and study of single spores remains as a possibility. Because only one of the four meiotic products is identified in such a case, this type of study has been called single strand analysis. Such an approach is necessary when the spores of a tetrad do not remain together as in the case of Phycomycetes (Table I-1). However, single strand analysis is often used in organisms having ordered or unordered tetrads, if a large number of spores can be obtained quickly and one wishes to avoid the tedious isolation of single asci or basidia.

The analysis of single meiotic products allows only the calculation of recombination value and therefore is particularly suited for establishing linkage (p. 165 f.) (PERKINS, 1953), for localizing genetic markers (p. 195 ff.), and for the determination of chromosome interference (p. 187 ff., Table IV-5).

At least two marker genes are required for these investigations. In *two-factor crosses two parental types* (P_1 and P_2) and *two non-parental* or *recombination types* (R_1 and R_2) i.e. a total of four genotypes, are obtained. With each additional factor, the number of different genotypes is doubled (Table IV-4).

In conclusion Table IV-5 summarizes the experimental possibilities which are discussed in this chapter and which are available for use by the geneticist.

Summary

1. A large number of fungi have been particularly useful for genetic investigation because the four products of each meiosis remain together as a tetrad and may be analyzed directly. Analysis of ordered tetrads is possible, however, only if the mechanism of nuclear division leads to a regular linear arrangement of spores in the ascus.

2. The amount of information which can be obtained is greatest from ordered tetrads, and least from single spores. The same is true with respect to the number of distinguishable tetrad types and genotypes in crosses involving a particular number of markers (Table IV-4).

3. Ordered, in contrast to unordered tetrads indicate the time of reduction of particular allelic pairs and their relation to the centromere. Isolation and analysis of unordered tetrads as well as single spores, on the other hand, require much less time and thus allow a more rapid solution of certain types of genetic problems.

4. The applicability of the three types of analysis to genetic investigations is indicated in Table IV-5.

II. Linkage

Linkage between genetic markers refers to the predominance of parental combinations among the offspring of crosses, i.e. the genetic markers do not recombine at random. The genes which are linked with one another constitute a *linkage group*. Unlinked genes belong to different linkage groups. Cytogenetic investigations have established the fact that the number of linkage groups corresponds to the number of chromosomes in the haploid complement (compare with Table I-1 and p. 167). The concepts are often used interchangeably because of the rigorous correlation between number of linkage groups and chromosome number.

Linkage of genes in a linkage group is not absolute. *Partial linkage,* which lies between the extremes of complete linkage and independent assortment, is the most common situation. If linkage is *complete,* only parental combinations occur among the progeny. With *independent assortment* parental and recombinant types occur in equal frequencies.

Existence of linkage may be determined experimentally by analyzing the meiotic products of a cross. The criteria commonly used to determine linkage or independent assortment of two genes are summarized in Table IV-6; a few examples are given in Table IV-7.

1. Recombination and crossing over

The breakdown of linkage is a result of crossing over; this does not invariably lead to recombination of the marker genes used in a particular cross. Therefore, the gene combinations and tetrad types which are of interest for the analysis of partial linkage are those which allow *direct* inferences about the number and type of crossovers that have taken place.

Depending upon the number of crossovers occurring simultaneously in a tetrad, single or multiple crossing over is said to occur. In the latter category double crossovers are particularly noteworthy. Two-, three-, and four-strand double crossovers can be distinguished on the basis of the number of chromatids which are involved (compare Fig. IV-8, IV-9, and IV-11).

In the following discussion the tetrad types will be designated by capital letters, e.g. A_1, B, D_2, P, R (compare Fig. IV-5). These symbols are also used in indicating the numbers of different tetrad types, e.g. $A_1 = 243$. Lower case letters are used to designate the frequencies of the tetrad types, e.g. a_1, b, d_2, p, r. By multiplying these values by 100, the percentage of each type may be obtained, e.g. $100 \times a_1 =$ percentage of the A_1 types. The recombination value gives the percentage of recombination and is 100 times greater than the corresponding recombination frequency (ϱ).

a) Ordered tetrads

1. *Analysis of crosses.* A maximum of 7 tetrad types are expected from the two-factor cross, $a^+b^+ \times ab$, if the meiotic products are ordered. This number is independent of the linkage or independent assortment of the markers; nor does it depend upon the amount of crossing over between the markers, if the genes are linked. On the other hand, the small number (7) of tetrad types corresponds to a significantly greater, theoretically unlimited number of crossing over configurations. Fig. IV-7 diagrams those which account most simply for the 7 types.

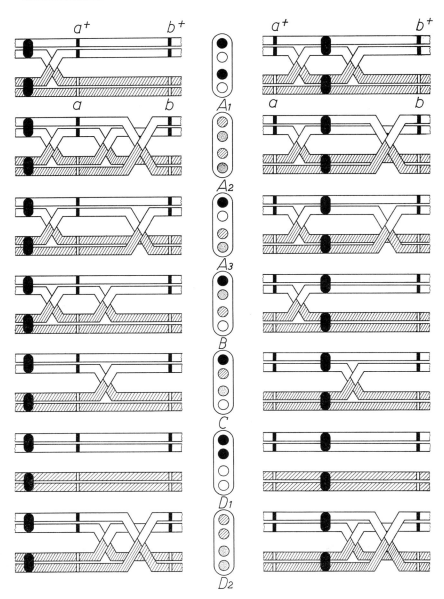

Fig. IV-7. Interpretation of the 7 tetrad types in terms of crossovers. Only the simplest crossover configurations are shown. *Left*. Both loci lie on the same side of the centromere, i.e. in the same chromosome arm. *Right*. The two loci are located on opposite sides of the centromere, i.e. in different arms of the chromosome. *Center*. The genotypes are labelled as in Fig. IV-5. The symbols for the chromosomes, centromeres, and marker genes are the same as in Fig. IV-2

This figure shows:

a) Only the D_1 type results from the *absence of crossing over*, regardless of whether the genes are on the same side (I) (Fig. IV-7 left) or on different sides (II) (Fig. IV-7 right) of the centromere.

b) A *single crossover* accounts for type C in both cases, type A in case I, and type B in case II.

c) The following types result from a *double crossover:* type D_2 (a four-strand double in a single region; case I and II); type B and A_3 (two- or three-strand double in case I); A_1, A_2, and A_3 (two-, four- or three-strand double in case II).

The ratios $a_1 : a_2 : a_3$ in case II and $a_3 : b$ in case I give information on the distribution of chromatids in double crossing over. The random expectations are $a_1 : a_2 : a_3 = 1 : 1 : 2$ and $a_3 : b = 1 : 1$. A deviation from these ratios indicates chromatid interference (p. 179ff.).

d) A *triple crossover* accounts for type A_2 of case I.

In calculating crossover frequencies one must remember that many possible crossover configurations are not shown in Fig. IV-7, even if no more than two crossovers per region are considered. For example, on the basis of random distribution of chromatids from a double crossover (2-str.: 3-str.:4-str. $= 1 : 2 : 1$; p. 179) if $d_2 \neq 0$, a few D_1 types (as a result of two-strand doubles), and double the number of C types (as a result of three-strand doubles) would be expected with the frequencies d_2 and $2d_2$ (see discussion in WHITEHOUSE, 1942, 1949; PERKINS, 1953, 1955; KUENEN, 1962a).

2. *Evidence for linkage.* By definition the stronger the linkage between two markers, the closer the frequency of parental types approaches one and the recombinant frequency approaches zero. The percentage of crossing over also approaches zero because recombination can only result when at least one crossover takes place between the two genes. Therefore, when genes are linked, D_2 types are always fewer than D_1 types ($d_1 > d_2$). Further, for the case shown on the left in Fig. IV-7, A_2 types are expected at a lower frequency than A_1 types ($a_1 > a_2$). Each inequality is a sufficient condition for establishing linkage (Table IV-6 and IV-7). Thus the following generalization holds: A significant deviation of the observed distribution of tetrads from the random is sufficient to establish linkage.

The tetrad distribution which is expected on the basis of random assortment of the pairs of markers, a^+/a and b^+/b, can be calculated. The postreduction frequencies of the two allelic pairs can be determined independently from the linkage calculations. This allows a classification consisting of four groups $A-D$: post $(a^+/a) \times$ post $(b^+/b) =$ frequency of group A $(= a$ with $a = a_1 + a_2 + a_3)$; post $(a^+/a) \times$ pre $(b^+/b) =$ frequency of group B $(= b)$; pre $(a^+/a) \times$ post $(b^+/b) =$ frequency of group $C (= c)$; pre $(a^+/a) \times$ pre $(b^+/b) =$ frequency of group $D (= d$ with $d = d_1 + d_2)$. Moreover, with random assortment the relationships $a_1 = a_2 = 1/2 a_3$ and $d_1 = d_2$ hold, so that calculation of the frequencies of all tetrad types is possible.

The equation $a_1 = a_2 = 1/2 a_3$ holds when the two genes are linked and lie on different sides of the centromere (Fig. IV-7 right) and chromatid interference does not occur (p. 179). This equation is not a sufficient, but a necessary condition for the non-linkage of two markers (p. 166).

The frequency of crossing over between two markers is a measure of the distance between them (p. 196). Accordingly in the case of linkage

Table IV-6. *Summary of criteria for linkage and independent assortment of two allelic pairs*

For explanation, see text. Application of the first four criteria in Table IV-7 and the last two in Table IV-2. Designations of the tetrad distributions are explained on p. 155 (unordered tetrads). (Expanded from KUENEN, 1962a)

ordered tetrads			unordered tetrads	single strands	inferences
tetrad distribution		map distance	tetrad distribution	recombination frequency	
$d_1 > d_2$	$a_1 > a_2$	$w = y - x$	$p > r$ $r:t < 1:4$ (L-distribution)	$\varrho < 0.5$	Markers linked and on the same side of the centromere
	$a_1 = a_2$	$w = x + y$			Markers linked and on different sides of the centromere
$d_1 = d_2$	$a_1 = a_2$	$w = y - x$ or $w = x + y$	$p = r$ $r:t \leq 1:4$ (N-, T-distribution)	$\varrho = 0.5$ $\varrho \leq x + y$	Linkage or non-linkage of markers cannot be determined
$d_1 = d_2$	$a_1 = a_2$	$w \neq y - x$ and $w \neq x + y$	$p = r$ $r:t > 1:4$ (F-distribution)	$\varrho = 0.5$ $\varrho > x + y$	Markers are not linked
$d_1 > d_2$	$a_1 > a_2$	$w \neq y - x$ and $w \neq x + y$	$p > r$ (L-distribution)	$\varrho < 0.5$	Markers are located on different chromosomes which assort in the parental combination at meiosis. Linkage is simulated (quasi-linkage: rare! p. 145)
$d_1 < d_2$	$a_1 < a_2$	$w \neq y - x$ and $w \neq x + y$	$p < r$ (R-distribution)	$\varrho > 0.5$	Markers located on different chromosomes which assort at meiosis in non-parental combinations (reverse linkage: rare! p. 145)

the distance w between two genes is equal either to the sum or the difference of the distances x and y between the genes and the centromere $(w = x + y)$ or either to the sum or the difference of the distances, x and y, between the genes and the centromere $(w = x + y$ or $w = y - x$ when $x \leq y)$ depending on whether the genes lie on different sides, or on the same side of the centromere (Fig. IV-7 right and left and Table IV-6 and IV-7).

With reference to Fig. IV-7, w represents the distance between a and b, x the distance between a and the centromere, and y the distance between b and the centromere. The distances x and y correspond to half their respective postreductional frequencies and the distance w to the frequency of recombinants or half the tetratype frequency (Table IV-7 and p. 196).

b) Unordered tetrads

1. *Analysis of crosses.* An analysis of unordered tetrads reveals a maximum of three types, namely the parental ditype (P), the non-parental

ditype, i.e. recombination ditype (R), and the tetratype (T) (Figs. IV-5 and IV-8). Certain crossover combinations fail to yield recombination of marker genes as in the case of ordered tetrads. The relation of crossover to tetrad type is shown in Fig. IV-8 for cases involving a maximum of two crossovers in the marked region.

Fig. IV-8 shows that in the simplest case parental ditypes result if there is no crossing over, tetratypes from a single crossover, and nonparental ditypes from a four-strand double. With random participation

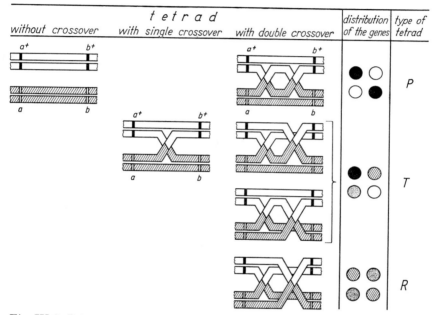

Fig. IV-8. Interpretation of the three types expected in unordered tetrads obtained from a two-factor cross with no more than 2 crossovers in the marked region. Genotypes represented as in Fig. IV-5

of chromatids in a double exchange as many two-strand as four-strand exchanges and twice as many three-strand exchanges are produced. The two- and three-strand exchanges lead to parental ditypes and tetratypes, respectively. Therefore, in calculating the frequency of crossing over the P and T classes may also include some types which have arisen through double crossing over. The frequency of these can be determined from the number of R types.

To illustrate, let us calculate the crossover frequency on the basis of the simplest assumptions from data taken from Table IV-7 (*Podospora anserina*, example I). From the values $P = 203$, $R = 5$, and $T = 407$ we obtain: $p = 0.33$, $r = 0.01$, and $t = 0.66$. Assuming a maximum of two crossovers in the marked region, the r value of 0.01 gives the frequency of four-strand

Table IV-7. *Examples of the application of the first three criteria described in Table IV-6*

object and cross	no. of asci analyzed	tetrad types (ordered tetrads)							tetrad types (unordered tetrads)			recombination value (%) (100 w)	1/2 post-reduction value (%) for	
		A_1	A_2	A_3	B	C	D_1	D_2	P	T	R		gene 1 (100 x)	gene 2 (100 y)
Podospora anserina $t_1 \times i$	615	67	1	6	2	399	136	4	203	407	5	33.9	6.2	38.5
Podospora anserina $m \times un$	1,036	20	24	47	76	692	166	11	186	815	35	42.7	8.1	37.8
Sordaria macrospora $j_2 \times vo_1$	593	50	41	90	93	212	48	59	98	395	100	50.2	23.1	33.1
Podospora anserina $t_1 \times m$	400	2	3	6	30	57	153	149	155	93	152	49.6	5.1	8.5

doubles (Fig. IV-8). Among the parental ditypes and tetratypes are some tetrads which have arisen from two- and three-strand double crossovers. In the absence of chromatid interference (p. 179f.) their frequencies are r and $2r$, respectively. Because every double crossover must be counted twice, crossover frequencies are obtained from P tetrads (r doubles): $2r = 0.02$, from T tetrads ($2r$ double crossovers and $t - 2r$ single crossovers): $4r + (t - 2r) = t + 2r = 0.66 + 0.02 = 0.68$ and from R tetrads (r double crossovers): $2r = 0.02$ (see Fig. IV-8). The frequency of crossing over between the two markers t_1 and i on the basis of the total tetrad frequencies is: $2r + t + 2r + 2r = t + 6r = 0.72$, i.e. 72 crossovers occur in every 100 tetrads. (For methods for calculating crossover frequencies more precisely see p. 195ff.)

2. *Evidence for linkage.* Parental and non-parental ditypes occur in equal frequency from a cross involving unlinked markers $(p = r)$. A statistically significant deviation from such a distribution is a sufficient condition for proof of linkage, provided the number of parental ditypes predominates $(p > r$, Table IV-6 and IV-7).

For such proof of linkage the rare cases in which an affinity between non-homologous chromosomes leads to simulated linkage (p. 144f.) are excluded. The above criterion for linkage has proved valuable in all other cases (RIZET and ENGELMANN, 1949; CATCHESIDE, 1951; PERKINS, 1953; BARRATT et al., 1954).

The absolute amount of crossing over rises with increasing distance between two markers and therefore the frequencies of tetratypes and recombination ditypes, t and r, also rise. The ratio $r:t$ cannot exceed the

See explanation in text and Table IV-6 (References: *S. macrospora:* HESLOT, 1958; *P. anserina:* KUENEN, 1962b).

evidence for linkage or independent assortment from analysis of						interpretations
ordered tetrads			unordered tetrads		single strands	
$D_1:D_2$	$A_1:A_2$	$100w:100(x+y)$ [+] / $100w:100(y-x)$ [-]	$P:R$	$R:T$	ϱ	
136:4 / $d_1 > d_2$	67:1 / $a_1 > a_2$	33.9:44.7 [+] / 33.9:32.3 [-] / $w \approx y - x$	203:5 / $p > r$	5:407 / $r:t < 1:4$	0.339 / $\varrho < 0.5$	Markers located on the same side of centromere
166:11 / $d_1 > d_2$	20:24 / $a_1 \approx a_2$	42.7:45.9 [+] / 42.7:29.7 [-] / $w \approx x + y$	186:35 / $p > r$	35:815 / $r:t < 1:4$	0.427 / $\varrho < 0.5$	Markers located on opposite sides of the centromere
48:59 / $d_1 \approx d_2$	50:41 / $a_1 \approx a_2$	50.2:56.2 [+] / 50.2:10.0 [-] / $w \approx x + y$	98:100 / $p \approx r$	100:395 / $r:t \approx 1:4$	0.502 / $\varrho \approx 0.5$ / $\varrho < 0.562$	Linkage or non-linkage of markers cannot be determined
153:149 / $d_1 \approx d_2$	2:3 / $a_1 \approx a_2$	49.6:13.6 [+] / 49.6: 3.4 [-] / $w \neq x + y$ / $w \neq y - x$	155:152 / $p \approx r$	152:93 / $r:t > 1:4$	0.496 / $\varrho \approx 0.5$ / $\varrho > 0.136$	Markers are unlinked

value of $1/4$ in the absence of interference (see Table IV-6), since $r(n)$ converges versus $1/6$ and $t(n)$ versus $2/3$ for $n \to \infty$ (see Table IV-8).

Explanation of Table IV-8: If there is no crossing over ($n=0$) in the marked region, only parental ditypes occur, i.e. $p(0)=1$, $t(0)=r(0)=0$. With a single crossover ($n=1$) only tetratypes result: $t(1)=1$, $p(1)=r(1)=0$. Double crossovers ($n=2$) lead to 25% each of parental and non-parental ditypes and 50% tetratypes: $p(2)=r(2)=1/4$, $t(2)=1/2$. (The results for $n=0$, 1, 2 are summarized in Fig. IV-8.)

The segregation into P, T, and R types for $n \geq 3$ can be derived from the tetrad type frequencies for $n-1$ crossovers; all tetrads which with $n-1$ crossovers yield parental and non-parental ditypes become tetratypes with an additional crossover, i.e. n crossovers. The tetratypes arising with $n-1$ crossovers become P, T, and R types in a $1:2:1$ ratio. The following relationships are derived from this: $p(n)=1/4\,t(n-1)$; $t(n)=p(n-1)+r(n-1)+1/4\,t(n-1)$; $r(n)=1/4\,t(n-1)$. From these relationships the frequencies given in the next to last line of Table IV-8 for n crossovers (with $n \geq 1$) can be calculated using the regression method.

The above relation $r:t < 1:4$ is not a sufficient condition for linkage, however, because it can be reconciled with independent assortment; the tetratype frequency may range in value from 0 to 1 independently of the frequency of the non-parental ditypes in the case of unlinked genes (WHITE-HOUSE, 1949; PERKINS, 1953).

As in the case of ordered tetrads, information about linkage and the location of the genes can be obtained by comparing the distances between each two of the three genes (p. 162f.). However, three-factor crosses are necessary to establish the order of loci with certainty.

Table IV-8. *The frequency of parental ditypes* (P), *non-parental ditypes* (R), *and tetratypes* (T) *for* $0, 1, 2, \ldots, n$ *crossovers between two markers*

The absence of chromatid interference is assumed for the calculation (p. 179ff.). $p(n)$, $t(n)$, and $r(n)$ designate the frequencies of P, T, and R tetrad types if n crossovers occur. In the last line are given the limiting values for the frequencies $n \to \infty$ (shown in the next to last line). See text for further explanation.

no. of crossovers	frequency of tetrad types		
	$p(n)$	$t(n)$	$r(n)$
0	1	0	0
1	0	1	0
2	$\frac{1}{4}$	$\frac{1}{2}$	$\frac{1}{4}$
3	$\frac{1}{8}$	$\frac{3}{4}$	$\frac{1}{8}$
4	$\frac{3}{16}$	$\frac{5}{8}$	$\frac{3}{16}$
5	$\frac{5}{32}$	$\frac{11}{16}$	$\frac{5}{32}$
$\vdots$	$\vdots$	$\vdots$	$\vdots$
n	$\frac{1}{6} + \frac{1}{3}(-\frac{1}{2})^n$	$\frac{2}{3} - \frac{2}{3}(-\frac{1}{2})^n$	$\frac{1}{6} + \frac{1}{3}(-\frac{1}{2})^n$
$n \to \infty$	$\frac{1}{6}$	$\frac{2}{3}$	$\frac{1}{6}$

If the distance between the first and second gene is designated as w_{1-2}, and the corresponding distances between the second and third, and the first and third as w_{2-3} and w_{1-3}, then linkage is indicated by either $w_{1-2} + w_{2-3} = w_{1-3}$ or $w_{1-2} - w_{2-3} = w_{1-3}$ or $w_{2-3} - w_{1-2} = w_{1-3}$. In the first case the sequence of the markers is $1 - 2 - 3$, in the second, $1 - 3 - 2$, in the third, $2 - 1 - 3$.

c) Single strands

1. *Analysis of crosses.* A two-factor cross yields four different genotypes either in independent assortment or partial linkage. These may be detected even if the products of meiosis cannot be analyzed by tetrads (p. 153f.). If recombinants occur, at least one crossover must have taken place in the marked region. No further conclusions can be drawn beyond this correlation between crossing over and recombination with a two-factor difference.

The particular combination of two linked genes depends upon the number of exchanges which take place in the region between them on the same strand. If an even number of exchanges occurs, parental types result (compare with Fig. IV-8: $P_1 = $ black, $P_2 = $ white); these also result, of course, if crossing over does not occur. On the other hand, an odd number of exchanges yields recombinant types (see in Fig. IV-8: $R_1 = $ hatched, $R_2 = $ stippled).

2. *Evidence for linkage.* With random assortment of two pairs of alleles the four possible genotypes occur in equal frequency. The frequency of recombinants (recombination frequency ϱ) equals 0.5 in the absence of linkage. By definition (p. 158f.) the value $\varrho < 0.5$ holds for linked genes. This is a sufficient condition for linkage (Table IV-6 and IV-7).

The condition, $\varrho < 0.5$, is not necessary because two markers may be linked with a recombination frequency of 0.5. In such a case linkage can only be established when an additional marker is found which gives a recombination frequency of less than 0.5 with each of the original markers (LUDWIG, 1938).

2. Linkage groups

If two genes, a and b, are linked, a third gene c must either be linked to both or assort independently. This relationship between three genes allows classification of markers into specific linkage groups.

The absence of linkage between two markers is shown by their independent assortment. As mentioned previously, this does not prove absence of linkage, however, because in some cases linkage may exist even with random assortment of markers (p. 161, p. 164f.). *Unconditional evidence for the absence of linkage, i.e. markers belonging to different linkage groups*, is obtained from recombination values which do not indicate a linear arrangement of three genes (Table IV-6 and IV-7, fourth example).

This requirement is significant for the analysis of ordered tetrads in that linkage of two genes is excluded only if either the sum or the difference of the two distances x and y between the genes and their centromeres are not equal to the "distance" w between the genes: $w \neq x + y$ and $w \neq y - x$ when $x \leqq y$ (Table IV-6 and IV-7). By the "distance between two unlinked genes" is meant the value that can be formally calculated from the recombination or tetratype frequencies (compare p. 196).

In order to establish the independence of different linkage groups, a marker is generally selected from each group and their non-linkage demonstrated. Centromere markers are particularly suited for this purpose (p. 155) because the sum of their distances from the centromere $(x + y)$ and especially the difference $(y - x)$ is less than the recombination frequency (ϱ) and therefore less than the "distance" (w) between them.

If three markers can be arranged linearly on the basis of their recombination values, it is *not possible* to decide between the alternatives of linkage or non-linkage even though random assortment occurs for at least two of them (Table IV-6 and IV-7). Nevertheless, in such borderline cases it is usually possible to classify them into the proper linkage group by introducing additional markers in further crosses.

"Quasi-linkage" and "reverse linkage" between markers of different linkage groups have been reported in rare instances (p. 144ff.). In these cases the markers are not assorted independently at meiosis; however, on the one hand, recombination frequencies below 0.5 occur to indicate linkage, and on the other, frequencies above 0.5 are obtained which cannot be reconciled with either linkage or independent assortment (Table IV-2 and IV-6).

In spite of certain difficulties in proving linkage or non-linkage (e.g. *Saccharomyces:* DESBOROUGH and LINDEGREN, 1959) it is generally possible to determine unequivocally the number of linkage groups in an organism. This assumes, however, that a sufficient number of genes are known.

The larger the number of markers which are known, the more likely the determination of the number of linkage groups will be accurate. If few markers are available, additional linkage groups may exist from which no genes have been identified. Further, a poorly marked linkage group may be mistaken for two separate groups, if the available markers are far enough apart to show 50% crossing over (compare Table IV-6 and IV-7).

The genes which are known for an organism in most cases fall into a relatively small and constant number of linkage groups. This is the number that is characteristic for the haploid chromosome complement (see Table I-1, Figs. IV-15 and IV-16).

3. Differences in recombination frequencies

The strength of linkage between markers, e.g. the recombination frequency, only remains constant in experiments which are carried out under the same conditions. For this reason it is important to know the sensitivity of the recombination process to variation in environmental conditions in order to set the genetic and environmental limits of the mechanism of recombination.

a) Genetic factors

Variations in the recombination frequencies between identical markers have been observed in a great variety of organisms in which different wild type stocks have been crossed under identical experimental conditions (Table IV-9 and IV-10). Crosses with the same or closely related strains have generally not shown significant differences in recombination frequencies.

Table IV-9. *Organisms in which different wild type stocks have shown different recombination frequencies and postreduction frequencies respectively under constant experimental conditions.* (From literature cited by FROST, 1961)

object	differences in		reference
	post-reduc-tion frequen-cies	recom-bination frequen-cies	
Neurospora crassa	+		TEAS, 1947; HOLLOWAY, 1954; NAKAMURA, 1961; STADLER and TOWE, 1962
	+	+	BARRATT et al., 1954; STADLER, 1956a, c; FROST, 1961
		+	MITCHELL and MITCHELL, 1954; FROST, 1955a, b; RIFAAT, 1956, 1958; MITCHELL, 1958; DE SERRES, 1958a; TOWE, 1958; PERKINS, 1959
Neurospora sitophila	+		WÜLKER, 1935; FINCHAM, 1951
Saccharomyces cerevisiae		+	DESBOROUGH and LINDEGREN, 1959
Schizophyllum commune		+	PAPAZIAN, 1950a, 1951; RAPER et al., 1958
Coprinus lagopus		+	DAY, 1958
Venturia inaequalis	+		BOONE and KEITT, 1956

Table IV-10. *Crosses between different wild type strains of Neurospora crassa*
Note that the postreduction values for the four markers used vary widely, if wild type stocks other than the L strains are employed. L = Lindegren stock; A = Abbott stocks (A_4 and A_{12}); E = Emerson stock (E^{5297}). (From data cited by FROST, 1961.)

marker (isolate no.)	linkage group	cross	no. of asci analyzed	postreduction value	P value for deviation from $L \times L$	reference*
mating type	I	$L \times L$	827	13.1 ± 1.2	—	1–5
		$A_4 \times L$	282	4.6 ± 1.2	< 0.01	2, 4, 6, 7
		$(A_4 \times L) \times A_4$	140	22.9 ± 3.6	< 0.01	4
		$E \times L$	134	27.6 ± 3.7	< 0.01	3, 4
		$(E \times L) \times (E \times L)$	169	32.5 ± 3.6	< 0.01	3
		$(A_4 \times L) \times A_{12}$	77	26.0 ± 5.0	< 0.01	4
al-1 (4637 T)	I	$L \times L$	298	11.1 ± 1.8	—	} 4 6
		$L \times A_4$	85	1.2 ± 1.2	< 0.01	
me-3 (36104)	V	$L \times L$	60	55.0 ± 6.4	—	} 2
		$L \times A_4$	19	26.3 ± 10.1	$0.02—0.05$	
aur (34508)	I	$L \times L$	270	52.2 ± 3.0	—	} 4
		$[(A_4 \times L) \times L] \times E$	122	76.2 ± 3.9	< 0.01	

* Reference: [1] LINDEGREN, 1932b, 1936b; [2] BUSS, 1944; [3] HOLLOWAY, 1953; [4] HOULAHAN et al., 1949; [5] REGNERY, 1947; [6] SRB, 1946; [7] DOERMANN, 1946.

This phenomenon has been studied particularly in *Neurospora crassa*. The discussion is therefore for the most part restricted to work on this organism (for literature see BARRATT, 1954; FROST, 1955b, 1961).

The origins of many of the strains of *N. crassa* which are currently being used for genetic investigation are to a large extent unknown. They are often the offspring of different wild type stocks of which the genome has become thoroughly mixed through a succession of crosses. LINDEGREN (1932a) used only two wild type stocks of opposite mating type at first. BEADLE and TATUM (1945) introduced additional wild type stocks, the "Abbott" and "Chilton" strains which had been isolated in Louisiana by E. V. ABBOTT and St. J. P. CHILTON. Through the crossing of the Abbott and Lindegren stocks, EMERSON and CUSHING (1946) and St. LAWRENCE (DE SERRES, 1958b) produced new "wild type" strains (see also FROST, 1955a).

Table IV-10 shows that crosses with the strains A_4, A_{12}, and E^{5297} (out of $A_{12} \times L$) yield different postreduction values for various markers than crosses between two "Lindegren" strains. These results agree with those of STADLER (1956a, c) and NAKAMURA (1961). Both workers demonstrated a pronounced variation in the postreduction frequencies by using different wild type strains. The analysis of thousands of asci for spore color markers yielded on the one hand postreduction values from 10—58% and on the other, values from 26—61% depending upon the particular cross.

Evaluation of the experimental data leads to the conclusion that *differences in recombination and postreduction frequencies result from genetic differences in the wild type strains employed.* The following facts support this view:

1. Crosses with *Aspergillus nidulans* have in no case shown varying recombination values in contrast to the results with *N. crassa* and other species. However, the mutant strains used are known to have been derived from a single wild type strain (PONTECORVO, 1953b).

WHITEHOUSE (1954) also failed to find differences in recombination frequencies in *Coprinus fimetarius*. He likewise used strains which were derived from a single wild type stock.

2. Through tetrad analysis an ascus from a cross between two wild type strains of *N. crassa* was found in which a segregation for at least one factor responsible for different postreduction frequencies occurred (FROST, 1955b).

Further investigation by FROST (1961) has shown that the number of factors involved must be at least three. The results of STADLER (1956a), STADLER and TOWE (1962) and RIFAAT (1958) argue for a larger number of factors which influence crossing over.

3. Different wild type strains of *N. crassa* exhibit cytogenetical and physiological differences. Similar differences exist between *N. crassa* and *N. sitophila*. These are more pronounced than between different races of the same species (FAULL, 1930; TATUM and BELL, 1946; SINGLETON, 1948; FINCHAM, 1950, 1951; HOLLOWAY, 1953; GARNJOBST, 1953; HIRSCH, 1954).

There is *no definitive theory of the mechanism by which genes control the recombination of the genetic material*. Two models have been proposed:

1. The *heterozygosis* resulting from crosses between different wild type strains *influences the effective pairing between homologous strands*. For example, *pairing may be inhibited*. Since pairing is a prerequisite of crossing over (p. 232ff.), in such a case heterozygosis would result in fewer crossovers than homozygosis.

This hypothesis is supported by the fact that the recombination frequency generally remains constant or increases in successive backcrosses with the same wild type strains but never decreases (STADLER, 1956a, c; TOWE, 1958; PERKINS, 1959; NAKAMURA, 1961; FROST, 1961). Such an increase can be attributed to a progressive nullification of mating restrictions.

The fact that the variation in recombination frequency is not limited to particular loci or chromosomal regions further supports this idea. It can be demonstrated between heteroalleles (p. 211), between genes and their centromeres, as well as in short regions distal or proximal to the centromere.

2. Particular regions of a chromosome are characterized by *interference* which either raises or lowers the frequency of crossing over (FINCHAM, 1950, 1951; FROST, 1955b, 1961; STADLER, 1956a; RIFAAT, 1958). This effect may be related to the hypothesis that a *specific enzyme controls the recombination process* (BRESCH, 1964). A mutated gene might lose the ability to synthesize the enzyme, so that a lowered crossing over frequency would result.

Differences in the postreduction frequencies of mating type genes in *N. crassa* and *N. sitophila* were explained by FINCHAM (1950, 1951) as an interference which in some way extended out from the centromere. Not all observations can be explained by such effects, however, since variations in recombination frequency have also been demonstrated for linked markers which are not close to the centromere.

b) Temperature

Cytological investigations have shown that meiotic processes are influenced by temperature changes (compare OEHLKERS, 1940a, b). It thus seemed pertinent to study the influence of temperature upon the recombination process by genetic methods. Experiments of this kind were first carried out on *Drosophila* (PLOUGH, 1917; STERN, 1926; GRAUBARD, 1934; SMITH, 1936). Later plant materials were also investigated. Some of the innumerable results obtained with fungi are presented graphically

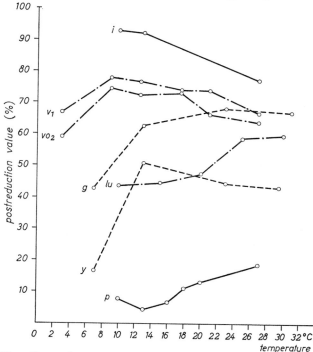

Fig. IV-9. Dependence of postreduction frequency on temperature. Each marker has been analysed at least at three different temperatures. o———o *Podospora anserina* (markers *i* and *p*, KEMPER and KUENEN, unpublished; see also Fig. IV-16). o—·—·—o *Sordaria macrospora* (markers v_1, vo_2 and *lu*, HESLOT, 1958; JOUSSEN, unpublished). o------o *Sordaria fimicola* (markers *y* and *g*, OLIVE, 1956)

in Figs. IV-9 and IV-10 and summarized in Table IV-11. The following points can be derived from this survey:

1. The postreduction frequency of many genes changes with a change in temperature (Figs. IV-9 and IV-10).

The first genetic investigations on the influence of temperature in fungi were carried out by HÜTTIG (1931, 1933a) on different species of *Ustilago*. HÜTTIG found differences of up to 40% in the postreduction frequencies of the mating type gene. Such a temperature influence has been repeatedly confirmed by other workers with a variety of organisms. However, no temperature effect on the recombination or postreduction frequency has been observed for certain loci; e.g. *Bombardia lunata* (ZICKLER, 1934a, b; 1937), *Neurospora sitophila* (WÜLKER, 1935), *Ascobolus stercorarius* (BISTIS, 1956), *Neurospora crassa* (STADLER, 1959b) and *Sordaria macrospora* (KUENEN, unpublished) (compare with the discussion of Fig. IV-10).

2. The change in postreduction frequency is not uniform for the different markers which have been studied. The postreduction values in some cases increase, in others decrease both as the temperature is raised and as it is lowered (Figs. IV-9 and IV-10).

Such changes of the postreduction frequencies in different directions have been demonstrated not only for markers of different organisms but also for markers of the same species and race. *Podospora anserina* and *Sordaria macrospora* are the best examples of this behavior. *Aspergillus nidulans* is an exception; in this ascomycete a similar decrease in recombination frequency was found in four different regions following an increase in temperature from 25° to 37° and 42° C (ELLIOTT, 1960a).

3. The dependence of postreduction frequency upon temperature is generally non-linear and a clearly defined maximum or minimum is rare (Fig. IV-9).

A *regular* decrease or increase in the postreduction and recombination frequency over a range of temperatures has been demonstrated in a few cases, e.g. in *Aspergillus nidulans* (ELLIOTT, 1960a), *Sordaria macrospora* (*lu*), and in *Podospora anserina* (*i*). Curves with a distinct minimum were established for three species of *Ustilago* by HÜTTIG. The latter investigator found the lowest postreduction frequency for those temperatures which were optimal for the germination of spores. Such a relationship between recombination extremes and optimal temperatures for germination as in *Ustilago* could not be confirmed for other organisms.

4. *A similar correlation between temperature and postreduction frequency is not generally found except in the case of markers in particular regions.* It can be shown that genetic markers, according to their position on the chromosome, show a change in postreduction frequency which is characteristic for all organisms when the temperature is lowered from the optimal to 15° C. Four regions may be distinguished on the basis of the kind of temperature effect (Fig. IV-10).

The optimal temperature, i.e. the temperature which has proven particularly favorable for growth and spore maturation, falls between 25—28° C for many Ascomycetes. Lowering the temperature to approximately 10 to 13° C generally leads to a marked unidirectional shift in postreduction frequency for genes located close to one another. Such a relationship between the position of the gene on the chromosome and the type of temperature effect can be graphically represented with temperature curves or maps (Fig. IV-10). Regions with the same "temperature sensitivity" are about 10 map units in length, i.e. they correspond to a postreduction value of about 20%. Since neighboring regions show opposite behavior with regard to the temperature effect, the temperature curve is a sine wave. Additional evidence that this type of curve reflects the true situation lies in the fact that postreduction frequencies of markers at the ends of the regions do not vary significantly (Fig. IV-10).

Since most of the data bearing on this problem are derived from experiments with *Podospora anserina* and *Sordaria macrospora* the temperature curves shown in Fig. IV-10 will be explained with reference to these examples. In both organisms three or four groups of markers may be distinguished depending upon the type of temperature-induced change of postreduction frequency. Each of these groups corresponds to one of the four regions shown in the figure. *Region I* (postreduction values up to about 25 %): decrease in frequencies with a lowering of temperature. *Region II* (postreduction values of 25 to 45 %): increase in frequency with a decrease in temperature. *Region III* (postreduction values from 45 to 65 %): decrease in postreduction frequencies. Markers known only for *Sordaria*. *Region IV* (postreduction values of 65 to 85 %): increase in postreduction frequencies. Certain *Sordaria* loci nevertheless show a significant decrease in postreduction frequency.

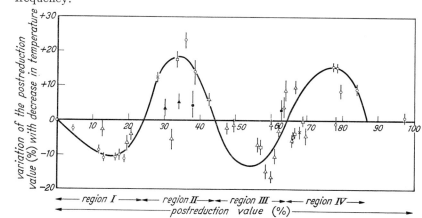

Fig. IV-10. Graphical representation of the relation between the position of the marker on the chromosome and the effect of temperature on the postreduction frequency. The postreduction value of a gene at a normal temperature (25—28° C) may be read on the abscissa. The numbers of the ordinate indicate the extent to which the postreduction frequency increases (+) or decreases (−) with a decrease in temperature (to 10—13° C) (average error taken into account in the representation). The regions I—IV designate segments of the chromosome with varying temperature sensitivity. Explanation in text. *References: Podospora anserina* (○): RIZET and ENGELMANN, 1949; MONNOT, 1953; KEMPER and KUENEN, unpublished. *Sordaria macrospora* (△): HESLOT, 1958; KEMPER, unpublished; JOUSSEN, unpublished; KUENEN, unpublished. *Sordaria fimicola* (□): OLIVE, 1956. *Ascobolus stercorarius* (×): BISTIS, 1956. *Neurospora sitophila* (●): WÜLKER, 1935. *Neurospora crassa* (▲): TOWE and STADLER, 1964 (see also MCNELLY and FROST, 1963)

Little is known about the control of localization of temperature effects on recombination. One can reason, however, that the *lower temperatures interfere with pairing* in the same way as heterogeneous regions do (p. 170). It is possible that the strands, which are twisted around one another on each side of the centromere become loosened in pairing and undergo less frequent crossing over in the vicinity of the centromere. As a result of their relaxation, the spirals become more closely associated at certain points so that pairing is more intimate and crossing over more frequent than at higher temperatures. If one assumes that the degree of relaxation and the points of close contact are fixed for each temperature range, the opposite effects of temperature on postreduction frequency in neighboring chromosomal regions can be explained.

Data from experiments with linked markers in *P. anserina* support this interpretation. A decrease in postreduction frequency of genes in the vicinity of the centromere and an increase in frequency of those lying more distantly were observed simultaneously (MONNOT, 1953; KEMPER, 1964). Such opposite effects are explained by MONNOT as the result of a temperature controlled shift of a single crossover within the chromosomes (compare also RIZET and ENGELMANN, 1949). Such a working hypothesis is also compatible with the idea of pairing inhibition at lower temperatures.

A further observation supports this view. One would expect that the curve will be pushed in the direction of the centromere at intermediate temperatures, i.e. temperatures between 13 and 25° C. Accordingly markers should exist which shift into another sensitivity range. Such markers have actually been found in *S. macrospora* (HESLOT, 1958) and *S. fimicola* (OLIVE, 1956).

The temperature sensitivity is particularly strong in the vicinity of the centromere in certain cases (*Neurospora crassa:* RIFAAT, 1959; TOWE and STADLER, 1964; *Drosophila:* Literature cited in LUDWIG, 1938). This observation is also compatible with the interpretation described above, namely that the relaxation process begins at the centromere.

The effect of temperature is more complex than is at first apparent. This is shown by the recent investigations of KEMPER (1964) on *Sordaria macrospora* and *Podospora anserina*. This author has demonstrated that *both the map distance and the interference value change with a temperature change* (p. 195f. and p. 187f.) (Table IV-11). There are also differences in the behavior of the two experimental organisms; in *P. anserina* the interference values drop with a decrease in temperature although the map distances increase in part, and decrease in part. In *S. macrospora*, on the other hand, the interference values generally increase only negligibly, while the map distances almost always decrease (compare with results in *Aspergillus nidulans*, ELLIOTT, 1960a).

One would expect that the interference values would either remain constant or change in the same direction as the map distances when the temperature is lowered. For example, if the number of crossovers is reduced by the effect of temperature, one would assume that both the absolute, as well as the relative number of multiple crossovers would decrease. It is possible that not only chromosome (p. 187ff.) but also chromatid interference (p. 179ff.) plays a role here (compare with discussion in BISTIS, 1956; KEMPER, 1964). Results of experiments which clarify this problem are not available.

If three dimensional mechanical models are devised for explaining chromosome interference (p. 193), changes in amount of interference may be attributed to shifts in the locations of close pairing between homologous strands which are induced by temperature changes. Another possible interpretation is that recombination results from the action of short-lived, temperature sensitive enzymes (p. 170) which are periodically synthesized during replication (BRESCH, 1964).

c) Chemicals and radiation

Repeated attempts to influence the meiotic, and particularly the mitotic recombination processes through radiation or chemicals have been made.

Table IV-11. *Data on the influence of temperature on the recombination process in Sordaria macrospora and Podospora anserina*

Map distances have been calculated from postreduction or tetratype frequencies (p. 197 ff.). The interference value indicates the strength of chromosome interference. It is inversely proportional to the interference (p. 187). Compare Figs. IV-9 and IV-10 and for the markers of *P. anserina*, Fig. IV-16 (C = centromere). (Data from KEMPER, 1964)

region	map distance	inter-ference value	map distance	inter-ference value	map distance	inter-fer-ence value	increase (+), decrease (−) or constancy (=) of	
							map dis-tance	inter-fer-ence value
Sordaria macrospora	28°		10°		5°			
C-pe	6.0	0.05	5.3	0.1	3.1	0.1	−	+
C-d	10.0	0.1	6.4	0.2	—	—	−	+
C-a	10.5	0.1	8.4	0.1	8.2	0.1	−	=
C-ire	17.6	0.3	14.7	0.4	—	—	−	+
C-lu	40.0	0.4	27.3	0.5	24.0	0.5	−	+
C-r_1	45.5	0.4	43.2	0.5	41.9	0.5	−	+
C-le	49.3	0.8	42.8	0.95	—	—	−	+
C-m	61.0	0.55	53.8	0.6	50.9	0.6	−	≈
C-pal	61.9	0.8	47.3	0.95	—	—	−	+
C-r	100	0.9	87.3	1.0	72.2	1.0	−	+
r_1-a	35.0	0.3	34.8	0.35	33.7	0.4	−	+
lu-pe	34.0	0.4	22.0	0.5	20.9	0.5	−	+
lu-r	60.0	0.6	60.0	0.5	48.2	0.6	−	≈
d-(C)-ire	27.6	0.9	21.1	0.7	—	—	−	−
lu-(C)-a	50.5	0.7	35.7	0.8	32.2	0.8	−	+
r_1-(C)-pe	51.5	0.6	48.5	0.7	45.0	0.7	−	+
Podospora anserina	27°		10°					
C-t_1	5.9	0.1	1.6	0.05			−	−
C-p	8.8	0.1	3.9	0.05			−	−
C-v_1	20.9	0.3	31.4	0.1			+	−
C-la	42.0	0.1	42.3	0.1			=	=
C-i	42.3	0.1	49.6	0.05			+	−
C-z	46.7	0.1	49.7	0.05			+	−
t_1-i	36.4	0.1	48.0	0.05			+	−
p-z	37.9	0.1	45.8	0.05			+	−
v_1-(C)-sp	21.0	0.35	31.5	0.1			+	−
v_1-(C)-la	62.9	0.6	73.7	0.6			+	=

Mitotic intrachromosomal recombination will be discussed in a later section (p. 204 ff.). Nevertheless, in order to emphasize the correlation with results of meiotic investigations, certain observations on the sensitivity of the mitotic exchange process to external influences will be mentioned here.

Some of the results of experiments dealing with this problem are summarized in Table IV-12. The following points are noteworthy:

Chemicals: An *increase in the frequency of crossing over* generally follows treatment with chemical agents. A correlation between concentration and activity has only been observed in a few cases (e.g. HÜTTIG, 1933 a, b). In *N. crassa* the crossover frequency is subject to deviations within regions of the same linkage group (PRAKASH, 1963 b). According to the author, this indicates that crossing over frequency is independent of the activity of the chelating agents which were used (Table IV-12). The variations are interpreted as differential effects of the two chemicals on the general intracellular distribution of ions. In this connection it is particularly noteworthy that the frequency of crossing over as well as the involvement of all four strands in double crossing over may be influenced by specific chemical agents. PRAKASH (1964) found a *chromatid interference shift from negative to positive in N. crassa* after treatment with ethylene diamine tetraacetic acid (EDTA); there were more two-strand than four-strand doubles in the controls while in the series treated with EDTA the four-strand doubles were more numerous than the two-strand doubles (for details see p. 181/183 and 190f.).

Experiments designed to influence the frequency of crossing over with chemical agents which are known to reduce the number of chromosomal breaks proved negative. No statistically significant change in postreduction frequency of the spore color gene in *Sordaria fimicola* could be established for any of the substances employed (cysteine, thiourea, 2-mercaptoethylamine (MILLER and BEVAN, 1964).

UV radiation: The number of crossovers generally increases following irradiation. FOGEL and HURST (1963) demonstrated a distinct *dependence of recombination on UV dose* in *S. cerevisiae.*

For example, in one case, these authors found 0.2 % recombination in the control, 1.1 % after 60 seconds, and 2.45 % after 120 seconds of UV irradiation. These percentages are based on the number of survivors with a survival of 83 % after 60, and 25 % after 120 seconds of irradiation.

A tentative interpretation can be made of the effects of irradiation on recombination in *Saccharomyces* and *Ustilago*. One can assume that an increase in the frequency of mitotic crossing over is correlated with the *inhibitory effect of UV radiation and certain chemicals* (e.g. mitomycin C) *on DNA replication*. An artificially induced delay in replication may *facilitate chromosome pairing* and thus increase the probability of recombination (HOLLIDAY, 1961 b, 1962 a, b, 1964 a; WILKIE and LEWIS, 1963; FOGEL and HURST, 1963; HURST and FOGEL, 1964). This hypothesis [like those proposed to explain the influence of genetic factors (p. 170) and temperature (p. 174) on recombination] suggests that the change in recombination frequency is the result of an alteration in the pairing behavior of homologous chromosomes.

Summary

1. The breakdown of linkage results from crossing over. The event always produces reciprocal products.

2. Linkage of genes, i.e. the occurrence of genes in particular groups, may be determined by using the criteria summarized in Table IV-6.

Table IV-12. *Examples of the alteration of recombination frequencies by means of chemical agents and radiation*

agent	object	somatic recombination	meiotic recombination	reference
Nitrogen mustard	*Penicillium chrysogenum*	+		MORPURGO and SERMONTI, 1959
	Aspergillus nidulans	+		MORPURGO, 1962b
Formalin	*Aspergillus nidulans*	+		FRATELLO et al., 1960
p-fluoro-phenylalanine	*Aspergillus nidulans*	+		MORPURGO, 1961
Diepoxybutane	*Aspergillus nidulans*	+		MORPURGO, 1963
Methyl bis-(β-chloro-ethyl)-amine (HN-2)	*Aspergillus nidulans*	+		MORPURGO, 1963
Mitomycin C	*Saccharomyces cerevisiae*	+		HOLLIDAY, 1964a
	Ustilago maydis	+		HOLLIDAY, 1964a
Ethylene diamine tetra-acetic acid (EDTA)	*Neurospora crassa*		+	PRAKASH, 1963b, 1964
8-oxychinoline	*Neurospora crassa*		+	PRAKASH, 1963b
Various alkali salts	*Ustilago avenae*		+	HÜTTIG, 1933a, 1933b
	Ustilago hordei		+	
	Ustilago decipiens		+	
Various urethanes	*Ustilago avenae*		+	HÜTTIG, 1933a, b
UV radiation	*Penicillium chrysogenum*	+		MORPURGO and SERMONTI, 1959
	Aspergillus nidulans	+		KÄFER and CHEN, 1964
	Aspergillus sojae	+		IKEDA et al., 1957; ISHITANI, 1956
	Saccharomyces cerevisiae	+		JAMES and LEE-WHITING, 1955; ROMAN and JACOB, 1958; FOGEL and HURST, 1963; WILKIE and LEWIS, 1963; HURST and FOGEL, 1964
	Ustilago maydis	+		HOLLIDAY, 1961b, 1962a, b; 1964a
X-radiation	*Aspergillus nidulans*	+		MORPURGO, 1962a
Gamma radiation (cobalt 60)	*Aspergillus nidulans*	+		KÄFER, 1963

3. The number of linkage groups corresponds to the number of chromosomes in the haploid complement. It is characteristic for each species. The independent assortment of different linkage groups may be established according to the criteria listed in Table IV-6.

4. The frequency of recombination between two markers, or the strength of linkage, remains constant under a given set of experimental conditions. However, if different wildtype strains are used or if the offspring are grown at different temperatures, the frequency of crossing over may often vary. Crossover frequency may also be altered with UV radiation, X- and gamma radiation as well as with chemical agents.

III. Interference

Interference is defined as the non-random distribution of crossovers among the four chromatids of the tetrad. Two types of interference are recognized: 1. By *chromatid interference we mean that the occurrence of a crossover either decreases or increases the probability that a second crossover will involve the same strand.* In the first case the interference is positive, while in the second, it is negative. In both cases chromatid interference leads to a non-random distribution of 2-, 3-, and 4-strand double crossovers (Fig. IV-11). 2. *Chromosome interference occurs if following a crossover, there is a less or greater than random chance that additional crossovers will occur in the vicinity of the first.* Here also the former is termed positive, the latter negative interference. Again in both cases the frequencies of multiple crossovers deviate from a random expectation. *Thus chromosome interference refers to the distribution of crossovers in a tetrad, while chromatid interference is concerned with the participation of individual strands in multiple crossing over.*

The phenomenon of interference can be investigated experimentally only if at least two regions are marked (i.e. at least three linked markers). Since both chromatid and chromosome interference have an influence on recombination frequency, single strand analysis (p. 158) does not allow differentiation of the two types of interference; such random analysis permits detection of a double crossover, but does not reveal the manner of its occurrence. Such information can only be obtained through tetrad analysis and by assuming that no sister strand exchanges (p. 185) take place.

Little is known regarding the genetic role of the centromere. The centromere is generally utilized as a marker. To consider the centromere equivalent to a gene does not always seem justified, however, since it has been repeatedly demonstrated that crossovers interfere with other crossovers in the same arm of a chromosome, but not in the arm on the opposite side of the centromere. This phenomenon can possibly be explained by assuming that the recombination process begins at the centromere and extends in both directions (compare with the modified polaron hypothesis, p. 241 f.).

1. Chromatid interference

a) Methods of determination

By definition chromatid interference exists if the distribution of double crossovers among the chromatids is non-random, i.e., if *the proportion of the double exchange types, 2-strand : 3-strand (type I) : 3-strand (type II) : 4-strand, deviates from a 1:1:1:1 ratio* (see Fig. IV-11, left). The difficulty in determining chromatid interference lies in the fact that double crossovers are not always unequivocally identifiable as such, since they must be inferred indirectly from chromatid recombinations.

A similar relationship exists between chromatid interference and chromatid recombination as between crossing over and the recombination of genetic markers. In both cases the cytological events are inferred on the basis of recombinational data, namely the number and type of crossovers in a region with multiple marker genes.

Fig. IV-11 (left) shows that 2-strand double crossovers lead to parental ditypes (P) for the two distal markers, and correspondingly, 3- and 4-strand doubles to tetratypes (T) and non-parental ditypes, i.e. recombination ditypes (R). The ratio $p:t:r = 1:2:1$ for chromatid recombination is expected in the absence of chromatid interference. A significant deviation from this ratio is a criterion for the occurrence of chromatid interference (p. 185).

The condition $p:t:r = 1:2:1$ is not sufficient, however, for the opposite conclusion, namely that no interference exists. *The absence of chromatid interference may be simulated* particularly by the following phenomena:

1. *Sister strand exchange:* Sister strand exchange accompanying a high positive or negative interference between non-sister chromatids may lead to a simulated random distribution (Fig. IV-11 right, and p. 185 ff.).

2. *Multiple crossing over:* The farther two markers lie apart, the higher is the probability of multiple crossing over between them; this may lead to the ratio $p:t:r = 1:2:1$ in spite of interference.

The influence of multiple exchanges on the proportion $p:t:r$ in long regions is not significant if it approximates the random expectation. For this reason the method of WHITEHOUSE (1956) for "correction" for widely separate markers is effective only if there is already a more or less pronounced deviation from a 1:2:1 ratio for the "uncorrected" data.

3. *Domain of interference:* Chromatid interference may no longer be detectable when long regions are used, since the end markers may lie outside the domain of interference. This may be the case particularly if the regions which are utilized are not adjacent.

4. *Opposing interferences:* Positive and negative interferences may cancel each other, if large regions are used for the analysis, or if the data from different regions are added together.

These sources of error indicate that it is essential to use closely linked markers (distance between markers not more than 10 map units) for the analysis of chromatid interference.

12*

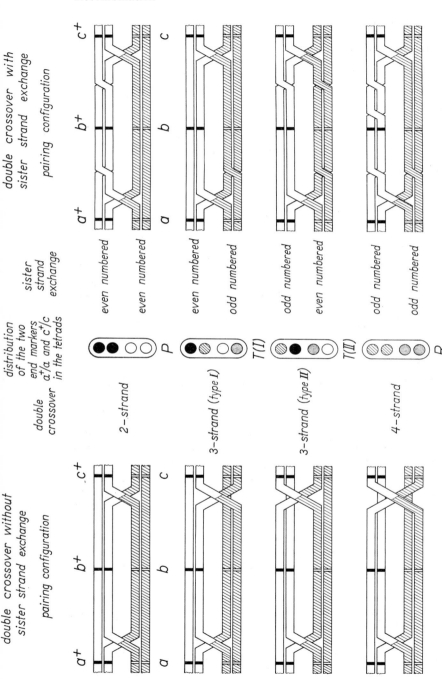

Fig. IV-11. The influence of double crossing over on the distribution of the marker genes a^+/a and c^+c. *Left:* The four possible double crossover types assuming that no sister strand exchange occurs. *Right:* Two-strand double crossing over between non-sister strands with simultaneous sister strand exchange. These crossover configurations lead to the same distribution of markers (*center*) as the four doubles on the left. In the same manner three- or four-strand crossing over in combination with sister strand exchanges result in all four types of tetrads. The genotypes are designated as in Fig. IV-5, chromosomes, centromeres

Since double crossing over occurs very rarely because of the positive chromosome interference (p. 190) which is usually present, the numbers necessary for a statistical confirmation are laborious to obtain. The data which have been published in the literature relating to short regions are frequently not extensive enough to allow statistical confirmation.

In certain cases an interpretation of the results is complicated by the fact that it is not possible to determine with certainty whether a rare tetrad type arose through double crossing over or as a result of cytological mechanisms of a secondary nature, e.g. an irregular slippage of nuclei (p. 148) or a non-random segregation of the centromere (p. 141 ff.). Such difficulties of interpretation can be avoided by marking the centromeres of different linkage groups (WHITEHOUSE, 1942, 1957a, b; PERKINS, 1955; HOWE, 1956).

b) Chromatid interference across the centromere

Data on chromatid interference pertaining to regions on different sides of the centromere are shown in Table IV-13. These data permit the following generalizations:

1. *Chromatid interference across the centromere can generally be detected only if relatively closely linked markers are used.* In most cases in which a significant deviation from a random distribution occurs, R tetrads are less frequent than P tetrads, and T tetrads are less frequent than P and R tetrads combined [$R:P < 1$ and $T:(P+R) < 1$]. This result is an indication of negative chromatid interference (4-strand < 2-strand).

Extensive analysis of two-factor crosses in *N. crassa* (about 58,000 tetrads) as well as in many other species has shown that four-strand double crossovers arise much more rarely than one would expect in the absence of chromatid and chromosome interference (PERKINS, 1962a). Only *Aspergillus nidulans* shows agreement with the expectation (STRICKLAND, 1958b).

2. Chromatid interference across the centromere is generally not observed [$R:P = 1$ and $T:(P+R) = 1$] when *longer regions* are used.

In the liverwort *Sphaerocarpus donellii* the deviation from a random distribution is likewise not significant, if widely separated markers are used (KNAPP, 1937, $P:T:R = 79:147:91$).

3. Different results are obtained if the centromere itself is not used as a marker, e.g. in the analysis of unordered tetrads (p. 153 ff.). If the region which includes the centromere is relatively short, no chromatid interference is generally detectable. In contrast, with longer segments of which the larger portion lies on the same side of the centromere as the adjacent marked segment, the deviation from a random distribution in certain cases is significant.

These results agree with the observation that in chromosome segments uninterrupted by the centromere fewer R than P tetrads arise (e.g. *S. cerevisiae*: SHULT and LINDEGREN, 1959; DESBOROUGH et al., 1960).

Not only can the frequency of crossing over be increased (PRAKASH, 1963b), but the type of chromatid interference influenced (p. 176) through treatment with ethylene diamine tetraacetic acid (EDTA). PRAKASH (1964) found a marked negative chromatid interference in control experiments with *N. crassa*, i.e. more two-strand than four-strand double crossovers (Table IV-13). In experiments with appropriate concentrations of EDTA three times more four-strand than two-strand doubles were demonstrated. This *reversal from a negative to a positive chromatid*

Table IV-13. *Chromatid interference across the centromere*

The data for each species are arranged and summarized according to the map distance between the end markers. The ratios $R:P$ and $T:(P+R)$ serve as measures of chromatid interference. The numbers within frames are those which deviate from a random distribution. See text for further details. For references for this and Tables IV-14, IV-17, and IV-18, see below.

object	map distance between end markers (in map units)	chromatid recombinations			$\dfrac{R}{P}$	$\dfrac{T}{P+R}$	reference [*]
		P (2-str.)	T (3-str.)	R (4-str.)			
Neurospora crassa	up to 10	25	8	6	☐0.2☐	☐0.3☐	4, 6, 17a
	11—20	96	69	47	☐0.5☐	☐0.5☐	1, 4, 5, 6, 7, 16, 17a
	21—30	97	191	85	0.9	1.0	9, 16, 17a
	31—40	31	66	32	1.0	1.0	16
	total:	249	334	170	0.7	0.8	
Neurospora sitophila	50	15	31	10	0.7	0.8	18, 19
Podospora anserina	41—50	11	30	11	1.0	1.4	24
	51—60	19	44	18	0.9	1.2	24
	total:	30	74	29	0.9	1.3	
Sordaria macrospora	41—50	18	21	20	1.1	☐0.6☐	21
	51—60	68	103	61	0.9	0.9	20
	61—70	46	88	45	1.0	1.0	21
	total:	132	212	126	0.9	0.8	
Sordaria fimicola	51—60	4	7	2	0.5	1.2	22
	61—70	8	28	8	1.0	1.8	22
	total:	12	35	10	0.8	1.6	
Saccharomyces cerevisiae	31—40	3	6	4	1.3	0.9	27, 28
	41—50	35	48	29	0.8	0.8	27, 28
	51—60	19	31	29	1.5	☐0.6☐	26, 28
	over 70	51	53	19	☐0.4☐	0.8	26
	total:	108	138	81	0.8	0.7	
Aspergillus nidulans	31—40	17	28	14	0.8	0.9	29
	41—50	21	26	23	1.1	☐0.6☐	29
	over 50	6	16	13	2.2	0.8	29
	total:	44	70	50	1.1	0.7	
Coprinus lagopus	up to 10	12	1	0	0	0.1	31

[*] *Reference to Tables IV-13, IV-14, IV-17, IV-18:*
Neurospora crassa: [1] LINDEGREN, 1936b; [2] LINDEGREN and LINDEGREN, 1937; [3] LINDEGREN and LINDEGREN, 1939; [4] LINDEGREN and LINDEGREN,

interference was explained as the effect of EDTA on the two-strand stage (PRAKASH, 1964) or the induction by EDTA of a high degree of breakage in sister strands (STRICKLAND, 1961); (see also section on the interpretation of intrachromosomal recombination, p. 231 ff.).

PRAKASH obtained in the controls the segregation 2-str:3-str:4-str = 53:24:15 (negative chromatid interference) for regions which lie on different sides of the centromere in linkage groups I and VI. After treatment with three different concentrations of EDTA (4×10^{-5}, 10×10^{-5} and 20×10^{-5}) the following values were obtained: 22:16:37 (weak interference), 16:33:56 and 31:22:59 (in the latter cases, positive interference). EVERSOLE and TATUM (1956) likewise observed a strong increase in four-strand double crossovers after the addition of EDTA to *Chlamydomonas*.

c) Chromatid interference within a chromosome arm

The most significant data from tetrad analysis of three-factor crosses bearing on this problem have been assembled in Table IV-14. The evaluation of these results is complicated by the fact that the centromere is frequently used as an end marker making it impossible to differentiate between P and R tetrads and two- and four-strand doubles (WHITE-HOUSE, 1942). The following can be concluded from Table IV-14:

1. *Negative chromatid interference within a chromosome arm can be detected in short as well as longer regions,* i.e. fewer four-strand than two-strand doubles occur $(R: P < 1)$.

More P than R tetrads may be observed if regions which are separated by a short or long interstitial segment are used for the analysis. The deviation from a $1:2:1$ ratio is not significant in the alga, *Chlamydomonas reinhardi* (EBERSOLD and LEVINE, 1959). Data from other materials, e.g. *Drosophila*, mice, and certain plants are presented in LUDWIG (1938), WHITE-HOUSE (1942), OWEN (1950), and CARTER (1954).

2. *Fewer three-strand double crossovers* than expected on a random basis were found in *S. cerevisiae* in contrast to *N. crassa* and *P. anserina*. As a matter of fact, the frequency of tetratypes is lower in *Saccharomyces* than the sum of the frequencies of the parental and non-parental ditypes $[T:(P+R) < 1]$.

The rare occurrence of tetratypes agrees with data of experiments in which two different chromosome arms were marked (p. 181). However, this contradicts the observations in other materials; statistical confirmation of the differences between observation and expectation has not been possible

1942; [5] HOULAHAN et al., 1949; [6] HOWE, 1954; [7] BARRATT et al., 1954; [8] PERKINS, 1956; [9] STADLER, 1956b; [10] GILES et al., 1957; [11] PERKINS, 1959; [12] PERKINS and ISHITANI, 1959; [13] MALING, 1959; [14] STRICKLAND et al., 1959; [15] STRICKLAND, 1961; [16] BOLE-GOWDA et al., 1962; [17] PERKINS, 1962b; [17a] PRAKASH, 1964.

Neurospora sitophila: [18] WÜLKER, 1935; [19] WHITEHOUSE, 1956.

Sordaria macrospora: [20] HESLOT, 1958; [21] JOUSSEN, personal communication.

Sordaria fimicola: [22] PERKINS et al., 1963; [17] PRAKASH, 1964.

Podospora anserina: [23] MONNOT, 1953; [24] KUENEN, 1962b.

Saccharomyces cerevisiae: [25] SHULT and LINDEGREN, 1956a, [26] SHULT and LINDEGREN, 1959; [27] HAWTHORNE and MORTIMER, 1960; [28] DESBOROUGH et al., 1960.

Aspergillus nidulans: [29] STRICKLAND, 1958b; [30] KÄFER, 1958.

Coprinus lagopus: [31] FINCHAM and DAY, 1963, p. 87.

Table IV-14. *Chromatid interference within a chromosome arm*

The data for each organism are arranged and summarized according to the map distance between the outer markers used for the tetrad analysis. The ratios $R:P$ and $T:(P+R)$ serve as measures of chromatid interference. The numbers in frames are those which deviate from a random distribution. For further explanation see text.

object	map distance between end markers (in map units)	chromatid recombinations				$\dfrac{R}{P}$	$\dfrac{T}{P+R}$	reference[*]
		P	T	R	P or R			
		(2-str.)	(3-str.)	(4-str.)				
Neurospora crassa	up to 10	49	59	36	6	0.7	0.6	4, 6, 15, 16, 17a
	11—20	42	52	27	4	0.6	0.7	1, 4, 17, 17a
	21—30	68	122	32	22	0.5	1.0	3, 5, 7, 8, 16, 17
	31—40	87	167	81	—	0.9	1.0	16, 17
	41—50	20	27	11	—	0.5	0.8	17
	51—60	21	41	15	—	0.7	1.1	17
	61—70	52	108	42	—	0.8	1.1	17
	71—80	18	47	15	—	0.8	1.4	17
	total:	357	623	259	32	0.7	1.0	
Sordaria fimicola	11—40	3	16	7	—	2.3	1.6	22
	51—60	5	10	4	—	0.8	1.1	22
	total:	8	26	11	—	1.4	1.4	
Podospora anserina	25—45	—	12	—	5	—	2.4	24
	46—50	—	20	—	16	—	1.3	24
	total:	—	32	—	21	—	1.5	
Saccharomyces cerevisiae	11—30	21	3	2	—	0.1	0.1	26, 28
	31—50	14	21	20	1	1.4	0.6	25, 27
	51—60	14	8	5	—	0.4	0.4	26
	total:	49	32	27	1	0.6	0.4	
Aspergillus nidulans	up to 20	19	30	16	—	0.8	0.9	29
	21—30	11	14	17	—	1.5	0.5	29
	total:	30	44	33	—	1.1	0.7	

* References follow Table IV-13.

in *N. crassa* or *P. anserina* nor in *Chlamydomonas reinhardi* (EBERSOLD and LEVINE, 1959) or *Sphaerocarpus donellii* (KNAPP, 1937).

Chromatid interference can be modified through treatments with specific chemical agents. PRAKASH (1964) was able to show on the basis of extensive investigations with *N. crassa* that the negative chromatid interference observed in controls (Table IV-14) could be transformed into positive interference under the influence of EDTA. This result agrees with that discussed in the previous section (p. 181/183) (see also p. 176).

PRAKASH obtained the following values for double crossovers in the arms of chromosomes I and VI: Controls: 2-str:3-str:4-str $= 30:7:6$, EDTA 4×10^{-5} 17:8:26; EDTA 10 $\times 10^{-5}$ 17:27:53; EDTA 20 $\times 10^{-5}$ 20:12:46. In the latter two cases a distinct shift to positive chromatid interference is apparent.

d) Chromatid interference and sister strand exchange

Up to now we have assumed that crossing over occurs only between non-sister strands. The close relationship between chromatid interference and sister strand exchange necessitates evaluation of this assumption. Exchange between sister strands can lead to chromatid recombination which masks or modifies genuine chromatid interference (p. 179). Up to the 1950's it was believed on the basis of cytological studies on ring chromosomes in *Drosophila* that exchange between sister strands did not occur (MORGAN, 1933; SCHWEITZER, 1935; other literature cited by LUDWIG, 1938, and SCHWARTZ, 1953, 1954). The question of the participation of chromatids in crossing over was not re-examined until 1953. Investigations on ring chromosomes were repeated, this time in maize (SCHWARTZ, 1953). The cytological picture at anaphase led, in contrast to the earlier study, to the conclusion that sister strand exchange did occur quite frequently (further experiments in *Drosophila*: SCHWARTZ, 1954; WELSHON, 1955). Apart from these results the autoradiographic studies on sister strand exchange in mitotic chromosomes makes the phenomenon probable[1] (TAYLOR, 1957, 1958; TAYLOR et al., 1957).

These results show that a discussion of chromatid interference cannot be restricted to interference between non-sister strands. Therefore, we have combined double crossing over between non-sister strands with the corresponding non-sister strand exchanges in Table IV-15 and Fig. IV-11 (right). It is apparent from these presentations that:

1. The ratio of chromatid recombination $p:t:r = 1:2:1$ is independent of sister strand exchange in the absence of chromatid interference (2-str:3-str:4-str $= 1:2:1$).

2. If chromatid interference and frequent sister strand exchanges (i.e. if the frequency of even numbered and odd numbered exchanges are statistically equal) do occur, a deviation from $p:t:r = 1:2:1$ cannot be demonstrated.

3. Chromatid interference without sister strand exchange never leads to a random distribution.

4. *A statistically significant deviation from $p:t:r = 1:2:1$ indicates in every case that chromatid interference is present. It is not possible to determine, however, whether or not sister strand exchanges also occur.*

The data of Tables IV-13 and IV-14 show that parental ditypes are more frequent than recombination ditypes $(p > r)$ and tetratypes less frequent than both ditypes together $(t < p + r)$. As seen from Table IV-16, these results can be explained by assuming negative, as well as positive chromatid interference:

---

[1] WOLFF (1964) recently discovered that when radioactive substances were used sister strand exchange did not occur spontaneously, but was induced by the incorporated isotopes.

On the one hand $p > r$ can be related to the effect of negative chromatid interference in the absence of sister strand exchange (even numbered — odd numbered); on the other hand this relationship can be attributed to the effect of positive chromatid interference, if one assumes that every exchange between non-sister strands is accompanied by a sister strand exchange on each side (odd numbered-odd numbered).

The relationship $t < p + r$ can likewise have a dual interpretation because of chromatid interference. If more three-strand than two- and four-strand doubles are produced, the number of sister strand exchanges in the two chromatid pairs must be different (even numbered-odd numbered). If the frequency of three-strand doubles is less than the frequency of the two other types, the corresponding number of sister strand exchanges must be the same (even numbered-even numbered or odd numbered-odd numbered).

Table IV-15. *The influence of sister strand exchange on chromatid recombination with double crossing over between non-sister strands*

Both pairs of sister strands of a tetrad are marked with 1—2 and 3—4 respectively (column 1). The effect of sister strand exchange in the presence of 2-strand double crossovers (column 3) can be seen in the right part of Fig. IV-11. Columns 4 to 6 may be interpreted in a similar way (see further in text and Table IV-16).

sister strands	sister strand exchange	chromatid recombination with			
		2-str.	3-str. (I)	3-str. (II)	4-str.
1—2 3—4	even numbered even numbered	P	T (I)	T (II)	R
1—2 3—4	even numbered odd numbered	T (I)	P	R	T (II)
1—2 3—4	odd numbered even numbered	T (II)	P	R	T (I)
1—2 3—4	odd numbered odd numbered	R	T (II)	T (I)	P

Table IV-16. *The influence of chromatid interference and sister strand exchange on chromatid recombination (for further explanation see text) (compare also Table IV-15)*

type of chromatid interference	chromatid recombination in sister strand exchange		
	1—2 even numbered 3—4 even numbered	1—2 even numbered 3—4 odd numbered	1—2 odd numbered 3—4 odd numbered
No interference 2-str. = 4-str. 3-str. = (2 + 4)-str.	$p = r$ $t = p + r$	$p = r$ $t = p + r$	$p = r$ $t = p + r$
Negative interference 2-str. > 4-str. 3-str. < (2 + 4)-str.	$p > r$ $t < p + r$	$p = r$ $t > p + r$	$p < r$ $t < p + r$
Positive interference 2-str. < 4-str. 3-str. > (2 + 4)-str.	$p < r$ $t > p + r$	$p = r$ $t = p + r$	$p > r$ $t > p + r$

Most authors attribute a deviation of $p:t:r$ from 1:2:1 in the above sense to negative chromatid interference. This is done frequently on the assumption that either no, or only rare sister strand exchange occurs. The negative interference observed in longer regions therefore does not fit well into this scheme (see also the discussion on p. 193 f. and MÖLLER, 1959).

2. Chromosome interference

a) Methods of determination

By definition (p. 178) chromosome interference occurs when the frequencies of n number of crossovers ($n = 1, 2, 3 \ldots$) deviate significantly from a random distribution. In order to detect this generally at least three linked markers are necessary. The four methods which can be used to determine chromosome interference are based on different premises.

1. *Coincidence.* The most commonly used measure of interference is the coincidence value, which represents the relation between the observed and the expected frequency of double crossovers. Coincidence is the reciprocal of interference. *With positive interference the coincidence value is less than 1, with negative interference, greater than 1. In the absence of interference the coincidence value equals 1.* With complete interference a value of 0 is obtained (no double crossovers). Coincidence can only be used for determining chromosome interference, however, if the absence of chromatid interference is assumed.

Other methods of determining interference are based on similar assumptions, e.g. use of the KOSAMBI factor (KOSAMBI, 1944), the WEINSTEIN constant (WEINSTEIN, 1955), or the interference value (KUENEN, 1962a). In all cases one calculates the theoretically expected frequency of double crossing over (i.e. the frequency with which a crossover is expected on a random basis *simultaneously* in each region) by multiplying the frequencies of the observed crossing over in two regions.

The advantage of this method lies in the fact that by a simple calculation not only the existence, but also the type of chromosome interference (positive or negative) can be determined. Nevertheless, the strength of interference cannot be established precisely for the following reasons: 1. The frequency of observed crossing over is determined in part by the interference and therefore cannot be used to calculate the theoretically expected double crossovers in the absence of interference (KUENEN, 1962a). 2. Further, the "real" crossover frequencies necessary for this calculation are inaccurate, because it is generally impossible to detect all crossovers in a region. This is particularly true if the marked regions are long.

2. *Poisson distribution.* The random distribution of an event (e.g. crossing over) can be determined approximately with the Poisson formula, $p(n) = \dfrac{e^{-x} x^n}{n!}$. In this equation $p(n)$ represents the expected frequency of trials (tetrads) with n events (crossovers), while x represents the overall average frequency of events (crossovers per tetrad). By comparing the observed distribution with that determined by this method, the existence of chromosome interference can be detected. *With positive interference $p(n)$ for $n \geq 2$ is always greater; with negative interference it is always less than the corresponding frequency of observed multiple crossing over.*

The same restrictions hold for determining the strength of interference as already pointed out in connection with the determination of coincidence values.

3. *Interference models.* The expected distributions of crossovers can be determined with the aid of models (p. 197ff.). *A comparison of the observed distribution with the distribution expected on the basis of the model*

not only indicates the existence, but also the degree of chromosome interference.

This manner of determining interference is particularly appropriate in tetrad analysis, since a better comparison of crossover distribution is possible than in single strand analysis (p. 197 ff.) with the same number of markers. Interference models for tetrad analysis have been developed by BARRATT et al. (1954), JOUSSEN and KEMPER (1960) and KUENEN (1962a) (p. 198 f.).

Computation of the degree of interference is also beset with difficulties in this method; the functional relation between the observed data and the strength of interference is so complex that an exact representation is not possible. This difficulty can be overcome, however, by determining the strength of interference with graphical methods (p. 198—200).

4. *Postreduction and tetratype values above 66.7%*. The theoretically expected maximum postreduction or tetratype value reaches 66.7% in the absence of chromosome or chromatid interference (MATHER, 1935; see in Table IV-8: $t(n) \rightarrow {}^2/_3$ for $n \rightarrow \infty$, and in Fig. IV-13: mapping function for $Q = 1.0$). *A value beyond this upper limit expected on the basis of a random distribution indicates positive chromosome and/or chromatid interference* (Table IV-19).

A theoretically expected maximum postreduction or tetratype frequency exists not only in the absence of interference, but also for each degree of interference (JOUSSEN and KEMPER, 1960; KUENEN, 1962a). It can be shown that with increasing distance between two markers the corresponding frequencies converge toward a constant maximum value (Fig. IV-13 and p. 198f). The latter is 66.7% without interference and 100% with complete interference; the other interference values range between these limits.

With similar chromosome interference in the so-called interference regions the maximum value is only attained through a "saturation" of crossing over. This means that every level of interference has a corresponding theoretically expected maximum crossover frequency (KUENEN, 1962a). A drop in the postreduction or tetratype frequencies following a maximum thus indicates a variable interference (e.g. in *Neurospora crassa*: PERKINS, 1956, 1962b; in *Podospora anserina*: KUENEN, 1962b).

b) Chromosome interference across the centromere

In general, chromosome interference across the centromere cannot be demonstrated (coincidence value $K = 1$, Table IV-17). A good agreement with a random expectation can be shown when extended regions are analyzed.

In addition to the authors cited in Table IV-17, others have also failed to observe a deviation from a random distribution, e.g. in *N. crassa*: STADLER (1956a), HOWE (1956), PERKINS (1959); in *S. cerevisiae*: SHULT and LINDEGREN (1956b). Mostly long regions were utilized for analysis. Most of the investigations on other plants and animals (e.g. maize, *Sphaerocarpus*, *Drosophila*) likewise show no chromosome interference across the centromere (Literature in LUDWIG, 1938, and OWEN, 1950).

A correlation between exchanges on the right and left of the centromere has been found, however, in certain instances (Table IV-17, see also Table IV-11).

1. *Positive interference* was observed in both *Podospora anserina* and *Sordaria macrospora* (Table IV-11) ($K < 1$). The degree of interference depends upon the length of the regions selected for investigation. Chromosome interference is no longer detectable with widely separated markers (Table IV-17).

Table IV-17. *Chromosome interference across the centromere*
(see also Table IV-11)

The data for each organism are arranged and summarized according to the map distance between end markers used in the analysis. The coincidence value (double exchanges observed/double exchanges expected) serves as a measure of chromosome interference. The values in frames are those which deviate from a random distribution. For further explanation see text.

object	map distance between end markers (in map units)	double exchanges		coinci-dence value	refer-ence[*]
		observed	expected		
Neurospora crassa	up to 10	11	2.3	4.8	1, 2
		30	30.4	1.0	16
	11—20	35	9.7	3.6	1, 2
		57	70.5	0.8	16, 17
	21—30	170	174.1	1.0	16
	total:	303	287.0	1.1	
Sordaria macrospora	41—50	38	32.6	1.1	21
	51—60	232	221.0	1.0	20
	61—70	91	91.2	1.0	21
	total:	361	344.8	1.0	
Sordaria fimicola	51—60	15	18.3	0.8	22
	61—70	50	52.8	0.9	22
	total:	65	71.1	0.9	
Podospora anserina	up to 20	8	32.3	0.2	24
	41—50	122	167.5	0.7	24
	51—60	155	152.7	1.0	24
	total:	285	352.5	0.8	
Saccharomyces cerevisiae	31—40	10	12.2	0.8	28
	41—50	115	124.3	0.9	27, 28
	51—60	79	86.9	0.9	26, 28
	over 70	123	136.2	0.9	26
	total:	327	359.6	0.9	
Aspergillus nidulans	31—40	108	96.5	1.1	29, 30
	41—50	72	59.9	1.2	29, 30
	51—60	122	121.6	1.0	29, 30
	61—70	40	48.9	0.8	29, 30
	over 70	47	36.6	1.3	29, 30
	total:	389	363.5	1.1	

[*] References follow Table IV-13.

Positive interference has also been found in *Drosophila* in rare instances (e.g. Pätau, 1941, in data of Gowen, 1919).

2. *Negative interference* has been found in *Neurospora crassa* in isolated cases if small regions in the vicinity of the centromere are analyzed ($K > 1$, Table IV-17).

PAPAZIAN (1952) also demonstrated negative interference in *N. crassa* from an analysis of the data of BUSS (1944). Further, negative interference has been observed in *Drosophila* when relatively small regions adjacent to the centromere have been studied (KIKKAWA, 1935; PÄTAU, 1941, from the data of GOWEN, 1919).

The degree of *positive* chromosome interference observed in *Podospora* depends upon the distance between the gene markers and between the markers and centromere. The increase in the coincidence value with increasing map distance is evidence that the sphere of influence of interference is limited. The mechanism of interference remains unknown. The role of the centromere in the recombination process likewise is unexplained (p. 178).

KIKKAWA (1933, 1935) assumed that the strands in the vicinity of the centromere fail to pair in a small fraction of the cases in order to explain negative interference in *Drosophila*. In such a case the coincidence value must be above 1, if one assumes that exchanges on two sides of the centromere are independent of one another. Thus, negative interference is simulated. This interpretation will be discussed in another connection (p. 218).

c) Chromosome interference within a chromosome arm

In contrast to the foregoing results, the data discussed here show that *within a region lacking a centromere, positive chromosome interference is most common.*

We shall consider first the results of experiments with multiple markers in which extremely closely linked markers were disregarded. The negative interference shown in such situations will be discussed in connection with genetic fine structure (p. 216ff.).

The most significant experimental results are summarized in Tables IV-18 and IV-19; these data give information on the occurrence and strength of interference within a chromosome arm (see also Table IV-11). Table IV-18 shows that the *coincidence value varies in different organisms and according to the location and size of the regions analyzed.* The coincidence value is generally smaller than 1, but in the case of markers lying far apart, values very close to 1 are frequently found, i.e. positive chromosome interference is usually detectable only within short regions. Further, *the postreductional and tetratype values which are greater than 66.7% indicate positive interference* (Table IV-19).

Doubts have repeatedly been expressed regarding the existence of such high postreduction and tetratype values (compare discussions of PERKINS, 1955; BISTIS, 1956). In certain cases a statistical test of the 66.7% limit is not possible, e.g. in *Glomerella cingulata* (WHEELER, 1956) and *Schizosaccharomyces pombe* (LEUPOLD, 1950). In other cases it is uncertain whether or not the cytological conditions for a statistically confirmed conclusion are available (e.g. *Ustilago:* HÜTTIG, 1931, 1933a, b). On the basis of the remaining data in Table IV-19 there is no doubt that the upper limit of 66.7% is exceeded.

High values have also been found for other organisms, e.g. for the algae, *Chlamydomonas reinhardi* and *C. eugametos* (HAWTHORNE cited by PERKINS, 1955, and GOWANS, 1960), for the bryophyte, *Sphaerocarpus donellii* (KNAPP, 1936, 1937, 1960) and for the fruit fly, *Drosophila melanogaster* (using attached X-heterozygotes; literature in PERKINS, 1955).

Table IV-18. *Chromosome interference within a chromosome arm*
(see also Table IV-11)

The data for each organism are arranged and summarized according to the map distance between the terminal markers used in the analysis. The coincidence value (double exchanges observed/double exchanges expected) serves as a measure of chromosome interference. The values in frames deviate from a random expectation. For further details see text.

object	map distance between end markers (in map units)	double exchanges		coincidence value	reference*
		observed	expected		
Neurospora crassa	up to 10	153	586.5	0.3	10, 15, 16
	11—20	40	93.4	0.4	11, 16
	21—30	140	201.9	0.7	3, 16, 17
	31—40	125	246.9	0.5	11, 13, 17
	41—50	32	37.9	0.8	3, 11, 14
	51—60	65	60.1	1.1	11, 12, 17
	61—70	239	243.0	1.0	17
	over 70	112	114.0	1.0	17
	total:	906	1583.7	0.6	
Sordaria macrospora	51—60	122	239.3	0.5	21
	61—70	120	117.8	1.0	21
	over 70	111	227.2	0.5	21
	total:	353	584.3	0.6	
Sordaria fimicola	11—40	27	32.9	0.8	22
	51—60	19	27.2	0.7	22
	total:	46	60.1	0.8	
Podospora anserina	21—30	8	50.9	0.2	23, 24
	41—50	15	129.0	0.1	24
	51—60	49	303.2	0.2	24
	total:	72	483.1	0.1	
Saccharomyces cerevisiae	11—20	8	3.0	2.7	28
	21—30	18	14.8	1.2	26, 28
	41—50	18	36.1	0.5	26, 27
	51—60	27	32.2	0.8	26
	over 60	21	22.3	0.9	25
	total:	92	108.4	0.8	
Aspergillus nidulans	11—20	93	92.1	1.0	29, 30
	21—30	35	43.0	0.8	29, 30
	total:	128	135.1	0.9	

* References follow Table IV-13.

It must be pointed out in interpreting the data of Table IV-19 that postreduction and tetratype values above 66.7% may result not only from positive chromosome interference, but also from chromatid interference. In the former case more single and fewer multiple crossovers occur, while in the latter, three-strand double crossovers are particularly frequent. Because it has been shown through tetrad analysis that three-strand crossing over generally occurs less frequently than expected (p. 181 and 183), it is very likely that values in excess of 66.7% should be attributed to chromosome interference.

Positive chromosome interference has not only been observed in the fungi, but in almost all other organisms which have been investigated genetically.

Table IV-19. *Species of fungi in which the postreduction and tetratype frequencies exceed 66.7%*

The tetratype values are based on data obtained in experiments with linked markers. Only maximum values found for each organism are included.

object	postreduc- tion value ($\pm$ standard deviation)	tetratype value ($\pm$ standard deviation)	reference
Neurospora crassa	—	72.1 ± 1.5	PERKINS, 1962b
Neurospora tetra- sperma	91.4 ± 4.6	—	HOWE, 1963
Podospora anserina	98.8 ± 1.2	92.0 ± 2.6	RIZET and ENGELMANN, 1949; MONNOT, 1953; KUENEN, 1962b
Sordaria macrospora	78.1 ± 1.3	—	HESLOT, 1958; KEMPER, 1964
Venturia inaequalis	91.1 ± 3.8	—	KEITT and BOONE, 1954
Saccharomyces cerevisiae	76.9 ± 2.7	83.3 ± 4.5	HAWTHORNE and MOR- TIMER, 1960; DESBO- ROUGH et al., 1960
Coprinus fimetarius	—	80.4 ± 3.8	QUINTANILHA, 1933
Ustilago hordei	96.2 ± 0.7	—	HÜTTIG, 1931
Ustilago avenae	86.0 ± 1.1	—	HÜTTIG, 1933a
Ustilago decipiens	98.7 ± 0.4	—	HÜTTIG, 1933a

Results of this kind are known, for example, for *Drosophila* (literature: LUDWIG, 1938; PARSONS, 1958) for corn (literature: LUDWIG, 1938; PARSONS, 1957) for *Sphaerocarpus donellii* (KNAPP and MÖLLER, 1955), and for *Chlamydomonas reinhardi* (EBERSOLD, 1956; EVERSOLE, 1956; EVERSOLE and TATUM, 1956; LEVINE and EBERSOLD, 1958; EBERSOLD and LEVINE, 1959).

The low frequency of recombination ditypes in two-factor crosses also argues for the existence of positive chromosome interference (PERKINS, 1962a). Only those fungi which are not included in Table IV-18 are mentioned here: *Neurospora sitophila* (WHITEHOUSE, 1948, 1956; FINCHAM, 1951), *Sordaria fimicola* (EL-ANI et al., 1961), *Glomerella cingulata* (WHEELER, 1956), *Venturia inaequalis* (BOONE et al., 1956; WILLIAMS and SHAY, 1957), *Ustilago maydis* (HOLLIDAY, 1961a, b).

Aspergillus nidulans is the only organism for which chromosome interference has not been demonstrated either for small or large intervals in spite of intensive analysis (Table IV-18).

The decrease in the amount of interference (increase in coincidence value) that is frequently observed with increasing distance between markers is explained by the assumption that the *"domain" of interference is limited*. The analysis of a 74 map unit region subdivided into five intervals in *N. crassa* (PERKINS, 1962b) supports this interpretation. *More* double crossovers (but fewer three- or four-strand doubles) were observed in the region as a whole than expected on the basis of a Poisson distribution. However, *fewer* doubles than expected were generally observed within each subdivision.

In contrast to most other organisms, *Podospora anserina* exhibits a high positive chromosome interference even when the distance between markers is great (Table IV-18).

The causes of gradual variation on the effect of interference in different organisms and in different chromosomes of the same organism are as little understood as the causes of interference. Attempts have been made to explain positive chromosome interference as the result of mechanical phenomena. It has been assumed that during prophase of meiosis when chiasmata form the chromosomes are more or less inflexible threads; the simultaneous occurrence of two immediately adjacent chiasmata is prevented [see also the explanation of negative chromatid and chromosome interference adjacent to the centromere proposed by MÖLLER (1959)]. Another hypothesis of positive chromosome interference considers crossovers to arise through a periodical production of enzymes during replication which remain active only a short time (p. 174). Such an explanation would also account for the high negative interference found within genes (p. 216ff.) (BRESCH, 1964).

3. The combined effect of chromatid interference, sister strand exchange, and chromosome interference on the exchange of genetic markers

The data from investigations on chromatid and chromosome interference were interpreted in more or less the same way in earlier years. The only recombination mechanism considered possible during these times was the classical crossing over (breakage and reunion; p. 234f.). Much greater caution is exercised today in explaining interference because of a greater awareness of the difficulties which confront any hypothesis. For example, thirty years ago sister strand exchange was out of the question as an explanation, because such exchange had been demonstrated as unlikely in *Drosophila* (p. 185). Sister strand exchange must be taken into account today both on theoretical grounds and also because experimental evidence for it has been found (p. 185). Models for the mechanism of recombination have been proposed which postulate sister strand exchanges among others (e.g. copy choice: p. 236f.).

We have attempted to show diagrammatically in Fig. IV-12 the complex interrelationships between chromatid and chromosome interference, and sister strand exchange in their effects on recombination frequencies.

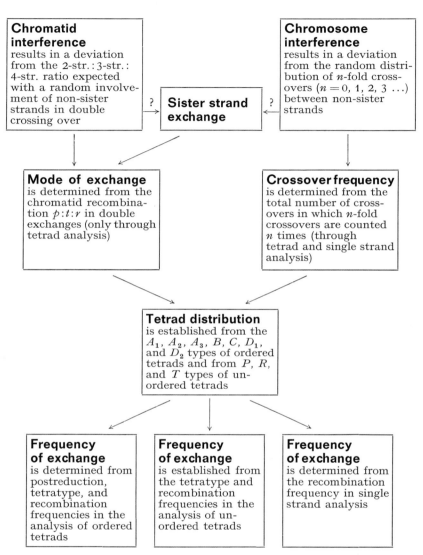

Fig. IV-12. The interaction of chromatid and chromosome interference and sister strand exchange and its effect on the recombination of markers. For explanation see text

One can see from this figure that chromatid interference and sister strand exchange determine the *type of exchange* while chromosome interference influences the *frequency of crossing over*. The *tetrad distribution* is established through the frequency and type of crossing over and in turn determines the postreduction, tetratype, and recombination frequencies. Whether or not chromatid and chromosome interference are restricted to non-sister strand crossing over is not known. Therefore, we have shown in the figure the possible influence of interference on recombination processes within sister chromatids.

Summary

1. Chromatid interference influences the type of exchange; it occurs when the relative frequencies of double exchanges deviate significantly from the random expectation of 2-str. : 3-str. (type I) : 3-str. (type II) : 4-str. = 1 : 1 : 1 : 1. Two-strand exchanges occur preferentially in all cases of chromatid interference, i.e. fewer four- than two-strand doubles are found. Moreover, three-strand doubles are less frequent than expected on a random distribution.

2. Chromosome interference has an influence on the frequency of exchange. The most useful genetic measure of this type of interference is the coincidence value. Chromosome interference across the centromere has been demonstrated only in rare cases. In contrast, a positive interference is usual within chromosome arms, i.e., centromere-free regions.

3. The relationships between chromatid and chromosome interference and sister strand exchange are diagrammed in Fig. IV-12.

IV. Chromosome maps

A graphical linear representation of chromosomes in which the markers of a linkage group are arranged according to the relative distances between them is called a chromosome or linkage map. The frequency of crossing over between two linked markers serves as a measure of the map distance between them. The construction of accurate linkage maps depends primarily upon the precision with which the number of crossovers can be determined from the recombination data.

The first linkage maps were constructed by MORGAN and his coworkers for *Drosophila melanogaster* (MORGAN, 1911a, b; reviewed in BRIDGES and BREHME, 1944). Maps for the other classical genetic materials (mouse, maize, peas, etc.) followed shortly. Mapping of markers in mosses, fungi, algae, bacteria, and viruses has only recently been undertaken.

1. Map units and map distance

Mapping involves the arrangement of markers at particular sites (loci) on a linear structure with specific distances between them. Such an arrangement is based on the following considerations:

1. *The frequency of exchange between two markers remains constant under the same experimental conditions* (p. 168). As an index of the frequency of exchange, the recombination ditype, as well as the tetratype, and postreduction frequency can be used. In the latter two cases tetrad analysis is required.

2. *With three linked markers, a, b, and c, the largest of the three exchange values, r_{a-b}, r_{a-c}, r_{b-c}, is always less than or equal to the total of both smaller values*, i.e. $r_{a-b}+r_{b-c}\geqq r_{a-c}$ if the exchange between a and c is more frequent than between the two other pairs of markers. In general, the relation, $r_{a-b}+r_{b-c}>r_{a-c}$, holds; only with close linkage or complete positive interference (p. 197f.) is $r_{a-b}+r_{b-c}=r_{a-c}$. An indisputable sequence for any three markers can be established on the basis of this relationship; further, the third marker always lies between the two which exhibit the greatest frequency of exchange. For example, the sequence a—b—c or c—b—a follows from $r_{a-b}+r_{b-c}\geqq r_{a-c}$.

Markers cannot be precisely located from the exchange frequencies, since the recombination and tetratype frequencies are not generally additive, i.e. the relation, $r_{a-b}+r_{b-c}>r_{a-c}$, holds in most cases. The center marker (b) can be located at two different loci, namely in relation to its adjacent marker on the "left" (a or c) or on the "right" (c or a).

3. *The usual lack of "additiveness" of the exchange frequencies is the result of multiple crossing over.* If crossovers occur simultaneously in each of two adjacent regions (e.g. a—b and b—c) a recombination between neighboring genes as well as a parental combination for the two end markers (a and c) occurs.

For example, tetratypes for adjacent genes and parental ditypes for the end markers result from two-strand crossing over (Fig. IV-11). In single strand analysis the types ab^+c and a^+bc^+ represent a double crossover, if one assumes that the recombination types were derived from the cross $a^+b^+c^+\times abc$ and the markers lie in the sequence a—b—c.

4. *The values required for exact mapping are the crossover frequencies, since these, in contrast to the exchange frequencies, are additive,* i.e. the sum of the crossover frequencies for two adjacent regions (e.g. a—b and b—c) is equal to the frequency of the crossing over which took place in the entire region (a—c).

If the markers lie close together so that the exchange frequencies are low, the recombination values may generally be added. By the addition of such small values in *Drosophila* MORGAN (1926) was able to locate exactly a large number of markers, some of which are far apart. The distance between two markers equivalent to an exchange value of 1 % later came to be called a Morgan unit.

The frequency of breaks occurring in a chromosome region can be used as a measure of distance between two markers instead of the crossover frequency (p. 148). The same dimension as the Morgan unit of classical genetics was selected for determining distances. Like the Morgan unit, the *map unit is equivalent to 1% breaks.*

Crossing over and breakage frequencies differ by a factor of two, since for each *crossover* in a tetrad only two of the four strands are involved in a *break*. The same relationship exists between postreduction or tetratype, and recombination frequencies (p. 198).

The mapping of markers thus involves the problem of determining the frequency of crossing over or breaks. Under the simplest assumptions these can be calculated from the corresponding exchange frequencies (p. 163 f.). These assumptions are inadequate for the exact determination of the number of crossovers. The following questions are pertinent in this connection; these can be answered only partially (see page references).

1. *What mechanism is responsible for the recombination of markers* (p. 234 ff.)? Are always two strands involved in an exchange? Does an exchange always lead to reciprocal recombinations (p. 222 ff.)?

2. *Which strands are involved in crossing over?* Does an exchange occur among any two strands or only between non-sister strands (p. 185)? Does strand involvement in double crossovers follow a random distribution or are some strands involved preferentially (p. 179 ff.)?

3. *How frequently does crossing over occur?* Are crossovers distributed at random along the total length of the tetrad or do they occur more or less frequently in certain regions? Are multiple crossovers randomly distributed or does a crossover influence exchanges in neighboring regions (p. 187 ff.)? Do sister strand and non-sister strand exchanges occur independently of one another or does interference occur between the two processes (assuming that sister strand exchange is a general phenomenon)?

4. *What is the relation between inter- and intragenic recombination processes?* Is the same, or are different mechanisms involved (p. 223 ff.)?

2. Interference models and mapping functions

The functional relationship between crossing over and the experimentally confirmed exchange of markers can be understood under simplified assumptions with the aid of mathematical interference models. The mathematical formulation of this relationship leads to mapping functions which allow a localization of markers and also permit a determination of the degree of chromosome interference. It has frequently been assumed in deriving interference models that no chromatid interference or sister strand exchange occurs. We know now that these assumptions are questionable (p. 185 and 193).

a) Single strand analysis

Since single strand or random analysis distinguishes only between parental and recombinant types (p. 158), empirical and theoretical mapping functions have been derived from which the distance between two genes can be determined from the frequency of recombination. The method developed by MORGAN (1926) which utilizes recombination between closely linked markers to determine the distance between more widely spaced genes (p. 196) can only be used when sufficient markers are known for an organism. In all other cases a direct proportionality exists between recombination frequency (ϱ) and breakage frequency (x) only when *positive chromosome interference is complete* ($\varrho = x$). HALDANE (1919, 1931) first set up a mapping function for *zero interference* utilizing the Poisson distribution (p. 187), namely $\varrho = \frac{1}{2}(1 - e^{-2x})$. The empirical and theoretical mapping functions derived for all degrees of positive interference lie between these two limits (complete positive interference and the absence of interference) (see BARRATT et al., 1954).

Empirical functions can only be established for organisms which have been investigated thoroughly genetically. In such materials by using a large number of markers the chromosome can be mapped in such detail that the detection of every crossover or break is highly probable. Empirical functions have been derived, for example, for *Drosophila* (LUDWIG, 1934), for the mouse (CARTER and FALCONER, 1951), and for certain plants (KOSAMBI, 1944) (literature in OWEN, 1950, and BARRATT et al., 1954).

Theoretical mapping functions and interference models have been developed and discussed under a variety of assumptions, as by RADEMACHER (1932), FISHER et al. (1947), OWEN (1950), SPIEGELMAN (1952), CARTER and ROBERTSON (1952), BARRATT et al. (1954), FISHER (1955), PERKINS (1955), PAYNE (1956, 1957), WALLACE (1957), PAPAZIAN (1960), KUENEN (1962a) and KEMPER (1964).

b) Tetrad analysis

Mapping functions are also used for localizing markers in organisms in which the meiotic products remain together as tetrads. They allow the calculation of crossing over frequencies from postreduction and tetratype frequencies. The postreduction values give the relative location of the centromere. Such mapping of the centromere was first carried out by LINDEGREN (1933) (p. 140). He chose one-half the postreduction value as the distance between the centromere and a particular marker (p. 196). One can determine the distance between two marker genes in the same way from the corresponding one-half of the tetratype frequency. However, this method generally leads to imprecise mapping, since the mapping function employed ($t = 2x$) holds only when *complete positive chromosome interference* exists (t = postreduction or tetratype frequency, x = breakage frequency, and $2x$ = crossover frequency). RIZET and ENGELMANN (1949) and PAPAZIAN (1951) independently derived a mapping function for *zero interference* from the Poisson distribution, namely $t = {}^2/_3 (1 - e^{-3x})$. Two limits exist, just as for single strand analysis, one for the absence of interference and one for complete interference; the empirical and theoretical mapping curves for all degrees of interference lie between these.

An interference model for *unordered* tetrads was first developed by BARRATT et al. (1954), and for *ordered* tetrads by KUENEN (1962a). Both models allow the derivation of mapping functions for different degrees of interference and the graphical and arithmetical determination of the degree of chromosome interference from the tetrad distributions. The assumptions for both models are essentially the same: 1. chromatid recombination $p:t:r = 1:2:1$ (p. 179), 2. a regular effect of chromosome interference between adjacent non-sister strand crossing over (p. 188ff.). In the model for unordered tetrads the tetratype frequency is represented as a function of the crossover frequency and the coincidence factor; the latter serves as a measure of interference. In the model for ordered tetrads in addition to these two values the maximum crossover frequency is taken into consideration, i.e. the frequency of crossing over in the longest interval of a linkage group in which chromosome interference is uniform (= interference region).

The mapping function, $x = -\frac{1}{3}Q^{-1} \ln (1 - \frac{3}{2}Qt)$ is derived with the aid of the interference model for ordered tetrads (KUENEN, 1962a). The function $\tau(x)$ used in Fig. IV-13 gives the same relationship for the particular case in which $t = \tau(x)$ (for the meaning of the symbols see Fig. IV-13). The interference value Q is a function of the coincidence factor and the maximum breakage frequency. Nevertheless, the interference value, which is analogous to the coincidence value (p. 187), can be calculated directly from the data of a two-factor cross.

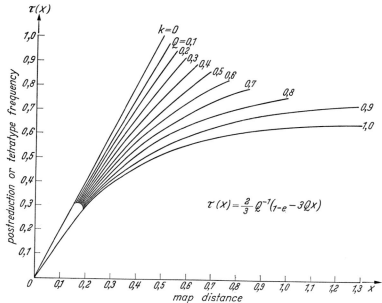

Fig. IV-13. Graphical representation of tetrad mapping functions for eleven different degrees of interference ($k = 0$; complete interference; $Q = 0.1$ to $Q = 0.9$: positive interference; $Q = 1.0$: no interference, i.e. distribution random). For explanation see text. (From KUENEN, 1962a)

An example of the calculation of map distances is shown below. The data have been selected from a cross with *Podospora anserina* ($t_1 \times i$ in Table IV-7, p. 164f.).

1. *Determination of the interference value Q.* It can be shown that the two markers t_1 and i lie on the same side of the centromere on the basis of the linkage criteria presented in Table IV-6 (compare with Table IV-7). In this instance the interference value is:

$$Q = \frac{a_3 + b}{(a_1 + a_2 + a_3 + b)(a_3 + b + c)}.$$

The small letters in this equation represent the frequencies of corresponding tetrad types A_1, A_2, A_3, B, C (compare Fig. IV-5, p. 151). Thus

$$Q = \frac{0.010 + 0.003}{(0.109 + 0.002 + 0.010 + 0.003)(0.010 + 0.003 + 0.649)}$$

$$= \frac{0.013}{0.124 \cdot 0.662} = \boxed{0.16}.$$

2. *Determination of the map distance between the centromere and the markers t_1 and i.* In the mapping function shown above t represents the post-reduction frequency for t_1 and i. Table IV-7 shows the values 0.124 (t_1) and 0.770 (i). Thus the map distance between the centromere and t_1 is:

$$x_1 = -\tfrac{1}{3} \cdot 0.16^{-1} \ln (1 - \tfrac{3}{2} \cdot 0.16 \cdot 0.124)$$
$$= -2.083 \ln 0.97024 = \boxed{0.0628}$$

and the map distance between the centromere and i:

$$x_2 = -\tfrac{1}{3} \cdot 0.16^{-1} \ln (1 - \tfrac{3}{2} \cdot 0.16 \cdot 0.770)$$
$$= -2.083 \ln 0.8152 = \boxed{0.4254} .$$

3. *Determination of the map distance between t_1 and i.* In this case t represents the frequency of the tetratypes. The value, $t = 0.662$ is obtained from $T = 407$ (Table IV-7). Therefore, the map distance is:

$$x_3 = -\tfrac{1}{3} \cdot 0.16^{-1} \ln (1 - \tfrac{3}{2} \cdot 0.16 \cdot 0.662)$$
$$= -2.083 \ln 0.84112 = \boxed{0.3631} .$$

The calculated map distances are additive. Since the centromere lies beyond the region marked by t_1 and i, the relationship $x_2 - x_1 = x_3$ holds, or in specific values: $0.4254 - 0.0628 = 0.3626 \approx 0.3631$. It is readily apparent in Fig. IV-13 that the points P_1 (0.0628/0.124), P_2 (0.4254/0.770), and P_3 (0.3631/0.662) lie between the two curves, $Q = 0.1$ and $Q = 0.2$; i.e. for the ordinate values (postreduction and tetratype frequencies) the calculated values on the abscissa (map distances) are found on the curve for $Q = 0.16$.

The mapping function developed by BARRATT et al. (1954), in contrast to KUENEN's, does not give a specific value for breakage frequency. Therefore, direct calculation of map distance cannot be carried out. Greater dependence must be placed on the graphical representation of the mapping function. The same shortcoming applies to the function derived by JOUSSEN and KEMPER (1960) for unordered tetrads.

SHULT and LINDEGREN (1956a, b) likewise describe mapping functions for unordered tetrads. Linked, as well as unlinked markers are considered. These authors investigated not only the effect of chromosome interference but also the effect of chromatid interference and sister strand exchange on the distribution of crossovers (compare also PERKINS, 1955).

All the models for mapping presented here are based on the unproven assumption that recombination of genetic material results from a "crossover-like" mechanism. Moreover, this is not the only unproven assumption, as we have already emphasized (p. 193f. and 197). However, if all processes which might influence mapping were taken into consideration the number of unknown values in a mathematical model would be too comprehensive. Such a model would be too complex and unwieldy to be of practical value in mapping genes.

3. Linkage groups and chromosome maps

In the fungi the mapping of markers by the "*Drosophila* method" of MORGAN (p. 196) is generally not feasible because the number of markers available is too few and the distance between them too great. Localization with the aid of exchange frequencies as a rule leads to imprecise linkage maps. For this reason the mapping functions described above are frequently used (p. 197f.). These permit a straightforward mapping of genes and centromeres if the degree of chromosome interference is known. Linkage maps have been worked out for many organisms (Table IV-20). Nevertheless, these are frequently fragmentary, especially when a particular organism has not been sufficiently analyzed genetically. In certain cases it can be shown from comparative genetic and cytological investigations that not all the linkage groups (chromosomes) of an organism are represented by marker genes; for example, in *Venturia inaequalis* there are seven chromosomes, but only six linkage groups have been identified (DAY et al., 1956; KEITT and BOONE, 1956). Maps for the seven chromosomes of two well-studied organisms, *Neurospora crassa* and *Podospora anserina* are shown in Figs. IV-14 and IV-15. The number of chromosomes determined cytologically for both organisms corresponds to the number of linkage groups (*N. crassa:* MCCLINTOCK, 1945; SINGLETON, 1953; BARRATT et al., 1954; *P. anserina:* FRANKE, 1962; KUENEN, 1962b).

Table IV-20. *Organisms in which genetic markers have been mapped:*

In certain cases the number of linkage groups is smaller than the number of chromosomes observed cytologically. This results from too few markers being known. For the same reason certain chromosome maps remain fragmentary (+).

object	reference
Neurospora crassa	BARRATT et al., 1954; MITCHELL and MITCHELL, 1954; STADLER, 1956a; PERKINS, 1959; PERKINS and ISHITANI, 1959; MALING, 1959; STRICKLAND et al., 1959; PERKINS et al., 1962; PERKINS and MURRAY, 1963
Podospora anserina	ESSER, 1956, and unpublished; KUENEN, 1962b
Sordaria macrospora	HESLOT, 1958; KEMPER, 1964
Sordaria fimicola (+)	EL-ANI et al., 1961; PERKINS et al., 1963
Glomerella cingulata (+)	WHEELER, 1956
Aspergillus nidulans	KÄFER, 1958; PONTECORVO and KÄFER, 1958; STRICKLAND, 1958b
Venturia inaequalis (+)	BOONE and KEITT, 1956; DAY et al., 1956
Saccharomyces cerevisiae	LINDEGREN, 1949a; SHULT and LINDEGREN, 1955, 1956b; LINDEGREN et al., 1959; DESBOROUGH and LINDEGREN, 1959; HAWTHORNE and MORTIMER, 1960; DESBOROUGH et al., 1960; PAPAZIAN and LINDEGREN, 1960; MORTIMER and HAWTHORNE, 1963; LINDEGREN et al., 1963; HWANG et al., 1963
Schizosaccharomyces pombe (+)	LEUPOLD, 1950, 1957, 1958
Schizophyllum commune (+)	RAPER and MILES, 1958
Coprinus lagopus	DAY, 1960; LEWIS, 1961; DAY and ANDERSON, 1961; PRÉVOST, 1962
Ustilago maydis (+)	HOLLIDAY, 1961a, b

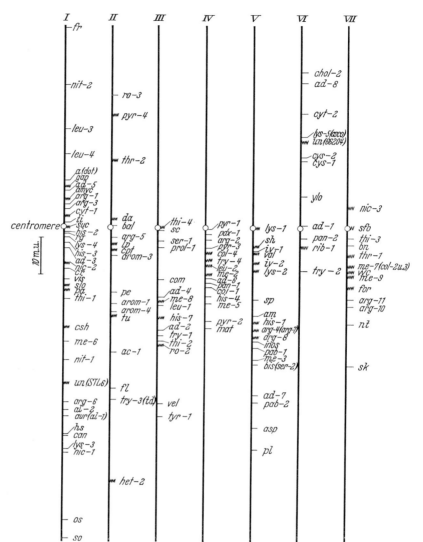

Fig. IV-14. Chromosome maps of the seven linkage groups of *Neurospora crassa*. The heavy vertical lines represent the chromosomes, the open circles the centromeres. The left arms of the chromosomes are oriented upward, the right downward. Genes whose positions have been confirmed are designated by the short horizontal straight lines. The short zig-zag lines represent genes the positions of which have not yet been determined precisely. Localization of markers is based mostly on the analysis of unordered tetrads or single spores, less frequently on the analysis of ordered tetrads; that is, the arrangement of genes into linkage groups is generally based upon the determination of map distances between markers and only in part on the determination of distances between markers and their centromeres. The first review of the mapping of markers was carried out by BARRATT et al. (1954). We have included here, as far as possible, all genes which are mentioned in the present volume (for references see Table IV-20). A compilation of all known genes arranged into the seven linkage groups (without mapping) can be found in BARRATT and OGATA (1964a, b)

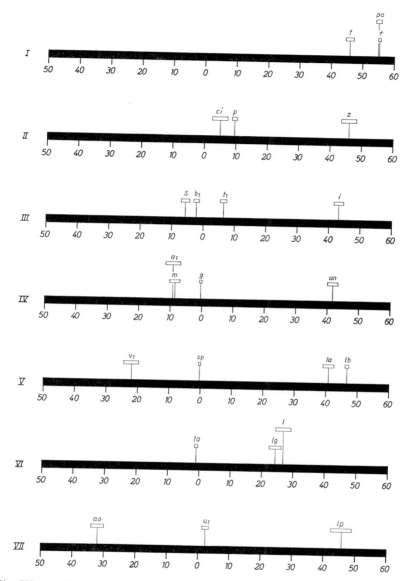

Fig. IV-15. Chromosome maps for the seven linkage groups of *Podospora anserina*. The chromosomes are represented by the heavy black lines. The mapping functions developed by KUENEN (1962a) were used for mapping the markers (Fig. IV-13). The centromeres were used as reference points (marked with 0) for locating the markers, since the localization is based entirely on analysis of ordered tetrads (postreduction frequencies). The scale shows the map distances from the centromeres to the genes, which are designated by the lines oriented upward. The margin of error is shown by the small rectangle. (From KUENEN, 1962b, supplemented by unpublished data of ESSER)

Summary

1. Genetic markers can be arranged linearly into linkage maps on the basis of recombination frequencies. The map distance between two markers is defined as the average number of breaks between these markers in one hundred single strands.

2. Map distances may be calculated with the aid of mapping functions directly from recombination frequencies, i.e. breakage or crossover frequencies. Such mapping functions have been derived for using data from single strand analysis as well as those from tetrad analysis.

V. Somatic recombination

As already mentioned in the introduction to this chapter (p. 136), recombinational events may occur in somatic cells as well as during meiosis. While interchromosomal somatic recombinations seem to result only from a redistribution of entire chromosomes over a number of successive irregular mitoses, intrachromosomal recombinations may involve meiosis-like phenomena (in rare cases) as well as mitotic processes (for references see Table II-6).

Following infection experiments with heterogenic dikaryons of *Puccinia graminis tritici* ELLINGBOE (1961) isolated a series of strains from the host plants which represented new combinations of individual markers. Since the exchange frequencies exceeded those known for mitotic recombination, ELLINGBOE's interpretation was that meiosis-like exchange phenomena took place during vegetative reproduction.

The results of WILKIE and LEWIS (1963) with *Saccharomyces* also suggest a partial meiotic behavior for certain nuclear divisions. These authors believe that this type of nuclear division is induced by UV radiation (p. 176). The observed frequency of recombination of linked genes argues against the conventional meiotic recombinational mechanism, namely, crossing over in the four-strand stage. WILKIE and LEWIS believe that crossing over in the two-strand stage is more probable.

Meiosis-like recombinations likewise seem to be involved in *Schizophyllum commune* (PARAG, 1962; ELLINGBOE and RAPER, 1962; ELLINGBOE, 1963, 1964; MIDDLETON, 1964).

Mitotic recombination was first observed by STERN (1936) in diploid heterozygous somatic cells of *Drosophila*; in the 1950's it was also discovered in fungi (*Aspergillus, Penicillium*) (see Table II-6 for references). *The significance of this discovery lies primarily in the fact that somatic recombination provides an alternative to sexual reproduction which allows organisms unable to reproduce sexually to produce new combinations of their genetic material* (see chapter on reproduction). Thus genetic investigation of the imperfect fungi becomes possible.

1. Mitotic crossing over

Intrachromosomal recombination in somatic cells had previously been assumed to occur by means of a mechanism similar to meiotic crossing over (STERN, 1936; PONTECORVO et al., 1953; PONTECORVO and ROPER,

1953; ROPER and PRITCHARD, 1955; PONTECORVO, 1958; PONTECORVO and KÄFER, 1958). This mechanism was referred to as mitotic crossing over. The assumption was not confirmed until the experiments of KÄFER (1961) on *Aspergillus nidulans* and of HOLLIDAY (1961 b) on *Ustilago maydis*. These workers analyzed diploid heterozygous nuclei which arose

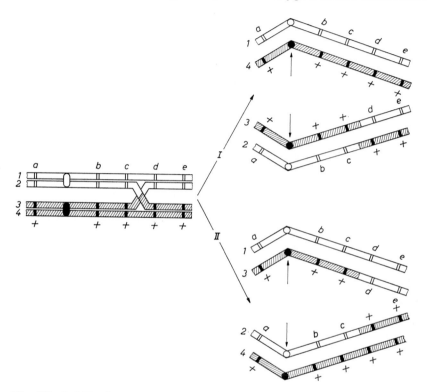

Fig. IV-16. Mitotic crossing over in a diploid somatic cell and the distribution of chromosomes in the subsequent mitosis. Above (I): One daughter nucleus receives the parental strands, the other the recombinant strands. Below (II): Each of the daughter nuclei receives one parental and one recombinant strand. Chromosomes, markers, and centromeres are represented as in Fig. IV-2. Explanation in text. (Adapted from PONTECORVO, 1958)

from the fusion of genetically different haploid nuclei within a heterokaryon during the parasexual cycle (p. 92 f.). They analyzed the two diploid daughter nuclei produced in the first mitotic division of a newly-arisen recombinant. In some cases each of the two nuclei contained one parental and one recombinant strand (Fig. IV-16, below), while in other cases the two reciprocal recombinant strands were in one nucleus and the two parental strands in the other (Fig. IV-16, above).

Recombination

These and other results led to the following conclusions: 1. *Mitotic, like meiotic crossing over occurs in the four-strand stage. 2. Only two of the four strands take part in any one exchange. 3. Mitotic recombination is reciprocal.*

Although mitotic recombination occurs in diploid nuclei in *A. nidulans, U. maydis* and certain other fungi, in *N. crassa* it very probably takes place in disomic $(n + 1)$ nuclei. A haploidization follows here as in the diploid nuclei (MITCHELL et al., 1952; PITTENGER, 1954, 1958; PITTENGER and COYLE, 1963).

Unexpected segregations may result in crosses, if mitotic recombination occurs shortly before meiosis (*N. crassa:* MITCHELL, 1963).

The spontaneous occurrence of mitotic recombination is rare. Therefore, analyses such as carried out by KÄFER and HOLLIDAY to prove the occurrence of reciprocal crossing over are tedious and time-consuming; for this reason agents which increase the frequency of crossing over have been used (p. 174ff.). Generally selective methods are employed to detect the rare haploid or diploid recombinants among the large number of non-recombinant cells. For selecting *diploid recombinants* markers are used which when homozygous produce a phenotype which is easily distinguished from the heterozygous parental type. This permits identification of a recombinational event by a sector (p. 93). A crossover between a marker and its centromere has one-half chance of yielding a nucleus homozygous for that marker, if chromatids assort at random (Fig. IV-16, below; selected marker, *e* or *d*).

PONTECORVO (1958) and PONTECORVO and KÄFER (1958) describe five different possibilities of selection which can be used relatively easily for genetic analyses: 1. Selection for genes with morphological effects (e.g. conidial color, markers *y* and *w*, Fig. IV-17). 2. Selection for recessive suppressor genes (e.g. marker *su-1-ad-20*). 3. Selection for recessive or semi-dominant resistance against toxic agents (e.g. *Acr-1* and *acr-2*). 4. Selection for genes which block the synthesis of particular compounds. 5. Selection for differential heterozygosis. Such a selection is possible only when two genes express different phenotypes in the *cis* and in the *trans* configuration (e.g. cd/c^+d^+ and cd^+/c^+d in Fig. IV-16, below).

Recombinations in haploid nuclei which arise through a stepwise loss of chromosomes from diploid nuclei (p. 93) can be detected by the mycelial sectors which develop from the division of the recombinant nuclei or by uninucleate conidia. Conidial or mycelial characters are frequently used as markers here also (e.g. color mutants). In this case, however, a single "selector marker" does not suffice, since the centromere cannot serve as a marker. Thus only recombinations in regions which are marked by two genes can be detected in haploid nuclei.

The frequency of mitotic crossing over can be increased appreciably with UV treatment (*Ustilago maydis:* HOLLIDAY, 1961b, *Saccharomyces cerevisiae:* FOGEL and HURST, 1963), X-radiation (*Aspergillus nidulans:* MORPURGO, 1962a), and certain chemical agents (*Aspergillus nidulans:* MORPURGO, 1962b, 1963) (p. 174ff.). In such cases selective methods for isolating recombinants may not be necessary.

Let us turn now to the question of how frequently mitotic crossing over occurs. KÄFER (1961), by using distal markers found that the percentage of homozygotes arising from mitotic recombination between the centromere and "selector marker" (e.g. *d* or *e* in Fig. IV-16) ranged from 0.05 to 0.07 for different chromosome arms of *A. nidulans*. Assuming that the distribution of chromosomes during mitosis occurs at random, i.e. that for each homozygous type (Fig. IV-16, below) there is a corresponding heterozygous type (Fig. IV-16, above) which is not detected because of selection, a *recombination frequency of 2% per mitosis is obtained*.

This calculation is based on the assumption that only fourteen approximately equal arms are present in the haploid chromosome complement ($n = 8$; in two chromosomes the centromeres are located terminally). With a frequency of homozygotes of 0,07 %, i.e. a recombination frequency of 2×0.07 or 0.14 % per chromosome arm, the value for the entire genome is obtained by multiplying by the number of chromosome arms ($14 \times 0.14 \approx 2$).

Double exchanges within an arm are observed only rarely, since for an average of 50 mitoses only one crossover occurs. Interference between crossovers in different arms of the same chromosome or different chromosomes could not be demonstrated (*A. nidulans:* KÄFER, 1961).

2. Chromosome maps

Mitotic, like meiotic recombination can be used to map chromosomes. Construction of chromosome maps involves three steps (PONTECORVO and KÄFER, 1958; KÄFER, 1958):

1. The linkage of markers is determined through analysis of haploid nuclei arising from the parasexual cycle.

2. The sequence of markers in the arms of chromosomes and the map distances between them are determined by comparison of crossover frequencies.

3. It is possible to determine which two arms belong to the same chromosome by combining steps 1 and 2; thus the position of the centromere can be established.

a) Linkage groups

Markers of the same linkage group almost always remain together during mitotic divisions and subsequent stepwise haploidization, because of the rarity of mitotic crossing over. In contrast, markers of different linkage groups combine at random during the reduction from diploid to haploid chromosome complement, i.e. half are parental combinations and half are recombinants. Therefore, *if two markers are linked, the recombination frequency is zero or near zero, while with non-linkage the recombination frequency is approximately 50%*.

The localization of a given marker is particularly easy if a partner strain is used in which the linkage groups are already known and each is marked by at least one gene. The analysis of uninucleate haploid conidia which are smaller than the diploid (e.g. *A. nidulans:* ROPER, 1952) reveals that the marker combines at random with the markers of all other chromosomes and only fails to show recombination with the markers of the chromosome on which it lies.

A large number of genes in *Aspergillus nidulans* have been arranged into eight linkage groups by this method (KÄFER, 1958). Two linkage groups have been identified for *Ustilago maydis* (HOLLIDAY, 1961a, b). These results agree with those from analysis of crosses and cytological investigations (e.g. ELLIOTT, 1960a). Linkage of markers has also been established in other fungi by analysis of mitotic recombination, e.g. *Penicillium expansum* (BARRON, 1962), *Saccharomyces cerevisiae* (FOGEL and HURST, 1963) and *Schizophyllum commune* (ELLINGBOE, 1964).

In *Aspergillus fumigatus*, however, only three out of twenty-six markers proved to be linked (STRØMNAES and GARBER, 1962). These investigators assumed that genetically controlled selection processes acting during the parasexual cycle interfere with the detection of linkage.

The analysis of recombination in mitosis rather than in meiosis has an advantage in that *linkage can always be proved unequivocally*. In the analysis of meiotic recombination linkage is frequently not excluded even if markers recombine at random (Table IV-6, p. 162 and IV-7, p. 164—165). Random assortment in mitotic recombination *always* indicates non-linkage.

b) Sequence of genetic markers

With a mitotic crossover in a heterozygous diploid nucleus, all markers which lie distal to the point of exchange become homozygous in half of the cases (Fig. IV-16, below); thus the sequence of markers can be determined from the homozygous-heterozygous relationship.

For example, if a chromosome arm is marked by *b*, *c*, *d*, and *e* as in Fig. IV-16, and *e* is the marker being selected (p.206), a crossover between *e* and the centromere gives a homozygote for *e*, thus allowing its selection (Fig. IV-16, below). All markers distal to the exchange site (e.g. *d*) will become homozygous, while all those between the centromere and the site of the crossover will be heterozygous. Since double crossovers are improbable (p. 207), one can conclude conversely that the markers that have been shown to be heterozygous lie closer to the centromere than those which occur in the homozygous condition.

Such a *determination of marker sequence is possible*, however, *only for individual chromosome arms* and not for entire chromosomes; a crossover on the opposite side of the centromere, i.e. in another chromosome arm (between marker *a* and the centromere in Fig. IV-16) cannot become homozygous for the "selector marker" (*e*). Thus by analyzing diploid mycelia or conidia one cannot differentiate between two arms of the same chromosome or two arms of different chromosomes (PONTECORVO and KÄFER, 1958). This difficulty can easily be circumvented, since the two arms with markers belonging to the same linkage group belong together (p. 207). *The position of the centromere is thereby fixed.*

For example, if the sequence, *centromere-a-b-c*, has been established in one arm and *centromere-d-e-f* in the other, then the centromere can only lie between *a* and *d*, assuming that all six markers are linked with one another (PONTECORVO and KÄFER, 1958).

c) Map distance between genetic markers

The frequency of crossing over is used as a measure of map distance between two markers just as in the case of "meiotic" chromosome maps. It is not possible to determine the absolute frequency of crossing over because selective methods are generally employed for the isolation of mitotic recombinants; therefore, *the map distances are relative*. The relative "mitotic" map distances of a chromosome arm shown in Fig. IV-17

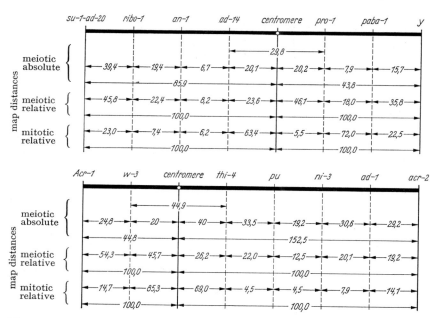

Fig. IV-17. Chromosome maps for linkage group I (above) and II (below) in *Aspergillus nidulans*. The chromosomes are represented as heavy horizontal lines. The distances between the markers (indicated by broken vertical lines) are based on the one hand on analysis of meiotic recombination (compare Fig. IV-14 and IV-15), and on the other, on analysis of mitotic recombination. For details see text. (Adapted from KÄFER, 1958)

indicate how often a marker distal to the centromere is homozygous and how often the proximal marker is heterozygous among 100 "selector homozygotes" (compare Fig. IV-16). *Map distances from different chromosome arms are not comparable.*

A comparison between "mitotic" and "meiotic" chromosome maps is possible within limits. The results with *Aspergillus nidulans* (Fig. IV-17) and *Ustilago maydis* (PONTECORVO and KÄFER, 1958; KÄFER, 1958, 1961, and HOLLIDAY, 1961 a, b) are informative on this point.

1. The sequence of markers is the same in the mitotic and meiotic chromosome maps (Fig. IV-17).

2. Crossing over is approximately one thousand times less frequent in mitotically dividing than in meiotically dividing nuclei.

3. The relative frequencies of mitotic crossing over in similarly marked regions differ sharply from the corresponding frequencies of meiotic crossing over (Fig. IV-17).

As shown in Fig. IV-17 mitotic crossing over is relatively more frequent than meiotic, for example, in the vicinity of the centromere (with the exception of the right arm of chromosome I). Approximately 60—70 % of all mitotic crossing over is concentrated in one interval in the left arm of chromosome I and in the right arm of chromosome II, while the same intervals show only one quarter of the meiotic crossing over.

We have seen that mitotic crossing over has many characteristics in common with meiotic crossing over. It remains to be seen whether the identity of the two mechanisms can be demonstrated with further experiments.

Summary

1. Intrachromosomal recombination occurs not only during meiosis, but in rare cases during nuclear divisions in vegetative cells. Such somatic recombinations generally take place during mitosis in diploid cells (mitotic recombination). In isolated instances they may also result from meiosis-like processes.

2. Mitotic recombination results from a mechanism which, like meiotic crossing over, leads to reciprocal segregations. It is thus designated as mitotic crossing over. Interference between crossovers has not been observed either in the same or in different chromosomes.

3. Mitotic recombination is rare. It is generally detected by the use of selective methods.

4. Genes may be localized on chromosomes on the basis of mitotic, as well as meiotic recombination. However, "mitotic" chromosome maps differ from "meiotic" in the relative frequencies of crossing over in similarly marked regions.

C. Intrachromosomal intragenic recombination

The results of recombination and mutation studies of the 1950's have led to a rejection of the classical concept of the gene (gene = the unit of recombination, mutation, and function). Such studies showed that *the functional unit may mutate at different sites which may recombine with one another*[1]. Such an extensive subdivision of the genetic material

---

[1] BENZER (1957) introduced the term *"cistron"* to designate the functional unit on the basis of his investigations of bacteriophage. The assignment of mutants to a cistron is accomplished by the cis-trans test as follows: If the individual defects of two mutants, a and b, are not alleviated in the trans configuration (ab^+/a^+b) when the two genomes are introduced into the same cell (in phage: after infection of a host cell with two types of phage;

became possible through the development of selective methods that enabled detection and quantitative determination of rare mutational and recombinational events.

I. Fine structure of the gene

1. Gene maps

The mutational sites of a gene may be localized and arranged in a linear sequence just as genes are mapped on chromosomes (p. 195 ff.). *By the term "gene map" is meant the graphical representation of small chromosome segments in which the mutational sites (alleles) of a particular functional unit (gene) are arranged according to the relative distances between them.* The recombination frequency between two sites is used as a measure of distance. Such recombination is *intragenic*, in contrast to the intergenic type, which occurs between mutational sites of different genes.

The different configurations of a gene are designated *alleles*, whether or not they produce identical or different phenotypes. If two alleles are different, one refers to *heteroalleles* (ROMAN, 1956). For proof of heteroallelism, i.e. the non-identity of alleles, the mutational, recombinational, and functional characteristics of the genetic material must be considered:

1. *Recombination.* If two mutants, a_1 and a_2, which have arisen independently of one another are crossed, wild types resulting from recombination between a_1 and a_2, are generally obtained at a low frequency (10^{-3} to 10^{-7}; Table IV-21). On the other hand, identical mutations are unable to recombine. Nevertheless, wild types may be expected from the cross of identical mutants also, but as a result of back mutation the frequency of which is generally much lower than that observed for recombination of heteroalleles.

2. *Back mutation.* Since every mutant allele has a characteristic reversion rate, differences in mutation rates are an indication of heteroallelism (p. 297 ff.). Identity of back mutation rates is, however, not a criterion for the identity of alleles, because non-identical alleles may exhibit the same reversion frequency. Heteroallelism is also indicated when reverse mutants show different physiological and biochemical properties (p. 401 ff.).

3. *Complementation.* Whether or not two mutations occupy the same site can be established by complementation tests (p. 386 ff.). For example, if a_1, but not a_2 complements a_3, then a_1 and a_2 must be heteroalleles. On the other hand, the same complementation behavior is not a criterion for identity of alleles.

---

in fungi: in heterokaryotic or heterozygous cells), the two mutants represent alterations in the same functional unit. The wild phenotype, or one close to it, is generally produced in the cis configuration (ab/a^+b^+). Because the term "cistron" is identical to the current concept of the gene (gene = functional unit), we have continued to use the term "gene" throughout this volume.

Recombination

Table IV-21. *Results of crosses between allelic auxotrophic mutants*

All crosses are shown in the form, $b(a_1 a_2^+)c^+ \times b^+(a_1^+ a_2)c$ (compare Fig. IV-19). a_1 and a_2 represent two heteroallelic markers of the a locus; b and c are markers on each side of the a locus. The sequence of the four markers is $b—(a_1—a_2)—c$ as shown in column 2. The combinations of end markers in the table refer to the prototrophic recombinants, $a_1^+ a_2^+$, (random analysis) isolated by means of selective methods. Data from reciprocal crosses or crosses the results of which are essentially in agreement have in some cases

object, linkage group and locus (a)	sequence of markers			map distance from end markers to gene a		frequency of proto-trophs $a_1^+ a_2^+$ ($\times 10^{-5}$)	num-ber of proto-trophs ana-lyzed
	region I $b—$	region II $(a_1—a_2)$	region III $—c$	$b—a$	$a—c$		
Neurospora crassa							
I nic-1	lys-3—(nic¹—nic²)—os			1	19	57	117
	nic³—nic²					55	73
	nic⁴—nic^x					4—25	107
IV me-2	try-4—(meα—meβ)—pan-1			6	4	4—8	545
	meα—meγ					4—73	2900
	meα—meδ					10—110	1181
	meβ—meγ					7—19	1564
	meβ—meδ					18—32	555
	meγ—meδ					10—37	1154
	meγ—meγ					1—7	890
V his-1	iv-2—(his²—his¹)—inos			20	6	1	82
	his¹—his³					3	134
	his²—his³					5	219
V pab-1	inos—(pab¹—pab⁵)—me-3			1	1	12	274
	pab¹—pab⁷					59	421
	pab⁵—pab⁷					29	254
V am	sp—(am²—am³)—inos			7	3	2	90
VI pan-2	try-2—(panB3—panB5)—ylo			8	3	90—630	1000
	panB3—panB3					20—120	860
	panB5—panB5					0.9—3	46
VI cys	lys-5—(cysA—cysB)—ylo			6	9	97—146	1637
	cysB—cysB					5—41	680
	cysA—cysA					0.9—2.5	39
Aspergillus nidulans							
I ad-8	y—(ad¹⁹—ad¹¹)—bi-1			0.2	6	8	268
	ad¹⁶—ad¹⁹					7	243
	ad¹⁶—ad¹¹					21	113
I paba-1	ad-9—(paba^x—paba^y)—y			0.5	16	0.06—0.09	125
						0.1—1.0	716
						1.2—5.1	476
						14—26	930
						40—51	347

been combined. The number of mutational sites designated by a particular symbol (column 2) are as follows: *nic-1*: $x = 3$; *me-2*: $\alpha = 7$, $\beta = 1$, $\gamma = 9$, $\delta = 2$; *pan-2*: $B\,3 = 6$, $B\,5 = 5$; *cys*: $A = 3$, $B = 4$; *paba-1*: $x + y = 14$. The data for the *paba* locus are arranged according to map distances. The calculation of the coincidence values K_1 and K_2 (second from the last and next to the last columns) and their significance are discussed on p. 217 and 218. Further explanation in text. Parallel experiments have been carried out in other organisms, e.g. yeast by KAKAR (1963).

number of prototrophs with				coincidence value		reference
parental combination of end markers and exchanges in region		recombination of end markers and exchanges in region				
I+II type P_1	II+III type P_2	II type R_1	I+II +III type R_2	K_1	K_2	
38	22	35	22	98.9	1.0	ST. LAWRENCE, 1956
22	28	12	11	79.3	0.6	
20	31	41	15	73.7	1.0	
64	235	194	52	39.8	0.9	MURRAY, 1963
331	1208	1025	336	48.3	1.0	
117	635	273	156	54.7	0.9	
335	549	455	225	60.0	0.8	
98	249	128	80	60.0	0.8	
209	555	235	155	56.0	0.7	
169	349	271	101	47.3	0.7	
21	28	13	20	20.3	0.8	FREESE, 1957a
41	43	29	21	13.1	0.7	
53	64	57	45	17.1	0.9	
81	46	136	11	495	0.6	FREESE, 1957b
124	97	179	21	617	0.5	
57	69	122	6	300	0.7	
40	19	18	13	890	0.7	PATEMAN, 1958, 1960a,b
257	242	364	137	62.2	0.9	CASE and GILES, 1958b
124	106	499	131	69.2	2.1	
1	26	10	9	89.1	1.2	
510	346	489	292	37.7	0.9	STADLER and TOWE, 1963
152	183	137	208	56.6	1.0	
18	8	9	4	19.0	0.6	
42	52	167	7	229.5	0.7	PRITCHARD, 1960
53	30	154	6	216.8	0.7	
16	25	71	1	77.5	0.3	
13	37	66	9	9.0	1.1	SIDDIQI, 1962
93	186	416	21	3.7	1.2	
58	137	268	13	3.4	0.6	
25	312	589	4	0.5	0.4	
3	104	236	4	1.4	1.8	

As we have indicated previously (p. 211), alleles within a gene have been mapped in the same way as genes within a chromosome. Two methods are most useful for such intragenic mapping (compare complementation maps, e.g. CASE and GILES, 1960, and p. 388ff.).

1. Mapping of mutational sites according to the distances between them. The frequency of wild type recombinants arising from the cross, $a_1 \times a_2$, serves as a measure of the distance between the two alleles, a_1 and a_2. Since auxotrophic mutants have been utilized for analysis in almost all

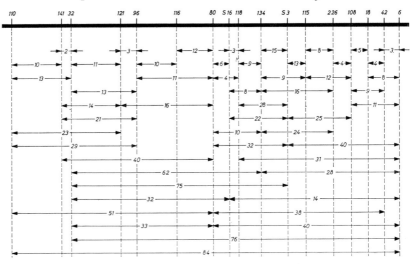

Fig. IV-18. Gene map of the *ad-8* locus of *Neurospora crassa*. Mutational sites are designated by isolation numbers. Map distances between mutational sites indicate the number of prototrophs among 10^5 surviving ascospores. (Adapted from ISHIKAWA, 1962)

cases, the recombinants are prototrophs and therefore easy to isolate (compare selective methods used for isolation of reverse mutants, p. 276f.). Mapping is based on the assumption that the wild types arise through a recombinational process and not by mutation, and further, that the probability of recombination increases with the distance between mutational sites. The units of distance used for constructing gene maps are not identical with those used for chromosome maps (p. 196f.). It has been shown from tetrad analysis that, in contrast to intergenic recombination, intragenic recombination does not necessarily lead to reciprocal products (p. 222ff.).

The gene map of the *ad-8* locus of *Neurospora crassa* (Fig. IV-18) shows that an *unequivocal linear arrangement of the mutational sites of the gene* is possible. On the other hand, the map distances are not always additive. The cause of the non-additivity is probably not double recombination in adjacent regions (as in intergenic recombination), since frequently in the case of three adjacent alleles with the sequence a_1, a_2, a_3, the sum of the two distances a_1—a_2 and a_2—a_3 is less than the distance a_1—a_3. An explanation of this discrepancy may be found by utilizing triply-marked genes (see further in CASE and GILES, 1958b, 1964).

2. *The mapping of mutational sites with respect to the distances to adjacent end markers* (for example, see CATCHESIDE et al., 1964). This method is based on the assumption that the genetic information of a chromosome is arranged in a linear fashion and that the DNA of individual genes does not extend laterally from the main strand. Two markers, *b* and *c*, which lie as close as possible to the left and right ends of the mutant gene, respectively, are generally used. Nearly all the prototrophic recombinants, $a_1^+ a_2^+$, from a cross, $b\,(a_1\,a_2^+)\,c^+ \times b^+\,(a_1^+\,a_2)\,c$ are expected to have the same combination of end markers; this is dependent upon the relative position of the sites a_1 and a_2 (Fig. IV-19, left; type R_1).

For example, if a_1 lies close to *b* and correspondingly, a_2 close to *c*, i.e. the sequence b—a_1—a_2—c (Fig. IV-19, left), then one would obtain $b^+(a_1^+a_2^+)c^+$ as the end marker combination, while if the sequence were the reverse (b—a_2—a_1—c), the combination $b(a_1^+a_2^+)c$.

The data assembled in Table IV-21 show that not only does the expected end marker combination R_1 occur, but that the remaining three combinations P_1, P_2 and R_2 (Fig. IV-19) are also relatively frequent. These results argue for a strong negative interference which will be discussed in the following section (p. 216ff.). In spite of this unexpected result, the mutational sites of a gene can be arranged in a linear fashion on the basis of different numbers of end marker recombinants: One of the two recombinant types (here R_1) is many times more frequent than the other (R_2), as Table IV-21 shows. This method of mapping is illustrated by the following example:

Three mutational sites, *1, 5* and *7*, are known for the *pab-1* locus of *N. crassa* (Table IV-21). The end markers *inos* and *me-3* both lie 0.9 map units from the *pab* gene. Since $R_1 > R_2$, the sequence in the first case is *inos-pab¹-pab⁵-me*, in the second case, *inos-pab¹-pab⁷-me*, and in the third, *inos-pab⁵-pab⁷-me*. These three observations indicate that the marker sequence is *inos-pab¹-pab⁵-pab⁷-me*. The same sequence is obtained from the results of the reciprocal crosses. The distances between the individual *pab* markers are obtained from the frequencies of the prototrophs. Since site *5* lies between *1* and *7*, the frequency of *pab¹⁺/pab⁷⁺* prototrophs is the highest (0.06) of the three values.

Another method of mapping alleles, which differs basically from the others, depends upon the mutation frequency of the individual alleles. This method was originally used by MULLER (1932) for mapping chromosomes and recently employed again for constructing gene maps in yeast (MANNEY and MORTIMER, 1964; MANNEY, 1964). It is based on the assumption that the chance of mutation depends upon the number of alleles in a gene and that this number is directly proportional to the length of the gene (p. 219).

MANNEY and MORTIMER used the mitotic reversion frequencies of 29 allelic markers for mapping in yeast; these were induced in heteroallelic diploid cells by X-irradiation. The intervals between each two alleles proved to be additive. These map units are obviously not identical with the "recombination units" of ordinary gene maps (p. 195).

Numerous maps of the mutational sites within genes have been constructed with the aid of the first of these methods during the last ten years (see Table IV-21). Agreement between the results of various investigations leads to the conclusion that the *genetic material of all organisms is linearly arranged down to its smallest unit.*

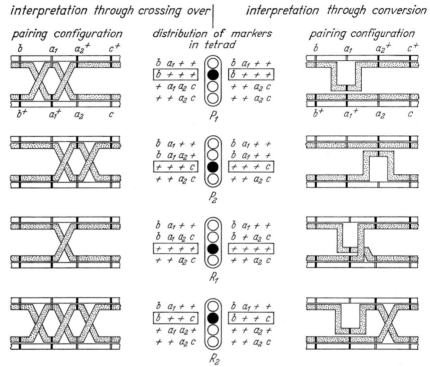

Fig. IV-19. The four end marker combinations, P_1, P_2, R_1, and R_2 resulting from simultaneous recombination of heteroalleles a_1 and a_2 (end markers $= b$ and c; compare Table IV-21); left: assuming that recombination is always reciprocal (crossing over); right: assuming that recombination is not reciprocal (conversion); center: the prototrophic recombinants $a_1^+ a_2^+$ are shown in the frames and as black spores. This illustration is of primary significance to section C. Further explanation at various places in the text, e.g. p. 215f., p.217 f., p. 228f.

2. Negative interference

One would not expect an intragenic exchange to be accompanied by another in close proximity, if there were positive interference, as with intergenic recombination. Contrary to this expectation the mapping of mutational sites (p. 211ff.) has revealed that *a recombination of two*

heteroalleles giving the wild type is often accompanied by additional recom-
binations which produce the "unexpected" end marker combinations
P_1, P_2 and R_2 (Fig. IV-19, left and Table IV-21). Such a correlation
of recombinations which is a sign of negative interference, was first
observed by PRITCHARD (1955) within the *ad-8* locus of *Aspergillus*
nidulans in an analysis of mitotic recombination (p. 204ff.) (see also
ELLIOTT, 1960b). The phenomenon was demonstrated in phage shortly
afterwards (CHASE and DOERMANN, 1958) and in meiotic recombination
in *N. crassa* and other fungi (references in Table IV-21 and for *Saccharo-*
myces, e.g. SHERMAN and ROMAN, 1963).

Negative, like positive chromosome interference can be measured by
the coincidence value (p. 187). Values well over 1 are obtained here
(K_1 in Table IV-21). With the assumption that intragenic as well as
intergenic recombination is reciprocal, the crossing over configurations
shown in Fig. IV-19, left, explain the observed genotypes. A comparison
of the data of Table IV-21 with the diagrams of these exchanges
strengthens the belief that principles other than those governing exchange
between different genes underlie the recombination process within a gene.

For example, one would expect the recombination type R_1 (Fig. IV-19,
left) to predominate, since recombination between two heteroalleles, a_1 and
a_2, leading to the wild type phenotype, simultaneously involves a recombination between the two end markers, *b* and *c* (e.g. for the *cys* locus of *Neuro-*
spora crassa in Table IV-21, observed: expected = 119:341.1). In contrast
the recombination type R_2 should occur at only a low frequency, since it
results from a triple crossover (29:1.7). Further, the two parental types P_1
and P_2 are expected at a frequency no higher than that corresponding to
the map distances between *b* and *a* or *c* and *a* (190:16.5 for P_1; 56:34.7
for P_2).

The coincidence value K_1 in Table IV-21 is given by the equation
$K_1 = \dfrac{r_2}{x \cdot y}$. Here r_2 represents the frequency of the R_2 genotypes (with
reference to the wild type recombinants) and *x* or *y* the frequency of recombination between *b* and *a* or *c* and *a* ($= {}^1/_{100}$ map distance *b—a* or *c—a*
respectively).

A more precise concept of intragenic recombination is obtained from
crosses in which more than two mutational sites are used to mark a gene.

For example, CASE and GILES (1958a, b) utilized three heteroalleles of
the *pan-2* locus in *N. crassa* in various crosses, particularly the double
mutant of the two distally situated sites, *B5* and *B36*, and a number of
single mutants (here indicated by *Bx*) that lie between *B5* and *B36*. With
this experimental design wild type offspring will arise in the cross *B5B36* × *Bx*
if exchanges between *Bx* and *B5* or *Bx* and *B36* occur simultaneously.
The expected frequency of wild type recombinants ($B5^+Bx^+B36^+$) can be
computed from the frequencies of the *pan-2* prototrophs for *B5* and *Bx* or
Bx and *B36*. The wild types arise from fifty to one hundred times more
frequently than expected theoretically with no interference. The high negative interference is even more pronounced with regard to the end markers.
The "rarer" of the two parental types is found at a frequency up to 20%
among the *pan-2* prototrophs even though its occurrence requires at least
four recombinations (each of which is a recombination between adjacent
markers).

These results appear to contradict those of DE SERRES (1956) with the
ad-3 locus of *N. crassa*. DE SERRES failed to find negative interference when
using several closely linked mutational sites. This is not a contradiction of

the results described above, however, because the *ad-3* locus has been shown to be structurally and functionally complex and is composed of at least two closely linked genes (*ad-3A* and *ad-3B*; see also p. 312) (DE SERRES, 1964).

All such results show that *recombinations within a particular gene are rare occurrences that tend to take place, however, close together*. The strength of negative interference, i.e. the size of the coincidence value (K_1) appears to depend more on the individual gene than on the distances between the mutational sites (Table IV-21).

For example, the coincidence values (K_1) for a specific locus lie in the same range, those for different loci in different ranges: e.g. *nic-1:* 70—100; *me-2:* 40—60; *his-1:* 10—20; *pab-1:* 300—500 etc.

A different, although related phenomenon bears on the question whether or not such closely adjacent exchanges are distributed at random within the gene. Data from Table IV-21 as well as the results of crosses involving three markers (e.g. CASE and GILES, 1958a, b) generally support a random distribution. A weak "positive interference" within the "negative interference region" can be demonstrated in only a few cases. The coincidence values K_2 in Table IV-21 lie for the most part between 0.6 and 1.

The coincidence values K_2 are computed as follows: a recombination between *b* and a_1 results in $a_1^+a_2^+$ prototrophs at a frequency p_2+r_2, recombination between a_2 and *c* at a frequency p_1+r_2. p_1, p_2 and r_2 indicate the frequency with which the corresponding end marker combinations P_1, P_2, and R_2 occur. With a random distribution simultaneous recombinations in regions *b*—a_1 and a_2—*c* are expected at a frequency $(p_1+r_2):(p_2+r_2)$. The coincidence value is then equal to:

$$K_2 = \frac{r_2}{(p_1+r_2)\cdot(p_2+r_2)}.$$

Assuming that recombinations within a gene interfere with one another ($K_2 = 1$), PAPAZIAN (1960) developed a model which allows calculation of the average number of recombinations within a gene. In addition to a random distribution, he assumed that (1) the average number of recombinations is the same for each gene and (2) recombinations in one gene do not interfere with recombinations in another.

PRITCHARD (1955, 1958) explained negative interference by assuming that recombinations do not occur randomly at sites along paired chromosomes, but take place preferentially or exclusively in "regions of effective pairing". He based this on the assumption that pairing is not uniform along the length of the chromosome, but is closer at those points where the homologous first came together by chance.

The excess of immediately adjacent recombinations can also be interpreted as localized effects of a specific enzyme, "recombinase" (BRESCH, 1964). This proposes that recombinations are enzymatically controlled (see also p. 170 and p. 174).

3. Recombinative capacity of genetic material

The observation that every gene consists of different mutational sites which can recombine with one another has raised the questions: To what extent is the genetic material divisible by recombination? At how many sites can the genetic information be recombined? How many recombinable mutational sites exist in the entire genome?

The number of mutational sites cannot be determined by mutation experiments alone, because not all parts of the genome are equally mutable. For example, sites exhibiting high mutability (called "hot spots", see p. 295) are known. The same limitation exists when we attempt to estimate the number of recombinable sites in the genome on the basis of recombination data. Such an approximation involves the unproven assumptions: (1) Recombination has the same probability of occurrence between all intra- as well as intergenic sites. (2) Mutational sites are uniformly distributed within the gene and the genes are uniformly distributed within the chromosome. On the basis of these assumptions an average number of mutational sites per gene is given in Table IV-22 (columns a—d).

The calculation involves the following steps:

1. *Map length of the genome.* An estimate must be accepted in this case, because the known number of map units (in *N. crassa* about 500; see Fig. IV-14) is generally smaller than the "actual" map distance of the whole genome; the number of known genes is relatively few. A uniform map distance of 800—1000 units has been estimated for the organisms listed in Table IV-22 (see PONTECORVO, 1958; FINCHAM and DAY, 1963).

2. *The minimal distance between two genes.* The lowest recombination frequency between two "adjacent" genes is in agreement for *N. crassa*, *Aspergillus nidulans*, and *Ascobolus immersus* at a value of between 0.01—0.05 % (DE SERRES, 1956, 1963, 1964; PRITCHARD, 1955, and LISSOUBA et al., 1962). An average value of 0.03 map units is listed in Table IV-22.

3. *Number of genes in a genome.* According to the assumptions of uniform recombinability and gene distribution within the genome, the number of genes is calculated by dividing the total map distance by the minimal map distance between genes. One obtains a value of 10^4—10^5 genes per genome in this way. With a distance of 0.03 map units between adjacent genes the number of genes is about 3×10^4. Since inter- and intragenic distances are not comparable (p. 211 ff.) the number of genes cannot be calculated by using "gene length" (as done by PONTECORVO, 1958) this is because determination of genome length is based on reciprocal recombination, while length of the gene is established from non-reciprocal recombination (p. 222 ff.) which involves other principles (p. 232 f.).

4. *Length of a gene.* By this is meant the map distance between two mutational sites which lie at the "ends" of the gene (Table IV-22, column a). This map distance is most often not identical with the highest recombination frequency determined for the gene, since intragenic recombination values are not additive (p. 214).

5. *Minimal distance between two mutational sites.* The lowest measurable recombination frequency within a gene is designated as the minimal distance (Table IV-22, column b). If only two heteroalleles are known for a gene, the length of the gene coincides with the "minimal value".

6. *Number of mutational sites per gene.* This calculation also depends upon the assumptions that the mutational sites within a gene are distributed uniformly and that recombination is equally probable at all sites within the gene. The number of mutational sites as shown in Table IV-22 has been determined by two methods, one by utilizing the minimal value (a/b; column c) within the *gene*, the other by using the minimal value ($a/b*$; column d) within the *genome*.

7. *Number of mutational sites within the genome.* This value is computed by multiplying the number of mutational sites of the gene by the number of genes in the entire genome.

Table IV-22. *Data from intragenic recombination used to calculate the genetic (columns c and d) and the chemical (columns e and f) dimensions of individual genes*

The arithmetic methods by which the values were obtained are given at the head of columns c—f. *b** in column d designates the minimum map distance of the organism; the smallest value in column b is designated by an asterisk (*) for each organism. The number, *N*, in columns e and f indicates the average value of nucleotide pairs per gene. *N* is calculated by dividing the number of nucleotide pairs per nucleus by the number of genes per genome (p. 221). The average number of genes for each of the organisms is assumed to be 3×10^4 (p. 219). The DNA content of *Ascobolus* is taken as equal to that of *Neurospora*, and that of *Schizosaccharomyces* as equal to that of *Saccharomyces*. The number of nucleotide pairs per haploid nucleus of *N. crassa* is estimated at 4.3×10^7 (HOROWITZ and MACLEOD, 1960) for *A. nidulans* at 4.1×10^7 (HEAGY and ROPER, 1952; PONTECORVO and ROPER, 1956), and for *S. cerevisiae* at 2.2×10^7 (OGUR et al., 1952). The average number of nucleotide pairs per gene (*N*) (columns e and f) obtained for *N. crassa* and *A. immersus* is 1.43×10^3, for *A. nidulans* 1.37×10^3, and for *S. pombe* 0.73×10^3. See text for further explanation.

object and gene	map distance between		average number of mutational sites per gene		average number of nucleotide pairs between adjacent mutational sites of a gene		reference
	end mutational sites	adjacent mutational sites					
	a	b	$\left(\frac{a}{b}\right)$	$\left(\frac{a}{b^*}\right)$	$\left(\frac{N}{c}\right)$	$\left(\frac{N}{d}\right)$	
			c	d	e	f	
Neurospora crassa							
cys	0.30	0.002	150	1,500	10	1	STADLER and TOWE, 1963
pan-2	0.82	0.002	410	4,100	3	0.3	CASE and GILES, 1958
pab-1	0.12	0.024	5	600	287	2	FREESE, 1957b
me-2	0.10	0.0004	250	500	6	3	MURRAY, 1960a, b, 196
his-1	0.01	0.002	5	50	287	29	FREESE, 1957a
pyr-3	0.034	0.001	34	170	42	8	SUYAMA et al., 1959
am	0.017	0.0002*	85	85	17	17	PATEMAN, 1960a, b
ad-8	0.17	0.004	42	850	34	2	ISHIKAWA, 1962
Aspergillus nidulans							
ad-8	0.16	0.05	3	5,300	456	0.3	PRITCHARD, 1955, 196
bi	0.10	0.04	2.5	3,300	547	0.4	ROPER, 1950
paba-1	0.051	0.00003*	1,700	1,700	1	1	ROPER, 1950; SIDDIQ 1962
pro-3	0.0001	0.0001	—	—	—	—	FORBES, 1956
Ascobolus immersus							
Serie 46	1.26	0.053	24	63,000	60	0.02	
Serie 19 A	0.002	0.00004	50	100	29	14	LISSOUBA et al., 1962
Serie 19 B + C	0.16	0.00002*	8,000	8,000	0.2	0.2	
Schizosaccharomyces pombe							
ad-2	0.04	0.015	3	250	244	3	LEUPOLD, 1957
ad-7	0.15	0.00024	625	940	1	1	LEUPOLD, 1957, 1961
ad-8	0.0013	0.00016*	8	8	90	90	LEUPOLD, 1961

Table IV-22 shows that the distances between each two adjacent mutational sites is generally on the order of 10^{-3} to 10^{-4} (in *Ascobolus* even 10^{-5}), while the dimensions of the gene itself may be $\leqq 1$ map unit. There are an estimated 100 to 1,000 mutational sites per gene. With 10^4 to 10^5 genes per genome, there are approximately 10^6 to 10^8 mutational sites in the entire chromosome complement. (With 3×10^4 genes, the mutational sites number between 3×10^6 and 3×10^7.)

4. DNA and recombination

As the carrier of genetic information, DNA is directly involved in recombination. Exactly how the DNA is recombined is as yet unknown, however. Nevertheless, the discovery of intragenic recombination has focused attention on *whether or not the subdivision of the genetic material by classical genetic means corresponds to a subdivision of the DNA*. The question whether a nucleotide is the smallest unit of recombination or whether the lower limit of recombination corresponds to a larger unit of molecular organization has been a particular subject of investigation. Studies dealing with this problem were first carried out with phage (BENZER, 1955, 1957; STREISINGER and FRANKLIN, 1956) and later on fungi (first with *A. nidulans*: PONTECORVO and ROPER, 1956). The average number of nucleotides between adjacent mutational sites within genes in certain organisms is shown in Table IV-22. *The calculation is based on the assumption that DNA content of a haploid nucleus (measured in nucleotide pairs) corresponds to the total map length of the genome:*

1. *DNA content per nucleus.* The DNA content of nuclei can be determined by biochemical methods. It has been shown that the amount of DNA is proportional to the number of chromosome complements (sets) (p. 314f.). For example, *A. nidulans* has 4.4×10^{-14} g. of DNA per nucleus for haploid nuclei and 9.0×10^{-14} g. for diploid (HEAGY and ROPER, 1952; PONTECORVO and ROPER 1956).

2. *Number of nucleotide pairs per nucleus.* Biochemical data (molecular weight etc.) allows a determination of the number of nucleotide pairs which corresponds to a specific DNA content. For example, a nucleus of *A. nidulans* has 4.1×10^7 nucleotide pairs (Table IV-22).

3. *Number of nucleotide pairs per gene.* Assuming that the DNA content of a haploid nucleus (measured in nucleotide pairs) corresponds to the genome of the nucleus, the average number of nucleotide pairs per gene can be determined by dividing the number of nucleotide pairs by the number of genes. With 4.1×10^7 nucleotide pairs per nucleus (*A. nidulans*) and 10^4 to 10^5 genes per genome (p. 219) a value of 400 to 4,000 nucleotide pairs per gene is obtained. This result agrees with the following consideration: If one assumes that a single gene determines a polypeptide chain of approximately 100 to 300 amino acids and that a triplet of base pairs codes for a single amino acid (p. 345), a gene is estimated to consist of 300 to 900 nucleotides.

4. *Number of nucleotide pairs between "adjacent" mutational sites.* The number of mutational sites (p. 219) and of nucleotide pairs per gene correspond. By dividing, a value is obtained for the number of nucleotides in the region between two adjacent mutational sites. If the mutational sites are assumed to be directly next to one another, i.e. lacking "spacers", the quotient indicates the dimension of a single mutational site (Table IV-22, columns e and f).

The calculation follows from the assumptions discussed in the previous section (p. 219):

1. Mutational sites and genes are uniformly distributed in the genome.
2. Intragenic recombinations have an equal probability of occurrence between all mutational sites and intergenic recombinations between all loci on the chromosome. Thus it is possible to make the following generalization (Table IV-22): *Mutational sites which are close together are separated by relatively few nucleotides.* A value of 1 to 3 nucleotides has been calculated in a number of cases. One can conclude, therefore, that *adjacent nucleotides are separable by recombination* (compare the recombination mechanism shown in Figs. IV-23 and IV-24).

Summary

1. Recombination may occur at many sites within a gene, i.e. the mutational sites of a gene are recombinable.
2. The mutational sites of a gene may be arranged into a linear gene map on the basis of the recombination frequencies. Genes, like linkage groups, have a unidimensional structure.
3. Intragenic recombination is a rare event. Such recombinations tend to occur in close proximity to each other (negative interference).
4. The number of mutational sites per gene is of the order of 10^2 to 10^3. A single gene has about 300 to 1,500 nucleotide pairs; this agrees with the expectation, if one assumes that a gene determines a polypeptide chain composed of about 100 to 300 amino acid residues. Adjacent recombinable sites are separated by relatively few, or possibly only a single nucleotide.

II. Non-reciprocal recombination

The regular 1:1 segregation of allelic pairs among the four meiotic products of a tetrad leads to the conception that the recombination process is exclusively reciprocal. Results in fungi and bacteriophage which showed conclusively that a non-reciprocal type of recombination also occurs were not forthcoming until 1955 (MITCHELL, 1955a, b; BRESCH, 1955). These experimental results shed a new light on earlier observations of abnormal tetrad segregations. Aberrant tetrad distributions had previously been explained as technical errors of isolation and thus essentially disregarded.

Abnormal tetrad segregations were reported almost forty years ago: BRUNSWICK (1926) in *Coprinus fimetarius*, BURGEFF (1928) in *Phycomyces blakesleeanus*, KNIEP (1928) in *Aleurodiscus polygonius*, DICKINSON (1928) in *Ustilago levis*, HANNA (1929) in *Ustilago zeae*, and von WETTSTEIN in the moss, *Funaria hygrometrica*, already in 1924. A few years later abnormal segregations were also found in certain Ascomycetes: ZICKLER (1934b) in *Bombardia lunata*, WÜLKER (1935) in *Neurospora sitophila*, LINDEGREN and LINDEGREN (1942) in *Neurospora crassa*, LINDEGREN (1949a, b, 1955), MUNDKUR (1949, 1950), WINGE and ROBERTS (1950, 1954a, b), ROMAN et al. (1951), LEUPOLD and HOTTINGUER (1954) in *Saccharomyces cerevisiae*.

It is surprising that in spite of non-reciprocity being observed relatively frequently, intensive analyses of tetrads were not carried out until the 1950's. For example, LINDEGREN and coworkers repeatedly found deviating tetrad segregations in the asci of *Saccharomyces cerevisiae*, which they explained as "allele-induced mutations" (LINDEGREN, 1949a, b, 1955; LINDEGREN et al., 1956). They adopted the term *conversion*, originally introduced by WINKLER (1930) to designate this phenomenon. WINKLER used the term to mean the mutation of a gene induced by the allele associated with it. He sought to explain by conversion not only the deviations from Mendelian segregations then known, but all recombinations of linked genes. While WINKLER took a position against the classical breakage-reunion hypothesis (p. 234ff.) with this conception, the Lindegren group limited the concept of conversion to the explanation of non-reciprocal recombinations. WINGE and ROBERTS (1950, 1954a, b, 1957), who also observed abnormal segregations in the asci of yeast strains, showed that non-reciprocity could arise by other processes, such as postmeiotic mitoses and subsequent diploidization or dikaryotization. Investigations of MUNDKUR (1949, 1950) and LEUPOLD and HOTTINGUER (1954) on the same material failed to reconcile the two hypotheses. The unexplained situation in *Saccharomyces* contributed to the failure to pursue this phenomenon further at the time.

1. Crossing over and conversion

Interest in abnormal tetrad segregations did not reawaken until the work of MITCHELL (1955a, b) on *N. crassa*. This author presented conclusive evidence that the unusual distribution of markers in the asci occurred through non-reciprocal recombination. The discovery stimulated a series of further studies in subsequent years on *Neurospora* (Table IV-23) and other species (e.g. *Ascobolus:* Table IV-23; *Sordaria:* OLIVE, 1956, 1959; EL ANI et al., 1961; KITANI et al., 1961; *Aspergillus:* STRICKLAND, 1958a; *Saccharomyces:* ROMAN, 1958; *Salpiglossis:* REIMANN-PHILIPP, 1955). The fact that rare non-reciprocal recombinations do occur was repeatedly confirmed.

The use of the term *"conversion"* for the mechanism underlying this process came to be generally accepted in order to distinguish it from crossing over which leads to reciprocal recombination. In spite of many attempts neither mechanism has been elucidated. Thus conversion and crossing over cannot be related to particular molecular phenomena. We will show in section D (p. 231ff.) the difficulties inherent in the interpretation of both processes on a molecular level. Although both terms were originally used to designate very specific molecular processes (p. 148 and p. 223), the above definitions of crossing over and conversion have become more and more widely accepted.

Aberrant tetrad segregations can be especially well observed with spore color markers. For example, Fig. IV-3 shows an ascus of *Sordaria macrospora* with 6 black and 2 white spores. Heteroallelic auxotrophic mutants have been employed exclusively for analyzing the phenomenon, at least in *Neurospora* (Table IV-23). In order to make the data in the table easier to understand, the method of proof of non-reciprocal recombination is explained below, using the classical example of MITCHELL:

Table IV-23. *Data from analyses of aberrant tetrads*

The mutational sites used as markers are designated as a_1, a_2 and for the corresponding wild type alleles a_1^+, a_2^+ in the column headings. Selective methods have generally been used for isolating aberrant tetrads, i.e. only tetrads with $a_1^+ a_2^+$ spores have been chosen (crosses of the $a_1 a_2^+ \times a_1^+ a_2$ type). In the few cases in which complete tetrad analyses have been carried out, some additional types of asci have been found (see next to last column): e.g. $2 a_1^+ : 6 a_1$ or $2 a_2^+ : 6 a_2$ or simultaneously $6 a_1^+ : 2 a_1$ and $2 a_2^+ : 6 a_2$ or $2 a_1^+ : 6 a_1$ and $6 a_2^+ : 2 a_2$ (see also data of THRELKELD, 1962b). For further details see text.

object, linkage group, and gene	mutational site a_1—a_2	recombination percentage between a_1 and a_2	number of asci analyzed	number of asci with reciprocal recombination (with double mutants) $4a_1^+:4a_1$ $4a_2^+:4a_2$	number of asci with non-reciprocal recombination (without double mutants)			reference*
					$\dfrac{6a_1^+:2a_1}{4a_2:4a_2}$	$\dfrac{4a_1:4a_1}{6a_2:2a_2}$	other segregation types	
Neurospora crassa								
IV: *pdx-1*	*pdx—pdxp*	0.17	4	—	1	2	1	1
VI: *pan-2*	*B 3—B 5*	0.27	19	2	5	6	6	2
	B 23—B 72	0.18	10	2	4	—	4	
	B 72—B 36	0.13	4	—	1	—	3	} 3
	B 23—B 36	0.30	11	2	4	—	5	
II: *td*	*3—11*	0.16	5	1	—	4	—	4
VI: *cys*	*c—t*	0.21	3	—	1	2	—	5
	17—64	0.11	28	—	16	12	—	
	4—38	0.14	6	—	5	1	—	} 6
	38—64	0.04	5	—	2	3	—	
	9—64	0.03	6	—	3	3	—	
Ascobolus immersus series 46	*188—63*	0.08	8	—	8	—	—	
	188—46	0.12	2	—	2	—	—	
	188—w	0.50	10	—	10	—	—	
	188—138	0.94	11	—	11	—	—	
	63—46	0.05	2	—	2	—	—	
	63—w	0.28	18	—	18	—	—	
	63—1216	0.29	5	1	4	—	—	7
	63—138	1.29	35	—	35	—	—	
	46—w	0.07	7	—	7	—	—	
	46—138	1.15	22	—	22	—	—	
	w—1216	0.09	18	—	18	—	—	
	w—138	0.83	15	—	15	—	—	
	1216—137	0.84	2	—	2	—	—	

* Reference: [1] MITCHELL (1955a); [2] CASE and GILES (1958b), [3] CASE and GILES (1964); [4] WEIJER (1959); [5] STADLER (1959a); [6] STADLER and TOWE (1963); [7] LISSOUBA et al. (1962).

MITCHELL studied a cross of two heteroallelic pyridoxine mutants of *N. crassa*. Among 585 asci analyzed, there were four which contained protrophic spores. Three of these four asci exhibited a segregation of six auxotrophs to two prototrophs. Further analysis of the six auxotrophic spores revealed that four of them carried only the one allele (*pdx*), while the remaining two spores carried only the corresponding heteroallele (*pdxp*). None of the pyridoxine auxotrophic spores were *pdx/pdxp*; the double mutant expected on the basis of the classical theory of crossing over did not appear. In this case the ascus showed the segregation 2 *pdx⁺ pdxp*: 4 *pdx pdxp⁺*: 2 *pdx⁺ pdxp⁺*, or if one writes the segregation for the two markers separately, 4 *pdx⁺*: 4 *pdx* and 6 *pdxp⁺*: 2 *pdxp*. The recombination between the two heteroalleles thus led to an abnormal 6:2 distribution for the *pdxp* marker.

The results of tetrad analyses compiled in Table IV-23 reveal that *intragenic recombination leads almost exclusively to aberrant, i.e. nonreciprocal tetrad segregations*. Reciprocal exchange, i.e. the appearance of the double mutant as well, can be demonstrated only in occasional cases.

However, more reciprocal recombinations are observed between tightly linked genes. Nevertheless, in part of these cases the selected markers belong, not to the same gene, but to different, although related functional regions (e.g. for the loci *ad-3A* and *ad-3B* of *N. crassa:* DE SERRES, 1956, 1963, 1964; GILES et al., 1957, see also p. 217). In other cases it has been possible to assign reciprocal and non-reciprocal recombinations to separate genetic regions through extensive ascus analysis (*Ascobolus immersus:* LISSOUBA et al., 1962). Fig. IV-20 shows a small chromosomal segment of *A. immersus* with mutational sites of different regions. Only sites *231* and *322* belong to the same region (for further discussion of this phenomenon, see p. 225f.).

Because of the relatively low frequency of *reciprocal recombination* within a gene one is tempted to interpret these rare cases as *independent events occurring together* and consider all intragenic recombination simply as a non-reciprocal process. This problem will be discussed in greater detail in section D (p. 232f.). At any rate, the 3:1 or 6:2 tetrads cannot be explained by the breakage and reunion of strands. The non-reciprocity could be due to one of the parental strains being replicated twice in a certain segment. Such an interpretation is shown by the pairing configuration in Fig. IV-19 on the right.

2. Polarized non-reciprocal recombination

The results of analyses of *Ascobolus immersus* and *Neurospora crassa* that show a "*preferential direction of recombination*" are critical for an understanding of recombination. RIZET, LISSOUBA, and coworkers, who first observed polarity in *Ascobolus*, used spore color mutants for their investigations. These are particularly well-suited for the analysis of aberrant tetrads, since segregation can be determined by direct observation of the spores (RIZET et al., 1960a, b; LISSOUBA and RIZET, 1960; LISSOUBA et al., 1962).

The mutants utilized by RIZET and coworkers arose spontaneously. Their spores do not form the brown violet pigment in the wall. By appropriate crosses it was possible to subdivide 2,000 white spored mutants into "series". All mutants in a series are extremely closely linked. The individual

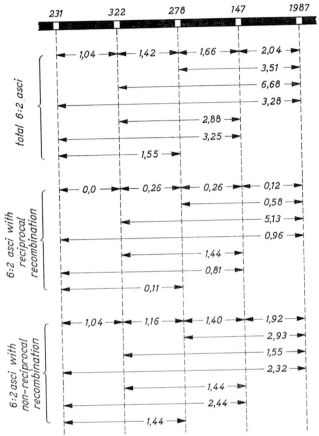

Fig. IV-20. Linear representation of a small chromosomal segment of *Asco-bolus immersus* as an example of reciprocal and non-reciprocal recombination between closely linked markers. The mutants used here belong to series 75; they differ from the wild type by possessing light colored spores (p. 269f.). The mutational sites are designated by isolation numbers. The map distances between mutational sites indicate the frequency of asci with 6 white : 2 black spores (measured in %). Below and center: these frequencies are classified according to whether they are conversion or crossover asci. (Adapted from LISSOUBA et al., 1962)

series lie at greater or less distances apart on the same chromosome or lie on different chromosomes. It was first assumed that each series corresponded to a gene, or possibly a small number of genes with a similar function. When mutants of the same series were intercrossed, asci with eight white spores were generally obtained. In rare cases, however, asci with 2 dark and 6 light spores were observed. Part of these asci resulted from reciprocal crossing over between the two mutational sites (one pair of wild-type spores : two pairs of single-mutant spores : one pair of double-mutant spores). The remaining asci lacked the double mutants. Therefore aberrant tetrads occurred in which one marker was present in "excess".

Table IV-24. *Results of an analysis of 45 asci with 3:1 segregations in Neurospora crassa*

Only the genotypes of the wild types arising from allelic recombination are shown. All asci arose from crosses of $lys\ cys^1\ cys^{2+}\ ylo^+ \times lys^+\ cys^{1+}\ cys^2\ ylo$. The cys^1 and cys^2 symbols designate heteroalleles of the cys locus (see Table IV-23); lys and ylo are distal markers with the following map distances: lys-6-cys-9-ylo (in map units). For explanation of the combination types of the distal markers see Fig. IV-19 (From data of Stadler and Towe, 1963).

genotype of *cys* recombinant	outer marker combination	double replication of	
		cys^{1+}	cys^{2+}
$lys^+\ cys^+\ ylo$	P_1	1	14
$lys\ cys^+\ ylo^+$	P_2	12	0
$lys^+\ cys^+\ ylo^+$	R_1	8	1
$lys\ cys^+\ ylo$	R_2	1	6
total:		22	21

The analysis of such aberrant tetrads has shown that when markers of the same series are used, the same one of the two markers is almost always the site of non-reciprocal (6:2) segregation while the other segregates normally (4:4) (Table IV-23). Further, the markers replicated in excess lie either always to the left or always to the right of the normally segregating locus. Rizet and coworkers concluded from such results that the *frequency of conversion shows a continuous increase from one end of the gene to the other*. They called such a region with a directed conversion a *"polaron"* (p. 233 and p. 240f.). More recent investigations have led to the assumption that the structural region which has been considered the gene until now, is identical with the polaron (Rossignol, 1964). The alleles of *series 46* (Table IV-23) have been shown to belong to the same functional unit with the help of the *cis/trans* test (see footnote p. 210). Binucleate heterokaryotic spores, which *Ascobolus immersus* produces in rare cases, were used for the test. (The spores are larger than normal and correspond to two uninucleate spores.)

Such a distinct correlation between the position of the marker within the gene and the site of the duplicate replication has only been observed for the genes of *Ascobolus*. It was recently shown that this fungus also possesses genes in which conversion can occur at mutational sites on the left as well as on the right (Rizet and Rossignol, 1963; Rossignol, 1964; Makarewicz, 1964). The intragenic markers used in crosses in *Neurospora crassa* and *Aspergillus nidulans* are always equally involved in conversion (Table IV-23 and Siddiqi and Putrament, 1963). Nevertheless, it was shown in both species by utilizing end markers that the recombination is polarized. Data on the *cys* locus of *Neurospora crassa* have been compiled in Table IV-24 as an example. These show that the end marker *ylo* to the right of the *cys* locus appears in the parental combination almost exclusively with the *cys* marker which occurs three times (*ylo cys*$^{1+}$ and *ylo*$^+$ *cys*$^{2+}$). In contrast, the end marker *lys* to the left of *cys* recombines at random with *cys*$^+$.

Twenty out of twenty-one asci (Table IV-24) in which the *cys*$^+$ recombinants carry the *ylo*$^+$ marker lying to the right of *cys*^{1} in the same strand are characterized by a duplicate replication of the *cys*^{1} allele. There is a similar correlation between *ylo* and the 3:1 segregation for *cys*^{2} in 20 out of 22 cases.

SIDDIQI and PUTRAMENT (1963) in their experiments with *Aspergillus nidulans* found that a recombination in the *paba*-1 locus more frequently accompanies an exchange in the distally adjacent than in the proximally adjacent region. These results with *Neurospora* and *Aspergillus* have led to the conclusion that the process underlying recombination proceeds from one end of the chromosome to the other and is related to the formation of regions of effective pairing (p. 218ff.). The observation that the distal site in the case of widely separated markers in *N. crassa* is replicated in duplicate more frequently than the proximal site supports this view.

Unfortunately, end markers were not used in *Ascobolus*. Thus we cannot say whether the recombination process proceeds outside of the marked regions of the gene in the same way as in *Neurospora* and *Aspergillus*. The polarity observed in all three cases leads to the assumption, however, that *intragenic non-reciprocal recombination always takes place during the continuously unidirectional replication* (compare the recombination models p. 239ff.).

3. Interference

We had concluded from the distribution of end markers in prototrophs arising through intragenic recombination (p. 216ff.) that recombinations within small regions tend to occur close together (negative interference). As pointed out in the last two sections, recombination within a gene is probably exclusively non-reciprocal. Thus the question can be asked: Do the observed segregations arise only from conversions or does the negative interference involve both intragenic, non-reciprocal (conversion) and intergenic, reciprocal recombination (crossing over)?

If one assumes that most or all of the wild-type recombinants in crosses between alleles arise through conversion (Fig. IV-19, right) and not through crossing over (Fig. IV-19, left), each of the four end marker combinations can only be explained through one exchange between the two heteroalleles and at least one in the two adjacent regions. *Such a correlation between adjacent recombinations leads to the idea that only a tiny segment is replicated in duplicate.* From the end marker combinations P_1, P_2 and R_2 (see Fig. IV-19, right) it is further apparent that within a limited region only one strand participates in the replication error, i.e. most copy errors are corrected in the immediately neighboring region. Results of experiments with three markers and particularly tetrad analysis not only confirm these conclusions, but show further that such a duplicate copying within a gene can occur more than once (e.g. CASE and GILES, 1958a, b; STADLER and TOWE, 1963; TOWE and STADLER, 1964). *Negative interference thus involves, at least in part, a clustering of intragenic conversions.*

Let us turn now to the second part of the question, namely, whether or not conversion and crossing over in close proximity are correlated. The results of tetrad analysis permit an answer to this question. Investigations of aberrant tetrads have established that an *intragenic non-reciprocal recombination is frequently* (up to 50%) *accompanied by a reciprocal exchange outside of the markers of the same gene* (e.g. in *N. crassa*; CASE and GILES, 1958a, b, 1964; STADLER and TOWE, 1963; in *S. fimicola:* EL-ANI et al., 1961; KITANI et al., 1961; KITANI, 1962). In these cases the *conversion strand*, i.e. the strand with the duplicate replication, *is usually involved in reciprocal crossing over at the same time.* Parallel results are known in the analysis of 5:3 asci in *S. fimicola* (p. 230f.). The end marker recombinants of the R_2 type (Fig. IV-19, right) which have been demonstrated through selective methods, also argue for such a correlation. Negative interference disappears rapidly if widely separated markers are used. Conversion and crossing over occur completely independent of one another in non-adjacent regions (STADLER, 1959b).

In conclusion, we would like to call attention to a result with *Sordaria fimicola* which suggests that *the duplicate replication is preferential for one of the two parental strands.* The aberrant tetrads arising from a cross of a gray spored mutant (g) and the wild type (g^+) showed a five fold greater segregation of $2\,g : 6\,g^+$ than $6\,g : 2\,g^+$ (OLIVE, 1956, 1959; EL-ANI et al., 1961; KITANI et al., 1961, 1962). The preferential copying of the g^+ marker may possibly result from the fact that the replication of the g strand is controlled by a replication-inhibiting site in the region of the g locus. This causes a template switch and the g^+ rather than the g marker is copied.

4. Conversion between half-chromatids

In addition to the 6:2 asci mentioned above, OLIVE and coworkers found 5:3 tetrads at about the same frequency in *Sordaria fimicola* (Fig. IV-21, above). In a few cases 7:1, 8:0, and irregular 4:4 segregations (Fig. IV-21, below) were observed. Similar results have been reported in *Ascobolus immersus* and in *N. crassa* in isolated cases (RIZET et al., 1960a, b; LISSOUBA, 1960; LISSOUBA et al., 1962; STADLER and TOWE, 1963; CASE and GILES, 1964).

A 5:3 ratio indicates that a pair of spores consists of genetically different members, even though their identity is expected, since their nuclei arise from mitosis of a single meiotic product (see Fig. IV-4). By using three markers, it could be excluded that this anomaly came about through the degeneration of the regular nuclei and their replacement by others which arose from additional postmeiotic mitoses. All three markers exhibited a 4:4 segregation. *5:3 tetrads require the presence of two subunits in each chromatid of a tetrad during the recombination process.* It is necessary to assume that recombination occurs before or during meiosis I and not during meiosis II. Conversions thus may possibly occur exclusively in the 8-strand stage (Fig. IV-21). 6:2 asci are accordingly explained by the simultaneous involvement of the two subunits in a conversion, although not necessarily at the same site (compare Figs. IV-23 and IV-24).

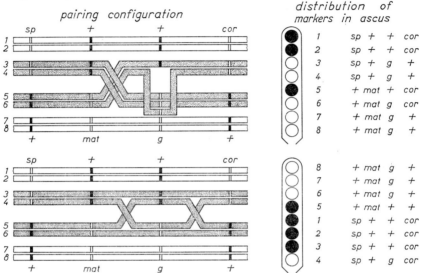

Fig. IV-21. Two aberrant tetrads of *Sordaria fimicola* as examples of recombination between half-chromatids. *Above:* ascus with a non-reciprocal half-chromatid recombination at the *g* locus and adjacent crossing over between the chromatids *3—4* and *5—6*. The color marker *g* segregates in a ratio of five gray spores (open circle) to three black spores (solid circle). *Below:* ascus with a double reciprocal half-chromatid recombination. During prereduction of the end marker *sp* the half chromatids *1—4* in the lower half of the ascus and the strands *5—8* in the upper half of the ascus have shifted. The strand distribution, which can be followed by the numbers next to the ascus, has led to the following segregation of color markers: *3* gray spores : *4* black spores : *1* gray spore. In both cases the template strands are white, the replicated strands stippled. The centromere is not included; it lies 46 map units to the left of the end marker, *sp*. (Adapted from KITANI et al., 1961)

Abnormal 4:4 asci (Fig. IV-21, below) in which not only one, but two pairs of spores are different (3—4 and 5—6) arise through a reciprocal exchange between the two subunits of different parental chromatids. This reciprocity must involve a crossover- or conversion-like double recombination (see also KITANI, 1962).

In agreement with the observations on the 6:2 asci (p. 229f.) OLIVE and coworkers found in analyses of 5:3 asci a pronounced correlation between conversion and crossing over in close proximity (compare e.g. Fig. IV-21, above).

Among 61 asci with 5:3 segregations 26 showed a crossover in a segment 3.4 map units in length lying to the right of g and 29 of the asci had a crossover in a region to the left of g only one map unit long. Reciprocal recombination was found simultaneously in both regions in 14 cases (23%). This value exceeds by 160 fold the figure expected on a random basis (0.14%). By use of a marker lying closer to the g locus (0.4 map units distant), it could be shown that crossing over occurs preferentially in the region immediately adjacent to the g locus when a conversion for g occurs.

Certain data from analyses of aberrant tetrads have led to the assumption that *the occurrence of non-reciprocal recombination* not only depends upon chance, but is also *genetically controlled*. In *Sordaria* (KITANI et al., 1961, 1962) and also *Ascobolus* (RIZET et al., 1960a, b) it has been observed that in crosses of particular mutants with the wild type some give no abnormal tetrads at all, while others produce only 6:2, but no 5:3 (or 7:1) asci. These results agree with those from analysis of reciprocal recombination, namely that different crossover frequencies may be caused by genetic differences between the wild-type strains utilized (p. 168 ff.). No explanation for this relationship has been forthcoming as yet. The absence of aberrant tetrads may possibly involve a pairing inhibition resulting from intragenic inversions.

Summary

1. The occurrence of aberrant tetrads (6:2 and 5:3) proves that non-reciprocal as well as reciprocal recombinations can occur. The non-reciprocal recombinations, which are called conversions, are found only within the gene.

2. A series of results supports the idea of a polarized recombination within genes. Intragenic conversion appears to take place during the continuously unidirectional replication.

3. Intragenic, non-reciprocal recombination (conversion) is correlated with reciprocal exchange (crossing over) in the immediate vicinity. The same strand frequently undergoes a conversion as well as a crossover.

4. The occurrence of 5:3 tetrads is evidence that each chromatid of a tetrad consists of at least two subunits which can recombine with one another during processes of recombination.

D. Possible explanations of intrachromosomal recombination

In the course of our discussion we have dealt with three different processes by which new combinations of the genetic material results.

1. *Assortment of entire chromosomes* at meiosis with the spindle mechanism providing for a *reciprocal* distribution of characters (p. 137 ff.).

2. The breakdown of linkage through *crossing over*. This process is always *reciprocal* (p. 159 ff.).

3. The separation of alleles of a gene through *conversion*, which leads to *non-reciprocal* recombination (p. 222 ff.).

In contrast to the spindle mechanism, crossing over and conversion produce changes within the chromosome. On the basis of the data given in sections B and C we will attempt an interpretation of the significance of crossing over and conversion in the following section.

I. Crossing over and conversion

The pairing of homologous chromosomes is necessary for both processes. This similarity raises the question *whether reciprocal and non-reciprocal recombination involve a single or two different mechanisms, one of which always produces crossovers, the other conversions.* This question must be considered in relation to one raised earlier, namely, *whether reciprocal or non-reciprocal recombination occurs at random throughout the genome or only at predetermined sites* at which either exclusively crossovers or conversions are possible (see also discussion in ST. LAWRENCE and BONNER, 1957; OLIVE, 1962; BERNSTEIN, 1962, 1964; WESTERGAARD, 1964).

An experimental approach to the solution of the *first problem* consists of comparing the frequencies of crossing over and conversion under different environmental conditions. While the temperature experiments of STADLER (1959b) with *Neurospora crassa* proved fruitless, the parallel researches of MITCHELL (1957) led to altered conversion frequencies without corresponding changes in the frequencies of crossing over. ROMAN and JACOB (1958) obtained similar results after treating yeast cells with UV radiation. Conversions observed in mitotically-dividing yeast cells also proved independent of crossing over when the experimental conditions were changed (ROMAN, 1956). *These results argue for the existence of two separate mechanisms.* Data from experiments with variously marked strains of *Neurospora* may also be interpreted similarly (STADLER, 1959a).

The strains used by STADLER were similar in that they carried two identical heteroallelic markers of the *cys* locus, although differing in other genes. A comparison of the crossing results showed that in certain cases (determined by the heterogeneity of the strains used) the frequency of crossing over between the two end markers rose four fold (compare p. 168 f.). In contrast the frequency of conversion remained constant or even decreased.

The assumption that two separate recombination mechanisms exist is further supported by the observation that conversion does not interfere with crossing over in regions that are well separated (STADLER, 1959b).

Other data can be interpreted to indicate crossing over and conversion as simply different expressions of the same mechanism. In this connection the fact that conversions are frequently accompanied by crossovers in the immediate vicinity is particularly noteworthy. The correlation exists not only for conversions between entire chromatids (6:2 or 3:1), but also for those between half-chromatids (5:3). The fact that the conversion strand has crossovers adjacent to the conversion region in many cases (p. 228f.) gives the impression that a causal relationship exists between the two recombination processes. It is possible that the correlation between conversion and crossing over is simply a result of particularly close pairing between certain regions (p. 218) so that both recombination processes are favored.

FOGEL and HURST (1963) observed that the frequency of gene conversion as well as mitotic crossing over in *Saccharomyces* is increased by UV irradiation. There is a distinct dependence upon dosage, e.g. following 0, 60 and 120 seconds of UV, respectively, conversion frequencies of 0.08%, 0.32%, 0.66% were obtained, and the crossover frequencies were 0.2%, 1.1%, 2.45%. The percentages are based on 83% and 24% survivors following 60 and 120 seconds of UV. The authors explain the correlation between conversion and crossing over through the assumption that both processes require the same preconditioning. Further, the UV is believed to promote the effective pairing of homologous chromosomes (p. 176).

Later we will show in a recombination model that the occurrence of both types of recombination is compatible with the idea of two distinct mechanisms (p. 240ff.).

Let us turn our attention to the *second question*, namely, whether crossovers and conversions occur anywhere in the genome or whether each phenomenon is localized at specific, but different sites. The results of aberrant tetrad analysis described in the previous section make it likely that *non-reciprocal* recombination takes place exclusively *within the gene*, while *reciprocal* recombination is only possible *between genes*. Because of the high frequency of non-reciprocal processes within genes (Table IV-22), one is inclined to interpret the rare exceptions as double conversions which yield reciprocal results.

According to a suggestion of RIZET and coworkers (references: p.225) regions with exclusively non-reciprocal recombinations (= polarons) alternate with regions in which only reciprocal recombinations occur (= linkage structures). Thus a polaron corresponds to the region that we have considered the gene. More recent investigations indicate that such an identity is highly probable (p. 235f.). However, such an interpretation leads one to conclude that if two closely-linked mutants of identical phenotype yield any reciprocal recombinants, they must be in separate genes. According to another suggestion (STADLER and TOWE, 1963) the ends of a gene are the sites of reciprocal recombination (p. 240ff.).

Discussion of both questions has shown that a series of results argues for the existence of two different recombination mechanisms with unlike sites of action. Conversion occurs only within the gene and leads to non-reciprocal recombination, while crossing over is only possible between different genes and produces reciprocal recombinations. Further work is needed to confirm the two-mechanism hypothesis.

II. Recombination mechanisms

The problem of recombination is not so much the *number* of mechanisms as the *kind* of mechanism, i.e. how are crossovers and conversions actually produced to recombine the genetic material. In addition to the classical breakage-reunion model, two others have been proposed: copy choice and partial replication. Only the latter two come into question in the case of both crossing over and conversion. The breakage-reunion mechanism can only explain reciprocal exchanges, since according to it recombination does not occur during replication (review and discussion, e.g. in BRESCH, 1964).

1. Breakage and reunion

In accordance with this classical idea, a recombination occurs when *two of the four chromatids of a tetrad break at exactly homologous sites and rejoin in a crossed manner* (breakage-reunion hypothesis; Fig. IV-22a). This is the concept underlying the chiasmatype hypothesis that. JANSSENS proposed in 1909 to explain the cross configurations (chiasmata) which can be seen between chromatids during the prophase of meiosis (p. 137). DARLINGTON (1930) used the hypothesis to explain crossing over; according to it, crossing over occurs during pachytene following completion of pairing and results in a chiasma which is visible in diplotene after chromatid pairs separate. Even though the longitudinal split in the chromosomes cannot be observed in pachytene, it is believed that the genetic material has already duplicated by this time. Tetrad analysis has shown (p. 140) that at least four strands are present at the time of recombination; two of these are involved in each crossover.

According to the breakage-reunion hypothesis, recombination takes place after replication is complete. Therefore, such a mechanism can only explain reciprocal recombinations and not conversions, since the locations of the breaks are always at exactly homologous positions on the two strands and only reciprocal chromatid segments are exchanged. A number of questions remain unanswered, however, even if the breakage-reunion mechanism is reserved for explaining reciprocal exchanges:

1. How do *breaks occur* in intact structures? What causes them?

2. How can the fact that *the breaks always occur simultaneously in the two strands* be explained? Does the first break induce the second or do both result simultaneously from the same cause? Why are the breaks limited to two strands? Do single breaks occur which are simply repaired if a second break fails to take place at the homologous site in another strand?

3. How can one explain *the strands breaking in precisely homologous sites*? Are there specific loci which are predestined for breakage so that a break occurs in each chromatid only at sites of the same specificity?

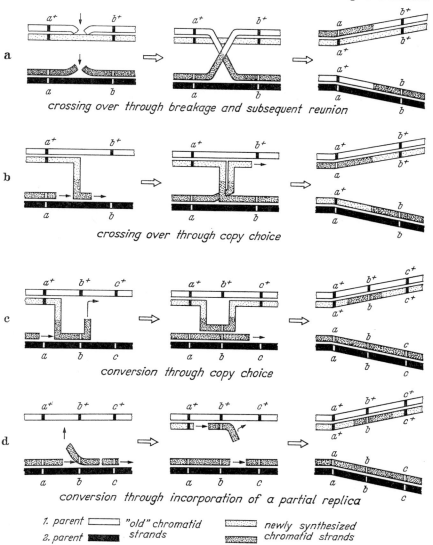

a crossing over through breakage and subsequent reunion

b crossing over through copy choice

c conversion through copy choice

d conversion through incorporation of a partial replica

1. parent ☐ "old" chromatid strands   ▨ newly synthesized chromatid strands
2. parent ▩

Fig. IV-22a—d. Hypothetical mechanisms of reciprocal and non-reciprocal recombination. Three phases of recombination are shown. (a) and (b): crossing over according to the breakage-reunion and copy choice mechanisms, respectively. (c) and (d): conversion according to the copy choice and the partial replication mechanism, respectively. Further explanation in text

4. How can *reunion of two broken chromatids* be explained? How do the broken ends of different strands join? Do broken ends fail to unite so that pieces are lost?

Satisfactory answers to these questions have yet to come forth.

2. Copy choice

Another theory of recombination was proposed in the early thirties by BELLING (1933). BELLING's idea was that *recombination occurs between new strands during replication* and not between intact strands after replication is complete (Fig. IV-22b, c). If the two parental strands contact each other during duplication, the replicating structures may switch from one template to the other. Reciprocal recombinations result from this scheme under normal circumstances even if the replication on both templates does not proceed synchronously (Fig. IV-22b). A replica which has a headstart on the other and at some point switches over to the yet unduplicated parental structure forces the "slower" replicating strand to make a template switch at the same point.

The Belling model, which was later called "copy choice" by LEDERBERG (1955), has certain advantages over the breakage-reunion hypothesis. It eliminates the need to account for breaks at exactly homologous sites on the chromatids, since with copy choice the replica shifts at precisely the site at which the other template is already occupied. In spite of this advantage, the Belling model was rejected at first because with tetrad analysis recombination could be demonstrated not only in two strands (i.e. those replicating) but in all four strands. BELLING attempted to resolve this difficulty by postulating *exchange between template and replicate structures*. This proposal was rejected during the thirties because sister strand exchanges were considered unlikely on the basis of cytological investigations on ring chromosomes (p. 185). In the interim it has become clear that sister strand exchanges *may* take place (p. 185). Nevertheless, copy choice loses its advantage over the breakage-reunion hypothesis because *it also requires breaks in two of the strands.*

If copy choice is considered from the standpoint of DNA and its semi-conservative replication (p. 132), new difficulties arise. According to current thinking, two sister strands cannot be considered template and replica, since each sister strand receives an "old" and a newly synthesized DNA helix. Sister strand exchange is not the same as an exchange between a template and replica, but between two double structures each consisting of an old and a new DNA strand (BRESCH, 1964). Such an exchange thus also requires breaks, at least in the old strands.

This theoretical requirement, namely breaks in already-replicated strands, has been confirmed experimentally in phage investigations (compare hypothetical recombination mechanism shown in Figs. IV-23 and IV-24). In phage *lambda* of *E. coli* it has been shown that genetic recombination without concurrent replication is possible (MESELSON and WEIGLE, 1961; KELLENBERGER et al., 1961; IHLER and MESELSON, 1963). Results from yeast also suggest that recombination may occur, not only during, but also after DNA synthesis (SHERMAN and ROMAN, 1963).

In spite of this disadvantage the copy choice mechanism has been revived during the last 10 years to explain two phenomena, namely, negative chromatid interference (1) and aberrant tetrad distributions (2).

1. Since the concept of chromatid interference is meaningful only if both sister strands of one parent can recombine with both of the other parent the copy choice mechanism is not strictly pertinent to the problem of chromatid interference; in this case only the two replica strands

undergo exchanges. Nevertheless, certain results concerning *negative chromatid interference* discussed earlier (p. 181 and p. 183) allow inferences on the problem of the recombination mechanism. Double crossovers have been observed to involve the same strands more frequently than expected in many cases. Such a favoring of two-strand doubles is to be expected on the basis of a copy choice mechanism. The fact that the recombinant strand resulting from a duplicate replication within a gene is also involved in a crossover likewise possibly supports copy choice (p. 228). A correlation between conversion and crossing over is particularly interesting in this connection because a recombination process to account for these results is only conceivable *during* replication.

2. We are concerned in this process with non-reciprocal recombination within a gene which leads to aberrant tetrads (p. 222ff.). Neither a 6:2 (or 3:1) nor a 5:3 tetrad can be explained by the exchange of segments between already replicated strands. The non-reciprocal nature of the 3:1 tetrads can only involve the two-fold duplication of a particular segment of one parental strand and the partial or complete failure of duplication of that segment in the other parental strand (Fig. IV-22c). The occurrence of 5:3 tetrads also demands a duplicate replication. However, in this case only a half-chromatid segment is copied. In such instances one is tempted to consider each of the two subunits of a chromatid as a single strand of DNA and a recombination (conversion) as resulting from semi-conservative replication (p. 131f.). The 6:2 tetrads remain unexplained under these circumstances; therefore conservative replication of at least a small region is probable.

The copy choice mechanism undoubtedly possesses advantages over the breakage-reunion hypothesis. Nevertheless, a series of unsolved problems still remain (compare discussion of WHITEHOUSE, 1963).

1. What induces the replicating strand to forsake its original template and switch to the other?

2. How does the shifting replica find the correct site in the other parental strand?

3. How are breaks in the already completely replicated strands explained in the case of sister strand exchanges (see breakage-reunion hypothesis p. 234f.)?

4. How is it possible that in certain cases whole chromatid strands and in others only subunits are replicated in duplicate?

5. How can copy choice be reconciled with semi-conservative replication? Is replication largely semi-conservative and only occasionally conservative in small regions where conversions are found?

Although many features of the copy choice mechanism remain puzzling, it has been used repeatedly in recent years in a modified form to explain the recombination process (p. 239ff.). Its advantage over the breakage-reunion hypothesis consists primarily in the fact that it can explain crossing over and conversion by relating recombination to replication.

3. Partial replication

The concept of a partial replication was originally developed by LURIA (1947) to explain recombinations observed in bacteriophages; some years later it was modified in certain details by HERSHEY (1952). In the modified form the hypothesis can also be applied to genetic phenomena in fungi (e.g. MITCHELL, 1955 a, b; BRESCH, 1964). In particular, it provides a possible interpretation of the occurrence of aberrant tetrads.

According to the partial replica mechanism, recombination occurs during and not following replication. To explain the recombination process in phage, BRESCH (1962) assumed that the synthesis of a daughter strand may begin at any point on the parental strand and proceeds in only one direction (Fig. IV-22d). If during the process one replicating segment meets another, it may slip under the other and separate it from the template. In general several such discarded pieces of genome may be completed to form intact strands.

The 3:1 tetrads in fungi can be interpreted by assuming that *a replicated segment formed along one of the two parental strands and separated from it becomes incorporated into the replica of the other chromatid during duplication* (Fig. IV-22d). Here as in copy choice a small segment of one of the parental structures is replicated twice; the same segment of the other parental strand is not copied. Nevertheless, in partial replication a continuous replica is formed through the *incorporation of the detached, autonomous partial replicas* and not as a result of switching from one template to the other. Such an interpretation of 3:1 tetrads is only made possible by assuming a conservative replication (p. 131), at least in small regions. The occurrence of 5:3 tetrads argues against semi-conservative replication if one proceeds from the related, but unproven assumption that the two subunits of a chromatid are identical to the two strands of the DNA double helix. The partial replication model also requires breaks in the two parental strands in so far as higher organisms are concerned, because only replicas recombine, as in copy choice, and the assumption that the replicas form the complete genome exclusively is hardly acceptable. With this limitation necessary for eukaryotic organisms, the basic advantage of the partial replication scheme is lost: a *recombination without breaks or template switch* through formation of a complete genome from partial replicas.

A further difficulty arises if one wishes to use the partial replication scheme to interpret reciprocal recombinations. In this instance one must postulate that the incorporation of a segment in the replica from one parental strand is correlated with the simultaneous incorporation of a homologous segment of replica into the other. It would not be necessary, however, that both partial replicas are identical. The correlation between conversion and crossing over (p. 228f.) in adjacent segments argues for different length segments with different genetic information at their ends. Such a mechanism is difficult to understand, however.

A number of questions also remain unanswered by the partial replication hypothesis. These overlap with those resulting from the discussion of copy choice (p. 237):

1. How can the incorporation of corresponding replicas in reciprocal recombination be explained?

2. How do breaks in "old" chromatids or half-chromatids arise (see copy choice)?

3. How can the occurrence of 3:1 tetrads be reconciled with semi-conservative replication (see copy choice)?

We see that on the one hand non-reciprocal recombination can only be explained by a recombination mechanism which occurs during and not following replication. This is the advantage of the copy choice and partial replication interpretations. On the other hand, all three mechanisms require breaks in intact strands. One comes away with the impression that *recombination is a complex process which may occur during synthesis of new strands as well as in intact strands after replication is complete.*

III. Recombination models

The recombination mechanisms described in the preceding section serve as a basis for models which have been developed to explain the experimental data. We wish to discuss two recombination models which are entirely different in their basic premises. One was proposed by STADLER and TOWE (1963), the other by WHITEHOUSE (1963)[1]. Both models base their explanation of recombination largely on the mode of replication and the structure of DNA. Their characteristics become particularly apparent in the distinctions between other, partially refuted models. For this reason we present first brief descriptions of three other hypotheses.

1. Switch hypothesis

Frequent template switch occurring in certain regions distributed at random along the chromosome. According to this model which was developed by FREESE (1957a, b) some years ago and since then has already been revised, crossing over and conversion are explained by a single mechanism. FREESE assumes that replication of a strand generally proceeds continuously along the same template. However, there are occasional regions of switching (randomly distributed along the chromosome) in which the replicating strand relinquishes its template and switches to the other homologue. He further postulated that within this very short region the switched replica may return to the original template and may even make three or more switches. The second replica does not

---

[1] Another recombination model was recently described and discussed in detail by HOLLIDAY (1964b). A brief reference to it is found in the footnote on p. 245. Note also the discussion in BERNSTEIN (1962, 1964) and WESTERGAARD (1964).

make all the reciprocal switches as those made by the first. The only requirement is that the second replica must emerge from the switch region on the unoccupied template to insure 2:2 segregation for the end markers.

With frequent template switches within short regions, the four combinations P_1, P_2, R_1, and R_2 of end markers (Fig. IV-19) are expected to appear in approximately equal frequencies among the $(a_1^+ a_2^+)$-recombinants, if the distance between a_1 and a_2 is small compared to the length of a switch region. According to the switch hypothesis with greater distances one should obtain more parental than recombinant types, although the same numbers of P_1 and P_2 types. While the results of FREESE (1957a, b) in *Neurospora crassa* appear to confirm this expectation (Table IV-21: p. 212/213) contradictory observations have been made repeatedly in the interim; these have led to the rejection of this model (Table IV-21; PRITCHARD, 1960; SIDDIQI, 1962; MURRAY, 1963; STADLER and TOWE, 1963).

2. Modified switch hypothesis

Frequent template switches in predetermined regions. In this model, which goes back to a proposal by STAHL (1961), reciprocal recombination of two markers may involve, as in the switch hypothesis, a multiple, non-reciprocal template alternation in a short region. However it is here postulated that the regions of switching are not randomly distributed throughout the chromosomes but restricted to specifically structured chromosomal segments; this feature distinguishes it from the switch hypothesis. A template shift is considered possible only by effective pairing in these predetermined regions. On the basis of these additional assumptions one would expect the parental end marker combinations P_1 and P_2 to always arise with differential frequency in allelic crosses, if one of the two heteroalleles lies close to the end of a switch region.

Although certain results which cannot be explained by the switch model are elucidated by the modified hypothesis, experimental data which contradict it also exist. According to the modified switch model one would expect that a mutational site lying at the end of a region of switching would give differential frequencies for the parental end marker combinations P_1 and P_2 (in the wild type recombinants) in crosses with other allelic markers. Such an expectation has not been confirmed in a number of cases, e.g. in *N. crassa* by STADLER and TOWE (1963) and in *Aspergillus nidulans* by SIDDIQI and PUTRAMENT (1963).

3. Polaron hypothesis

A single template switch in predetermined regions with polarized replication. In this recombination model proposed by RIZET and coworkers for *Ascobolus immersus* special regions in which a template switch may occur are also postulated. This model differs from the two previously described in three points:

1. A replica changes its template generally only once within a region of switching; it remains on the new template from the point of switching to the end of the region; the second replica likewise does not leave

its template. The region between the switch point and the end of the region of switching is copied in duplicate in one parental strand and not at all in the other parental strand.

2. The replication of strands always goes in a specific direction. The polarity is expressed in the double copying which always involves only one and the same side of the switching region (= polaron, p. 227) i.e. always the distally situated mutational site (p. 225 f.).

3. Reciprocal recombination can only occur between two different polarons in so-called "linkage structures". A non-reciprocal template switch cannot be "corrected" by a shift of the homologous replica in the same polaron.

The discovery (Tables IV-23 and IV-24) that both mutational sites within a gene have more or less equal chance of being copied in duplicate through a unilateral template switch, unlike the observation in *Ascobolus* that the same site is always the one which is copied in duplicate, argues against the generalization of this concept to other organisms. This fact could be reconciled with the polaron model by assuming that both mutational sites lie in different polarons; however, with such an interpretation reciprocal recombinations between the two heteroalleles would also be expected. This has not been observed (compare discussion of STADLER, 1963; and STADLER and TOWE, 1963).

4. Modified polaron hypothesis

Template switch with frequent return to the original template in predetermined regions with polarized replication. STADLER and TOWE (1963) developed this model for interpreting their results with *Neurospora crassa*; it differs from the polaron model in two points.

1. The authors assume — in contrast to the polaron hypothesis — that the switched replica very frequently returns to the original template while still within the polaron; further, the second switch of the replica occurs immediately adjacent to the first. Moreover, single switches are also possible; these are particularly frequent when the alleles are far apart. In this case the distal mutational site is copied in duplicate preferentially.

The following experimental results can be explained on the basis of this modification:

(a) Non reciprocal recombinations only occur within genes.

(b) Even with very closely-linked mutational sites a duplicate copying of either site is possible, if the replica returns to the original template.

(c) In the case of widely separated sites, only one, the distal site, is frequently or exclusively copied twice. According to STADLER and TOWE the different results in *Ascobolus* and *Neurospora* result from the fact that the relative frequency of single and double switches differs in the two organisms.

2. STADLER and TOWE, assuming a polarized replication, further postulate that at the distal end of the polaron there are equal chances that the two replicating strands will either switch templates or will continue to copy their original templates. Single switches by which the double copying continues to the end of the polaron (compare polaron hypothesis) are "corrected" at these sites, i.e. the switched replica

returns either to the original template or continues to copy the new template while the second replica switches to the other template at this point. A reciprocal crossover at the end of the polaron is possible by double as well as single switches.

Making the additional assumption permits the explanation of the following results:

(a) Reciprocal recombinations do not occur within a gene (p. 222 ff.).

(b) A correlation exists between intragenic non-reciprocal recombination and reciprocal segregation of end markers (p. 228 ff.).

(c) The conversion strand is frequently one of the two crossover strands (p. 229).

(d) The end markers which are proximal in relation to the direction of replication are linked with the allelic markers (p. 227 f.).

(e) The distally situated end markers are distributed at random among the wild-type recombinants, i.e. they are not linked with the allelic markers (p. 227f.).

The modified polaron hypothesis represents a formal genetic model of recombination which explains the bulk of the experimental data. However, on the molecular level there are certain difficulties which we will discuss briefly.

On the one hand, replication is known to be semi-conservative (see chapter on replication); on the other, the existence of 3:1 tetrads argues for a conservative mode of replication (p. 132). Further, it has been proved that all four strands of a tetrad may undergo recombination (p. 140). The copy choice mechanism explains only recombinations between replica strands, in semi-conservative replication only exchanges between subunits (= DNA single strands?) of the four chromatids present at meiosis. To explain these apparently contradictory facts, STADLER and TOWE assume that the bulk of the genetic material is replicated semi-conservatively prior to synapsis. Nevertheless, small regions (= polarons) exist in which replication can only occur later and in the conservative manner. The authors' reason that in these segments a template switch may lead to non-reciprocal recombination, assuming that the structures are tightly paired.

In light of these considerations STADLER and TOWE have proposed that recombination involves the following steps:

1. After synapsis the chromatids separate except at the points at which replication has not occurred.

2. The old strands of a chromatid break at the ends of a non-replicated region. The broken piece is incorporated into the "sister chromatid".

3. The synthesis of the fourth, not yet replicated chromatid segment proceeds only from the proximal end (polarization). During the process the replicas may switch to other templates and copy them for a short region (conversion). If only one of the two new strands being synthesized switches to the other template, a 5:3 segregation can occur.

4. The free, or distal end of the unreplicated chromatid can either attach to the approaching replica (without crossing over) or unite with one of the two non-sister chromatids (with crossing over). In the latter case a chromatid consisting of an "old" and a "new" strand must break at the distal end of the polaron.

5. After the union the tetrad opens and the chromatids separate from one another as far as the next non-replicated region. Here the process may repeat itself.

STADLER and TOWE's model of partial conservative replication, like all the other models, does not circumvent breaks in intact structures. A polarized replication is only required in the regions which are synthesized after synapsis, i.e. for the regions with conservative replication. STADLER and TOWE assume that the direction of replication in a polaron depends upon the relative position of the latter in the chromosome. This assumption is based on data from *N. crassa* and *A. nidulans*; these show that replication always begins from that side of the polaron on which the centromere lies (*N. crassa: me* locus, MURRAY, 1960a, b, 1963; *cys* locus, STADLER and TOWE, 1963; *A. nidulans: paba-1*-locus, SIDDIQI and PUTRAMENT, 1963). On the basis of these results it can be assumed that replication of polarons proceeds in the direction which is determined by the separation of the chromatid pairs; this begins at the centromere and moves in both directions from it.

5. Hybrid DNA hypothesis

Formation of hybrid DNA segments following breaks in two DNA single strands and subsequent replication in the region of the breaks. The molecular mechanism proposed by STADLER and TOWE is not the only one which is reasonably compatible with the data on recombination and replication. WHITEHOUSE (1963) developed a model with a wholly different molecular basis giving greater emphasis to the structure and mode of replication of DNA. According to this model the following events are postulated (Fig. IV-23; compare also Fig. IV-24):

1. A single DNA strand breaks in each of two DNA double helices (= chromatids?). The breaks need not to be in precisely homologous positions as in the breakage-reunion mechanism, but can occur at neighboring sites (Fig. IV-23 a).

2. In this case the two broken strands separate from their sister strands in the region of the breaks (Fig. IV-23 b) and become associated in most cases through pairing of complementary bases (Figs. IV-23 c and IV-24).

3. The spaces arising in this region of each of the double helices as a result of strand separation become filled by the extension of the intact strands through replication (Fig. IV-23 d).

4. The newly synthesized DNA single strands separate from their templates (Fig. IV-23 e), come together, and remain associated because of base pairing (Figs. IV-23 f and IV-24). Crossing over results from the double union between the separated portions of the strands of the two double helices. The DNA molecule is therefore hybrid in the crossover region.

5. The strands not involved in the crossover, i.e. each of the single strands that remain intact in the two DNA double helices, are now separated (Fig. IV-23 g) and attach to the DNA overlaps of the other two single strands. Eventually the spaces which arise through the breaks are filled in by newly added nucleotides (Fig. IV-23 h).

16*

WHITEHOUSE shows that breaks at homologous sites in the DNA single strands can also lead to hybrid DNA double strands by an additional replication. An "old" and a "new" strand and not two "old" and two "new" strands, lie next to one another in contrast to the recombination model shown in Fig. IV-23.

WHITEHOUSE's explains the occurrence of 6:2 and 5:3 tetrads without difficulty (Fig. IV-24). A base pair of one parent may be replaced by an allelic base pair from the other parent through loss of single nucleotides and the supplementary replication in the crossover region.

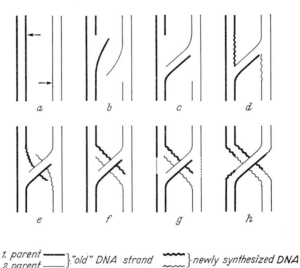

1. parent ——— }"old" DNA strand ∿∿∿ }newly synthesized DNA
2. parent ———

Fig. IV-23a—h. The eight phases of the recombination process according to the hybrid DNA hypothesis. Of the four DNA double helices only those two directly involved in recombination are shown. For the sake of clarity the spirals are omitted. The newly synthesized DNA in the region of the breaks (arrow in *a*) is shown with wavy lines and the extra DNA resulting from the supplementary synthesis in the old strands by dotted lines (phase *g*). Further explanation in text. (Adapted from WHITEHOUSE, 1963)

A 6:2 segregation may occur if a base pair is exchanged (for example, A—T for G—C; see Fig. IV-24, center) and if such an exchange involves an alteration of the genetic information. In most cases, as WHITEHOUSE assumes, the bases are complementary in the crossover region and a complete pairing between the detached segments of DNA will occur (see Fig. IV-23, c and f). In rare cases non-complementary bases (i.e. DNA segments with different genetic information) may be involved in the crossover region (Fig. IV-24, shown by two asterisks). This may lead to a 5:3 or 3:5 segregation of genetic markers.

The correlation between conversion and crossing over can be explained by the Whitehouse model (p. 228f.). The polarity observed in conversion

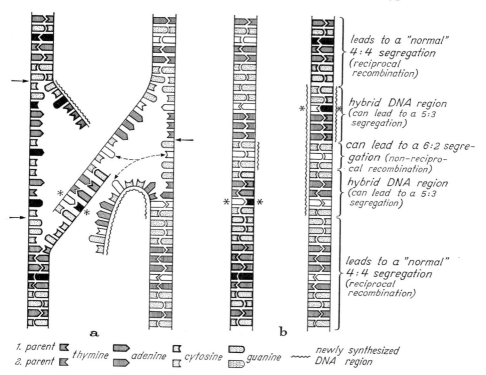

leads to a "normal"
4:4 segregation
(reciprocal
recombination)

hybrid DNA region
(can lead to a 5:3
segregation)

can lead to a 6:2 segre-
gation (non-recipro-
cal recombination)

hybrid DNA region
(can lead to a 5:3
segregation)

leads to a "normal"
4:4 segregation
(reciprocal
recombination)

a   b

1. parent ▌ thymine ▶ adenine ▌ cytosine ▦ guanine ～～ newly synthesized
2. parent ▌   ▶   ▌   ▦   ～～ DNA region

Fig. IV-24a and b. Interpretation of reciprocal and non-reciprocal recombination according to the hybrid DNA hypothesis. (a) Formation of
crossover corresponding to the phase of recombination illustrated in Fig.
IV-23, e. The sites of the DNA where breaks of single strands will occur in a
later stage of recombination (see Fig. IV-23g, h) are marked by arrows.
(b) The final result of the recombination process showing both DNA strands
at the completion of crossing over. The DNA strands which have not parti-
cipated in crossing over have been omitted in (a) and (b). Therefore, a 5:3
or 6:2 segregation appears as a 3:1 or 4:4 segregation respectively in this
figure. The same symbols for DNA are used here as in Fig. III-2. Bases or
base pairs for which both parents differ are printed in white or black respectively. Sites of non-complementary bases are marked by two asteriks.
Neither the spiral structure nor the portions of the nucleotides other than
the bases are shown. See Figs. III-1 and III-2 for further explanation of the
manner of representation. (Adapted from WHITEHOUSE, 1963)

also fits into the scheme through the formation of the hybrid DNA segments
(HASTINGS and WHITEHOUSE, 1964; WHITEHOUSE and HASTINGS, 1965).

The Whitehouse model differs from that of STADLER and TOWE in
the following features:

1. The replication of DNA is *entirely semi-conservative*.

2. A recombination is not produced as a result of a replication error,
but occurs *through breaks in already completed strands*. Complementary
strands are *not required to break at homologous points*.

3. In the breakage region a *supplementary semi-conservative replication* runs its course. This is essential for the "repair" of the breaks.

4. Reciprocal and non-reciprocal recombinations arise from the *formation of hybrid DNA segments*. Both the "old" and "new" single strands of DNA participate directly in recombination.

The concept that recombination is in some way linked with replication is taken into consideration in both models. Further, breaks in the DNA strands are postulated, although how and for what reason these breaks occur remains unexplained. We refrain from a commitment to this model, because of its speculative nature[1].

The recombination process will not be understood in detail until molecular rearrangements of genetic material are elucidated. A more precise knowledge of chromosome structure and the replication of DNA will undoubtedly contribute to a clarification of the relation between recombination and replication and thus an understanding of the recombination process.

Summary

1. The results of numerous investigations argue for the concept that two different recombination mechanisms with different action domains exist. One of these, conversion, is only possible within a gene and leads to non-reciprocal recombination. The other, crossing over, takes place only between genes and gives reciprocal recombinations.

2. The mechanism of recombination is unknown. Three hypotheses have been proposed to explain the recombination process. The breakage-reunion hypothesis postulates that recombination takes place after replication is complete by the breakage and reunion of two chromatids. The copy choice mechanism requires that recombination occurs during replication between the newly forming strands; the replica formed on one template strand switches to the template strand of the other parent. According to the mechanism of partial replication, recombination occurs during and not following replication by the incorporation of detached, autonomous partial replicas. The advantage of the latter two mechanisms over the breakage-reunion scheme is that the concurrence of recombination and replication explains both crossing over and conversion.

3. Discussion of certain recombination models reveals that a molecular interpretation of the process will remain unsatisfactory until the manner of DNA replication and the molecular structure of the chromosomes are fully understood.

---

[1] HOLLIDAY (1964b) recently proposed a model on the basis of which non-reciprocal recombination likewise occurs in hybrid DNA regions. It differs from the Whitehouse model in that a pair of non-complementary bases in the hybrid DNA region can be exchanged for unbound bases which "fit," this exchange being favored by lack of pairing. Such a base exchange is comparable to the replacement of base analogues when new mutants arise (HOLLIDAY, 1962b; see also Fig. V-6 in the chapter on mutation). It can lead to a 6:2 segregation unless a subsequent base exchange produces a reversion.

Literature

Apirion, D.: Formal and physiological genetics of ascospore colour in *Aspergillus nidulans*. Genet. Res. **4**, 276—283 (1963).

Barratt, R. W.: A word of caution: genetic and cytological effects of Abbott stocks in *Neurospora crassa*. Microbial. Genet. Bull. **11**, 5—6 (1954).

— D. Newmeyer, D. D. Perkins, and L. Garnjobst: Map construction in *Neurospora crassa*. Advanc. Genet. **6**, 1—93 (1954).

—, and W. N. Ogata: *Neurospora* stock list, second revision. Neurospora Newsletter **5**, 24—82 (1964a).

— — First supplement to *Neurospora* stock list, second revision as published in Neurospora Newsletter 5. Neurospora Newsletter **6**, 29—40 (1964b).

Barron, G. L.: The parasexual cycle and linkage relationships in the storage rot fungus *Penicillium expansum*. Canad. J. Bot. **40**, 1603—1613 (1962).

Beadle, G. W., and E. L. Tatum: *Neurospora*. II. Methods of producing and detecting mutations concerned with nutritional requirements. Amer. J. Bot. **32**, 678—685 (1945).

Belling, J.: Crossing over and gene rearrangement in flowering plants. Genetics **18**, 388—413 (1933).

Bennett, J. H.: Modes of tetrad formation. Amer. Naturalist **90**, 195—199 (1956).

Benzer, S.: Fine structure of a genetic region in bacteriophage. Proc. nat. Acad. Sci. (Wash.) **41**, 344—354 (1955).

— The elementary units of heredity. In: W. D. McElroy and B. Glass (edits.), The chemical basis of heredity, pp. 70—93. Baltimore 1957.

Bernstein, H.: On the mechanism of intragenic recombination. I. The *r II* region of bacteriophage *T 4*. J. theor. Biol. **3**, 335—353 (1962).

— On the mechanism of intragenic recombination. II. *Neurospora crassa*. J. theor. Biol. **6**, 347—370 (1964).

Bistis, G.: Studies on the genetics of *Ascobolus stercorarius* (Bull.) Schrot. Bull. Torrey bot. Club **83**, 35—61 (1956).

—, and L. S. Olive: Ascomycete spore mutants and their use in genetic studies. Science **120**, 105—106 (1954).

Bohn, W.: Einige Untersuchungen über die Tetradenaufspaltung bei Basidiomyceten. Z. indukt. Abstamm.- u. Vererb.-Lehre **67**, 435—445 (1933).

Bole-Gowda, B. N., D. D. Perkins, and W. N. Strickland: Crossing over and interference in the centromere region of linkage group I of *Neurospora*. Genetics **47**, 1243—1252 (1962).

Boone, D. M., and G. W. Keitt: *Venturia inaequalis* (Cke) Wint. VIII. Inheritance of colour mutant characters. Amer. J. Bot. **43**, 226—233 (1956).

— J. F. Stauffer, M. A. Stahmann, and G. W. Keitt: *Venturia inaequalis* (Cke) Wint. VII. Induction of mutants for studies on genetics, nutrition and pathogenicity. Amer. J. Bot. **43**, 199—204 (1956).

Bresch, C.: Zum Paarungsmechanismus von Bakteriophagen. Z. Naturforsch. **10b**, 545—561 (1955).

— Replication and recombination in bacteriophage. Z. Vererbungsl. **93**, 476—490 (1962).

— Klassische und molekulare Genetik. Berlin-Göttingen-Heidelberg: Springer 1964.

Bridges, C. B., and K. S. Brehme: The mutants of *Drosophila melanogaster*. Carnegie Inst. Wash. Publ., Washington (D.C.) **1944**, 552.

Brieger, F.: Die genaue Bestimmung des Zeitpunktes der Mendelspaltung. Züchter **5**, 34—44 (1933).

Brunswick, H.: Die Reduktionsteilung bei den Basidiomyceten. Z. Bot. **18**, 481—498 (1926).

Burgeff, H.: Variabilität, Vererbung und Mutation bei *Phycomyces blakesleeanus* Bgff. Z. indukt. Abstamm.- u. Vererb.-Lehre **49**, 26—94 (1928).

Buss, H. R.: The genetics of methionineless mutants of *Neurospora crassa*. Ph. D. Thesis, Stanford University 1944.

CARTER, T. C.: A search for chromatid interference in the male house mouse. Z. indukt. Abstamm.- u. Vererb.-Lehre **86**, 210—223 (1954).
—, and D. S. FALCONER: Stocks for detecting linkage in the mouse, and the theory of their design. J. Genet. **50**, 307—323 (1951).
—, and A. ROBERTSON: A mathematical treatment of genetical recombination, using a fourstrand model. Proc. roy. Soc. B **139**, 410—426 (1952).
CASE, M. E., and N. H. GILES: Evidence from tetrad analysis for both normal and aberrant recombination between allelic mutants in *Neurospora crassa*. Proc. nat. Acad. Sci. (Wash.) **44**, 378—390 (1958a).
— — Recombination mechanisms at the *pan-2*-locus in *Neurospora crassa*. Cold Spr. Harb. Symp. quant. Biol. **23**, 119—135 (1958b).
— — Comparative complementation and genetic maps of the *pan-2* locus in *Neurospora crassa*. Proc. nat. Acad. Sci. (Wash.) **46**, 659—676 (1960).
— — Allelic recombination in *Neurospora:* Tetrad analysis of a three-point cross within the *pan-2*-locus. Genetics **49**, 529—540 (1964).
CATCHESIDE, D. G.: The genetics of microorganisms. 223 pp. London: Pitmen 1951.
— A. P. JESSOP, and B. R. SMITH: Genetic controls of allelic recombination in *Neurospora*. Nature (Lond.) **202**, 1242—1243 (1964).
CHASE, M., and A. H. DOERMANN: High negative interference over short segments of the genetic structure of bacteriophage *T 4*. Genetics **43**, 332—353 (1958).
COLSON, B.: The cytology and morphology of *Neurospora tetrasperma* DODGE. Ann. Bot. (Lond.) **48**, 211—224 (1934).
DARLINGTON, C. D.: A cytological demonstration of „genetic" crossing over. Proc. roy. Soc. B **107**, 50—59 (1930).
DAY, P. R.: The genetics of *Coprinus lagopus*. Rep. John Innes hort. Instn. **49**, 16—18 (1958).
— The structure of the *A* mating type locus in *Coprinus lagopus*. Genetics **45**, 641—650 (1960).
—, and G. E. ANDERSON: Two linkage groups in *Coprinus lagopus*. Genet. Res. **2**, 414—423 (1961).
— D. M. BOONE, and G. W. KEITT: *Venturia inaequalis* (CKE) WINT. XI. The chromosome number. Amer. J. Bot. **43**, 835—838 (1956).
DEMEREC, M.: The fine structure of the gene. In: J. M. ALLEN (edit.) The molecular control of cellular activity, pp. 167—177. New York-Toronto-London 1962.
DESBOROUGH, S., and G. LINDEGREN: Chromosome mapping of linkage data from *Saccharomyces* by tetrad analysis. Genetica **30**, 346—383 (1959).
— E. E. SHULT, T. YOSHIDA, and C. C. LINDEGREN: Interference patterns in family *y-1* of *Saccharomyces*. Genetics **45**, 1467—1480 (1960).
DICKINSON, S.: Experiments on the physiology and genetics of the smut fungi. Cultural characters. I. Their permanence and segregation. Proc. roy. Soc. B **102**, 174—176 (1928).
DODGE, B. O.: Spore formation in asci with fewer than eight spores. Mycologia (N.Y.) **20**, 18—21 (1928).
DOERMANN, A. H.: Investigations of the lysine-requiring mutants of *Neurospora crassa*. Ph. D. Thesis, Stanford University 1946.
DOWDING, E. S.: *Gelasinospora*, a new genus of Pyrenomycetes with pitted spores. Canad. J. Res. **9**, 294—305 (1933).
—, and A. BAKERSPIGEL: Poor fruiters and barrage mutants in *Gelasinospora*. Canad. J. Bot. **34**, 231—240 (1956).
EBERSOLD, W. T.: Crossing over in *Chlamydomonas reinhardi*. Amer. J. Bot. **43**, 408—410 (1956).
—, and R. P. LEVINE: A genetic analysis of linkage group I of *Chlamydomonas reinhardi*. Z. Verbungsl. **90**, 74—82 (1959).
EL-ANI, A. S., L. S. OLIVE, and Y. KITANI: Genetics of *Sordaria fimicola*. IV. Linkage group I. Amer. J. Bot. **48**, 716—723 (1961).

ELLINGBOE, A. H.: Somatic recombination in *Puccinia graminis tritici*. Phytopathology **51**, 13—15 (1961).
— Illegitimacy and specific factor transfer in *Schizophyllum commune*. Proc. nat. Acad. Sci. (Wash.) **49**, 286—292 (1963).
— Somatic recombination in dikaryon *K* of *Schizophyllum commune*. Genetics **49**, 247—251 (1964).
—, and J. R. RAPER: Somatic recombination in *Schizophyllum commune*. Genetics **47**, 85—98 (1962).
ELLIOTT, C. G.: The cytology of *Aspergillus nidulans*. Genet. Res. **1**, 462—476 (1960a).
— Non-localised negative interference in *Aspergillus nidulans*. Heredity **15**, 247—262 (1960b).
EMERSON, S.: Meiotic recombination in fungi with special reference to tetrad analysis. In: W. J. BURDETTE (edit.), Methodology in basic genetics, pp. 167—208. San Franzisco 1963.
—, and J. E. CUSHING: Altered sulphonamide antagonism in *Neurospora*. Fed. Proc. **5**, 379—389 (1946).
ESSER, K.: Die Incompatibilitätsbeziehungen zwischen geographischen Rassen von *Podospora anserina* (CES.) REHM. I. Die genetische Analyse der Semi-Incompatibilität. Z. indukt. Abstamm.- u. Vererb.-Lehre **87**, 595—624 (1956).
— Die Incompatibilitätsbeziehungen zwischen geographischen Rassen von *Podospora anserina* (CES.) REHM. II. Die Wirkungsweise der Semi-Incompatibilitäts-Gene. Z. Vererbungsl. **90**, 29—52 (1959a).
— Die Incompatibilitätsbeziehungen zwischen geographischen Rassen von *Podospora anserina* (CES.) REHM. III. Untersuchungen zur Genphysiologie der Barragebildung und Semi-Incompatibilität. Z. Vererbungsl. **90**, 445—456 (1959b).
—, u. J. STRAUB: Genetische Untersuchungen an *Sordaria macrospora* AUERSW., Kompensation und Induktion bei genbedingten Entwicklungsdefekten. Z. Vererbungsl. **89**, 729—746 (1958).
EVERSOLE, R. A.: Biochemical mutants of *Chlamydomonas reinhardi*. Amer. J. Bot. **43**, 404—407 (1956).
—, and E. L. TATUM: Chemical alteration of crossing over frequency in *Chlamydomonas*. Proc. nat. Acad. Sci. (Wash.) **42**, 68—73 (1956).
FAULL, A. F.: On the resistance of *Neurospora crassa*. Mycologia (N.Y.) **22**, 288—303 (1930).
FINCHAM, J. R. S.: Genetic and biochemical studies in *Neurospora*. Ph. D. Thesis, Cambridge University 1950.
— A comparative genetic study of the mating-type chromosomes of two species of *Neurospora*. J. Genet. **50**, 221—229 (1951).
—, and P. R. DAY: Fungal genetics. Oxford 1963.
FISHER, R. A.: A class of enumerations of importance in genetics. Proc. roy. Soc. B **136**, 509—520 (1950).
— The experimental study of multiple crossing over. Caryologia (Firenze) **6**, Suppl. 227—231 (1955).
— M. F. LYON, and A. R. G. OWEN: The sex chromosome in the house mouse. Heredity **1**, 355—365 (1947).
FOGEL, S., and D. D. HURST: Coincidence relations between gene conversion and mitotic recombination in *Saccharomyces*. Genetics **48**, 321—328 (1963).
FORBES, E. C.: Recombination in the *pro* region of *Aspergillus nidulans*. Microbial Genet. Bull. **13**, 9—11 (1956).
FRANKE, G.: Die Zytologie der Ascusentwicklung von *Podospora anserina*. Z. indukt. Abstamm.- u. Vererb.-Lehre **88**, 159—160 (1957).
— Versuche zur Genomverdoppelung des Ascomyceten *Podospora anserina* (CES.) REHM. Z. Vererbungsl. **93**, 109—117 (1962).
FRATELLO, B., G. MORPURGO, and G. SERMONTI: Induced somatic segregation in *Aspergillus nidulans*. Genetics **45**, 785—800 (1960).

FREESE, E.: The correlation effect for a histidine locus of *Neurospora crassa*. Genetics **42**, 671—684 (1957a).

— Über die Feinstruktur des Genoms im Bereich eines *pab*-Locus von *Neurospora crassa*. Z. indukt. Abstamm.- u. Vererb.-Lehre **88**, 388—406 (1957b).

FROST, L. C.: A possible interpretation of the cytogenetic effects of Abbott stocks in *Neurospora crassa*. Microbial Genet. Bull. **12**, 7—9 (1955a).

— The genetics of some wild-type and mutant strains of *Neurospora crassa*. Ph. D. Thesis, Cambridge University 1955b.

— Heterogeneity in recombination frequencies in *Neurospora crassa*. Genet. Res. **2**, 43—62 (1961).

GARNJOBST, L.: Genetic control of heterocaryosis in *Neurospora crassa*. Amer. J. Bot. **40**, 607—614 (1953).

GILES, N. H., C. W. H. PARTRIDGE, and N. J. NELSON: The genetic control of adenylo-succinase in *Neurospora crassa*. Proc. nat. Acad. Sci. (Wash.) **43**, 305—317 (1957).

GOLDSCHMIDT, R.: Prä- oder Postreduktion der Chromosomen? Die Lösung eines alten Problems. Naturwissenschaften **19**, 358—362 (1932).

GOWANS, C. S.: Some genetic investigations on *Chlamydomonas eugametos*. Z. indukt. Abstamm.- u. Vererb.-Lehre **91**, 63—73 (1960).

GOWEN, J. W.: A biometrical study of crossing over. Genetics **4**, 205—250 (1919).

GRAUBARD, M. A.: Temperature effect on interference and crossing over. Genetics **19**, 83—94 (1934).

HALDANE, J. B. S.: The combination of linkage values, and the calculation of distances between the loci of linked factors. J. Genet. **8**, 299—309 (1919).

— The cytological basis of genetical interference. Cytologia (Tokyo) **3**, 54—65 (1931).

HANNA, W. F.: Studies in the physiology and cytology of *Ustilago zeae* and *Sorosporium reilianum*. Phytopathology **19**, 415—442 (1929).

HASTINGS, P. J., and H. L. K. WHITEHOUSE: A polaron model of genetic recombination by the formation of hybrid deoxyribonucleic acid. Nature (Lond.) **201**, 1052—1054 (1964).

HAWTHORNE, D. C., and R. K. MORTIMER: Chromosome mapping in *Saccharomyces*: centromere linked genes. Genetics **45**, 1085—1110 (1960).

HAYES, W.: The genetics of bacteria and their viruses. Studies in basic genetics and molecular biology. Oxford 1964.

HEAGY, F. C., and J. A. ROPER: Deoxyribonucleic acid content of haploid and diploid *Aspergillus* conidia. Nature (Lond.) **170**, 713 (1952).

HERSHEY, A. D.: Reproduction of bacteriophage. Int. Rev. Cytol. **1**, 119—134 (1952).

HESLOT, H.: Contribution à l'étude cytogénétique et génétique des Sordariacées. Rev. Cytol. Biol. véget. **19**, Suppl. **2**, 1—235 (1958).

HIRSCH, H. M.: Temperature-dependent cellulase production by *Neurospora crassa* and its ecological implications. Experientia (Basel) **10**, 180—182 (1954).

HOLLIDAY, R.: The genetics of *Ustilago maydis*. Genet. Res. **2**, 204—230 (1961a).

— Induced mitotic crossing over in *Ustilago maydis*. Genet. Res. **2**, 231—248 (1961b).

— Effect of photoreactivation on ultra-violet-induced segregation of heterozygous diploids. Nature (Lond.) **193**, 95—96 (1962a).

— Mutation and replication in *Ustilago maydis*. Genet. Res. **3**, 472—486 (1962b).

— The induction of mitotic recombination by mitomycin C in *Ustilago* and *Saccharomyces*. Genetics **50**, 323—335 (1964a).

— A mechanism for gene conversion in fungi. Genet. Res. **5**, 282—304 (1964b).

HOLLOWAY, B. W.: Heterocaryosis in *Neurospora crassa*. Ph. D. Thesis, California Institute of Technology 1953.
— Segregation of the mating-type locus in *Neurospora crassa*. Microbial Genet. Bull. **10**, 15—16 (1954).
HOROWITZ, N. H., and H. MACLEOD: The DNA content of *Neurospora* nuclei. Microbial Genet. Bull. **17**, 6—7 (1960).
HOULAHAN, M. B., G. W. BEADLE, and H. G. CALHOUN: Linkage studies with biochemical mutants of *Neurospora crassa*. Genetics **34**, 493—507 (1949).
HOWE, H. B.: Crossing over in the first (sex) chromosome of *Neurospora crassa*. Genetics **39**, 972—973 (1954).
— Crossing over and nuclear passing in *Neurospora crassa*. Genetics **41**, 610—622 (1956).
— Markers and centromere distance in *Neurospora tetrasperma*. Genetics **48**, 121—131 (1963).
— Sources of error in genetic analysis in *Neurospora tetrasperma*. Genetics **50**, 181—189 (1964).
HÜTTIG, W.: Über den Einfluß der Temperatur auf die Keimung und Geschlechtsverteilung bei Brandpilzen. Z. Bot. **24**, 529—577 (1931).
— Über physikalische und chemische Beeinflussung des Zeitpunktes der Chromosomenreduktion bei Brandpilzen. Z. Bot. **26**, 1—26 (1933a).
— Über den Einfluß von Außenbedingungen auf die Chromosomenreduktion. Züchter **5**, 243—249 (1933b).
HURST, D. D., and S. FOGEL: Mitotic recombination and heteroallelic repair in *Saccharomyces cerevisiae*. Genetics **50**, 435—458 (1964).
HWANG, Y. L., G. LINDEGREN, and C. C. LINDEGREN: Mapping the eleventh centromere in *Saccharomyces*. Canad. J. Genet. Cytol. **5**, 290—298 (1963).
IHLER, G., and M. MESELSON: Genetic recombination in bacteriophage λ by breakage and joining of DNA molecules. Virology **21**, 7—10 (1963).
IKEDA, Y., C. ISHITANI, and K. NAKAMURA: A high frequency of heterozygous diploids and somatic recombination induced in imperfect fungi by ultra-violet light. J. gen. appl. Microbiol. (Tokyo) **3**, 1—11 (1957).
ISHIKAWA, T.: Genetic studies of *ad-8* mutants in *Neurospora crassa*. I. Genetic fine structure of the *ad-8* locus. Genetics **47**, 1147—1161 (1962).
ISHITANI, C.: A high frequency of heterozygous diploids and somatic recombination produced by ultra-violet light in imperfect fungi. Nature (Lond.) **178**, 706 (1956).
ITO, T.: Genetic study on the expression of the color factor of the ascospore in *Sordaria fimicola*. I. Segregation of the dark- and lightcolored ascospore. Res. Bull. Obihiro Zootech. Univ. Ser. I **3**, 223—230 (1960).
JAMES, A. P., and B. LEE-WHITING: Radiation-induced genetic segregations in vegetative cells of diploid yeast. Genetics **40**, 826—831 (1955).
JANSSENS, F. A.: La theorie de la chiasmatypie. Nouvelle interprétation des cinèses de maturation. Cellule **25**, 389—411 (1909).
JOLY, P.: Données récentes sur la génétique des champignons supérieurs (Ascomycètes et Basidiomycètes). Rev. Mycol. (Paris) **29**, 115—186 (1964).
JOUSSEN, H., u. J. KEMPER: Ein neues Interferenzmodell zur Aufstellung von Tetraden-Kartierungsfunktionen. Z. Vererbungsl. **91**, 350—354 (1960).
KÄFER, E.: An 8-chromosome map of *Aspergillus nidulans*. Advanc. Genet. **9**, 105—145 (1958).
— The processes of spontaneous recombination in vegetative nuclei of *Aspergillus nidulans*. Genetics **46**, 1581—1609 (1961).
— Radiation effects and mitotic recombination in diploids of *Aspergillus nidulans*. Genetics **48**, 27—45 (1963).
—, and A. CHEN: UV-induced mutations and mitotic crossing-over in dormant and germinating conidia of *Aspergillus*. Microbial. Genet. Bull. **20**, 8—9 (1964).
KAKAR, S. N.: Allelic recombination and its relation to recombination of outside markers in yeast. Genetics **48**, 957—966 (1963).

Recombination

KAPLAN, R. W.: Genetik der Mikroorganismen. Fortschr. Bot. **22**, 293—315 (1960).
KEITT, G. W., and D. M. BOONE: Induction and inheritance of mutant characters in *Venturia inaequalis* in relation to its pathologenicity. Phytopathology **44**, 362—370 (1954).
— — Use of induced mutations in the study of host-pathogen relationships. Genetics in Plant Breeding. Brookhaven Symp. in Biol. **9**, 209—217 (1956).
KELLENBERGER, G., M. L. ZICHICHI, and J. J. WEIGLE: Exchange of DNA in the recombination of bacteriophage. Proc. nat. Acad. Sci. (Wash.) **47**, 869—878 (1961).
KEMPER, J.: Temperaturabhängigkeit der Rekombinations- und Interferenzwerte bei *Sordaria macrospora* AUERSW. Diss. Math. Naturwiss. Fak. Univ. Köln 1964.
KIKKAWA, H.: Crossing over in the males of *Drosophila virilis*. Proc. Imp. Acad. (Tokyo) **9**, 535—536 (1933).
— Biological significance of coincidence in crossing over. Jap. J. Genet. **11**, 51—59 (1935).
KITANI, Y.: Three kinds of transreplication in *Sordaria fimicola*. Jap. J. Genet. **37**, 131—146 (1962).
— L. S. OLIVE, and A. S. EL-ANI: Transreplication and crossing over in *Sordaria fimicola*. Science **134**, 668—669 (1961).
— — — Genetics of *Sordaria fimicola*. V. Aberrant segregation at the *g* locus. Amer. J. Bot. **49**, 697—706 (1962).
KNAPP, E.: Zur Genetik von *Sphaerocarpus*. (Tetradenanalytische Untersuchungen.) Ber. dtsch. bot. Ges. **54**, 58—69 (1936).
— Crossing over und Chromosomenreduktion. Z. indukt. Abstamm.- u. Vererb.-Lehre **73**, 409—418 (1937).
— Tetrad analysis in green plants. Canad. J. Genet. Cytol. **2**, 89—95 (1960).
—, u. E. MÖLLER: Tetradenanalytische Auswertung eines Dreipunktversuches bei *Sphaerocarpus donellii* AUST. Z. indukt. Abstamm.- u. Vererb.-Lehre **87**, 298—310 (1955).
KNIEP, H.: Die Sexualität der niederen Pflanzen. Jena: Gustav Fischer 1928.
KOSAMBI, D. D.: The estimation of map distance from recombination values. Ann. Eugen. (Lond.) **12**, 172—175 (1944).
KUENEN, R.: Ein Modell zur Analyse der crossover-Interferenz. Z. Vererbungsl. **93**, 35—65 (1962a).
— Crossover- und Chromatiden-Interferenz bei *Podospora anserina* (CES.) REHM. Z. Vererbungsl. **93**, 66—108 (1962b).
LEDERBERG, J.: Recombination mechanisms in bacteria. J. cell. comp. Physiol. (Suppl. 2) **45**, 75—107 (1955).
LEUPOLD, U.: Die Vererbung von Homothallie und Heterothallie bei *Schizosaccharomyces pombe*. C. R. Lab. Carlsberg, Sér. Physiol. **24**, 381—480 (1950).
— Physiologisch-genetische Studien an adeninabhängigen Mutanten von *Schizosaccharomyces pombe*. Ein Beitrag zum Problem der Pseudoallelie. Schweiz. Z. allg. Path. **20**, 535—544 (1957).
— Studies on recombination in *Schizosaccharomyces pombe*. Cold Spr. Harb. Symp. quant. Biol. **23**, 161—170 (1958).
— Intragene Rekombination und allele Komplementierung. Arch. Klaus-Stift. Vererb.-Forsch. **36**, 89—117 (1961).
—, and H. HOTTINGUER: Some data on segregation in *Saccharomyces*. Heredity **8**, 243—258 (1954).
LEVINE, R. P., and W. T. EBERSOLD: Gene recombination in *Chlamydomonas reinhardi*. Cold Spr. Harb. Symp. quant. Biol. **23**, 101—109 (1958).
LEWIS, D.: Genetical analysis of methionine suppressors in *Coprinus*. Genet. Res. **2**, 141—155 (1961).
LINDEGREN, C. C.: The genetics of *Neurospora*. I. The inheritance of response to heat treatment. Bull. Torrey bot. Club **59**, 85—102 (1932a).

LINDEGREN, C. C.: The genetics of *Neurospora*. II. Segregation of the sex factors in asci of *Neurospora crassa*, *N. sitophila* and *N. tetrasperma*. Bull. Torrey bot. Club **59**, 119—138 (1932b).
— The genetics of *Neurospora*. III. Pure bred stocks and crossing over in *Neurospora crassa*. Bull. Torrey bot. Club **60**, 133—154 (1933).
— A six-point map of the sex chromosome of *Neurospora crassa*. J. Genet. **32**, 243—256 (1936a).
— The structure of the sex chromosome of *Neurospora crassa*. J. Hered. **27**, 251—259 (1936b).
— Chromosome maps of *Saccharomyces*. Proc. of the 8th Internat. Congr. of Genet. (Hereditas Suppl. Vol.) 338—355 (1949a).
— The yeast cell, its genetics and cytology. St. Louis: Educational Publishers Ltd. 1949b.
— Non-mendelian segregation in a single tetrad of *Saccharomyces* ascribed to gene conversion. Science **121**, 605—607 (1955).
—, and G. LINDEGREN: Non-random crossing over in *Neurospora*. J. Hered. **28**, 105—113 (1937).
— — Non-random crossing over in the second chromosome of *Neurospora crassa*. Genetics **24**, 1—7 (1939).
— — Locally specific patterns of chromatid and chromosome interference in *Neurospora*. Genetics **27**, 1—24 (1942).
— — R. B. DRYSDALE, J. P. HUGHES, and A. BRENES-POMALES: Genetical analysis of the clones from a single tetrad of *Saccharomyces* showing non-mendelian segregation. Genetica **28**, 1—24 (1956).
— — E. E. SHULT, and S. DESBOROUGH: Chromosome maps of *Saccharomyces*. Nature (Lond.) **183**, 800—802 (1959).
— — —, and Y. L. HWANG: Centromeres sites of affinity and gene loci on the chromosomes of *Saccharomyces*. Nature (Lond.) **194**, 260—265 (1962).
— — — — Chromosome maps of *Saccharomyces*. Microbial Gen. Bull., Suppl. to No **19** (1963).
—, and E. E. SHULT: Non-random assortment of centromeres with implications regarding random assortment of chromosomes. Experientia (Basel) **12**, 177 (1956).
LISSOUBA, P.: Mise en évidence d'une unité génétique polarisée et essai d'analyse d'un cas d'interférence négative. Ann. Sci. nat. Bot. **44**, 641—720 (1960).
— J. MOUSSEAU, G. RIZET, and J. L. ROSSIGNOL: Fine structure of genes in the ascomycete *Ascobolus immersus*. Advanc. Genet. **11**, 343—380 (1962).
—, et G. RIZET: Sur l'existence d'une unité génétique polarisée ne subissant que des échanges non réciproques. C. R. Acad. Sci. (Paris) **250**, 3408—3410 (1960).
LUDWIG, W.: Über numerische Beziehungen der crossover-Werte untereinander. Z. indukt. Abstamm.- u. Vererb.-Lehre **67**, 58—95 (1934).
— Über die Häufigkeit von Prä- und Postreduktion. Z. indukt. Abstamm.- u. Vererb.-Lehre **73**, 332—346 (1937a).
— Faktorenkoppelung und Faktorenaustausch bei normalem und aberrantem Chromosomenbestand. Leipzig: Georg Thieme 1938.
LURIA, S. E.: Reactivation of irradiated bacteriophage by transfer of self reproducing units. Proc. nat. Acad. Sci (Wash.) **33**, 253—263 (1947).
MAKAREWICZ, A.: First results of genetic analysis in series 726 of *Ascobolus immersus*. Acta Soc. Bot. Pol. **33**, 1—8 (1964).
MALING, B.: Linkage data for group IV markers in *Neurospora*. Genetics **44**, 1215—1220 (1959).
MANNEY, T. R.: Action of a super-suppressor in yeast in relation to allelic mapping and complementation. Genetics **50**, 109—121 (1964).
—, and R. K. MORTIMER: Allelic mapping in yeast using X-ray-induced mitotic reversion. Science **143**, 581—582 (1964).
MATHER, K.: Reductional and equational seperation of the chromosomes in bivalents and multivalents. J. Genet. **30**, 53—78 (1935).

Recombination

MATHIESON, M. J.: Polarized segregation in *Bombardia lunata*. Ann. of Bot., N. S. **20**, 623—634 (1956).

McCLINTOCK, B.: *Neurospora*. I. Preliminary observations of the chromosomes of *Neurospora crassa*. Amer. J. Bot. **32**, 671—678 (1945).

McELROY, W. D., and B. GLASS (edits.): The chemical basis of heredity. Baltimore 1957.

McNELLY, C. A., and L. C. FROST: The effect of temperature on the frequency of recombination in *Neurospora crassa*. Genetics **48**, 900 (Abstr.) (1963).

MESELSON, M., and J. J. WEIGLE: Chromosome breakage accompanying genetic recombination in bacteriophage. Proc. nat. Acad. Sci. (Wash.) **47**, 857—868 (1961).

MICHIE, D.: Affinity: A new genetic phenomenon in the house mouse. Evidence from distant crosses. Nature (Lond.) **171**, 26—27 (1953).

— Affinity. Proc. rov. Soc. B **144**, 241—259 (1955).

MIDDLETON, R. B.: Sexual and somatic recombination in common-*AB* heterokaryons of *Schizophyllum commune*. Genetics **50**, 701—710 (1964).

MILLER, M. W., and E. A. BEVAN: Radio-protective chemicals and genetic recombination in *Sordaria fimicola*. Nature (Lond.) **202**, 716 (1964).

MITCHELL, H. K.: Crossing over and gene conversion in *Neurospora*. In: W. D. McELROY and B. GLASS (edits), The chemical basis of heredity, pp. 94—113. Baltimore 1957.

MITCHELL, M. B.: Aberrant recombination of pyridoxine mutants of *Neurospora*. Proc. nat. Acad. Sci. (Wash.) **41**, 215—220 (1955a).

— Further evidence of aberrant recombination in *Neurospora*. Proc. nat. Acad. Sci. (Wash.) **41**, 935—937 (1955b).

— Genetic recombination in *Neurospora*. Genetics **43**, 799—813 (1958).

— Detailed analysis of a *Neurospora* cross. Genetics **44**, 847—856 (1959).

— Ascus formation and recombinant frequencies in *Neurospora crassa*. Genetics **45**, 507—517 (1960a).

— Evidence of non-random distribution of ascus classes in fruiting bodies of *Neurospora crassa*. Genetics **45**, 1245—1251 (1960b).

— Indications of pre-ascus recombination in *Neurospora* crosses. Genetics **48**, 553—559 (1963).

— Phenotype distributions in asci of *Neurospora crassa*. Amer. J. Bot. **51**, 88—96 (1964).

—, and H. K. MITCHELL: A partial map of linkage group D in *Neurospora crassa*. Proc. nat. Acad. Sci. (Wash.) **40**, 436—440 (1954).

— T. H. PITTENGER, and H. K. MITCHELL: Pseudowild types in *Neurospora crassa*. Proc. nat. Acad. Sci. (Wash.) **38**, 569—580 (1952).

MÖLLER, E.: Über Chromatideninterferenz. Z. Vererbungsl. **90**, 409—420 (1959).

MONNOT, F.: Sur la localisation du gène *S* sur quelques particularités du crossing over chez *Podospora anserina*. C. R. Acad. Sci. (Paris) **236**, 2330—2332 (1953).

MORGAN, L. V.: A closed x-chromosome in *Drosophila melanogaster*. Genetics **18**, 250—283 (1933).

MORGAN, T. H.: An attempt to analyse the constitution of the chromosomes on the basis of sex-linked inheritance in *Drosophila*. J. exp. Zool. **11**, 365—415 (1911a).

— Random segregation versus coupling in mendelian inheritance. Science **34**, 384 (1911b).

— The theory of the gene. New Haven: Yale University Press 1926.

MORPURGO, G.: Somatic segregation induced by p-fluorophenylalanine. Aspergillus News Letter **2**, 10 (1961).

— Increased frequency of somatic crossing over by X rays in *Aspergillus nidulans*. Microbial Genet. Bull. **18**, 18—20 (1962a).

— Quantitative measurement of induced somatic segregation in *Aspergillus nidulans*. Sci. Rep. Ist. sup. Sanità (Roma) **2**, 324—329 (1962b).

MORPURGO, G.: Induction of mitotic crossing over in *Aspergillus nidulans* by bifunctional alkylating agents. Genetics **48**, 1259—1263 (1963).

—, and G. SERMONTI: Chemically-induced instabilities in a heterozygous diploid of *Penicillium chrysogenum*. Genetics **44**, 137—152 (1959).

MORTIMER, R. K., and D. C. HAWTHORNE: Chromosome maps of *Saccharomyces*. Microbial Gen. Bull., Suppl. to No **19** (1963).

MULLER, H. J.: Further studies on the nature and causes of gene mutations. Proc. 6th Intern. Congr. Genet. **1**, 213—214 (1932).

MUNDKUR, B. D.: Evidence excluding mutations, polysomy, and polyploidy as possible causes of non-mendelian segregations in *Saccharomyces*. Ann. Missouri bot. Gard. **26**, 259—280 (1949).

— Irregular segregations in yeast hybrids. Curr. Sci. **19**, 84—85 (1950).

MURRAY, N. E.: The distribution of methionine loci in *Neurospora crassa*. Heredity **15**, 199—206 (1960a).

— Complementation and recombination between methionine-2 alleles in *Neurospora crassa*. Heredity **15**, 207—217 (1960b).

— Polarized recombination and fine structure within the *me-2* gene of *Neurospora crassa*. Genetics **48**, 1163—1184 (1963).

NAKAMURA, K.: An ascospore color mutant of *Neurospora crassa*. Bot. Mag. (Tokyo) **74**, 104—109 (1961).

OEHLKERS, F.: Meiosis und crossing over. Biol. Zbl. **60**, 337—348 (1940a).

— Meiosis und crossing over. Cytogenetische Untersuchungen an *Oenothera*. Z. indukt. Abstamm.- u. Vererb.-Lehre **78**, 157—168 (1940b).

OGUR, M., S. MINCKLER, G. LINDEGREN, and C. C. LINDEGREN: The nucleic acids in a polyploid series of *Saccharomyces*. Arch. Biochem. **40**, 175—184 (1952).

OLIVE, L. S.: Genetics of *Sordaria fimicola*. I. Ascospore colour mutants. Amer. J. Bot. **43**, 97—107 (1956).

— Aberrant tetrads in *Sordaria fimicola*. Proc. nat. Acad. Sci. (Wash.) **45**, 727—732 (1959).

— Mechanisms of genetic recombination in the fungi. In: G. DALLDORF (edit.), Fungi and fungous diseases. Springfield 1962.

OWEN, A. R. G.: The theory of genetical recombination. Advanc. Genet. **3**, 117—157 (1950).

PÄTAU, K.: Cytologischer Nachweis einer positiven Interferenz über das Centromer. (Der Paarungskoeffizient I.) Chromosoma (Berl.) **2**, 36—63 (1941).

PAPAZIAN, H. P.: Physiology of the incompatibility factors in *Schizophyllum commune*. Bot. Gaz. **112**, 143—163 (1950a).

— A method of isolating the four spores from a single basidium in *Schizophyllum commune*. Bot. Gaz. **112**, 139—140 (1950b).

— The incompatibility factors and a related gene in *Schizophyllum commune*. Genetics **36**, 441—459 (1951).

— The analysis of tetrad data. Genetics **37**, 175—188 (1952).

— Cluster model of crossing over. Genetics **45**, 1169—1175 (1960).

—, and C. C. LINDEGREN: A study of irregular quadruplets in *Saccharomyces*. Genetics **45**, 847—854 (1960).

PARAG, Y.: Studies on somatic recombination in dikaryons of *Schizophyllum commune*. Heredity **17**, 305—318 (1962).

PARSONS, P. A.: Genetical interference in maize. Nature (Lond.) **179**, 161—162 (1957).

— Genetical interference in *Drosophila spp.* Nature (Lond.) **182**, 1815—1816 (1958).

PATEMAN, J. A.: Aberrant recombination at the *am* locus in *Neurospora crassa*. Nature (Lond.) **181**, 1605—1606 (1958).

— High negative interference at the *am* locus in *Neurospora crassa*. Genetics **45**, 839—846 (1960a).

— Inter-relationships of the alleles at the *am* locus in *Neurospora crassa*. J. gen. Microbiol. **23**, 393—399 (1960b).

PAYNE, L. C.: The theory of genetical recombination: a general formulation for a certain class of intercept length distribution appropriate to the discussion of multiple linkage. Proc. roy. Soc. B **144**, 528—544 (1956).
— The theory of genetical recombination: effect of changing the $^{1}/_{4}\chi^{2}_{4}$ intercept length distribution. Heredity **11**, 129—139 (1957).
PERKINS, D. D.: Biochemical mutants in the smut fungus *Ustilago maydis*. Genetics **34**, 607—626 (1949).
— The detection of linkage in tetrad analysis. Genetics **38**, 187—197 (1953).
— Tetrads and crossing over. J. cell. comp. Physiol. **45**, 119—149 (1955).
— Crossing over in a multiply marked chromosome arm of *Neurospora*. Microbial Genet. Bull. **13**, 22—23 (1956).
— New markers and multiple point linkage data in *Neurospora*. Genetics **44**, 1185—1208 (1959).
— The frequency in *Neurospora* tetrads of multiple exchanges within short intervals. Genet. Res. **3**, 315—327 (1962a).
— Crossing over and interference in a multiply marked chromosome arm of *Neurospora*. Genetics **47**, 1253—1274 (1962b).
— A. S. EL-ANI, L. S. OLIVE and Y. KITANI: Interference between exchanges in tetrads of *Sordaria fimicola*. Amer. Naturalist **97**, 249—252 (1963).
— M. GLASSEY, and B. A. BLOOM: New data on markers and rearrangements in *Neurospora*. Canad. J. Genet. Cytol. **4**, 187—205 (1962).
— , and C. ISHITANI: Linkage data for group III markers in *Neurospora*. Genetics **44**, 1209—1213 (1959).
— , and N. E. MURRAY: New markers and linkage data. Neurospora Newsletter **4**, 26—27 (1963).
PITTENGER, T. H.: The general incidence of pseudo-wild types in *Neurospora crassa*. Genetics **39**, 326—342 (1954).
— Mitotic instability of pseudo-wild types in *Neurospora*. Proc. 10th Intern. Congr. Genet. **2**, 218 (1958).
— , and M. B. COYLE: Somatic recombination in pseudowild-type cultures of *Neurospora crassa*. Proc. nat. Acad. Sci. (Wash.) **49**, 445—455 (1963).
PLOUGH, H. H.: The effect of temperature on crossing over in *Drosophila*. J. exp. Zool. **24**, 147—210 (1917).
PONTECORVO, G.: Mitotic recombination in the genetic systems of filamentous fungi. Proc. 9th Intern. Congr. Genet. 1953a.
— The genetics of *Aspergillus nidulans*. Advanc. Genet. **5**, 141—238 (1953b).
— Mitotic recombination in the genetic systems of filamentous fungi. Caryologia, Vol. suppl. 1—9 (1954).
— Trends in genetic analysis. New York: Columbia University Press 1958.
— , and E. KÄFER: Genetic analysis based on mitotic recombination. Advanc. Genet. **9**, 71—104 (1958).
— , and J. A. ROPER: Genetic analysis without sexual reproduction by means of polyploidy in *Aspergillus nidulans*. J. gen. Microbiol. **6**, VII (1952).
— — Diploids and mitotic recombination. Advanc. Genet. **5**, 218—233 (1953).
— — Resolving power of genetic analysis. Nature (Lond.) **178**, 83—84 (1956).
— — L. M. HEMMONS, K. D. MACDONALD and A. W. J. BUFTON: The genetics of *Aspergillus nidulans*. Advanc. Genet. **5**, 141—238 (1953).
— , and G. SERMONTI: Parasexual recombination in *Penicillium chrysogenum*. J. gen. Microbiol. **11**, 94—104 (1954).
PRAKASH, V.: Parental and non-parental association of centromeres in *Neurospora crassa*. 11th Intern. Congr. Genet. **2**. The Hague, 1963a (in press).
— Effects of chelating agents on crossing over in *Neurospora crassa*. Genetica **34**, 121—151 (1963b).
— Chromatid interference in *Neurospora crassa*. Genetics **50**, 297—321 (1964).

Prévost, G.: Etude génétique d'un basidiomycète: *Coprinus radiatus* Fr. ex Bolt. Thèse Fac. de Science, Université Paris 1962.

Pritchard, R. H.: The linear arrangement of a series of alleles of *Aspergillus nidulans*. Heredity **9**, 343—371 (1955).

— Recombination and negative interference in *Aspergillus nidulans*. Proc. 10th Intern. Congr. Genet. **2**, 223—224 (1958).

— Localized negative interference and its bearing on models of gene recombination. Genet. Res. **1**, 1—24 (1960).

— Mitotic recombination in fungi. In: W. J. Burdette (edit.), Methodology in basic genetics, pp. 228—246. San Francisco 1963.

Prud'homme, N.: Recombinations chromosomique extra-basidales chez un basidiomycète «*Coprinus radiatus*» Ann. Génét. **4**, 63—66 (1963).

Quintanilha, A.: Le problème de la sexualité chez les Basidiomycètes. Recherches sur le genre „*Coprinus*". Bol. soc. Broteriana **8**, 1—99 (1933).

Rademacher, L.: Mathematische Theorie der Genkopplung unter Berücksichtigung der Interferenz. J.ber. schles. Ges. vaterl. Kultur **105**, 83—92 (1932).

Raper, J. R., M. G. Baxter, and R. B. Middleton: The genetic structure of the incompatibility factors in *Schizophyllum commune*. Proc. nat. Acad. Sci. (Wash.) **44**, 889—900 (1958).

—, and P. G. Miles: The genetics of *Schizophyllum commune*. Genetics **43**, 530—566 (1958).

Regnery, D. C.: A study of the leucineless mutants of *Neurospora crassa*. Ph. D. Thesis, California Institute of Technology 1947.

Reimann-Philipp, R.: Genetische Untersuchungen an den Tetraden einer höheren Pflanze (*Salpiglossis variabilis*). Z. indukt. Abstamm.- u. Vererb.-Lehre **87**, 187—207 (1955).

Rifaat, O. M.: Genetical studies on the mating-type chromosome of *Neurospora crassa*. Ph. D. Thesis Cambridge 1956.

— A possible inversion in the mating-type chromosome of *Neurospora crassa*. Genetica **29**, 193—205 (1958).

— Effect of temperature on crossing over in *Neurospora crassa*. Genetica **30**, 312—323 (1959).

Rizet, G., et C. Engelmann: Contribution à l'étude génétique d'un ascomycète tetrasporé: *Podospora anserina*. Rev. Cytol. Biol. végét. **11**, 201—304 (1949).

— P. Lissouba et J. Mousseau: Sur l'interférence négative au sein d'une série d'allèles chez *Ascobolus immersus*. C. R. Soc. Biol. (Paris) **11**, 1967—1970 (1960a).

— — — Les mutations d'ascospore chez l'ascomycète *Ascobolus immersus* et l'analyse de la structure fine des gènes. Bull. Soc. franç. Physiol. Vég. **6**, 175—193 (1960b).

—, et J. L. Rossignol: Sur la dissymmétrie de certaines conversions et sur la dimension de l'erreur de copie chez l'*Ascobolus immersus*. Revista Biol. **3**, 261—268 (1963).

Roman, H.: Studies of gene mutation in *Saccharomyces*. Cold Spr. Harb. Symp. quant. Biol. **21**, 175—185 (1956).

— Sur les recombinaisons non réciproques chez *Saccharomyces cerevisiae* et sur les problèms posés par ces phénomènes. Ann. Génét. **1**, 11—17 (1958).

— Genic conversion in fungi. In: W. J. Burdette (edit.), Methodology in basic genetics, pp. 209—227. San Francisco 1963.

— D. C. Hawthorne, and H. C. Douglas: Polyploidy in yeast and its bearing on the occurrence of irregular genetic ratios. Proc. nat. Acad. Sci. (Wash.) **37**, 79—84 (1951).

—, and F. Jacob: A comparison of spontaneous and ultraviolet-induced allelic recombination with reference to the recombination of outside markers. Cold Spr. Harb. Symp. quant. Biol. **23**, 155—160 (1958).

ROPER, J. A.: A search for linkage between genes determing vitamin requirements. Nature (Lond.) **166**, 956 (1950).
— Production of heterozygous diploids in filamentous fungi. Experientia (Basel) **8**, 14—15 (1952).
—, and R. H. PRITCHARD: The recovery of the complementary products of mitotic crossing over. Nature (Lond.) **175**, 639 (1955).
ROSSIGNOL, J. L.: Phénomènes de recombinaison intragénique et unite fonctionnelle d'un locus chez l'*Ascobolus immersus*. Thèse Fac. de Science, Univ. Paris 1964.
RYAN, F. J.: Crossing over and second division segregation in fungi. Bull. Torrey bot. Club **70**, 605—611 (1943).
SCHWARTZ, D.: Evidence for sister-strand crossing over in maize. Genetics **38**, 251—260 (1953).
— Studies on the mechanism of crossing over. Genetics **39**, 692—700 (1954).
SCHWEITZER, M. D.: An analytical study of crossing over in *Drosophila melanogaster*. Genetics **20**, 497—527 (1935).
SERMONTI, G.: Analysis of vegetative segregation and recombination in *Penicillium chrysogenum*. Genetics **42**, 433—443 (1957).
SERRES, F. J. DE: Studies with purple adenine mutants in *Neurospora crassa*. I. Structural and functional complexity in the *ad-3* region. Genetics **41**, 668—676 (1956).
— Recombination and interference in the *ad-3* region of *Neurospora crassa*. Cold Spr. Harb. Symp. quant. Biol. **23**, 111—118 (1958a).
— Studies with purple adenine mutants in *Neurospora crassa*. III. Reversion of x-ray-induced mutants. Genetics **43**, 187—206 (1958b).
— Studies with purple adenine mutants in *Neurospora crassa*. V. Evidence for allelic complementation among *ad-3 B* mutants. Genetics **48**, 351—360 (1963).
— Genetic analysis of the structure of the *ad-3* region of *Neurospora crassa* by means of irreparable recessive lethal mutations. Genetics **50**, 21—30 (1964).
SHAW, J.: Asymmetrical segregation of mating type and two morphological mutant loci in *Sordaria brevicollis*. Bull. Torrey bot. Club **89**, 83—91 (1962).
SHERMAN, F., and H. ROMAN: Evidence for two types of allelic recombination in yeast. Genetics **48**, 255—261 (1963).
SHULT, E. E., and S. DESBOROUGH: The application to tetrad-analysis-data from *Saccharomyces*, of principles for establishing the linear order of genetic factors. Genetica **31**, 147—187 (1960).
— — and C. C. LINDEGREN: Preferential segregation in *Saccharomyces*. Genet. Res. **3**, 196—209 (1962).
—, and C. C. LINDEGREN: The determination of the arrangement of genes from tetrad data. Cytologia (Tokyo) **20**, 291—295 (1955).
— — A general theory of crossing over. J. Genet. **54**, 343—357 (1956a).
— — Mapping methods in tetrad analysis. I. Provisional arrangement and ordering of loci preliminary to map construction by analysis of tetrad distribution. Genetica **28**, 165—176 (1956b).
— — Orthoorientation: A new tool for genetical analysis. Genetica **29**, 58—72 (1957).
— — A survey of genetic methodology from mendelism to tetrad analysis. Canad. J. Genet. Cytol. **1**, 189—201 (1959).
SIDDIQI, O. H.: The fine genetic structure of the *paba-1* region of *Aspergillus nidulans*. Genet. Res. **3**, 69—89 (1962).
—, and A. PUTRAMENT: Polarized negative interference in the *paba-1* region of *Aspergillus nidulans*. Genet. Res. **4**, 12—20 (1963).
SINGLETON, J. R.: Cytogenetic studies of *Neurospora crassa*. Ph. D. Thesis, California Inst. of Technology 1948.
— Chromosome morphology and the chromosome cycle in the ascus of *Neurospora crassa*. Amer. J. Bot. **40**, 124—144 (1953).

SMITH, F. H.: Influence of temperature on crossing over in *Drosophila*. Nature (Lond.) **138**, 329—330 (1936).

SPIEGELMAN, S.: Mapping functions in tetrad and recombinant analysis. Science **116**, 510—512 (1952).

SRB, A. M.: Ornithine-arginine metabolism in *Neurospora* and its genetic control. Ph. D. Thesis, Stanford University 1946.

STADLER, D. R.: A map of linkage group VI of *Neurospora crassa*. Genetics **41**, 528—543 (1956a).

— Double crossing over in *Neurospora*. Genetics **41**, 623—630 (1956b).

— Heritable factors influencing crossing over frequency in *Neurospora*. Microbial. Genet. Bull. **13**, 32—34 (1956c).

— Gene conversion of cysteine mutants in *Neurospora*. Genetics **44**, 647—655 (1959a).

— The relationship of gene conversion to crossing over in *Neurospora*. Proc. nat. Acad. Sci. (Wash.) **45**, 1625—1629 (1959b).

— Observations on the polaron model for genetic recombination. Heredity **18**, 233—242 (1963).

—, and A. M. TOWE: Genetic factors influencing crossing over frequency in *Neurospora*. Genetics **47**, 839—846 (1962).

—, — Recombination of allelic cysteine mutants in *Neurospora*. Genetics **48**, 1323—1344 (1963).

STAHL, F. W.: A chain model for chromosomes. J. Chim. Phys. **56**, 1072—1077 (1961).

— The mechanics of inheritance. Englewood Cliffs, New Jersey: Prentice-Hall, Inc. 1964.

STENT, G. S.: Molecular biology of bacterial viruses. San Francisco and London 1963.

STERN, C.: An effect of temperature and age on crossing over in the first chromosome of *Drosophila melanogaster*. Proc. nat. Acad. Sci. (Wash.) **12**, 530—532 (1926).

— Somatic crossing over and segregation in *Drosophila melanogaster*. Genetics **21**, 625—730 (1936).

ST. LAWRENCE, P.: The *q* locus of *Neurospora crassa*. Proc. nat. Acad. Sci. (Wash.) **42**, 189—194 (1956).

—, and D. M. BONNER: Gene conversion and problems of allelism. In: W. D. McELROY and B. GLASS (edits.), The chemical basis of heredity, pp. 114—122. Baltimore 1957.

STREISINGER, G., and N. FRANKLIN: Mutation and recombination at the host range genetic region of phage *T 2*. Cold Spr. Harb. Symp. quant. Biol. **21**, 103—109 (1956).

STRICKLAND, W. N.: Abnormal tetrads in *Aspergillus nidulans*. Proc. roy. Soc. B **148**, 533—542 (1958a).

— An analysis of interference in *Aspergillus nidulans*. Proc. roy. Soc. B **149**, 82—101 (1958b).

— A rapid method for obtaining unordered *Neurospora* tetrads. J. gen. Microbiol. **22**, 583—585 (1960).

— Tetrad analysis of short chromosome regions of *Neurospora crassa*. Genetics **46**, 1125—1141 (1961).

— D. D. PERKINS, and C. C. VEATCH: Linkage data for group V markers in *Neurospora*. Genetics **44**, 1221—1226 (1959).

STRØMNAES, Ø., and E. D. GARBER: Heterocaryosis and the parasexual cycle in *Aspergillus fumigatus*. Genetics **48**, 653—662 (1962).

SURZYCKI, S., and A. PASZEWSKI: Non-random segregation of chromosomes in *Ascobolus immersus*. Genet. Res. **5**, 20—26 (1964).

SUYAMA, Y., K. D. MUNKRES, and V. W. WOODWARD: Genetic analyses of the *pyr-3* locus of *Neurospora crassa*: the bearing of recombination and gene conversion upon intraallelic linearity. Genetica **30**, 293—311 (1959).

SWIEZYNSKI, K. M.: Somatic recombination of two linkage groups in *Coprinus lagopus*. Genetica Polonica **4**, 21—36 (1963).

17*

TATUM, E. L., and T. T. BELL: Neurospora. III. Biosynthesis of thiamin. Amer. J. Bot. 33, 15—20 (1946).

TAYLOR, J. H.: The time and mode of duplication of the chromosomes. Amer. Naturalist 91, 209—221 (1957).

— The organization and duplication of genetic material. Proc. 10th Intern. Congr. Genet. 1, 63 (1958).

— P. S. WOODS, and W. L. HUGHES: The organization and duplication of chromosomes as revealed by autoradiographic studies using tritium-labeled thymidine. Proc. nat. Acad. Sci (Wash.) 43, 122—127 (1957).

TEAS, H. J.: The biochemistry and genetics of threonine-requiring mutants of Neurospora crassa. Ph. D. Thesis, California Inst. of Technology 1947.

THRELKELD, S. F. H.: Effect of 5-bromouracil on ascus patterns in some Neurospora crosses. Nature (Lond.) 193, 1108—1109 (1962a).

— Some asci with nonidentical sister spores from a cross in Neurospora crassa. Genetics 47, 1187—1198 (1962b).

— Pantothenic acid requirement for spore color in Neurospora crassa. Canap J. Genet. and Cytol. 7, 171—173 (1965).

TOWE, A. M.: Factors influencing crossing over in Neurospora. Microbial Genet. Bull. 16, 31—32 (1958).

—, and D. R. STADLER: Effects of temperature on crossing over in Neurospora. Genetics 49, 577—583 (1964).

WALLACE, M. E.: Affinity, a new genetic phenomenon in the house mouse. Evidence within laboratory stocks. Nature (Lond.) 171, 27—28 (1953).

— The use of affinity in chromosome mapping. Biometrics 13, 98—110 (1957).

— Experimental evidence for a new genetic phenomenon. Phil. Trans. B 241, 211—254 (1958a).

— New linkage and independence data for ruby and jerker in the mouse. Heredity 12, 453—462 (1958b).

— An experimental test of the hypothesis of affinity. Genetica 29, 243—255 (1959).

— A possible case of affinity in tomatoes. Heredity 14, 275—283 (1960a).

— Possible cases of affinity in cotton. Heredity 14, 263—274 (1960b).

— Affinity: evidence from crossing inbred lines of mice. Heredity 16, 1—23 (1961).

WEIJER, J.: Aberrant recombination at the td locus of Neurospora crassa and its mendelian interpretation. Canad. J. Genet. Cytol. 1, 147—160 (1959).

WEINSTEIN, A.: Unraveling the chromosomes. J. cell. comp. Physiol. 45, Suppl. 2, 249—269 (1955).

WELSHON, W. J.: A comparative study of crossing over in attached x-chromosomes of Drosophila melanogaster. Genetics 40, 918—936 (1955).

WESTERGAARD, M.: Studies on the mechanism of crossing over. I. Theoretical considerations. C. R. Lab. Carlsberg, Sér. Physiol. 34, 359—405 (1964).

WETTSTEIN, F. v.: Morphologie und Physiologie des Formenwechsels. Z. indukt. Abstamm.- u. Vererb.-Lehre 33, 1—236 (1924).

WHEELER, H. E.: Linkage groups in Glomerella. Amer. J. Bot. 43, 1—6 (1956).

—, and C. H. DRIVER: Genetics and cytology of a mutant, dwarf-spored Glomerella. Amer. J. Bot. 40, 694—702 (1953).

WHITEHOUSE, H. L. K.: Crossing over in Neurospora. New Phytologist 41, 23—62 (1942).

— Genetics of ascomycetes. Ph. D. Thesis, Cambridge Univ. 1948.

— Multiple-allelomorph heterothallism in the fungi. New Phytologist 48, 212—244 (1949).

— Mapping chromosome centromeres by the analysis of unordered tetrads. Nature (Lond.) 165, 893 (1950).

— Analysis of unordered tetrads segregating for lethal or other epistatic factor. Nature (Lond.) 172, 463—464 (1954).

WHITEHOUSE, H. L. K.: The use of loosely-linked genes to estimate chromatid interference by tetrad analysis. C. R. Lab. Carlsberg, Sér. Physiol. **26**, 407—422 (1956).
— Use of tetratype frequencies for estimating spindle overlapping at the second division of meiosis in "ordered" tetrads. Nature (Lond.) **179**, 162—163 (1957a).
— Mapping chromosome centromeres from tetratype frequencies. J. Genet. **55**, 348—360 (1957b).
— A theory of crossing over by means of hybrid deoxyribonucleic acid. Nature (Lond.) **199**, 1034—1040 (1963).
—, and J. B. S. HALDANE: Symmetrical and asymmetrical reduction in ascomycetes. J. Genet. **47**, 208—212 (1946).
—, and P. J. HASTINGS: The analysis of genetic recombination on the polaron hybrid DNA model. Genet. Res. (in press) (1965).
WILKIE, D., and D. LEWIS: The effect of ultraviolet light on recombination in yeast. Genetics **48**, 1701—1716 (1963).
WILLIAMS, E. B., and J. R. SHAY: The relationship of genes for pathogenicity and certain other characters in *Venturia inaequalis* (CKE) WINT. Genetics **42**, 704—711 (1957).
WINGE, Ö., and C. ROBERTS: Non-mendelian segregation from heterozygotic yeast asci. Nature (Lond.) **165**, 157—158 (1950).
— — On tetrad analysis apparently inconsistent with mendelian law. Heredity **8**, 295—304 (1954a).
— — Causes of deviations from 2:2 segregations in the tetrads of monohybrid yeasts. C. R. Lab. Carlsberg, Sér. Physiol. **25**, 285—329 (1954b).
— — Remarks on irregular segregations in *Saccharomyces*. Genetica **28**, 489—496 (1957).
WINKLER, H.: Die Konversion der Gene. Jena: Gustav Fischer 1930.
WOLFF, S.: Are sister chromatid exchanges sister strand crossovers or radiation-induced exchanges? Mutation Research **1**, 337—343 (1964).
WÜLKER, H.: Untersuchungen über Tetradenaufspaltung bei *Neurospora sitophila* SHEAR et DODGE. Z. indukt. Abstamm.- u. Vererb.-Lehre **69**, 210—248 (1935).
ZICKLER, H.: Das Sichtbarwerden der Mendelspaltung im Ascus von *Bombardia lunata*. Ber. dtsch. bot. Ges. **52**, 11—14 (1934a).
— Genetische Untersuchungen an einem heterothallischen Ascomyceten (*Bombardia lunata*, nov. spec.). Planta (Berl.) **22**, 573—613 (1934b).
— Die Vererbung des Geschlechts bei dem Ascomyceten *Bombardia lunata* ZCKL. Z. indukt. Abstamm.- u. Vererb.-Lehre **73**, 403—408 (1937).

References

which have come to the authors' attention after conclusion of the German manuscript

A IV/V

BERG, C. M.: Biased distribution and polarized segregation in asci of *Sordaria brevicollis*. Genetics **53**, 117—129 (1966).
CHEN, K. C.: Evidence for the genetic control of asymmetrical segregation in *Sordaria brevicollis*. Genetics **50**, 240—241 (1964).
—, and L. S. OLIVE: The genetics of *Sordaria brevicollis*. II. Biased segregation due to spindle overlap. Genetics **51**, 761—766 (1965).
SCOTT-EMUAKPOR, M. B.: Random segregation in *Neurospora*. Genetica **36**, 407—411 (1965).

A VI

COWAN, J. W., and D. LEWIS: Somatic recombination in the dikaryon of *Coprinus lagopus*. Genet. Res. **7**, 235—244 (1966).
GARBER, E. D., and L. BERAHA: Genetics of phytopathogenic fungi. XIV. The parasexual cycle in *Penicillium expansum*. Genetics **52**, 487—492 (1965).

Recombination

HOFFMANN, G. M.: Untersuchungen über die Heterokaryosebildung und den Parasexualcyclus bei *Fusarium oxysporum*. III. Paarungsversuche mit auxotrophen Mutanten von *Fusarium oxysporum* f. *callistephi*. Arch. Mikrobiol. **56**, 40—59 (1967).

MALUZYNSKI, M.: Recombination in crosses between biochemical mutants of *Coprinus lagopus*. Acta Soc. Bot. Bol. **35**, 191—199 (1966).

B I, 1

COOKE, F.: Recombination values in *Neurospora*. A pitfall in statistical analysis and application of a correction factor. Canad. J. Genet. Cytol. **8**, 733—736 (1966).

HOWE, H. B.: Vegetative traits associated with mating type in *Neurospora tetrasperma*. Mycologia **56**, 519—525 (1964).

PATEMAN, J. A., and B. T. O. LEE: Segregation of polygenes in ordered tetrads. Heredity **15**, 351—361 (1960).

B II, 1

HOWE, H. B.: Determining mating type in *Neurospora* without crossing tests. Nature (Lond.) **190**, 1036 (1961).

—, and P. HAYSMAN: Linkage group establishment in *Neurospora tetrasperma* by interspecific hybridization with *N. crassa*. Genetics **54**, 293—302 (1966).

MITCHELL, M. B.: A model predicting characteristics of genetic maps in *Neurospora crassa*. Nature (Lond.) **205**, 680—682 (1965a).

— An extended model of periodic linkage in *Neurospora crassa*. Canad. J. Genet. Cytol. **7**, 563—570 (1965b).

B II, 2

BEGUERET, J.: Sur la répartition en groupes de liaison de gènes concernant la morphologie des ascospores chez le *Podospora anserina*. C. R. Acad. Sci. (Paris) **264**, 462—465 (1967).

McCULLY, K. S.: The use of p-fluorophenylalanine with "master strains" of *Aspergillus nidulans* for assigning genes to linkage groups. Genet. Res. **6**, 352—359 (1965).

B II, 3

BOUCHARENC, M., J. MOUSSEAU et J. L. ROSSIGNOL: Sur l'action de la température sur la fréquence des recombinaisons réciproques et non réciproques au sein du locus 75 de l'*Ascobolus immersus*. C. R. Acad. Sci. (Paris) **262**, 1589—1592 (1966).

CAMERON, H. R., K. S. HSU, and D. D. PERKINS: Crossing over frequency following inbreeding in *Neurospora*. Genetica **37**, 1—6 (1966).

ESPOSITO, R. E., and R. HOLLIDAY: The effect of 5-fluorodeoxyuridine on genetic replication and mitotic crossing over in synchronized cultures of *Ustilago maydis*. Genetics **50**, 1009—1017 (1964).

GRIFFITHS, A. J. F., and S. F. H. THRELKELD: Internuclear effects on prototroph frequencies of some crosses in *Neurospora crassa*. Genetics **54**, 77—87 (1966).

HOLLIDAY, R.: Induced mitotic crossing-over in relation to genetic replication in synchronously dividing cells of *Ustilago maydis*. Genet. Res. Camb. **6**, 104—120 (1965).

JANSEN, G. J. O.: UV-induced mitotic recombination in the $paba_1$ region of *Aspergillus nidulans*. Genetica **35**, 127—131 (1964).

— UV-induced mitotic recombination in the $paba_1$ cistron of *Aspergillus nidulans*. Diss. Utrecht 1966.

KWIATKOWSKI, Z.: Studies on the mechanism of gene recombination in *Aspergillus*. I. Analysis of stimulating effect of the removal of some metallic ions on mitotic recombination. Acta microbiol. pol. **14**, 3—13 (1965).

KWIATKOWSKI, Z., and K. GRAD: A comparison of the ultraviolet effect on the mitotic recombination in two different cistrons of *Aspergillus nidulans*. Acta microbiol. pol. **14**, 15—18 (1965).

LAVIGNE, S., and L. C. FROST: Recombination frequency and wild-type ancestry in linkage group I of *Neurospora crassa*. Genet. Res. Camb. **5**, 366—378 (1964).

MCNELLY-INGLE, C. A., B. C. LAMB, and L. C. FROST: The effect of temperature on recombination frequency in *Neurospora crassa*. Genet. Res. **7**, 169—183 (1966).

PARRY, J. M., and B. S. COX: Photoreactivation of ultraviolet induced reciprocal recombination, gene conversion and mutation to prototrophy in *Saccharomyces cerevisiae*. J. gen. Microbiol. **402**, 235—241 (1965).

PRAKASH, V.: Intra-chromosomal position interference in *Neurospora crassa*. Genetica **35**, 287—322 (1964).

SCOTT-EMUAKPOR, M. B.: Genetic recombination in *Neurospora crassa* and *N. sitophila*. Genet. Res. Camb. **6**, 216—225 (1965).

WOLFF, S., and F. J. DE SERRES: Chemistry of crossing-over. Nature (Lond.) **213**, 1091—1092 (1967).

ZIMMERMANN, F. K., R. SCHWAIER, and U. V. LAER: Mitotic recombination induced in *Saccharomyces cerevisiae* with nitrous acid, diethylsulfate and carcinogenic, alkylating nitrosamides. Z. Vererbungsl. **98**, 230—246 (1966).

B III, 1

BAKER, W. K., and J. A. SWATEK: A more critical test of hypotheses of crossing over which involve sister-strand exchange. Genetics **52**, 191—202 (1965).

SCOTT-EMUAKPOR, M. B.: Interference studies in *Neurospora crassa* and *N. sitophila*. Genet. Res. Camb. **6**, 226—229 (1965).

B III, 2

PRAKASH, V.: Intra-chromosomal position interference in *Neurospora crassa*. Genetica **35**, 287—322 (1964).

B IV, 3

BARRY, E. G.: Chromosome aberrations in *Neurospora* and the correlation of chromosomes and linkage groups. Genetics **55**, 21—32 (1967).

CHEN, K. C.: The genetics of *Sordaria brevicollis*. I. Determination of seven linkage groups. Genetics **51**, 509—517 (1965).

HOWE, H. B., and P. HAYSMAN: Linkage group establishment in *Neurospora tetrasperma* by interspecific hybridization with *N. crassa*. Genetics **54**, 293—302 (1966).

MORTIMER, R. K., and D. C. HAWTHORNE: Genetic mapping in *Saccharomyces*. Genetics **53**, 165—173 (1966).

PASZEWSKI, A., S. SURZYCKI, and M. MANKOWSKA: Chromosome maps in *Ascobolus immersus* (Rizet's strain). Acta Soc. Bot. Pol. **35**, 181—188 (1966).

B V, 1

COYLE, M. B., and T. H. PITTENGER: Mitotic recombination in pseudowild types of *Neurospora*. Genetics **52**, 609—625 (1965).

HOFFMANN, G. M.: Untersuchungen über die Heterokaryosebildung und Parasexualcyclus bei *Fusarium oxysporum*. III. Paarungsversuche mit auxotrophen Mutanten von *Fusarium oxysporum* f. *callistephi*. Arch. Mikrobiol. **56**, 40—59 (1967).

JOHNSTON, J. R., and J. M. MACKINNON: Spontaneous and induced mitotic recombination in diploid and tetraploid *Saccharomyces*. Abstract, 2nd Internat. Symp. on Yeasts, Bratislava 1966.

MALUZYNSKI, M.: Recombination in crosses between biochemical mutants of *Coprinus lagopus*. Acta Soc. Bot. Pol. **35**, 191—199 (1966).

Recombination

PRUD'HOMME, N.: Somatic recombination in *Coprinus radiatus*. In: Incompatibility in fungi (K. ESSER and J. R. RAPER, eds.), p. 48—52. Berlin-Heidelberg-New York: Springer 1965.

C I, 1

LEUPOLD, U., and H. GUTZ: Genetic fine structure in *Schizosaccharomyces*. Proc. XIth Int. Congr. Genetics, The Hague 1964, p. 31—35.
SMITH, B. R.: Genetic control of recombination. I. The recombination-2 gene of *Neurospora crassa*. Heredity 21, 481—498 (1966).

C I, 2

BARRICELLI, N. A., and K. WOLFE: Localized negative interference, some of its manifestations and biological function. Z. Vererbungsl. 96, 307—312 (1965).
BAUSUM, H. T., and R. P. WAGNER: "Selfing" and other forms of aberrant recombination in isoleucine-valine mutants of *Neurospora*. Genetics 51, 815—830 (1965).
EMERSON, S., and C. C. C. YU-SUN: Gene conversion in the pasadena strain of *Ascobolus immersus*. Genetics 55, 39—47 (1967).
HARTLEY, M. J., and W. J. WHITTINGTON: Possible effect of mitotic recombination on gene conversion and negative interference. Nature (Lond.) 209, 698—700 (1966).
PEES, E.: Polarized negative interference in the *lys-51* region of *Aspergillus nidulans*. Experientia (Basel) 21, 514 (1965).
PUTRAMENT, A.: Mitotic recombination in the *paba 1* cistron of *Aspergillus nidulans*. Genet. Res. Camb. 5, 316—327 (1964).
SMITH, B. R.: Genetic control of recombination. I. The recombination-2 gene of *Neurospora crassa*. Heredity 21, 481—498 (1966).

C II, 1

BAUSUM, H. T., and R. P. WAGNER: "Selfing" and other forms of aberrant recombination in isoleucine-valine mutants of *Neurospora*. Genetics 51, 815—830 (1965).
EMERSON, S.: Quantitative implications of the DNA-repair model of gene conversion. Genetics 53, 475—485 (1966).
HARTLEY, M. J., and W. J. WHITTINGTON: Possible effect of mitotic recombination on gene conversion and negative interference. Nature (Lond.) 209, 698—700 (1966).
HOLLIDAY, R.: Studies on mitotic gene conversion in *Ustilago*. Genet. Res. Camb. 8, 323—337 (1966).
JESSOP, A. P., and D. G. CATCHESIDE: Interallelic recombination at the *his-1* locus in *Neurospora crassa* and its genetic control. Heredity 20, 237—256 (1965).
MAKAREWICZ, A.: Colourless mutants in *Ascobolus immersus* with alternative phenotypes. Acta Soc. Bot. Pol. 35, 175—179 (1966).
MOUSSEAU, J.: Sur les variations de fréquence de conversion au niveau de divers sites d'un même locus. C. R. Acad. Sci. (Paris) 262, 1254—1257 (1966).
PASZEWSKI, A., and S. SURZYCKI: "Selfers" and high mutation rate during meiosis in *Ascobolus immersus*. Nature (Lond.) 204, 809 (1964).
PICARD, M.: La structure d'un locus complexe chez l'Ascomycete *Podospora anserina*. Théses Fac. Sci. d'Orsay, Univ. de Paris 1966.
PUTRAMENT, A.: Mitotic recombination in the *paba 1* cistron of *Aspergillus nidulans*. Genet. Res. Camb. 5, 316—327 (1964).
RIZET, G., et J. L. ROSSIGNOL: Sur la dimension probable des échanges réciproques au sein d'un locus complexe d'*Ascobolus immersus*. C. R. Acad. Sci. (Paris) 262, 1250—1253 (1966).

C II, 2

KRUSZEWSKA, A., and W. GAJEWSKI: Recombination within the *Y* locus in *Ascobolus immersus*. Genet. Res. Camb. **9**, 159—177 (1967).
PEES, E.: Polarized negative interference in the *lys-51* region of *Aspergillus nidulans*. Experientia (Basel) **21**, 514 (1965).

C II, 3

PUTRAMENT, A.: Mitotic recombination in the *paba 1* cistron of *Aspergillus nidulans*. Genet. Res. Camb. **5**, 316—327 (1964).

D

BAKER, W. K., and J. A. SWATEK: A more critical test of hypotheses of crossing over which involve sister-strand exchange. Genetics **52**, 191—202 (1965).
EMERSON, S.: Quantitative implications of the DNA-repair model of gene conversion. Genetics **53**, 475—485 (1966).
UHL, C. H.: Chromosome structure and crossing over. Genetics **51**, 191—207 (1965).
WHITEHOUSE, H. L. K.: Crossing-over. Sci. Progr. **53**, 285—296 (1965).
— An operator model of crossing-over. Nature (Lond.) **211**, 708—713 (1966).
— A cycloid model for the chromosome. J. Cell Sci. **2**, 9—22 (1967).

Chapter V

Mutation

Mutations are discontinuous, heritable alterations of the genetic material. They do not result from sexual or parasexual processes; this is in contrast to recombination through which new genotypes are produced by the reassortment of entire chromosomes or by the exchange of chromosome segments. Mutations arise *spontaneously* as well as through the action of *mutagenic agents.*

All mutations other than those which are extrachromosomal (discussed in the final chapter p. 439 ff.) are localized in the chromosomes or their equivalents (e.g. in bacteria). Three kinds of nuclear mutations are generally recognized:

1. *Point mutations: structural changes in single sites within a gene.* They do not alter the linkage relations with markers on the same chromosome. The locus of the mutation is called the mutational site. Through additional mutation in the same region a point mutation may become so altered that the original phenotype is again produced (reverse or back mutation). On the molecular level a point mutation involves a change in the smallest building blocks of DNA, including even the replacement of a single DNA base by another.

Every change in a gene was formerly called a gene mutation. Since we now know (p. 210 ff.) that a gene consists of many, rather than a single mutational site, "point mutation" is used to designate the change described above.

2. *Chromosomal mutations: structural alterations in the chromosomes through rearrangement or loss of chromosome segments.* Such mutations lead to changes in recombination values. They come about through breaks in a chromosome. The fragments are either lost in subsequent divisions or may rejoin in new combinations. Mutants of this type cannot revert by means of a single event.

3. *Genome mutations: changes in the number of intact chromosomes or chromosome sets per nucleus.* The structure of the chromosomes remains unaltered.

Genome mutations can generally be recognized in the fungi by analysis of cytological preparations just as in the higher organisms. They often result in a change in radiation sensitivity; for example, haplonts show a linear (single event) dose-effect curve, while such a curve is non-linear (multiple event) for X-ray inactivation of polyploids (p. 280 ff. and 282). Rearrangement or loss of very small chromosomal fragments whose size lie below the limit of resolution of the light microscope cannot be detected microscopically. Thus in order to differentiate between point and chromosomal mutations, both of which lead to a change in chromosomal structure, genetic criteria are generally employed, namely, segregation of characters in the F_1 generation, linkage relations with markers, and capacity to revert (compare p. 306 ff.).

Since recent advances in our understanding of mutation are based on experiments with mutagenic agents, the latter deserve particular attention; results with phage and bacteria must also be considered in this connection.

General references: DEMEREC (1955), GOLDSCHMIDT (1955), SWANSON (1957), KAPLAN (1957, 1959, 1962a), STRAUB (1958), ZAMENHOF (1959, 1963), STRAUSS (1960), LASKOWSKI (1960b), RÖBBELEN (1960, 1962, 1963), SAGER and Ryan (1961), Errera (1962), Schull (1962), Auerbach (1962, 1964), Stent (1963), Stahl (1964), Bresch (1964), Hayes (1964).

Other references may be found in the volumes edited by McElroy and Glass (1957) and by Stubbe (1960, 1962) as well as the Cold Spring Harbor Symposia volumes **16** (1951), **21** (1956), and **23** (1958).

A. Point mutations

Point mutations have been particularly useful for investigating a large number of genetic problems for two main reasons:

1. The analysis of recombination requires point marking of the genetic material (see chapter on recombination).

2. Point mutations permit the analysis of changes in the DNA base sequence (see interpretations of chemical mutagenesis, p. 291 ff., and the genetic code, p. 344 ff.).

Not all mutations are equally suited for genetic experiments. The study of mutation requires types of mutants which are unequivocally recognizable and can be easily selected and isolated. These conditions are not fulfilled by most allelic differences in wild populations, even though geographic races in general differ in numerous, although often not strongly divergent characters. Until the discovery of biochemical mutants genetic investigations were primarily limited to morphological and lethal mutations in the higher organisms. *Morphological mutations* are heritable alterations that are visibly different from the wild type phenotype and thus are relatively easy to isolate.

Such mutations are not always advantageous in fungi. The number of morphological characters is very limited. It follows that, when two or more morphological mutations occur together they are often difficult or impossible to identify (e.g. sterility in combination with spore color).

By *lethal mutation* we mean an heritable change in the genetic material which leads to a functional defect resulting in the death of the organism.

If a lethal mutation is recessive, the lethal effect may be masked by the normal allele in a diplont or heterokaryon. In spite of such experimental means of studying lethal recessives even in the usually haploid fungi such mutations are not well suited for investigating genetic problems of a general nature.

Biochemical mutations, in contrast to morphological and lethal mutations, fulfill all experimental requirements. Auxotrophic and resistance mutants fall into this category. *Auxotrophic mutants* have lost the capacity to synthesize certain metabolites essential for life (*deficiency mutants*); they are able to grow only if such substances are provided in the culture medium. Auxotrophs behave like lethal mutants under certain circumstances (on minimal medium); however, they are viable on complete medium (p. 271 ff.). *Resistance mutants* have, in contrast to auxotrophs, gained a capability; they are able to grow in the presence of substances which are toxic to the wild type.

Biochemical mutants are easily identified even if several mutations occur in the same nucleus. Such mutants can be isolated relatively quickly and in appreciable numbers by selective techniques (p. 273 ff.).

In considering point mutations we differentiate between "forward" and "reverse" or "back" mutation. A *forward mutation* is simply any alteration of a particular mutational site in a wild type gene.

The concept of "wild type" is by no means unambiguous especially if different geographic races are considered, which may possess a differential genetic endowment (p. 169). In the present discussions we mean by "wild type" the unmutated original strain used for the particular mutation experiments.

A mutation within the same gene that has previously undergone a forward mutation is called a *back* or *reverse mutation* provided the second mutation results in the reestablishment of the wild phenotype. The mutational site in such a case need not to be identical to that of the forward mutation (p. 401 ff.). Sometimes a back mutation may be simulated by mutation of a suppressor gene at another functional unit or locus (p. 397 f.).

I. Methods for the isolation and characterization of point mutants

1. Morphological mutants

Although higher plants and animals possess a diversity of form and color which is suitable for genetic investigation, similar morphological characters in the fungi are limited in number on account of their relatively simple morphology. The deficiency of morphological characters is particularly apparent in the yeasts, which are unicellular. In fungi with hyphae mutations are known which affect mycelial characters (color, growth form and rate), fruiting body characters (form and size), spores, and conidial characters (color, form, size, and arrangement) (compare Table II-1). Spore and conidial characters have proved especially suited for genetic studies (p. 148 f.). Such mutations are expressed in a single cell, separable from the parent, and in the ideal case, uninucleate (e.g. conidia of *Aspergillus nidulans*, p. 274).

The method generally used for the production and isolation of morphological mutants is as follows: Spores or conidia of a fungus are irradiated or treated with mutagenic chemicals and transferred to a culture medium. A large percentage of the cells are killed by the treatment; the remainder germinate to produce mycelia. Any mutations which have occurred can usually be recognized directly. The main disadvantage of this method is that no selection is possible. *Spontaneous* morphological mutations among spores and conidia are much easier to detect than those affecting mycelial characters. For example, RIZET and coworkers have developed an elegant method for the isolation of such mutants in *Ascobolus immersus* (RIZET et al., 1960a, b; LISSOUBA, 1960; LISSOUBA et al., 1962).

269

When mature, the eight spores are ejected from the ascus. They can be collected on an agar plate by inverting the latter over the mature culture. Over a thousand separate asci per plate can be surveyed for spontaneous mutations in this way. Since the eight spores of an ascus usually remain together, 4:4 segregations for spore form or color can be recognized, thus indicating that one of the parent nuclei had mutated prior to karyogamy. About two thousand spontaneous mutants were isolated in a relatively short time by employing this technique.

The detection of mutants is often complicated because of dominance in the heterokaryon. This is a problem if multinucleate macroconidia of *N. crassa* or ascospores which become multinucleate at maturity (e.g. *Podospora anserina, Sordaria macrospora*) are used for the production of mutants. If a mutation occurs in only one nucleus of such a reproductive cell and if it is recessive or if it reduces the division rate of the nucleus, the mutation may remain undetected. In such cases the heterokaryotic mycelium exhibits the wild phenotype. Under these circumstances the dosage of the mutagenic agent must be high enough to make the survival of more than one nucleus per spore unlikely. The probability of detecting recessive mutations is thus increased. If one is not concerned with obtaining quantitative data on the relation between dosage and mutation rate, treatment of multinucleate cells leads to the same result as treatment of uninucleate cells. Methods have been worked out in *Neurospora crassa* to avoid dominance in the heterokaryon by utilizing mutants which produce only uninucleate microconidia (*fluffy:* LINDEGREN and LINDEGREN, 1941; *peach-microconidial:* TATUM et al., 1950).

2. Lethal mutants

The origin and inheritance of lethal mutations can only be investigated if they are (1) *recessive* and (2) *lead to a balanced heterozygote or a heterokaryotic cell*. Since the lethal effect is suppressed by the dominant wild type allele in such cells, special methods are necessary to recognize lethal mutations and to study their mode of inheritance.

ATWOOD and MUKAI have described a method by which nuclei with lethal mutations can be maintained for indefinite periods in a heterokaryon through a "symbiosis" with normal nuclei. By using microconidia the viability or lethality of single nuclei can be tested (ATWOOD, 1950; ATWOOD and MUKAI, 1953a, b, 1954a, b).

These workers used a heterokaryon of *N. crassa* in which one nuclear type carried the marker, *or,* and the other the *me* and *amyc* markers. *or* homokaryons are auxotrophic for ornithine, homokaryons of *me amyc* for methionine; the latter undergo yeast-like cell-formation and grow very slowly. The heterokaryon is prototrophic by complementation of non-allelic defects (p. 386f.).

Macroconidia from such a heterokaryon were irradiated and sown on minimal medium. The resulting mycelia were all heterokaryotic, since neither homokaryon could grow on minimal. The heterokaryons were isolated and placed on minimal medium; after a few days they produced conidia. Conidia from each isolate were then placed on minimal plus methionine. The methionine supplement enabled *me* homokaryons to grow.

In general ATWOOD and MUKAI observed morphologically normal (heterokaryotic: *or + me amyc*) and also slow growing mycelia (homo-karyotic: *me amyc*) according to expectation. However, the latter type was completely absent in rare cases. This unexpected result was explained by the fact that the *me amyc* nuclei in the irradiated conidia already possessed a recessive lethal mutation. The mutation (designated *l*) may have occurred spontaneously prior to conidial formation or may have been induced in the macroconidia themselves. In either case a viable heterokaryon with *or* or *me amyc l* nuclei is produced.

Conidia which are formed from such a heterokaryon yield only normal mycelia on a methionine supplemented minimal medium, if they contain *both* nuclear types. The two remaining conidial types (homokaryotic for *or* or *me amyc l*) failed to survive, either because of the auxotrophy for ornithine or of the lethal mutation. Fifty-eight mutants were recovered from 2,764 isolated in this manner. Only two out of 26 non-allelic mutants proved to be viable on complete medium when homokaryotic. All others had lost essential functions.

This method developed by ATWOOD and MUKAI was later frequently modified, e.g. by DE SERRES and OSTERBIND (1962), ROYES (1962) and MORROW (1964). Other organisms have also proved suitable for investigating recessive lethal mutations. For example, in *Podospora anserina* viable heterokaryotic ascospores arise through postreduction of a lethal factor (MARCOU, 1963; compare Fig. IV-6). In contrast to the haploid Eumycetes *N. crassa* and *P. anserina*, in which lethal mutations are only viable in balanced heterokaryons, the diploidy of certain yeasts offers a further possibility for the analysis of recessive lethals. Since copulation between spores of different mating types in yeasts frequently occurs in the ascus (p. 12), viable diploid cells heterozygous for a lethal may arise, e.g. in *Saccharomyces ludwigii:* WINGE and LAUTSEN (1939), WINGE (1947).

3. Biochemical mutants

Because of the significance of auxotrophic mutants for genetic investigation, most methods of isolation have been developed for this type of mutant. Resistance mutants have received much less attention.

Literature: BEADLE and TATUM (1945), FRIES (1947, 1948a, b), PONTE-CORVO (1949), REAUME and TATUM (1949), RYAN (1950), ROBERTS (1950), LEDERBERG and LEDERBERG (1952).

a) Classical method

Since methods for isolating biochemical mutants quickly by selective means were originally unknown, the much more tedious technique of identifying the few mutants among a large number of cells was employed. BEADLE and TATUM (1945) were the first to isolate a series of auxotrophs in *Neurospora crassa* in this way; the mycelia which developed from surviving conidia after irradiation with ultraviolet (p. 278 ff.) were crossed with the wild type and the spores of a resulting ascus tested for auxotrophy in the manner described below. If an auxotrophic mutation had occurred in a conidium, four spores of the ascus would be auxotrophic and four prototrophic. A significant reduction in labor came with the elimination of crossing *before* the test. Only those mycelia for which a growth test has already been demonstrated a defect are crossed with the wild type. In this simplified method, which is shown in Fig. V-1, wild type conidia or spores which have been treated with mutagenic agents are shown on complete medium (a medium which contains all substances that auxotrophs might be expected to require for growth).

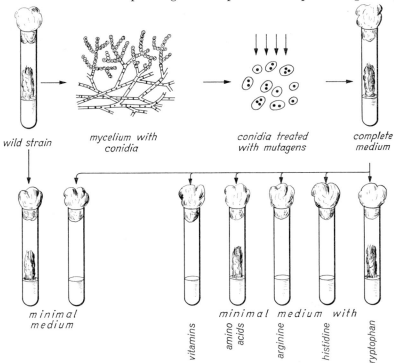

wild strain   mycelium with conidia   conidia treated with mutagens   complete medium

minimal medium   minimal medium with   vitamins   amino acids   arginine   histidine   tryptophan

Fig. V-1. Diagram of the classical method for the isolation and characterization of auxotrophic mutants. See further in text. (Adapted from BEADLE, 1945)

A transfer is made to minimal medium from each surviving mycelium. The minimal allows only those to grow which are able to synthesize their own essential metabolites (amino acids, vitamins etc.). If this capacity is lost through mutation for any of the individual growth requirements, the mycelial transfer is unable to grow and is thus identified as a mutant. The substance for which the deficiency mutant is auxotrophic is determined by a series of special tests. For this minimal media are used, each of which has been supplemented with individual vitamins, or amino acids, etc. (Fig. V-1). The test series is generally so arranged that the number of auxotrophic possibilities decreases with each additional test series.

To explain the technique of successive testing let us assume that a mutant has been found in the first test which can only grow on minimal if a series of amino acids are added (Fig. V-1). The next test series is designed to determine for which of the amino acids auxotrophy exists. A series of minimal media is prepared each of which is supplemented by a different amino acid (e.g. arginine, histidine, tryptophan etc., Fig. V-1). For example, if the mutant grows only on a medium which includes tryptophan, then tryptophan synthesis must be blocked at some point.

The number of mutants which can be recovered directly decreases when multinucleate conidia or spores are used. If only one of the two or more nuclei has undergone mutation, the fungus is able to synthesize all essential substances through the presence of the wild type nuclei and can thus grow on minimal medium. In order to avoid such a heterokaryotic condition in quantitative studies, mutant strains which produce only uninucleate microconidia can be used (e.g. in *N. crassa*) (p. 270).

By using a technique developed by PONTECORVO (1949) cells of a mutant can be tested simultaneously for different requirements on a petri dish of minimal medium (*auxanography*).

A thin layer of agar containing minimal medium is poured into a petri dish. A second layer of minimal agar containing a suspension of the auxotrophic cells under test is poured over the first. After the agar has hardened small quantities of different growth substances are added at separate points on the surface. Growth occurs only in those zones in which the particular mutant has been provided with the necessary growth factors.

A special auxotrophy for adenine can be recognized directly in *N. crassa* as well as in *Schizosaccharomyces pombe*, because the mutants in question produce a red pigment. The pigment is a derivative of a precursor of adenine (*ad-3* locus of *N. crassa:* DE SERRES and KØLMARK, 1958; *ad-6* and *ad-7* locus of *S. pombe:* LEUPOLD, 1958; GUTZ, 1961, 1963).

The classical method has the disadvantage of not allowing selective isolation of particular auxotrophic types. Selection techniques which have been developed to meet the requirements of modern genetics are described briefly in the following sections.

b) Replica plating technique

LEDERBERG and LEDERBERG (1952) originally developed this method for bacteria. Soon afterwards it was used for yeasts (*Saccharomyces cerevisiae:* TAKAHASHI, 1959) and smut fungi (*Ustilago maydis:* HOLLIDAY, 1956). Fungi which produce colonies composed of single cells and fail to form hyphae are best suited to this technique.

A cylindrical wood block covered with sterile velveteen serves as a stamp. By lightly touching it to a plate of complete medium on which colonies are growing and then to plates of various test media, a few cells of each mutant strain are transferred to the test media. The latter consist of minimal medium with various supplements, so arranged as to allow quick identification of the mutants under test.

Let us suppose that a large number of auxotrophic mutants are to be tested for requirement for the five substances, A, B, C, D and E. The mutant colonies are replicated on the following five different test media: minimal with (1) A + B + C + D, (2) A + B + C + E, (3) A + B + D + E, (4) A + C + D + E, (5) B + C + D + E. If the replicated mutant cells are not auxotrophic for any of the five substances, the pattern of the colonies will be identical on the test media and the complete medium. However, if a strain cannot grow on one of the five test media (e.g. minimal $+ A + B + D + E$) a colony which is present on the complete medium will be missing at the same location on the replica plate. Therefore, this isolate is auxotrophic for the substance that was *not* provided in the particular test medium (in our example, substance C).

The use of the replica plating technique has not remained restricted to unicellular organisms. For example, *Aspergillus nidulans* can be induced to form only microconidia by adding 0.08% sodium oxycholate to the medium. This substance is not mutagenic and does not influence the viability of conidia or ascospores. Up to eight replica platings can be made from the resulting microcolonies with moist, closely cropped velveteen (MACKINTOSH and PRITCHARD, 1963). ROBERTS (1959) used another type of stamp, consisting of a plate of closely set steel needles. ROBERTS was able to transfer conidia from low growing mycelia of *A. nidulans* to other plates if the needle points were first coated with moist agar. MALING (1960) transferred conidia of *N. crassa* by using velveteen or filter paper stamps. She used mutant strains which grew densely and produced abundant conidia. REISSIG (1956) allowed spores of *N. crassa* to grow on thin paper which lay in contact with a complete agar medium. As soon as the hyphae grew through the paper into the agar, the paper was transferred to a test medium. The hyphae remaining in the paper gave the same pattern on the test plate as on the complete medium of the original plate.

c) Layering technique

This technique, first described by REAUME and TATUM (1949) is particularly well suited for the isolation and characterization of auxotrophic mutants in yeast. Irradiated haploid cells are plated in an appropriate dilution on minimal medium. After a few days' incubation during which the surviving prototrophic cells have given rise to small colonies, a layer of complete agar is poured over the original medium. Cells which have lost the capacity to synthesize specific growth factors through mutation are now able to grow with the supplemented medium and produce colonies in the second layer. These can be isolated and tested for auxotrophy. If specific auxotrophic mutants are desired (e.g. for amino acid deficiencies) all essential metabolites except amino acids are added to the initial layer of minimal medium. The cells which are auxotrophic for amino acids are unable to grow and are then recovered from the second layer of agar.

Selection for particular mutant types can also be utilized for fungi which produce hyphae if a microscopic examination of the treated and plated spores is carried out instead of the layering technique. However, a portion of the ungerminated spores or weak germlings which are isolated in this way may not grow on complete medium, because some cells are killed as a result of the mutagenic treatment. This method was used with success in *N. crassa* by LEIN et al. (1948) and in *Venturia inaequalis* by BOONE et al. (1956).

d) Filter technique

Both the replica plating and the layering technique were originally developed for single celled organisms which produce colonies (bacteria, yeasts). An agar medium was employed in both cases. Since most of the fungi which are genetically interesting produce a mycelium, application of these techniques involved some difficulties. Thus, the method first developed by FRIES (1947) in *Ophiostoma multiannulatum* for hyphae-producing fungi which can be cultivated in liquid media is all the more noteworthy. *The technique is one of the best of the selective methods. It consists of the elimination of all prototrophic mycelia from a liquid minimal medium containing a suspension of treated spores or conidia through repeated filtration.* The hyphae of germinated prototrophic spores in minimal medium are held back by sterile filters (cotton: WOODWARD et al., 1954; gauze: CATCHESIDE, 1954). If the filtration is repeated at definite time intervals, the auxotrophs, i.e. the ungerminated spores as well as those killed by the mutagenic treatment are concentrated. The suspension of ungerminated cells is distributed on an agar medium supplemented by those substances for which auxotrophy is sought. Many of the mycelia growing on this medium nevertheless prove to be prototrophic, since some auxotrophic cells anastomose and prototrophic heterokaryons result. In other cases spores the germination of which has been delayed by the mutagenic treatment develop. Nevertheless, under favorable conditions the yield of particular auxotrophic mutants reaches 10% or more.

In order to prevent fusion of auxotrophic cells as much as possible, the conidial suspension is shaken or aerated during incubation so that extended contact between cells is avoided. A further increase in yield is possible through the choice of a favorable time period between the beginning of the experiment and the last filtration. If the time interval is too long, the prototrophic spores the germination of which has been delayed are eliminated by the last filtration but most of the auxotrophic spores may have died by that time. If the time interval is too short many of the prototrophic spores are not yet germinated. WOODWARD et al. (1954) carried out the first filtration of *Neurospora* conidia 18 hours after the beginning of the experiment and further filtrations at three and six hour intervals during the second 18 hour period and later at 6 to 12 hour intervals. The entire period for the procedure (between two and four days) depends upon the degree of delayed growth caused by the mutagenic treatment.

The application of the filter technique is limited to particular types of auxotrophs. Mutants which can grow with minute amounts of vitamins or other substances remain undetected as a rule. Traces of such growth factors are always present in a suspension as a result of lysis of dead spores. Other mutants (e.g. inositol auxotrophs) that die quickly in minimal medium likewise cannot be recovered by this method.

Concentration of auxotrophic mutants by repeated filtration was effective not only for *Ophiostoma* and *Neurospora* but also for *Saccharomyces cerevisiae* (TAKAHASHI, 1959) and *Coprinus lagopus* (DAY and ANDERSON, 1961). In order to use the technique for yeast TAKAHASHI used a mutant strain in which the cells fail to separate from one another after division and form a primary mycelium.

e) Double mutant method

FRIES (1948a, b) developed a method for concentration of mutants in *Ophiostoma multiannulatum* based on the fact that certain double auxotrophic mutants frequently survive longer in minimal medium than the corresponding single mutants. PONTECORVO et al. (1953) utilized this method for a strain of *Aspergillus nidulans* auxotrophic for biotin. Conidia of the mutant were irradiated, plated on minimal medium, and covered with a layer of the same medium. After a period of about 100 hours during which it was found most of the biotin auxotrophs had died, a layer of complete medium was added. The cells which had not yet starved to death produced mycelia.

About 60% of the isolates were auxotrophic for biotin as well as a second substance. LESTER and GROSS (1959) even obtained up to 80% double mutants with an inositol auxotroph of *Neurospora crassa*. The physiological basis of the selective advantage of the double mutants is unknown in both *Aspergillus* and *Neurospora*.

An increased viability of double auxotrophs was also observed by MIT-CHELL and MITCHELL (1950) in *Neurospora*, in *Ophiostoma* by FRIES (1953) and in *Saccharomyces* by ROMAN (1955). The increase in these cases involves an additional mutation in the same function, namely a double auxotrophy for adenine and guanine.

The isolation of specific mutants through selection of double mutants is the basis of the method described by REISSIG (1960). From the investigations of MITCHELL and MITCHELL (1952) on *Neurospora crassa* it was known that certain mutations within the *pyr-3* locus suppress the effect of the *arg-2* auxotroph. Thus the double mutant *pyr-3 arg-2* can grow without an arginine supplement (p. 400f.). The suppressor effect of *pyr-3* for *arg-2* was used by REISSIG for isolating many *pyr-3* mutants. He plated irradiated conidia of an *arg-2* strain on minimal to which he added pyrimidine but no arginine. Analysis of the mycelia growing on this medium yielded two different types of mutants: some mutations were reversions of *arg-2* to *arg-2⁺* while the remainder were mutations at the *pyr-3* locus.

f) Method for isolating reverse mutants

Reversion of auxotrophs to prototrophs is one of the most intensively studied types of mutation; reversions are particularly suited for quantitative determination of the frequency of mutation because of the simple isolation technique. In order to obtain back mutants a large number of cells (e.g. 10^8) of an auxotrophic strain is plated on minimal medium. Since only cells which have undergone mutation to prototrophy can grow, reverse mutation can be determined even if their frequency is very small (e.g. 10^{-7}).

For precise quantitative analysis the controls must establish that all mutations to prototrophy are recovered with the isolation technique. For a control prototrophic cells in known number are added to the auxotrophs and their capacity to grow is determined. The necessity for such a control experiment arises out of a discovery by GRIGG (1957) in *Neurospora crassa*;

he found that growth of prototrophic conidia is suppressed in the presence of high concentrations of auxotrophic conidia. A second difficulty in reversion experiments arises from the fact that in certain cases the prototrophs do not result from back mutation but from mutation of a suppressor gene (e.g. GILES, 1951 in *Neurospora crassa*, see also p. 397 ff.).

g) Method for isolating resistance mutants

A technique similar to that used for determining back mutation is used for isolating mutants that are characterized by their resistance to specific chemicals. ROPER and KÄFER (1957) isolated spontaneous mutations for acriflavine resistance in *Aspergillus nidulans* by plating conidia on an acriflavine medium in large numbers. A concentration of the toxic substance was selected that suppressed germination of the wild type conidia completely. Liquid cultures can be utilized in the same manner.

MORPURGO (1962), using a similar technique, obtained mutants of *A. nidulans* which were resistant to 8-azaguanine and p-fluorophenylalanine. LEBEN et al. (1955) selected antimycin-resistant mutants in the same way in *Venturia inaequalis*.

This experimental approach failed to give the expected results with *Neurospora crassa* in a number of cases. Mutants of the same type could, however, be obtained through UV irradiation (HOWE and TERRY, 1962). Resistance mutants were found for the following substances: acriflavine, cyclohexamide, sodium azide, sodium benzoate, and sodium caprylate. In *Penicillium chrysogenum* mutants which were resistant to 8-azaguanine were produced by mustard gas treatment (ARDITTI and SERMONTI, 1962).

FUERST and SKELLENGER (1958) developed a method which not only yields resistance mutants but also offers the possibility of testing the influence of specific substances (e.g. adenine, thiamine) on the inhibition of growth by antibiotics (details of this method can be found in the original paper).

Summary

1. The fungi possess relatively few morphological characters that are useful for genetic investigation because of their simple morphological structure. Since selective methods of isolation can generally not be used, obtaining large numbers of morphological mutants is laborious. Single mutants are relatively easy to identify; frequently this is not the case, however, if more than one mutation occurs in the same nucleus.

2. A lethal mutation can only be detected if it is recessive and occurs with its wild type allele in a heterokaryon or in a heterozygous cell. Lethals are generally not very suitable for genetic experiments because of these limitations.

3. In contrast to morphological and lethal mutations, biochemical mutants satisfy most of the experimental requirements. Their greatest advantage lies in the fact that they can be easily and quickly isolated in large numbers through selective techniques. The filter technique has proved especially useful as a method of isolation. Reverse mutations from auxotrophy to prototrophy are particularly suitable for quantitative studies of mutation frequency because of the ease of isolating them.

II. Radiation-induced mutagenesis

That high energy radiation induces changes in the genetic material has been known since the classical experiments of MULLER (1927, 1928). *Ionizing radiation* (X-rays, alpha, beta, and gamma radiation from radioactive substances) and the shorter wave lengths of *ultraviolet* have proved particularly effective. *For production of mutations in fungi X-rays and UV have been used primarily* (Table V-1). The table shows that morphological and biochemical mutations may be induced by non-ionizing as well as ionizing radiation. Further, radiation causes both forward and reverse mutation. The effects of the two types of radiation differ, however, as will be discussed in the following section. Point as well as chromosomal mutations are induced in both cases (p. 306 ff.).

Literature: BLAU and ALTENBURGER (1922), DESSAUER (1922), NADSON and FILLIPOV (1925, 1928), TIMOFEEFF-RESSOVSKY et al. (1935), KNAPP et al. (1939), KNAPP and SCHREIBER (1939), HOLLAENDER and EMMONS (1941), TIMOFEEFF-RESSOVSKY and ZIMMER (1947), SOMMERMEYER (1952), HOLLAENDER (1954, 1955, 1956), LEA (1956), MITCHELL et al. (1956), RAJEWSKY (1956), ZIMMER (1956), KAPLAN (1956, 1957), BACQ and ALEXANDER (1958), WALLACE and DOBZHANSKY (1959), FRITZ-NIGGLI (1959, 1962), HARRIS (1961), STUBBE (1962), RIEGER and BÖHME (1962), SZYBALSKI and LORKIEWICZ (1962), WOLFF (1963).
Literature review from 1896 to 1955: SPARROW et al. (1958).

Table V-1. *Examples of the mutagenic effect of ionizing (X-ray and gamma) and ultraviolet radiation*

object	kind of mutation				reference
	forward mutation	reverse mutation	morphological mutation	biochemical mutation	
1. X-radiation					
Aspergillus terreus	+		+		STAPLETON et al., 1952; STAPLETON and HOLLAENDER, 1952
Glomerella cingulata	+		+	+	MARKERT, 1952, 1956
Neurospora crassa	+			+	LINDEGREN and LINDEGREN, 1941; SANSOME et al., 1945; DE SERRES and KØLMARK, 1958
		+		+	GILES, 1951; KØLMARK, 1953; DE SERRES, 1958
Ophiostoma multiannulatum	+			+	ZETTERBERG, 1961, 1962
Podospora anserina	+		+		ESSER, unpublished
Schizosaccharomyces pombe		+		+	GUTZ, 1961, 1963
Sordaria macrospora	+		+		HESLOT, 1958

Table V-1 (Continued)

object	kind of mutation				reference
	forward mutation	reverse mutation	morphological mutation	biochemical mutation	

2. Gamma radiation

Saccharomyces cerevisiae		+		+	DUPIN, 1963
Schizosaccharomyces pombe		+		+	HESLOT, 1962

3. Ultra violet radiation

Aspergillus nidulans	+		+		APIRION, 1963; KÄFER and CHEN, 1964
		+		+	KILBEY, 1963a
Glomerella cingulata	+		+	+	MARKERT, 1952, 1953
Neurospora crassa	+			+	LINDEGREN and LINDEGREN, 1941; BEADLE and TATUM, 1945; REISSIG, 1960, 1963a, b; HOWE and TERRY, 1962
		+		+	GILES, 1951, 1959; KØLMARK, 1953, 1956; MALLING et al., 1959; VAHARU, 1961; ISHIKAWA, 1962; KILBEY, 1963b
Ophiostoma multiannulatum	+			+	ZETTERBERG and FRIES, 1958; ZETTERBERG, 1960a, 1961, 1962
Penicillium chrysogenum		+		+	AUERBACH and WESTERGAARD, 1960
Penicillium italicum	+		+	+	BERAHA et al., 1964
Penicillium digitatum	+		+	+	BERAHA et al., 1964
Saccharomyces cerevisiae	+		+		WILKIE, 1963
	+			+	PITTMAN et al., 1963; COSTELLO et al., 1963
		+		+	DUPIN, 1963; HAEFNER and LASKOWSKI, 1963; HAEFNER, 1964a, b
Schizosaccharomyces pombe		+		+	HESLOT, 1960, 1962; LEUPOLD, 1961; CLARKE, 1962, 1963
Ustilago maydis	+	+		+	HOLLIDAY, 1961, 1962
Venturia inaequalis	+		+		BOONE et al., 1956
Verticillium albo-atrum	+			+	BUXTON and HASTIE, 1962

1. Dose-effect relationship

Knowledge of the nature of the relationship between radiation dose and mutation frequency is essential for an interpretation of the radiation-induced mutation process. A graphical representation of the relationship between dose and effect gives the so called *dose-effect curves* (Figs. V-2 and V-3).

In a quantitative determination of dosage dependence one must remember that the fraction of mutated cells in a population of cells (frequency

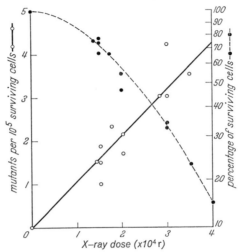

Fig. V-2. Mutagenic effect of X-rays. Relationship between X-ray dose and frequency of *inos+* back mutations in microconidia of *Neurospora crassa*. (Adapted from GILES, 1951)

of mutants) is not necessarily identical with the frequency of mutations induced by radiation (mutation rate). The number of mutants may increase not only as a result of new mutations but also because of the preferential survival of spontaneous mutants which arose before irradiation. A high percentage of cells are generally killed by irradiation and mutants may often be more susceptible to killing than non mutants. Participation of strongly selective factors is especially likely in the case of non-linear dose-effect curves. The degree of selection and the true mutation rate can be determined by experiments with mixtures of mutants (KAPLAN, 1953).

The *nature of the dose-effect curves* in the case of ionizing radiation is essentially the same in fungi as in other organisms. In the *middle range* the rate of induced mutation in the surviving nuclei is *directly proportional to the dose* (Fig. V-2). A *linear dose-effect relationship* also exists with *low* dosages, although the data are few and inconsistent; they are mostly from bacterial experiments (SPENCER and STERN, 1948; BONNIER and LÜNING, 1949; DEMEREC et al., 1952, 1953, 1959, 1960; DEMEREC, 1959; DEMEREC and SAMS, 1960). The dose-effect curves in general reach *saturation* levels with *high* doses (e.g. NEWCOMBE and McGREGOR, 1954); they rarely drop after reaching a maximum.

Linear dose-effect relationships have been observed for *N. crassa* for forward as well as reverse mutation of different auxotrophic loci after X-irradiation (SANSOME et al., 1945; GILES, 1951). Similar results were obtained for morphological mutations by STAPLETON et al. (1952) and STAPLETON and HOLLAENDER (1952) in *Aspergillus terreus* and by MARKERT (1956) in *Glomerella cingulata*. HESLOT (1962) found linear as well (for *arg-1*) as non-linear dose-effect curves (for *leuc-3*) following irradiation of *Schizosaccharomyces pombe* with gamma rays from a cobalt 60 source (see also mutation production by ^{32}P-disintegration in spores of *Aspergillus nidulans*: STRIGINI et al., 1963).

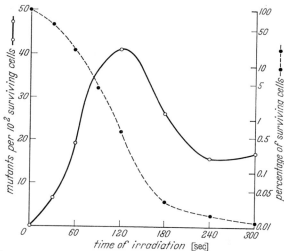

Fig. V-3. Mutagenic effect of UV-rays (see also Figs. V-8 and V-9). Relationship between the time of irradiation and the frequency of forward mutation in conidia of *Glomerella cingulata*. (Adapted from MARKERT, 1953)

With UV radiation non-linear dose-effect curves are frequently obtained in contrast to the effect of ionizing radiation. Here the maximum usually is followed by a decrease (Figs. V-3, V-8, and V-9).

Near the point of origin the dose-effect curves are often more or less proportional to both the number of radiation-killed conidia and the number of surviving radiation-induced mutants (e.g. KØLMARK, 1953; BUXTON and HASTIE, 1962). Such a parallel proportionality suggests that the death of a cell results from a lethal mutation (p. 268). The results of NORMAN (1954) also support this conclusion; he demonstrated that the death of *Neurospora* macroconidia through UV irradiation involves the nucleus. Nevertheless, ATWOOD (1950) demonstrated that with X-radiation true lethal mutations play a significant role at low doses only.

Large doses of UV and X-radiation lead to inactivation of the irradiated cells. This phenomenon has been investigated intensively particularly in yeasts. Radiation-induced inactivation of *haplonts* results predominantly from induction of *recessive* lethal mutations, while inactivation of *polyploid* strains can be attributed to the induction of *dominant* lethals (LASKOWSKI, 1960a; LASKOWSKI and STEIN, 1960; STEIN and LASKOWSKI, 1960; HAEFNER, 1964a). The latter mutations are brought about mostly by chromosomal breaks and resulting aberrations (p.306ff.). However, the inactivation effect of radiation may also involve in part a disturbance or inhibition of essential cellular processes without lethal mutation or other changes in the hereditary material.

In order to investigate the dependence of radiation resistance on ploidy level in detail, tri-, tetra-, penta-, and hexaploid strains were produced from a largely homozygous isogenic stock of *Saccharomyces* (LASKOWSKI, 1962a). With UV irradiation haploid as well as polyploid yeast strains showed the same radiation sensitivity. Such an identical sensitivity was not observed with X-radiation. The corresponding dose-effect curves revealed a clear decrease in radiation resistance with increasing ploidy level ($n \geqq 2$). The increase in radiation sensitivity is believed to be primarily the result of induction of chromosome aberrations (p. 306ff.) (LATARJET and EPHRUSSI, 1949; CALDAS and CONSTANTIN, 1951; MORTIMER, 1952; WARSHAW, 1952; ZIRKLE and TOBIAS, 1953; LUCKE and SARACHEK, 1953; SARACHEK and LUCKE, 1953; BEAM, 1955; POMPER and ATWOOD, 1955; WEINFURTNER and VOERKELIUS, 1955; LASKOWSKI and STEIN, 1960; STEIN and LASKOWSKI, 1960; LASKOWSKI, 1960a, b, 1962a, b; HAEFNER and LASKOWSKI, 1963; LASKOWSKI and HAEFNER, 1963; HAEFNER, 1964a). Nevertheless, for ploidy levels greater than one, genes which control the synthesis of amino acids also seem to have an influence on the resistance to X-radiation. This is also true for alpha radiation (LASKOWSKI, 1962a).

Other radiation experiments with *Saccharomyces cerevisiae* have shown that an increased radiation resistance occurs during meiosis when gamma rays are utilized (DUPIN, 1963). Such a resistance was not observed with UV radiation. The investigator concluded that UV and ionizing radiation affect different cellular mechanisms.

2. Theories of radiation effect

Two alternative theories have been developed to explain the fact that mutations occur in only a small fraction of the irradiated cells.

1. *Variability theory:* This assumes that individual cells differ in their sensitivity to radiation. A series of decisive observations argues against this interpretation and thus the second theory is more likely to be correct (for literature see ZIMMER, 1956; KAPLAN, 1956, 1957).

2. *Target theory: A mutation in the cell is produced by a single microphysical event in which one, or on occasion, a few "targets" are hit.* The effect of radiation is statistical, analogous to unaimed "shots" so that it is a matter of chance in which cell of an irradiated population of cells a mutation takes place (BLAU and ALTENBURGER, 1922; DESSAUER, 1922; for other literature see p. 278).

The fact that with moderate doses of ionizing radiation *a linear relationship between dose and mutation* rate is frequently observed favors the *target theory* (Fig. V-2). According to the target theory the cells behave like molecules in a first order reaction, i.e. a *single hit* is sufficient to produce a mutation. The *non-linear curves* found in the case of UV irradiation (p. 281) are interpreted as a result of *multiple hits*. The small energy content of the UV quanta as compared to ionization may be responsible for the necessity for numerous hits (KAPLAN, 1957).

3. Phases of the mutation process

Most dose-effect curves show a deviation from a single hit curve with higher doses. This may result in the curves approaching a saturation level (X-ray, p. 280) or after reaching a maximum decreasing again (UV light Fig. V-3). The cause of the drop-off may be found in the heterogeneity of the irradiated cell population.

For example, MARKERT (1953) interpreted his results with *Glomerella cingulata* in this way. He concluded that a part of the irradiated cells is resistant to mutation (with or without lethal effect). With high doses the percentage of mutations among the surviving cells decreases again.

Other observations, however, support the idea that a portion of the changes induced by radiation undergoes reversion as a result of further hits. On the basis of these and other experiments with bacteria, it is considered certain that the mutation process is not over in fractions of seconds, but runs through several phases before the hereditary changes are completed (for literature see HARM and STEIN, 1956; KAPLAN, 1957). Recent results (for literature see KAPLAN, 1962a) suggest that at least three successive stages may be involved:

1. *Initiating phase.* A mutation may be initiated by a single absorbed quantum of radiation or an ionization, if this primary physical event takes place in a particularly sensitive cell structure (target). This primary change occurs mostly in the DNA itself (compare also p. 290ff.).

The specific effect of certain mutagens on specific mutational sites (p. 294ff.) argues for the direct mutability of DNA. Such specificity is easy to understand if one assumes that the premutational change occurs in the specifically structured DNA of the individual genes.

HAAS and DOUDNEY believe, in contrast, that the primary change induced by UV is not in the DNA, but in precursors of RNA (HAAS and DOUDNEY, 1957, 1959; DOUDNEY and HAAS, 1958, 1959, 1960). According to these authors these altered precursors lead to a synthesis of a modified RNA. The actual mutation i.e. the ultimate change in the DNA itself would then depend first on the transfer of the premutation in the RNA to the DNA at the time of replication of the latter. This hypothesis has been repeatedly disputed (e.g. KAPLAN, 1962a; RIEGER and BÖHME, 1962), because it is based on the unproved idea that the genetic information can be transferred from RNA to DNA as well as from DNA to RNA (p. 342).

2. *Premutational phase.* After UV irradiation DNA synthesis is at first interrupted. It continues only after the onset of protein and RNA synthesis (KELNER, 1953; KANAZIR and ERRERA, 1956; HAROLD and ZIPORIN, 1958; DOUDNEY, 1959; DRACULIC and ERRERA, 1959). By inhibiting protein synthesis (e.g. with chloramphenicol) the time of new DNA synthesis can be postponed. This moment seems to mark the end of the premutational phase (WITKIN, 1959; LIEB, 1960; WEATHERWAX and LANDMAN, 1960).

If protein synthesis is blocked experimentally, a sharp decrease in mutational yield after UV and X-irradiation is observed (WITKIN, 1956, 1959, 1961; DOUDNEY and HAAS, 1958; KADA et al., 1960, 1961). This observation is explained by assuming an increased probability of reversion from the premutational condition as a result of lengthening the premutational phase (KIMBALL et al., 1959; WITKIN, 1961; compare experiments of VAHARU, 1961, with *N. crassa*).

A deeper insight into the premutational phase comes from photo-reversion experiments (p. 285). The mutagenic activity of UV can frequently be reduced by a post-radiation treatment with visible light (e.g. KELNER, 1949; KAPLAN and GUNKEL, 1960; ZELLE et al., 1958). The results of these experiments show that UV-induced mutation does not lead directly to a stable alteration in the genetic material (review by DULBECCO, 1955; JAGGER, 1958, 1960; for its significance at the molecular level, see p. 285).

Investigations through which the influence of particular factors during irradiation is determined have made significant contributions to the characterization of radiation lesions leading to mutation: *oxygen* (e.g. ALPER and HOWARD-FLANDERS, 1956; GLASS and METTLER, 1958; HOWARD-FLANDERS, 1958, 1959), *water content* (e.g. STAPLETON and HOLLAENDER, 1952; KAPLAN and KAPLAN, 1956; METZGER, 1960), *temperature* (e.g. ZAMENHOF and GREER 1958; KAPLAN, 1962b; HAINZ and KAPLAN, 1963; KILBEY, 1963b). The results of these experiments are interpreted to indicate that a labile reversible change precedes the completion of a mutation.

3. *Completion phase.* This phase begins with the new DNA replication. The alterations in the DNA which have persisted up to this time remain permanently and apparently cause mutation as a result of copying errors during replication (compare NAKADA et al., 1960, and p. 291 ff.).

All irradiation experiments have shown that sensitivity to radiation depends upon the stage of development and other physiological conditions.

For example, budding yeast cells are more resistant to radiation than old cells in the resting stage (OSTER, 1934a, b, c; OSTER and ARNOLD, 1934; BEAM et al., 1954; SARACHEK, 1954a, b). This is understandable since the genetic material has already duplicated at the beginning of the budding process (OGUR et al., 1953). Dried cells of *Saccharomyces* (DUNN et al., 1948) and dry spores of *Aspergillus* (STAPLETON and HOLLAENDER, 1952) have likewise proved more resistant to X-rays than normally hydrated cells. A structural alteration in the DNA produced by desiccation may be responsible (KAPLAN, 1957).

The effect of mutagenic agents may also be genetically influenced (ZETTERBERG, 1962); this was observed in experiments with uracil (u) and methionine (m) auxotrophs of *Ophiostoma multiannulatum*. For example, after UV irradiation the double auxotroph, $u\ m$, showed a significant decrease in reversion rate for $u \rightarrow u^+$ compared with the single mutant, u. The survival rate remained completely unaffected. A similar effect was also found following X-irradiation and after treatment with nitrogen mustard, dimethyl sulfate and N-nitroso-N-methylurethane [compare with the increased viability of double auxotrophs (p. 276) and the simulated mutagen specificity in the use of strains with multiple markers (p. 299f.)].

4. Chemical basis of radiation-induced mutation

Irradiation experiments with UV light of different wave lengths show that the action spectrum parallels the absorption spectrum of DNA (in fungi: HOLLAENDER and EMMONS, 1941; in higher plants: KNAPP and SCHREIBER, 1939; KNAPP et al., 1939; STADLER and UBER, 1942). This supports the hypothesis that *UV-induced mutations may result from a direct photochemical alteration of nucleic acid* (p. 128). Recent results of WILKIE (1963) with yeast cultures maintained under aerobic and anaerobic conditions can be interpreted in the same way.

Dose-effect curves for the induction of *"petite"* mutations (p. 442ff.) by UV light were determined for haploid and diploid strains of *Saccharomyces cerevisiae*. WILKIE obtained single hit curves for anaerobic yeast cultures and multiple hit curves for aerobic cultures; this was true for both haploid and diploid strains. The number of light quanta absorbed per cell was used as a measure of dose. Of the three wave lengths tested (245, 265, and 280 mμ) 265 mμ proved most effective. Further, the intensity of radiation and the temperature during irradiation had no influence on the mutation frequency. WILKIE concluded that the energy absorbing molecules were nucleic acid. The single and multiple hit curves are explained by assuming that in cells growing anaerobically only a single nucleic acid matrix is present while in aerobically growing cells a number of identical matrices are present.

A number of research groups using bacteria and viruses have investigated the *chemical alteration of DNA in vitro* in attempts to explain the irradiation effect on a molecular basis (BEUKERS et al., 1958, 1959, 1960; BEUKERS and BERENDS, 1960; WACKER et al., 1960, 1961; SCHOLES and WEISS, 1960). From the results of these experiments one can conclude that the photochemical reaction produced under the influence of UV can take place in the DNA molecule. At least a portion of the UV-induced premutations are dimers of thymine of DNA. The formation of such dimers by UV in vitro and in vivo has been demonstrated (WACKER et al., 1960, 1961). Photoreversion of mutations (p. 284) results in part from the splitting of the dimer into monomers by a specific photosensitive enzyme (RUPERT, 1962) and in part from other photochemical processes involving DNA (KAPLAN, 1963). Good evidence of the direct attack of UV on DNA is the fact that the replacement of thymine by 5-bromouracil (p. 291) in the DNA leads to an increased radiation sensitivity (GREER and ZAMENHOF, 1957; LORKIEWICZ and SZYBALSKI, 1960).

Summary

1. In fungi the mutagenic effect of X- and UV radiation has been extensively investigated. Both point and chromosomal mutations result from irradiation.

2. The frequency of mutations increases in a linear fashion up to saturation with increasing doses of X-rays. UV irradiation leads in part to a non-linear dose-effect curve and almost always to a drop with high doses.

3. The radiation effect as shown by dose-effect curves can be interpreted with the aid of the target theory. According to this theory, a single microphysical event (hit) can produce a mutation. A single hit generally suffices to produce a mutation with X-ray (linear dose-effect curves = single hit curves). With UV irradiation multiple hits are necessary to some extent (non-linear dose-effect curves = multiple hit curves).

4. Radiation-induced mutation is believed to involve at least three phases, the initiating, premutational and completion phase, to yield the final hereditary change; the evidence for this comes primarily from the results of experiments with bacteria. Hypotheses of the chemical basis of the mutation process are discussed, particularly for UV. In the latter case UV-induced premutational dimers of thymine occur in the DNA.

III. Chemical mutagenesis

1. Effect of mutagenic chemicals

Specific chemicals induce known chemical alterations of DNA in vitro; e.g. hydroxylamine specifically changes cytosine; nitrite alters other bases. Furthermore, 5-bromouracil is incorporated into DNA in place of thymine (p. 291). An analysis of these phenomena gives promise of an insight into the chemical basis of the mutation process.

A large portion of our knowledge of the mutagenic effect of chemicals in fungi comes from the work of WESTERGAARD and coworkers. These investigators studied the mutagenic effect of a large number of chemicals in *Neurospora crassa*. These are listed in Table V-2. In addition other investigators have tested other substances partly in *Neurospora*, partly in other organisms. Between a quarter to a third of these chemicals proved to be mutagenic. Such mutagens which are not listed in Table V-2 are listed in Table V-3.

Literature: KØLMARK and WESTERGAARD (1949, 1952, 1953); JENSEN et al. (1949, 1950, 1951); KØLMARK (1953, 1956); KØLMARK and GILES (1955); WESTERGAARD (1957, 1960); MALLING et al. (1959); AUERBACH and WESTERGAARD (1960); KØLMARK and KILBEY (1962); AUERBACH et al. (1962a, b).

For other references see STUBBE (1960) and also Table V-3.

Table V-2 and Figs. V-4 and V-5 yield information on the effect of chemical mutagens.

1. *Each of the tested substances produces a characteristic mutation rate.* For example, diepoxybutane gives from two to three hundred times more reversions than formaldehyde (Table V-2). A comparison with corresponding radiation experiments shows that certain chemicals produce a significantly higher yield of mutations than can be induced by irradiation (compare Table V-4).

For example, diepoxybutane induces 25 to 30 times more *ad-3+* prototrophs in *N. crassa* (Table V-4) and even 100 to 300 times more *ar-3+* prototrophs in *E. coli* than UV or X-rays (WESTERGAARD, 1960; and GLOVER, 1956). The survival rate generally amounts to 50—100% under optimal conditions (for maximum yield of mutations).

2. *Reverse mutation rates and survival rates are not correlated with one another.* Chemicals with the same mutagenic effect may show either a strong or a weak lethal effect; for example, p-N-di(β-chlorethyl) phenylalanine and propylene oxide induce about the same number of reverse mutants ($21—22/10^6$) while the percentage of surviving macroconidia ranges from 26 to 100 (Table V-2). In comparison, radiation,

Table V-2. *The effect of different mutagenic chemicals on the adenine auxotroph of N. crassa.* (From WESTERGAARD, 1957)

The concentration of the mutagen and the duration of the treatment were adjusted to give the optimal yield of reverse mutants. Since the macroconidia which were used are multinucleate (on the average 2—3 nuclei per conidium), one can assume that the frequency of reversions per nucleus exceeds the values given in the last column of the table by some multiple.

mutagenic agent	concentration in moles per liter	duration of treatment in minutes	percentage of surviving macroconidia	reverse mutants per 10^6 treated macroconidia
Diepoxybutane	0.2	40	56	85
Dimethyl sulfate	0.005	30	44	64
Epichlorhydrin	0.15	45	42	56
Chlorethylmethane sulfonate (CB 1506)	0.1	13	58	51
Glycidol	0.5	60	26	34
p-N-di (β-chlorethyl) phenyl-alanine CB 3025, L-form	0.03	40	100	22
Propylene oxide	0.5	60	27	21
Diethyl sulfate	0.04	40	68	18
Ethyl methane sulfonate (CB 1528)	0.1	12.5	14	17
Ethylene oxide	0.025	15	63	17
Ethylene imine	0.05	30	75	16
Hydrogen peroxide + formaldehyde	0.06 + 0.3	30	20	4.3
Di (β-chlorethyl) methyl amine (nitrogen mustard)	0.0025	25	60	3.4
1.2-monoepoxybutane	0.2	40	47	3.2
p-N-di (β-chlorethyl) phenyl-alanine CB 3026, D-form	0.03	120	90	3.1
Epibromohydrin	0.08	45	40	2.5
Monochlor (β-chlorethyl) dimethylamine	0.005	60	80	1.7
Tertiary butylhydroperoxide	0.09	30	50	1.5
Diepoxypropylether	0.1	20	65	0.7
Triethylene melamine (TEM)	0.02	50	55	0.6
Diazomethane	0.03	40	20	0.6
Trimethylphosphate	0.2	40	96	0.5
Hydrogen peroxide	0.2	45	10	0.4
Formaldehyde	0.01	180	80	0.3

particularly X-rays, generally kills more cells than chemical agents, in spite of the weaker mutagenic effect.

3. *The percentage of reverse mutations among the surviving cells generally increases when the duration of the treatment is extended or the concentration of the mutagen increased* (Figs. V-4 and V-5). The nature of the curves suggests a kind of target mechanism of action for certain mutagenic agents similar to that of the mutagenic effect of radiation (p. 282): for example, single hit curves in *E. coli* (DEMEREC et al., 1952, 1953), multiple hit curves in *N. crassa* (Fig. V-4 and KØLMARK and GILES, 1955; KØLMARK, 1956). In the first case a single molecule is sufficient to initiate the mutation process, while in the latter, multiple hits are

Table V-3. *Survey of the most important chemicals and organisms from the standpoint of their use in the induction of mutations*

Only those substances are listed which have not been included in Table V-2.

mutagenic agent	object	reference
Nitrous acid	*Neurospora crassa*	BARNETT and DE SERRES, 1963; REISSIG, 1963a
	Schizosaccharo-myces pombe	GUTZ, 1961, 1963; HESLOT, 1962; CLARKE, 1962, 1963
	Aspergillus nidulans	SIDDIQI, 1962; APIRION, 1963
N-nitroso-N-methyl urethane	*Ophiostoma multi-annulatum*	ZETTERBERG, 1960b, 1961, 1962
	Saccharomyces cerevisiae	MARQUARDT et al., 1964
	Colletotrichum coccodes	LOPRIENO et al., 1964
N-nitroso-N-ethyl urethane	*Saccharomyces cerevisiae*	MARQUARDT et al., 1964
	Colletotrichum coccodes	LOPRIENO et al., 1964
Various nitrosamines and nitrosamides	*Saccharomyces cerevisiae*	MARQUARDT et al., 1964
β-methylcholanthrene	*Neurospora crassa*	TATUM et al., 1950
β-propiolactone	*Neurospora crassa*	SMITH and SRB, 1951
	Aspergillus nidulans	KILBEY, 1963a
Caffeine	*Ophiostoma multi-annulatum*	FRIES, 1950; ZETTERBERG, 1960a
	Saccharomyces cerevisiae	NAGAI, 1962
	Schizosaccharo-myces pombe	HESLOT, 1962
Tri (β-chlorethyl) methylamine (nitrogen mustard)	*Saccharomyces cerevisiae*	REAUME and TATUM, 1949
	Neurospora crassa	McELROY et al., 1947; TATUM et al., 1950
n-butylchlorethyl sulfide	*Neurospora crassa*	STEVENS and MYLROIE, 1953
Di (β-chlorethyl)-sulfide	*Neurospora crassa*	HOROWITZ et al., 1946
Acriflavin Pararosaniline Pyronin B Pyronin Y Acridine red 3B	*Saccharomyces cerevisiae*	NAGAI, 1962
Various sulfate esters Various epoxides Various ethylene amines Various β-halogenethyl-amines	*Schizosaccharo-myces pombe*	HESLOT, 1962
5-bromodeoxyuridine 5-fluorodeoxyuridine	*Neurospora crassa*	ISHIKAWA, 1962; SCOTT, 1964
2,6-diaminopurine	*Saccharomyces cerevisiae*	PITTMAN et al., 1963
2-aminopurine	*Neurospora crassa*	BROCKMAN and DE SERRES, 1963

necessary. Nevertheless, one must remember with such an interpretation that neither the time-dependence nor the concentration curves are unequivocal.

The nature of the curves is not only determined by the conditions of the experiment (duration of treatment, concentration of mutagen), but also by other factors in the cell which are not always identifiable, e.g. penetration and rate of accumulation. Thus non-linear curves may result which simulate a multiple hit relationship.

The concentration and time-dependence curves frequently rise to a maximum and then drop off (e.g. HESLOT, 1962). Such curves can be

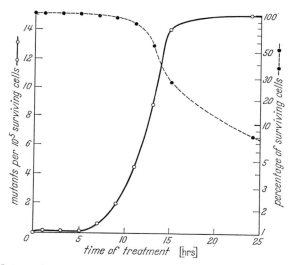

Fig. V-4. Mutagenic effect of chlorethylmethane sulfonate (compare Table V-2). The relationship between length of treatment and the frequency of *ad-3+* reversions of macroconidia of *Neurospora crassa* at a concentration of 0.1 mole/liter. (Adapted from WESTERGAARD, 1957)

interpreted as indicating that the final result of chemically-induced mutation is preceded by a premutational stage (p. 283) — as in the case of UV irradiation.

Experiments with substances called base analogues, which are very similar to particular components of DNA and are therefore incorporated into the molecule during replication, give further insight into the mechanism of mutation. Investigations on the effectiveness of such mutagenic agents have been carried out primarily on viruses and bacteria (also chemical treatment of nucleic acid in vitro, e.g. tobacco mosaic virus: MUNDRY and GIERER, 1958). Because of technical difficulties the fungi have proved less suitable for such work. Nevertheless, they have also recently been used in isolated instances. *Treatment of fungi with base analogues* also induces mutations (Table V-3; for interpretation of results see p. 291 f.).

Furthermore, experiments on the *effect of nucleic acids on the mutation process* in *Neurospora crassa, Schizosaccharomyces pombe,* and *Saccharomyces cerevisiae* have been carried out (OPPENOORTH, 1960; VAHARU, 1961; SHAMOIAN et al., 1961; SHOCKLEY and TATUM, 1962; FRITZ-NIGGLI, 1963). In certain cases an increase in the mutation frequency has been observed if RNA or DNA isolated from wild or mutant strains has been included in the culture medium.

The reversion rate of a leucine auxotroph (leucineless) of *Neurospora crassa* rose if RNA was added to the nutrient following irradiation with low

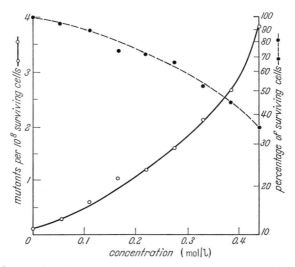

Fig. V-5. Mutagenic effect of 1,4 butane sulfonate. The relation between concentration of mutagen and the frequency of *arg-1+* reversions among cells of *Schizosaccharomyces pombe* with a one hour treatment. (Adapted from HESLOT, 1962)

UV doses (VAHARU, 1961). The relative frequency of prototrophic reversions increased with X-irradiation in the presence of RNA in the same way in a nitrite- and in a UV-induced adenine auxotroph of *S. pombe* (FRITZ-NIGGLI, 1963). Qualitative differences in the effect of two different RNA preparations were shown in the nitrite-induced mutant: When wild type RNA was used, reversions of the *adenine* locus as well as mutations of suppressor genes were obtained. After addition of RNA isolated from nitrite-induced suppressor mutants only suppressor mutations were found.

A mixture of RNA/DNA from a mycelial extract of wild type *Neurospora* was added along with uridine to a conidial suspension of a pyrimidine auxotroph (SHAMOIAN et al., 1961). After addition of the nucleic acid auxotrophic conidia showed a growth comparable to the prototrophic control in only three out of eight attempts. Unfortunately a genetic analysis was not carried out so that a conclusion about the mechanism of action is not possible. Neither do the results of SHOCKLEY and TATUM (1962) provide a clear conception of the effect of DNA preparations on *N. crassa*. In these experiments biochemical and morphological mutants were tested with partially purified DNA as well as extracts from a morphological mutant. An increase in the reversion rate (up to ten fold) was observed in only a few cases. Autoclaved extracts had the same effect as those not autoclaved. OPPENOORTH (1960) obtained similar results with *S. cerevisiae*.

The increase in mutation rate following addition of DNA was interpreted by OPPENOORTH, SHAMOIAN et al., and other workers as a *transformation effect* of the nucleic acid. This conclusion appears hasty, since the results do not provide convincing evidence of such a mechanism. *The phenomenon of genetic transformation*, well-established for bacteria, *remains uncertain in fungi* and higher organisms.

2. Interpretations of chemical mutagenesis

For interpretation of the effect of chemical mutagens on DNA several hypotheses are under discussion. The relevant experiments have been done almost exclusively on viruses. From our knowledge about structure and replication of DNA (p. 129ff.) one should expect that even small alterations of the base sequence can represent mutations. Such alterations may consist of substitutions of one nucleotide for another (FREESE and coworkers), or insertions or deletions of nucleotides (BRENNER and coworkers). Some of the possibilities are illustrated in Fig. V-6.

Literature: FREESE (1959a, b, c); BAUTZ and FREESE (1960); FREESE, BAUTZ and BAUTZ-FREESE (1961); FREESE, BAUTZ-FREESE and BAUTZ (1961); BAUTZ-FREESE (1961); ORGEL and BRENNER (1961); BRENNER et al. (1958, 1961); LERMAN (1963); FREESE (1963); ZAMENHOF (1963).

a) Exchange between bases

A base (e.g. guanine) may be split off from its nucleotide chain through the effect of a specific mutagen (e.g. ethyl methane sulfonate). In the replication of the complementary strand at the point opposite this gap another base, e.g. thymine, may be incorporated. This pairs with its complementary base, adenine, at the subsequent replication. In case the incorporated base is not complementary to the one which was split off, such an exchange leads to a permanently replicating alteration of the DNA molecule, i.e. a mutation (Fig. V-6a). The deletion of the base is thus a premutational change, which with subsequent replication of DNA, has a specific probability of becoming a mutation.

The alkylating reactions in particular play a mutagenic role here, since diethyl sulfate and ethylethane sulfonate act in a similar way (Table V-2).

b) Incorporation of base analogues instead of normal bases

A base (e.g. adenine) pairs with a synthetic base analogue during replication (e.g. 5-bromouracil) instead of with the naturally-occurring complementary base (thymine). Such incorporation of base analogues into DNA results from the structural similarity of analogues and natural bases. It does not lead to a mutation immediately, but only after an incorporation or pairing error during subsequent DNA replication (Figs. V-6b and V-6c).

19*

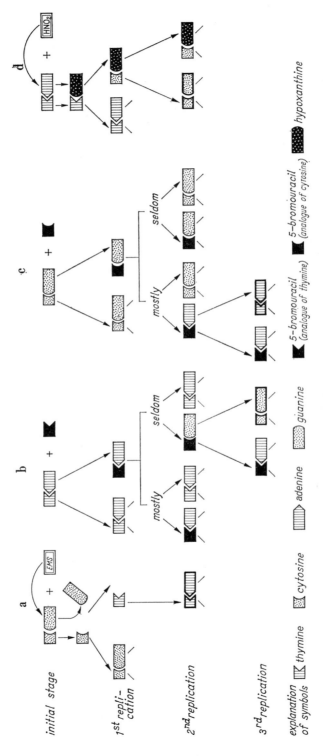

Fig. V-6a—d. Hypothetic mechanisms for interpretation of the mutagenic effect of chemical agents. a) Exchange of the base pair cytosine-guanine versus the base pair thymine-adenine as a result of the splitting off of guanine by ethyl methane sulfonate. b) Exchange of thymine-adenine for guanine-cytosine as a result of the incorporation of 5-bromouracil in place of thymine with a subsequent replication error. c) Exchange of cytosine-guanine for adenine-thymine as a result of incorporation of 5-bromouracil with altered pairing specificity in place of cytosine with the subsequent incorporation error. d) Exchange of thymine-adenine for cytosine-guanine as a result of deamination of adenine to hypoxanthine by nitrous acid (HNO₂). The symbols in heavy outline in each case designate the first occurrence of a new base pair, i.e. the completion of mutation. Further explanation in text

explanation of symbols: thymine · cytosine · adenine · guanine

5-bromouracil (analogue of thymine) · 5-bromouracil (analogue of cytosine) · hypoxanthine

initial stage

1st replication

2nd replication

3rd replication

In addition to 5-bromouracil (BU) 2-aminopurine (AP) is a mutagenic base analogue. Neither involves a splitting off or chemical alteration of bases (see below). Since BU is generally incorporated in place of thymine, it pairs mostly with adenine. As long as such pairing continues, no mutant phenotype is produced. The BU-containing DNA remains unmutated. In rare cases, however, the pairing specificity of BU is so altered as a result of a tautomeric shift or an ionization that it pairs with guanine instead of adenine (*replication error*, Fig. V-6b). The result is the transition T-A → BU-A → BU-G → C-G (C = cytosine, T = thymine, A = adenine, G = guanine). Since there is a possibility of a pairing error in subsequent replications as well, such BU-A nucleotide chains continue to give rise to C-G-containing DNA strands at a low frequency. Although as a rule BU replaces thymine, in exceptional cases it can also be incorporated in place of cytosine (*incorporation error*, Fig. V-6c). Since it almost always pairs with adenine in the next DNA replication, the transition C-G → BU-G → BU-A → T-A gives a mutated strand (compare also RUDNER, 1960).

In a similar way AP mutations may arise through an error resulting from incorporation of guanine instead of adenine and also through a replication error which results from pairing with cytosine instead of thymine (NAKADA et al., 1960; STRELZOFF, 1961, 1962).

Base analogues have also been used successfully as mutagens for fungi (Table V-3). Such substances have a highly specific effect (p. 295).

c) Chemical alteration of bases

A base (e.g. adenine) may become so altered by the effect of a specific mutagen (e.g. nitrous acid) that the new form (e.g. hypoxanthine) has a different pairing specificity; at the next replication it pairs with another base (cytosine). In the subsequent replication the latter pairs with the complementary base (guanine). This exchange of base pairs (guanine-cytosine for adenine-thymine) does not result in a completed mutation until the second replication (Fig. V-6d).

The mutagenic effect of *nitrous acid* can be related to the oxidative deamination of adenine and cytosine, the 6-amino groups of which are changed into hydroxyl groups; these are further transformed into keto groups by tautomeric shifts (SCHUSTER, 1960; VIELMETTER and SCHUSTER, 1960). The products of these deaminations are hypoxanthine and uracil, respectively; these, in contrast to the original bases exhibit a pairing behavior like guanine and thymine. Thus in the next replication they pair with cytosine and adenine, and these subsequently with guanine and thymine.

Hydroxylamine, as well as nitrous acid, leads to an altered pairing affinity. It opens a double bond in the cytosine ring. In the subsequent DNA replication the altered cytosine pairs with adenine instead of guanine.

While deamination of adenine and cytosine leads to a change in pairing specificity, deamination of guanine results in its failure to pair with other bases. As a result, the deamination product, xanthine, interferes with DNA replication. Xanthine is either replaced by the original base or causes the death of the individual concerned.

Nitrous acid has repeatedly proved to have a specific mutagenic effect in fungi (Table V-3). The results of SIDDIQI (1962) with *Aspergillus nidulans* in particular support the assumption that nitrous acid induces the same changes in fungal DNA as in phage and bacterial DNA. SIDDIQI found that half and quarter sectors with altered phenotypes segregated in mycelia arising from treated conidia of *Aspergillus nidulans*. Such sectoring was interpreted as the effect of nitrous acid on only one base of a base pair so that only one of the two newly replicated strands of DNA carries the mutation. Results of experiments on spontaneous mutation in *E. coli* allow the same conclusion (RYAN and KIRITANI, 1959).

d) Deletion and insertion of bases

Single nucleotides may either be deleted or added through the effect of specific mutagens (acridines, e.g. proflavine) at the time of DNA replication. The mutation does not involve a base exchange, but a deletion or insertion of one or more bases. If one assumes that each three adjacent nucleotide pairs determine an amino acid (see chapters on replication and function) an insertion or deletion of a base pair in a triplet results in a lateral shift of the grouping of the bases in threes; thus the overall information of the gene is changed. As a result a greatly modified protein appears (discussion in KAPLAN, 1962a, and LERMAN, 1963). In contrast, exchange of a base leads to a change in the protein at only one site, because only a single base pair of one triplet and not the entire grouping into threes is altered (ROSEN, 1961). It remains to be seen if such a mechanism of action can be confirmed for acridine (compare Table V-3).

On the basis of these considerations it is clear that treatment with mutagenic chemicals involves initially a premutational stage, which has a specific probability of later becoming a hereditary modification (p. 283 f.). The time between incorporation of base analogues and the completion of the mutation can be interpreted as the premutational stage.

In *E. coli* the fixation of caffeine-induced resistance mutations has been shown to require several cell divisions (KUBITSCHEK and BENDIGKEIT, 1958, 1961). Caffeine has also been used with success as a mutagen in fungi (Table V-3). An interpretation of caffeine-induced mutagenesis is still difficult, because it is not known whether or not caffeine is incorporated into DNA as a purine analogue, inducing mutation in this manner.

Summary

1. A large number of chemical substances have proved to be mutagenic. The individual mutagens differ considerably in their potency. Mutation and inactivation rates are not correlated with each other.

2. In general the frequency of mutations among surviving cells increases if the length of treatment is increased or the concentration of the mutagen is raised. In the cases of some mutagens the nature of the concentration and time-dependence curves suggests an action similar to that of the target theory and analogous to the effect of radiation.

3. Chemically-induced mutations are currently believed to arise mostly through changes in the base sequence of DNA. According to this view, bases may exchange with other bases, nucleotides may be added or deleted, or entire segments of the DNA strand may be shifted or deleted.

IV. Specific mutability

The direct and specific effect of certain agents on particular building blocks of DNA raises the important question, whether or not the specificity of mutagens on the molecular level is correlated with an effect on specific genes or on particular mutational sites of a gene. The answer to this question has been sought in numerous experiments on viruses, bacteria, and fungi. Such a specific mutagenic effect has indeed been observed frequently. Nevertheless the specificity cannot be equated with directed mutation. With our current knowledge of DNA structure the latter effect is not expected from any mutagens.

1. Intragenic specificity

By intragenic *specificity* we mean the *induction of mutations by a mutagen at preferred sites within a gene* (KAPLAN, 1962a). This is also called intra-locus specificity (AUERBACH and WESTERGAARD, 1960). *Intragenic specificity has been demonstrated for forward as well as reverse mutation.*

a) Forward mutation

The experiments of LEUPOLD (1961) and GUTZ (1961, 1963) on *Schizosaccharomyces pombe* serve as an example of specificity of a mutagen on forward mutation. The results, which are summarized in Fig. V-7 reveal the following:

1. *Mutations which have arisen independently of one another within a gene are not randomly distributed throughout the gene.* In addition to innumerable sites at which only very few mutations occur highly mutable sites, known as *"hot spots"*, have been observed.

In the *ad-7* locus of *S. pombe* out of 152 UV-induced mutations 39 ($=26\%$) and of 89 induced by nitrous acid, 26 ($=29\%$) occurred at one mutational site (Fig. V-7). A similar situation is found in *N. crassa*. All 15 mutations of the *ad-8* locus which were induced by base analogues were concentrated at a single site (ISHIKAWA, 1962). In contrast, experiments on the same organism with UV, X-ray, and nitrous acid did not reveal any hot spots; the number of mutants investigated may have been too small.

In bacteriophage *T4* 39% of the 753 independently isolated spontaneous mutations in the gene *rIIA* occurred at a hot spot; in the neighboring gene *rIIB* this was true for 60% of 855 such mutants (BENZER, 1959, 1961).

It is likely that a continuous graduation exists between "normal" mutational sites and hot spots, as the work on phage has shown. The existence of hot spots might suggest that all such mutations represent identical alterations in the DNA. However, this is not correct; BENZER demonstrated that mutants of a single site differ in their spontaneous reversion rate. Such a differential behavior is understandable, since a nucleotide can take on four different allelic configurations.

2. *The distribution of mutants throughout the gene depends* upon the gene itself as well as on *the type of mutagen utilized.* Dependence upon the mutagen is explained by structural differences within the gene. Each mutagen acts on specific parts of the DNA. The mutational spectra obtained with UV and X-irradiation in *S. pombe* are similar; on the other hand, treatment with nitrous acid gives an entirely different distributional pattern. Sites are affected which do not appear among

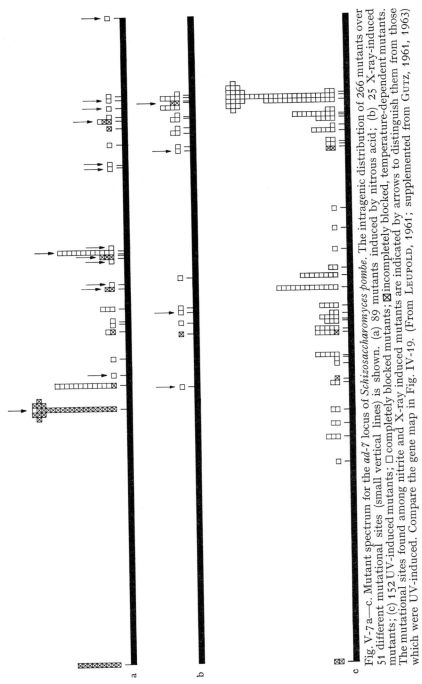

Fig. V-7a—c. Mutant spectrum for the *ad-7* locus of *Schizosaccharomyces pombe*. The intragenic distribution of 266 mutants over 51 different mutational sites (small vertical lines) is shown. (a) 89 mutants induced by nitrous acid; (b) 25 X-ray-induced mutants; (c) 152 UV-induced mutants; □ completely blocked mutants; ⊠ incompletely blocked, temperature-dependent mutants. The mutational sites found among nitrite and X-ray induced mutants are indicated by arrows to distinguish them from those which were UV-induced. Compare the gene map in Fig. IV-19. (From LEUPOLD, 1961; supplemented from GUTZ, 1961, 1963)

radiation-induced mutants; these are designated in Fig. V-7 by arrows. The differential distribution of mutations within the *ad-7* locus of *S. pombe* permit one to conclude that there are specific differences in the effect of radiation and nitrous acid (GUTZ, 1961, 1963). This may result from the fact that UV and X-radiation attack mainly the pyrimidines, while nitrous acid induces mutations primarily by chemical alteration of the purine base, adenine (LASKOWSKI, 1960b).

These differences were not pronounced in the *ad-8* locus of *N. crassa*; the number of mutants was too small, however, for a conclusive statement (ISHIKAWA, 1962).

This kind of comparative analysis of different mutant spectra was first carried out for the *rII* cistron of the bacteriophage *T4* and represents at present the largest system of mutations arising independently of one another. It includes spontaneous mutations as well as those produced by a variety of chemical mutagens (BENZER and FREESE, 1958; FREESE, 1959b; BENZER, 1959, 1961).

b) Reverse mutation

Results of experiments on reverse mutation also support the idea of intragenic specificity of mutagens. *Such specificity has been demonstrated in viruses, bacteria, and fungi* (BENZER, 1961; RUDNER and BALBINDER, 1960, and GILES, 1951; KØLMARK, 1953; MALLING et al., 1959; HESLOT, 1962). Certain of the findings on the *ad-3* and the *inos* loci of *N. crassa* have been summarized in Table V-4 and graphically presented in Figs. V-8 to V-10. These show that the effect of mutagens is similar for forward as well as reverse mutation. The following conclusions can be drawn from the table and figures:

1. *Different mutational sites of the same or different genes revert with varying frequencies if a particular mutagen is used* (Table V-4). These differences are independent of the length of treatment (Figs. V-8, V-9, V-10).

Table V-4. *The relative specificity of mutagenic agents in their effect on an inositol and adenine double auxotroph of Neurospora crassa*

The effectiveness of the mutagens is compared under "optimal conditions" (see Table V-2). (From WESTERGAARD, 1960.)

mutagenic agent	reverse mutants per 10^6 treated macroconidia		ratio of
	$inos^+$	ad^+	$ad^+ : inos^+$
Ethyl methane sulfonate	11.3	17.4	1.5
Diethyl sulfate	4.3	16.8	4
Dimethyl sulfate	3.4	64.0	19
Chlorethylmethane sulfonate	0.3	51.0	190
Diepoxybutane	0.2	89.0	445
Bromethylmethane sulfonate	0.04	152.0	3,800
Ultra-violet light	7.1	3.5	0.5
X-rays	0.2	3.2	16
Spontaneous	0.02	0.2	10

Intragenic specificity is explained as in the case of forward mutation by the fact that genes are not uniform in structure and are particularly sensitive at certain sites. Sites of low mutability exist along with those that are highly mutable. It is unlikely that all mutations at a hot spot involve identical changes. In some cases it has been shown that reverse mutation has not re-established the original molecular configuration (p. 401 ff.).

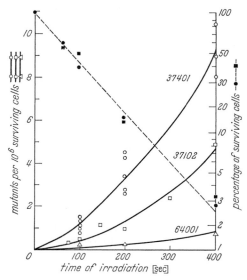

Fig. V-8. Mutagenic effect of UV radiation on three inositol auxotrophs of *N. crassa* which arose independently of one another. Isolate No. 37401: ○ and ●; No. 37102: □ and ■; No. 64001: ▵. The relation between the duration of radiation and the frequency of *inos*+ reversions among microconidia is shown. (Adapted from GILES, 1951)

As shown by Fig. V-8, the curves for the three inositol mutants of *N. crassa* diverge with time of UV irradiation. With the same length of treatment one of the three mutants (37401) reverted twice as frequently as a second (37102) and about nine times as frequently as a third (64001). The mutants behaved oppositely with X-irradiation (GILES, 1951; compare also 1959). KØLMARK (1953) obtained similar results in experiments with the double mutant *inos ad-3* of *N. crassa*. The *inos* marker reacted only very weakly to X-rays while the *ad* marker response was about 16 times greater (Table V-4). With UV radiation the two genes reacted in the reverse fashion (Tables V-4 and Fig. V-9). The differential effect was particularly evident after treatment with chlorethylmethane sulfonate, diepoxybutane, and bromethylmethane sulfonate (Table V-4). AUERBACH et al. (1962a, b) likewise found a gene-specific sensitivity toward particular mutagens in *N. crassa*.

Differential reversion rates were also obtained in combined treatments of UV light with formaldehyde and with hydrogen peroxide. In both cases the reversion rate of the *ad* allele was markedly increased while that of the *inos* allele remained essentially unchanged or was slightly reduced (MALLING et al., 1959). The two loci also behave oppositely if the treatment with formaldehyde precedes, and if it follows the UV irradiation. KøLMARK and AUERBACH made a similar observation on the same material with pre- and post-treatment with diepoxybutane (see AUERBACH and WESTERGAARD, 1960). Hydrogen peroxide did not give such a differential reversion rate (MALLING et al., 1959).

WESTERGAARD and his coworkers interpreted the differential effect of pre- and post-treatment with the hypothesis that inos⁺ reversions result from a direct radiation effect while reversions of the ad allele are caused by chemical processes.

HESLOT (1962) found similar relationships for the two genes *arg-1* and *leuc-3* of *Schizosaccharomyces pombe* after treatment with gamma rays of cobalt 60, UV, and several chemical mutagens.

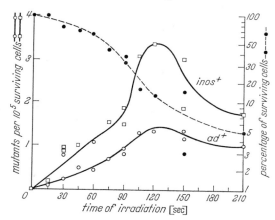

Fig. V-9. Mutagenic effect of UV radiation on an inositol-adenine double auxotroph of *Neurospora crassa* (compare Fig. V-10). The relation between the duration of radiation and the frequency of inos⁺ (○) and ad⁺ (□) reversions in macroconidia is shown. (Adapted from KøLMARK, 1956)

2. *One and the same mutational site* (= region in which a forward mutation has occurred) *reacts differently to different mutagens* (Table V-4, Figs. V-9 and V-10). This result corresponds to that found for the mutagen-specific mutation spectrum of forward mutation.

From the data of GILES (1951) and from data of Table V-4, it is apparent that the mutagens employed have very different effects on both the *ad* and the *inos* alleles. *ad* reverts about 280 times more frequently with ethylmethane sulfonate than with bromethylmethane sulfonate; *inos* shows only a nine times stronger reaction in the reverse fashion. The effect of UV and X-rays on the *inos* allele is likewise very different while no differences were found for the *ad* allele (Table V-4). In order to compare the mutation rates radiation doses with the same killing effect were chosen. HESLOT (1962) also found a differential reaction of mutational regions to different mutagens in *S. pombe*.

The interpretation of experiments on mutagenic specificity in strains with multiple markers (e.g. WESTERGAARD and coworkers) demands caution. This is shown by the work of CLARKE (1962, 1963) with an adenine and methionine double auxotroph of *S. pombe*. The apparent mutagenic specificity observed in this material (UV, nitrous acid) was actually simulated to a considerable extent by the suppression of ad^+ revertants in the presence of methionine in the selection medium. This suppression was very pronounced with UV-induced reverse mutations, but only weak in those induced by nitrous acid.

A clue to the *interpretation of intragenic specificity* is found in the "hot spot" investigations on *T4* phage (supplemented by results from fungi) and reverse mutation experiments on phage and bacteria. One is led to the following concepts:

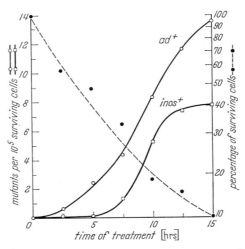

Fig. V-10. Mutagenic effect of diethyl sulfate on an inositol and adenine double auxotroph mutant of *N. crassa* (compare Fig. V-9). The relation between the duration of treatment and the frequency of *inos*+ (□) and *ad*+ (○) reversions among macroconidia at a concentration of 0.07 mol/liter is shown. (Adapted from KØLMARK, 1956)

1. A hot spot probably consists of a single base pair, at least one member of which is particularly sensitive to a specific mutagen (see hypotheses of chemical mutagenesis, p. 291 ff.).

2. The high mutability of a hot spot is not only determined by the nature of this single base, but also by the influence of the neighboring bases on it (RUDNER and BALBINDER, 1960; BENZER, 1961; KAPLAN et al., 1963). This is logical because the same base recurs at many other sites of the gene.

3. Certain metabolic processes of the cell not involving the nucleotides of DNA may be related to the existence of hot spots. The directing role of metabolism may derive from the fact that particular segments of DNA may be sensitized or desensitized to a mutagen by specific metabolic products. Furthermore certain premutative changes of DNA may be converted selectively to mutations; this may take place, seldom or even never (compare WINKLER, 1963).

2. Intergenic specificity

Intergenic or inter-locus specificity (KAPLAN, 1962a; AUERBACH and WESTERGAARD, 1960) of a mutagen designates the phenomenon in which *the total frequency of mutations per gene induced by a mutagen differs for different genes* (WINKLER, 1963). Such differences have been demonstrated in phage and bacteria and, in isolated cases, in fungi (e.g. *Ophiostoma*). The degree of expression of intergenic specificity depends mostly upon the nature of the mutagen. This permits the conclusion that the different mutation rates cannot be explained merely by differences in the length of genes. The causes of inter- and intragenic specificity may possibly be the same.

In order to confirm the heterogeneity statistically, extensive data are necessary. The total number of mutations must be large. Moreover, the detection of intergenic specificity depends upon the number of genes in each of the comparable mutant groups. These cannot include too many kinds of mutations, otherwise the classification becomes too broad, i.e. not too many genes with hot spots should be included within the comparison, since possible differences between genes may compensate one another and thus may not be detected. Since the mutation experiments with fungi have not been carried out on an extensive enough scale, it is not surprising that investigations of the problem in this group have given negative results.

Experiments of this type have been carried out on a number of fungi. Various mutagenic agents have been tested, for example, on *Venturia inaequalis:* UV radiation, nitrogen mustard (BOONE et al., 1956), *Neurospora crassa:* UV, X-radiation, nitrogen mustard (BEADLE and TATUM, 1945; HOROWITZ et al., 1946), *Schizosaccharomyces pombe:* UV radiation, diethyl sulfate, isopropylmethane sulfate and ethylene imine (HESLOT, 1960, 1962), *Penicillium chrysogenum:* UV light and diepoxybutane (AUERBACH and WESTERGAARD, 1960). None of these investigations showed a clear difference in the spectrum of morphological and auxotrophic mutations.

The results with *Penicillium* are particularly noteworthy in this connection because two mutagens were used the strong elective effect of which was known from the experiments on the double mutant *ad inos* of *Neurospora crassa* (Table V-4). After UV irradiation 4.1%, and after treatment with diepoxybutane, 3.9% biochemical mutants were obtained. Further classification of the mutants gave 0.8% (UV) and 0.6% (diepoxybutane) for vitamin, 2.6% and 2.5% for amino acid, 1% specifically for methionine auxotrophy in both cases (AUERBACH and WESTERGAARD, 1960).

However, a certain intergenic specificity was observed in *Ophiostoma multiannulatum* (ZETTERBERG, 1961). UV and X-radiation induced approximately the same number of auxotrophs for amino acids and nucleic acid components and somewhat fewer for vitamins and fatty acids in this fungus. In contrast amino acid deficiency mutants were obtained almost exclusively following treatment with N-nitroso-N-methylurethane.

Distinct differences were found in the type of spectrum when these mutants were classified: auxotrophs for methionine, arginine, lysine, and histidine arose in approximately equal frequency from methylurethane treatment while radiation produced predominantly methionine but no histidine mutants.

Results of LASKOWSKI (1960a) with yeasts likewise suggest an intergenic specificity. STEIN and LASKOWSKI (1958, 1959) showed with the help of a mathematical model that the number of genes undergoing mutation to a recessive lethal condition was four times as great after X-radiation than after UV radiation (compare also STEIN, 1962). According to LASKOWSKI, these results argue for the sensitivity of specific genes to specific mutagenic agents. AUERBACH and WESTERGAARD (1960) explain the differential effects of radiation found by LASKOWSKI by assuming that X-rays cause greater damage to DNA than UV and therefore produce more lethal mutations.

Results of experiments with *Drosophila* also allow such an interpretation; significantly fewer lethal and more morphological mutants were observed following treatment with certain chemical mutagens than after X-irradiation (FAHMY and FAHMY, 1956). Again ionizing radiation produces more change and damage to the chromosomes than chemical agents. Experiments with *Vicia faba* (e.g. REVELL, 1959; RIEGER and MICHAELIS, 1960) have shown that most chemical mutagens induce more chromosome breaks in heterochromatic than in euchromatic regions. AUERBACH and WESTERGAARD (1960) have therefore proposed an explanation of the intergenic specificity of mutagens in which it is assumed that genes that mutate preferentially after treatment with chemicals are localized in or directly adjacent to heterochromatic chromosome segments. The testing of this hypothesis is not possible in fungi because the chromosomes are too small to distinguish hetero- and euchromatic segments.

Summary

1. A gene always includes a few highly mutable sites or "hot spots" as well as a large number of relatively stable regions. Accordingly, the mutational sites of the same or different genes exhibit different reversion rates when specific mutagens are employed.

2. Treatment with different mutagenic agents generally yields different mutation spectra for a gene. The same mutational region exhibits different reversion rates with different mutagens.

3. Highly mutable sites as well as mutagen-specific mutational spectra show that the mutational sites within a gene react differently to different mutagens (intragenic specificity). Further, in materials which have been intensively enough studied, it has been shown that the total frequency of mutation induced by a mutagen per gene is gene specific, i.e. different genes show different mutation rates (intergenic specificity).

V. Spontaneous mutations

Changes in the hereditary material which arise in the absence of treatment with a mutagen are called spontaneous mutations. The concept of spontaneous change needs a more precise definition. We know that certain mutagens greatly increase the yield of mutants. On the other hand, some radiation (e.g. visible light) and chemicals (e.g. hydrogen peroxide, organic peroxides) either do not increase the spontaneous rate appreciably or cause only a barely detectable increase (e.g. MARQUARDT et al., 1963). One can visualize mutagenic agents arranged into a series according to the degree of effectiveness as judged by mutation rate at a specific radiation intensity or concentration of mutagen. At one end of this series falls a group of "natural mutagens" (WESTERGAARD, 1957) the effect of which cannot be distinguished by a rate significantly different from that of spontaneous mutation.

1. Mutation rates

The frequency of spontaneous mutation is generally very low. A direct determination of mutation rate (= average number of mutations per nucleus per unit of time) *is ordinarily not possible,* because neither the time of mutation, the number of nuclei (e.g. in growing mycelia), the viability, nor the rate of division of the mutated nuclei is known (discussion in KAPLAN, 1957 and SCHULL, 1962).

The complicating effect of the selective advantage or disadvantage of mutated in comparison with non-mutated nuclei can be avoided to a large extent, if one-celled organisms (e.g. bacteria, yeast) and proper techniques are used. To discuss these methods is beyond the scope of this treatment; two of the best known methods are the null culture (LURIA and DELBRÜCK, 1943; LEA and COULSON, 1949) and the papillar method (RYAN et al., 1955).

The determination of spontaneous mutation rate in hyphae-producing fungi is much more difficult and unreliable than in bacteria and yeasts. A mutation may manifest itself only after numerous nuclear divisions, i.e. a long time after it has arisen. If a mutation occurs spontaneously in a nucleus of a growing, multinucleate mycelium, such a change is generally not detected until a conidium or hyphal tip containing the mutated nucleus happens to be isolated and a new mycelium allowed to develop from it.

For example, when an unusually large number of mutant conidia are observed in a mycelium of *N. crassa*, one can conclude that the mutation occurred at approximately the time the culture began to grow. On the other hand, if only a few mutant conidia are found, it may be assumed that the mutation has taken place in the conidia or a few nuclear divisions prior to conidial formation. Since the time of occurrence of the mutation is rarely accurately determinable, the frequency of such mutants gives only an approximation of the frequency of mutation per conidium. The determination of mutation rates in multinucleate conidia, as in macroconidia of *N. crassa*, is further complicated by dominance in the heterokaryon (p. 270).

Because of such difficulties it is understandable why *few data exist on spontaneous mutation rates in fungi.* This lack of information is particularly apparent with respect to spontaneous *forward mutation.* The data on such rates are generally inaccurate; they range *from 10^{-7} to 10^{-6}* per nucleus and nuclear division for auxotrophic and other mutations.

WOODWARD (1956), using selective methods, found 5 citrulline and about 80 glutamic acid auxotrophs among 10^6 macroconidia of *N. crassa*. These values give little precise information on the mutation rate of an individual gene or a single mutational site, since a mutation to citrulline or glutamic acid auxotrophy can occur at many different sites in the genetic material. Moreover, because multinucleate macroconidia were utilized, it is probable that the frequency of mutation per nucleus may be even higher. TATUM et al. (1950) screened approximately 3,000 microconidia of *N. crassa* for spontaneous mutations to auxotrophy. They found only a single mutation. More precise data on the frequency of the mutation from *ad-3+* to *ad-3* (*N. crassa*) can be found in BARNETT and DE SERRES (1963). Rates here vary between 3.8×10^{-7} and 1.6×10^{-5}. In *Aspergillus nidulans* frequencies between 1.1 and 5.7×10^{-6} were observed for mutations of suppressor genes (*Su-meth-1*). A mutation rate for spontaneous "*colonial*" mutations of about 10^{-7} per nucleus and nuclear division was found in homo- and heterokaryons of *Schizophyllum commune* using the null culture technique (DICK and RAPER, 1961).

While investigations of forward mutation allow only inaccurate estimates of mutation rates, more precise values can be obtained from reverse mutation experiments. As Table V-5 shows, *the frequencies for reversion lie between 10^{-9} and 10^{-7}*. These mutation frequencies come to only $1/_{100}$ of the corresponding forward mutation. This difference is explained by the fact that all mutations within a gene or even different genes are included in the forward mutation rate. In the reverse mutation rate only mutations in the region of a specific forward mutation are counted except for the suppressor mutations which may occur.

Moreover, it can be seen from Table V-5 that *the mutation frequencies for different mutations within a gene are often quite different*. The radiation-induced *inos* mutant *JH-5202* and the nitrous acid induced mutant *2-017-0137* of the *ad-3A* locus in *N. crassa* are noteworthy because of their particularly high reversion rates (compare "hot spots", p. 295). BARNETT and DE SERRES (1963), who pursued this phenomenon in the *ad-3* locus, found that the prototrophs were unstable and produced *ad-3* types that were phenotypically identical to the original mutants the defects of which lay in the same gene. Suppressor or mutator genes could be excluded as a cause. Thus one is led to believe that in addition to stable mutations, highly mutable alleles arise in rare cases; these do not become stable until they first undergo a reversion or a forward mutation to another allele.

2. The mutation process

A series of results from experiments with phage, bacteria and fungi shows that spontaneous as well as induced mutations go through premutational stages. AUERBACH (1959) found that macroconidia of *N. crassa* which were held at cool temperatures exhibited a reduced mutability; after the temperature was raised, normal mutability was quickly reestablished. The capacity for such a rapid recovery was explained by the assumption that in the cold some kind of change in the DNA occurs spontaneously the completion of which requires higher temperatures (p. 284).

In one strain of *N. crassa* AUERBACH determined the frequency of recessive lethal mutations using a technique developed by ATWOOD and MUKAI (1953a, b) (p. 270f.) by which the rate was established each week for different temperatures. At 30° C the mutation rate increased linearly from 0 to 9% within 27 weeks. This corresponds to a weekly increase of 0.3%.

Table V-5. *Examples of the frequency of spontaneous reversions in different organisms*

In *N. crassa* mostly macroconidia (multinucleate) were used for the investigations. The corresponding values from analysis of microconidia (uninucleate) are shown in brackets.

object	gene	designation of mutant	number of reverse mutants per 10^6 survivors	reference
Neurospora crassa	*inos*	JH-5202	15.0 (15.0—726.3)	
		64001	0.1 (0.05)	
		37401	0.02 (0.01)	
		46316	0.02 (0.01)	GILES, 1951
		37102	0.01 (0.01)	
		JH-2626	0.01 (0.01)	
		89601	0.002 (< 0.007)	
	ad-3	38701	0.008—0.06	JENSEN et al., 1951
		38701	0.1 —0.29	KØLMARK and WESTERGAARD, 1953; KØLMARK, 1956
		38701	0.1 —0.6	MALLING et al., 1959
		2-017-0137	4.3	
		74A-OR-1	0.2	
		74A-OR-2	0.05—0.1	
		74A-OR-11	0.01—0.06	BARNETT and DE SERRES, 1963
		74A-OR-10	0.01—0.05	
		74A-OR-12	0.0	
		74A-OR-20	0.0	
Ophiostoma multiannulatum	*ur*	2657	4.7—100 [a]	
		2626	0.3 —9.6 [a]	
		2615	0.01—2.9	ZETTERBERG and FRIES, 1958; ZETTERBERG, 1960a
		2505	0.02—3.1 [a]	
		2614	< 0.01—1.4	
	me	2668	0.01—0.11	
		2511	< 0.01—0.26	
	hy	2494	0.01—2.5 [a]	
Saccharomyces cerevisiae	*ar-8*		0.93	
	met-2		0.42	MAGNI, 1963
	hi-1		0.02	
	hi-1	10118	0.22	
		10284	0.003	MAGNI et al., 1964
	can^r	10284	0.037	
		1060	0.004	
	iv	M-23	2.6 —3.2	
		M-2	0.18—0.26	
		M-13	0.12—0.46	KAKAR et al., 1964
		M-12	0.02—0.12	
		M-24	0.0 —0.02	
Schizosaccharomyces pombe	*arg-1*		5.0 —11.0	
	ur-1		0.2 —0.64	HESLOT, 1962
	leuc-3		0.02—0.03	
Ustilago maydis	*ad-1*		0.1	
	inos-2		0.01	HOLLIDAY, 1962
	inos-3		0.005	

[a] Values in some experiments lie above $100/10^6$.

At 4° C the increase per week came to only about a third (0.1%) of that at the higher temperature. The rate climbed to only 2.3% in 22 weeks. When the cooled conidia were placed in a temperature of 30° C after 24 weeks the mutation rate reached the same value within 4 weeks as that after continuous high temperature (8—11%).

The experiments of AUERBACH with *N. crassa* and those of RYAN and coworkers with *E. coli* show further that spontaneous mutations or at least certain premutational changes which are reversible may arise without DNA synthesis (RYAN, 1957; RYAN and KIRITANI, 1959; RYAN et al., 1959, 1961). Although it has been shown conclusively that in resting cells of *E. coli* mutations may occur in the absence of DNA replication, a similar conclusion for *N. crassa* depends upon the assumption that no DNA synthesis takes place in dry conidia. The completion of mutations in this case would not be possible until germination of the spores, i.e. at the time of DNA replication. Thus we are led to the conclusion *that changes (premutations) can occur in nonreplicating DNA and these are later completed usually during replication, or rarely, in its absence* (RYAN and coworkers). Little is known about the cause of spontaneous mutation. Hypotheses have been proposed but none are generally applicable, nor have any been confirmed.

Summary

1. The frequency of spontaneous mutation is generally very low. Precise data on mutation rates are available only for reversions of auxotrophy to prototrophy. They lie in the order of magnitude of 10^{-7} to 10^{-9} per nucleus and nuclear division.

2. A spontaneous mutation apparently involves premutational stages just as in the case of an induced mutation. Spontaneous mutation is assumed to involve changes in the DNA base sequence. Little is known about the cause of spontaneous mutation.

B. Segmental mutations

Segmental mutations differ from point mutations in the linear extent of the change in the chromosome. They can only be demonstrated cytologically in particularly favorable cases. The success of such cytological studies depends upon the size of the mutation as well as on the size the chromosome. Because of their small chromosomes, fungi are poorly suited for cytological investigation. This disadvantage is partially offset by the *possibility of investigating chromosomal mutations by tetrad analysis*. Not only is differentiation between point and segmental mutations possible with this genetic method but also between translocations, inversions, and deletions.

Literature: STRAUB (1958), SHARMA and SHARMA (1960), DAVIDSON (1960), WOLFF (1960), RÖBBELEN (1960, 1963), KIHLMAN (1961), EVANS (1962), RIEGER and MICHAELIS (1962).

I. Translocations

A translocation is a structural alteration of a chromosome characterized by the insertion of a chromosome segment at another position in the same or a different chromosome. Segments between two different chromosomes may also be exchanged; in such a case we speak of *reciprocal* translocation.

At synapsis cross configurations ordinarily form (Fig. V-11). Only in rare cases does meiosis lead to chromosome complements in which exactly a single set of genes is represented. For example, if one assumes that no crossing over between the centromere and the point of translocation has occurred (Fig. V-11 a) and that the centromeres segregate to the poles in anaphase I at random (p. 141 ff.), asci with 8 viable ascospores are obtained in only one third of the cases. Such asci arise if the centromeres lying diagonally opposite one another go to the same pole (compare Fig. V-11 a, left). All other asci contain *spores with incomplete chromosome complements.* Such spores can be recognized in the asci by their hyaline appearance. They are not viable.

Crossing over in the acentric arms does not introduce additional complications. On the other hand, crossovers between the centromere and translocation point prevent a separation of the chromatids and lead, among other things, to four viable and four inviable spores (Fig. V-11 b—d). Certain of the viable spores carry a duplication for a portion of the genetic information.

On the basis of the numerical distribution of the ascus types reproduced in Fig. V-11, the following questions can be answered:

1. Is a lethal point or a chromosomal mutation involved? In the former case only asci with 4 viable: 4 inviable spores are expected, in the latter, asci with 8 viable or 8 inviable spores are also found.

2. If the mutation is chromosomal, is it a translocation (Fig. V-11), an inversion (Fig. V-12), or a deletion (p. 311 f.). The relatively frequent appearance of asci with 8 inviable spores is evidence for a translocation (e.g. compare Figs. V-11 and V-12).

3. In the event of a translocation how far does the translocation point lie from the centromere? The closer the translocation lies to the centromere the fewer the crossovers between the two points, i.e. the more rarely do asci with 4:4 segregations occur.

4. How large is the translocated segment? The distance between the centromere and translocation point (see point 3) indicates the extent of the chromosome arm involved in the translocation.

Translocations and inversions (compare Fig. V-12) were demonstrated in *N. crassa* and *Sordaria macrospora* by means of tetrad analysis (McCLINTOCK, 1945; and HESLOT, 1958; ESSER and STRAUB, 1958). For example, only 10% normal asci were found in perithecia of *S. macrospora* from crosses made with the self-fertile mutant, *cr*. The remaining asci contained either 8 hyaline spores incapable of germination (15—17%) or four black and four hyaline spores in characteristic arrangements (73—75%). The perithecia arising from selfing of the *cr* mutant yielded, in contrast, asci with 8 normal spores (ESSER and STRAUB, 1958). This observation is evidence for a translocation, since the frequency of asci with 8 inviable spores exceeds the frequency of normal asci.

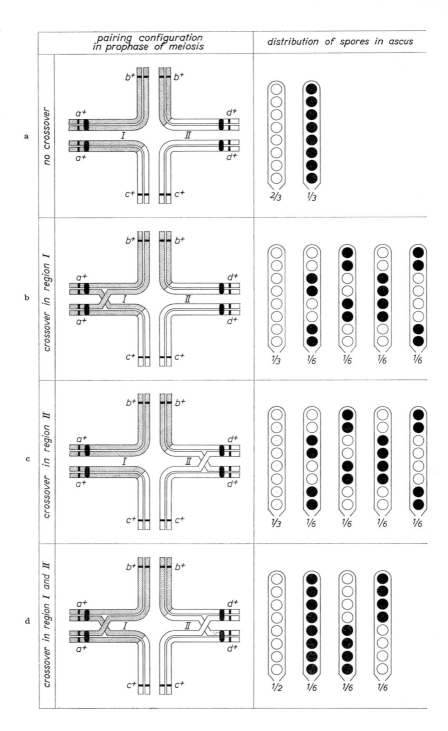

In *N. crassa* segmental mutations were demonstrated not only through tetrad analysis but also by cytological investigation in exceptional cases (McClintock, 1945).

Translocations have also been shown in strains of *Aspergillus nidulans* after treatments to induce mutation (Tector and Käfer, 1962; Käfer, 1962, 1963a, b). Tetrad analysis cannot generally be used with this material because the spores of the tetrad separate too readily. Therefore, Käfer developed a method which permits differentiation of translocations from other chromosomal mutations and their localization in the chromosome.

Since translocations are characterized by rearrangement of the linear structure of the chromosome, they lead to *distinctive alterations in gene sequence and linkage relationships*. The type of change permits an estimate of the location and size of the translocation.

II. Inversions

By inversion is meant an intrachromosomal rearrangement in which a segment of a chromosome is removed, rotated 180°, and reinserted. The pairing of two homologous chromosomes one of which possesses an inverted segment leads to the loop formation shown in Fig. V-12. *Inversions can also be recognized cytologically because of bridge formation in anaphase I and II.* Bridges arise from crossing over in the inverted segment; this leads to dicentric chromatids (chromatids with two centromeres) and to acentric chromatid fragments (compare double headed arrows in Fig. V-12). Such bridges will break at some point between the two centromeres, because the latter go to opposite poles. This results in *deletions* and *partial duplications*. In the first instance hyaline inviable *spores with incomplete chromosome complements* are formed. In the second case spores are formed which carry a duplication for part of the genetic information. Such spores are phenotypically normal and may produce a mycelium.

Since an ascus contains all four products of meiosis, asci with a segregation of 4 inviable to 4 normal spores result from a single crossover in the inverted segment (Fig. V-12b). With double crossover asci with 8 normal or 8 inviable spores are also found (2-strand or 4-strand double crossover in Fig. V-12c and e; 3-strand double crossover leads to the same result as a single crossover: Fig. V-12d). On the basis of the numerical distribution of individual ascus types inversions can be distinguished from translocations by the rare occurrence of asci with eight inviable spores (compare Figs. V-11 and V-12). Moreover, this frequency allows a determination of the size of the inverted segment.

Fig. V-11 a—d. The distribution of inviable (open circle) and viable spores (solid circle) expected in the asci of an eight-spored ascomycete with a reciprocal translocation. The four chromatids and chromatid segments of one pair of homologues are stippled, those of the other pair are blank. Centromeres and marker genes are shown as in Fig. IV-2. Only a single crossover between the centromere and translocation point is considered. See further in text. (Adapted from Heslot, 1958)

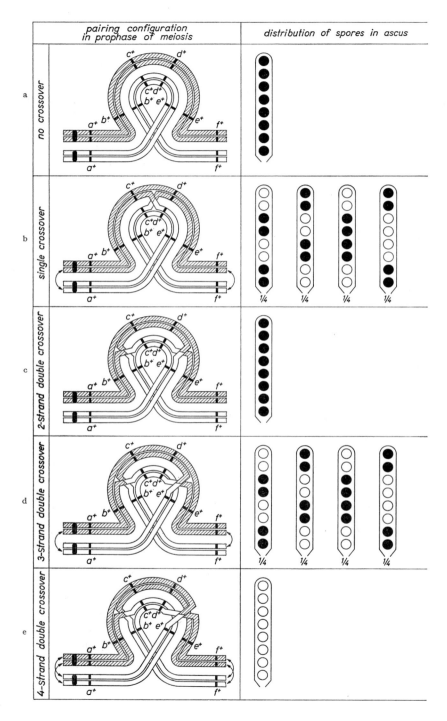

	pairing configuration in prophase of meiosis	distribution of spores in ascus
a	no crossover	
b	single crossover	¼ ¼ ¼ ¼
c	2-strand double crossover	
d	3-strand double crossover	¼ ¼ ¼ ¼
e	4-strand double crossover	

We have considered only paracentric inversions in Fig. V-12. In these the centromere lies outside of the inversion. Details regarding the more involved pericentric inversions (centromere within the inverted segment) can be obtained from HESLOT (1958).

Since crossing over in an inversion almost always leads to incomplete genomes (Fig. V-12), the analysis of viable cells gives only a *very small crossover frequency* between markers which lie within or on different sides of the inversion. Inversions are frequently recognized by such characteristic "crossover suppression" (SINGLETON, 1964).

III. Deletions

When intercalary or terminal chromosome segments are lost during the nuclear cycle, a deletion is said to occur. Such losses almost always have a lethal effect in the homokaryon. Larger deletions lead to the death of a cell even in the heterozygous condition, even though the missing genetic information is present in the homologous chromosomes.

Asci with inviable spores result from deletions just as from translocations and inversions. With a deletion, however, — in contrast to other types of chromosomal mutation — a part of the genome is missing so that no asci with eight normal spores are produced. *The absence of normal asci is thus an indication of a deletion.* Moreover, a deletion can be detected by altered recombination frequencies of marker genes which lie close to the deleted segment.

GILES (1955) found mutants in *N. crassa* following X-irradiation which showed atypical linkage relationships and never reverted. Similar results have also been observed in bacteria (KAUDEWITZ, 1959a, b, c; DEMEREC, 1960).

As we have already pointed out earlier (p. 306), point and chromosomal mutations are quantitatively different chromosomal alterations. Some chromosomal mutations encompass a major portion of a chromosome. Others, mostly deletions, are much smaller and include only part of a gene; thus they more or less approach the dimensions of point mutations. Such chromosomal mutations are called "*block mutations*". They have been observed in phage and bacteria as well as in fungi. In contrast to point mutations, *block mutations do not yield wild type recombinants with adjacent homologous mutational sites*. The length of a block mutation can be determined from the location of point mutations with which no wild type recombinants form. It usually does not exceed the length of a gene. Another difference between point and block mutations is

Fig. V-12a—e. Distribution of inviable (open circle) and viable (solid circle) spores in asci of an eight-spored ascomycete with a paracentric inversion. The sister chromatids of a chromosome are shaded, those of the homologue unshaded. The dicentric or acentric chromatid fragments which result from crossing over and which belong together are indicated by the double headed arrows. A maximum of two crossovers in the inverted segment is considered. Centromeres and marker genes are shown as in Fig. IV-2. See further in text. (Adapted from HESLOT, 1958) that most of the latter do not undergo reversion (SUYAMA et al., 1959; ISHIKAWA, 1962; HAWTHORNE, 1963; DE SERRES, 1964).

For example in *N. crassa* 10 block mutants were found among 70 *ad-8* mutants tested; none of these reverted even following UV treatment. The length of these block mutants lay between 0.03 and 0.4 map units (ISHIKAWA, 1962). DE SERRES (1964) investigated the occurrence of block mutations in the *ad-3* region of *N. crassa*. This region contains two genes (*ad-3 A* and *ad-3 B*) that are separated by a segment of unknown genetic function. DE SERRES showed that the irreparable recessive lethal mutations were actually block mutations which were longer than single genes and in some cases even longer than the entire *ad-3* region. On the other hand the *ad-3* mutations for which compensation is possible by the addition of adenine, result from smaller changes within *ad-3 A* and *ad-3 B*.

Reversions have been observed in the case of certain block mutations. GRIGG and SERGEANT (1961) developed the following hypothesis to explain this rare phenomenon: Reversions can be induced in loci which consist of a sequence of similar segments arising through unequal sister strand exchange. Thus, the number of similar segments increases in one strand and decreases in the other. The authors were able to show that such a hypothesis agreed, to a large extent, with the observations.

The occurrence of block mutations might suggest that point and chromosomal mutations (at least deletions) may involve essentially the same alterations in DNA. However, results of experiments with mutagenic agents argue against such an interpretation. In general chemicals are only mutagenic in producing point mutations, probably through a change in the base sequence of the DNA (p. 291 ff.); they very rarely induce chromosomal mutations.

Block mutations in *N. crassa* have occurred spontaneously and have been induced only through X-ray and UV radiation and not through chemicals such as nitrous acid and base analogues (ISHIKAWA, 1962).

Mutagenic specificity has also been observed in *Salmonella*. About 40% of all spontaneous and UV-induced *cys* mutations proved to be deletions; in contrast, 2-aminopurine induced only point mutations (DEMEREC, 1960).

The results with phage *T 4* differ from those with *Neurospora* and *Salmonella* in that nitrous acid induced both chromosomal and point mutations. Moreover, it was shown that sites exist in the gene where breaks occur preferentially to give rise to deletions (TESSMAN, 1962). These sites are analogous to the "hot spots" of point mutation (p. 295).

The causes of deletion are not fully understood. Results with phage and bacteria have led to the idea that deletion of a sizeable segment of a gene may occur as a result of a loop in the DNA during replication (e.g. DEMEREC, 1960). Such an explanation is however only possible for spontaneous or chemically induced small deletions. For the mutagenic effect of X-rays it is assumed that the primary damage, which is more or less reversible, occurs directly or indirectly with the passage of the ionizing particle. The occurrence of such damage in pairs allows recombination if both changes take place close enough together in time and space (RIEGER and BÖHME, 1962).

Summary

1. In the chromosomes of fungi chromosomal mutations can generally be distinguished from point mutations only through breeding experiments. Cytological methods usually fail to give definitive information.

Tetrad analysis is the best method of determining the kind of chromosomal mutation (translocation, inversion, deletion) by observing the number and distribution of inviable spores in the asci.

2. X-rays and UV have proved especially effective in inducing chromosomal mutations. Chemical agents rarely produce large alterations in the chromosomes. Chromosomal mutations may also occur spontaneously.

3. The occurrence of block mutations shows that small deletions may take place within a gene.

C. Genome mutations

Genome mutations are distinguished from chromosomal and point mutations in that the *structure of individual chromosomes remains unaltered.* They involve simply more or fewer chromosomes per nucleus. *Most genome mutations arise from abnormalities in mitosis and meiosis.* They are classified into the following types:

1. *Euploidy: Multiplication of complete chromosome complements.* This may occur spontaneously in rare cases but can also be induced by radiation, heat shock, and certain chemical agents (particularly colchicine). The effect of colchicine is to interfere with formation of the spindle apparatus which is necessary for the distribution of the chromosomes. As a result, the already duplicated chromosomes which normally go to the two daughter nuclei become included in a single nucleus; thus the chromosome complement is doubled (*polyploidy*). The number of chromosome sets in the nucleus is referred to as di-, tri-, tetra-, pentaploid, etc. If the multiplication of chromosome sets involves the genome of a single species, we speak of *autopolyploidy.* An increase in the number of chromosome sets involving genomes of different species is called *allopolyploidy.*

EMERSON and WILSON (1949) obtained an allopolyploid by crossing polyploid strains of *Allomyces arbuscula* and *A. javanicus.*

2. *Aneuploidy: Addition or loss of individual chromosomes.* Individuals with extra or missing chromosomes may arise through non-disjunction, i.e. the failure of homologues to separate from each other during meiosis. Moreover, in polyploids with odd numbers of chromosome sets, e.g. triploids, individual chromosomes may be lost because of the limitation of pairing opportunities. The loss of a chromosome is usually lethal.

Aneuploid nuclei were found to occur regularly in *Aspergillus nidulans* as a result of non-disjunction in somatic cells (p. 146). They represent intermediate stages in the process by which diploid nuclei are reduced to haploid (KÄFER, 1961, 1963b). Aneuploidy also occurs occasionally in yeasts (COX and BEVAN, 1962).

Literature: STRAUB (1950, 1958), RÖBBELEN (1960, 1962, 1963), RIEGER (1963).

I. Methods for determining polyploidy

Methods for determining polyploidy involve cytological, biochemical or genetic criteria (compare Table V-6).

1. Cytological methods

1. *Chromosome counts.* This is a matter of observing microscopic preparations of mitotic and meiotic divisions. In this way the existence as well as the degree of polyploidy can be established. Because the small size of their chromosomes makes fungi poorly suited for cytological study, methods other than chromosome counts have been developed for determining ploidy level.

2. *Measurement of nuclear volume.* Many investigations in higher and lower plants as well as animals have shown that diploid nuclei are approximately double the volume of haploid nuclei.

For instance KNIEP (1930) and SÖRGEL (1936, 1937) showed that the nuclear volumes in gametophytes and sporophytes of *Allomyces arbuscula* exist in a 1:2 ratio. Their assumption that the sexual generation is haploid and the asexual diploid was later confirmed by the cytological investigations of EMERSON (1941, 1950), SPARROW (1943), EMERSON and WILSON (1949), and WILSON (1952).

SOST (1955) determined the volumes of haploid, diploid, tetraploid, and octoploid nuclei in *A. arbuscula*. The values corresponded roughly to the ratio 1:2:4:8. The ploidy level was confirmed for each case by chromosome counts. Similar quantitative relationships between the number of chromosome sets and nuclear volume have also been found in other fungi (e.g. *Podospora anserina*: FRANKE, 1962).

3. *Determination of cell form and size.* As in the higher plants in which the volume of uninucleate cells increases 1.6 to 2.5 times with a doubling of the genome (STRAUB, 1950), individual cells of fungi such as gametes, conidia, spores, and yeast cells show an enlargement with an increase in the number of genomes (*gigas* forms).

SOST (1955) found that the cell volumes of haploid and diploid meiospores in *Allomyces arbuscula* showed the relationship 1:2.1 and the volumes of diploid and octoploid zoospores the ratio 1:3.6. SANSOME (1946, 1949) obtained the ratio 1:1.4 for the volumes of haploid and diploid conidia in *Penicillium notatum*. FRANKE (1962) found that the volumes of multinucleate ascospores with diploid nuclei in *Podospora anserina* showed an increase of 1.5 to 1.7 times over those with haploid nuclei (for method see e.g. BAUCH, 1942c, 1953).

The form as well as the size may differ in haploid and diploid cells. Yeast is the best known example of this phenomenon. The haploid cells of *Saccharomyces cerevisiae* are spherical, while the diploid are larger and ellipsoidal (see Fig. I-2).

2. Biochemical method

From the fact that DNA is a part of the chromosome one would expect *the DNA content of the nucleus to increase with the level of polyploidy.* The truth of this assumption has been shown in extensive investigations on polyploid yeasts, molds, and rusts.

OGUR et al. (1952) determined the DNA content of haploid, diploid, triploid, and tetraploid *Saccharomyces* cells: $n:2n:3n:4n = 2.4:4.8:6.5:9.9$ ($\times 10^{-14}$ g per nucleus). HEAGY and ROPER (1952) and PONTECORVO and ROPER (1956) found the ratio for haploid and diploid nuclei in *Aspergillus nidulans* $n:2n = 4.4:9.0$, ISHITANI et al. (1956) in *Aspergillus sojae* $n:2n = 7.3:14.6$, HOLLIDAY (1961) in *Ustilago maydis* $n:2n = 19.0:39.5$.

3. Genetic method

This method involves the *detection of abnormal segregations among the meiotic products*. Well-marked strains are necessary for genetic analysis. Triploid and tetraploid zygotes which have arisen from the fusion of a diploid with a haploid, or another diploid nucleus lead in part to segregations other than 2:2 in the meiotic tetrads. The frequency of individual segregational types (2:2, 3:1, 1:3, 4:0, 0:4) depends on several factors:

1. *Dominance of the alleles involved:* e.g. a^+ dominant over a or a over a^+.

2. *Numerical distribution of the two alleles in the zygotic nucleus,* e.g. in a tetraploid zygote with dominance of a^+ over a: simplex type a^+aaa, duplex type a^+a^+aa, triplex type $a^+a^+a^+a$.

3. *Frequency of exchange between the locus and its centromere.* The influence of recombination on the type of segregation is explained by the pairing behavior of the three or four homologous chromosomes within the same nucleus. For any specific region of the chromosome, pairing occurs only between two homologues in spite of the presence of three or four homologues, although a particular chromosome may pair with different homologues in its various regions. While individual chromosome segments in triploids may often lack a pairing partner, in a tetraploid cell all four chromosomes may be paired with one another. The division figures which characteristically result are either two bivalents, a quadrivalent or more rarely, a trivalent with an univalent (as can be seen microscopically in the first meiotic metaphase of higher plants). Depending upon the kind of pairing and the site of exchange, different combinations of alleles occur among the four diploid meiotic products, yielding different phenotypes.

From a tetraploid zygotic nucleus of the duplex type a^+a^+aa (with a^+ dominant over a) without exchanges 2:2 and 4:0 segregations occur in the ratio of 1:2. If an exchange occurs between the locus and its centromere, 3:1 and 4:0 segregations are obtained. Correspondingly in a triplex (or simplex) type without exchanges 4:0 (or 2:2 in a simplex type) segregations, with exchange between the centromere and locus, tetrads of the 3:1 (3:1) and 4:0 (2:2) types occur (compare FINCHAM and DAY, 1963, Table 17, p. 117).

Genetic analysis was employed particularly in determining polyploid cells of yeast (mostly diploid and tetraploid) (e.g. *Saccharomyces:* LINDEGREN and LINDEGREN, 1951); ROMAN et al., 1951, 1955; *Schizosaccharomyces:* LEUPOLD, 1956a). This method has also proved useful for other organisms, namely *Penicillium notatum* (SANSOME, 1946, 1949) and *Aspergillus nidulans* (PONTECORVO and ROPER, 1952; ROPER, 1952). ROMAN et al. (1955) and LEUPOLD (1956b, c) developed formulae which given specific assumptions allow the determination of the frequencies of different segregational types (4:0, 3:1, 2:2) of triplex, duplex, and simplex tetrads.

II. Spontaneous and induced polyploidy

Relatively little has been reported on the induction of polyploidy in fungi in contrast to the higher plants. Table V-6 gives a survey of organisms which have been used for experiments to produce polyploid strains. The *mutagenic agents* employed for polyploidization belong primarily to the group of *spindle poisons* (e.g. colchicine, camphor, acenaphthene), plant *growth substances* (e.g. indole and naphthalene acetic acid), and *carcinogens* (e.g. benzpyrene and methylcholanthrene). In rare cases UV and radium have induced polyploidy. Polyploid nuclei also arise spontaneously. Proof of polyploidy is in most cases based on several of the criteria discussed in the preceding paragraphs.

In so far as the higher plants are concerned heat and cold shocks may lead to an increase in the number of chromosome sets. In fungi temperature treatments have given mostly negative results. SUBRAMANIAM and RANGANATHAN (1947) reported a case of tetraploid gigas strains in a brewer's bottom yeast induced by cold shock. However, convincing proof of polyploidy was lacking.

The mechanism of action of the polyploidy-inducing mutagens is known in only a few cases. Following treatment with colchicine the spindle fibers necessary for normal separation of the chromosomes often are not formed so that a doubling of the chromosome complement results (p. 313). A similar action is assumed for camphor and acenaphthene. The cause of the mutagenic action of growth substances and carcinogens is largely unknown.

While the addition of colchicine always produced polyploidy strains in *Allomyces*, *Penicillium*, and *Podospora*, such treatments in yeast and *Sordaria macrospora* were unsuccessful (e.g. THAYSEN and MORRIS, 1943; BAUCH, 1941a, b, 1942a, b, 1953; and HESLOT, 1958).

It has been shown in a number of fungi that *polyploid strains regularly undergo a reduction to the haploid condition* after a time. The reductional process, which can be followed by nuclear measurements, takes place unusually rapidly in *Podospora anserina* (FRANKE, 1962). This peculiarity is explained by the fact that the mycelium possesses perforated cross walls so that with a reduction of diploid nuclei to the haploid condition the more rapidly dividing haploid nuclei migrate from cell to cell and quickly become predominant.

The phenomenon of "somatic reduction" was first observed by PONTECORVO et al. (1953) and later by KÄFER (1961) in *Aspergillus nidulans* as well as in *Neurospora crassa* by SANSOME (1956). The haploidization process goes through different stages of aneuploidy (p. 146f.). Results with yeast (BAUCH, 1953) also reveal such intermediate stages. The chromosomes of a nucleus appear to undergo haploidization more or less independently of one another.

The polyploidy experiments on *Allomyces arbuscula* carried out by SOST (1955) deserve particular attention. These investigations answer the question whether or not alternation of generations can be influenced experimentally in this phycomycete. SOST was able to produce diploid and tetraploid gametophytes and tetraploid and octoploid sporophytes through colchicine treatments. The polyploid gametophytes and sporo-

Table V-6. *Examples of induction of polyploidy in fungi*

mutagenic agent	object	method for establishing polyploidy						reference
		cytological				bio-che-mi-cal	ge-ne-tic	
		nu-clear size	cell size	cell form	chro-mo-some num-ber			
Camphor	*Achlya spec.*	+			+			SANSOME and HARRIS, 1963
	Aspergillus nidulans		+		+	+	+	PONTECORVO and ROPER, 1952; ROPER, 1952; HEAGY and ROPER, 1952
	Aspergillus oryzae		+		+		+	ISHITANI et al., 1956
	Aspergillus sojae		+		+	+	+	ISHITANI et al., 1956
	Neurospora crassa	+	+				+	SANSOME, 1956
	Penicillium spec.		+					KOSTOFF, 1946
	Penicillium notatum	+	+				+	SANSOME, 1946
	Phytophtora cactorum	+			+			SANSOME and HARRIS, 1963
	Podospora anserina	+	+		+			FRANKE, 1962
	Pythium debaryanum	+			+			SANSOME and HARRIS, 1963
	Saccharomyces cerevisiae and other yeasts		+	+				BAUCH, 1941 a, b, 1953; SKOVSTEDT, 1948; SUBRAMANIAM, 1945; RANGANATHAN and SUBRAMANIAM, 1950
Colchicine	*Allomyces arbuscula*	+	+		+			SOST, 1955
	Penicillium notatum	+	+					GORDON and MCKEHNIE, 1945
	Podospora anserina	+	+		+			FRANKE, 1962
Acenaph-thene	*Aspergillus oryzae*		+		+		+	ISHITANI et al., 1956
	Aspergillus sojae		+		+	+	+	ISHITANI et al., 1956
	Penicillium spec.		+					KOSTOFF, 1946

Table V-6 (Continued)

mutagenic agent	object	cytological				bio-che-mi-cal	ge-ne-tic	reference
		nu-clear size	cell size	cell form	chro-mo-some num-ber			
	Saccharomyces cerevisiae and other yeasts		+	+				BAUCH,1941a, b, 1953; SUBRAMANIAM, 1945; SUBRAMANIAM and RANGANATHAN, 1948
Naphthalene acetic acid, benzpyrene, methylcholanthrene	Saccharomyces cerevisiae		+	+				BAUCH, 1953
Radium	various yeasts		+	+				BAUCH, 1944
UV radiation	Aspergillus oryzae		+		+		+	ISHITANI et al., 1956
	Aspergillus sojae		+		+	+	+	ISHITANI et al., 1956
Spontaneous	Aspergillus oryzae		+		+		+	ISHITANI et al., 1956
	Aspergillus sojae		+		+	+	+	ISHITANI et al., 1956
	Cyathus stercoreus				+			LU and BRODIE, 1962; LU, 1964
	Penicillium notatum	+	+				+	SANSOME, 1949
	Saccharomyces cerevisiae		+	+		+	+	LINDEGREN and LINDEGREN, 1951; OGUR et al., 1952
	Schizosaccharomyces pombe		+				+	LEUPOLD, 1956a, b

phytes clearly show that no fixed relationship exists between chromosome number and type of generation. Under normal conditions alternation of generations in *Allomyces arbuscula* is strictly correlated with the nuclear phase (KNIEP, 1930; SÖRGEL, 1936, 1937). The results of EMERSON and WILSON (1949), who produced diploid, triploid, and tetraploid gametophytes of *Allomyces arbuscula* and *A. javanicus* also support this view.

Summary

1. The methods commonly used to establish polyploidy are based on cytological, biochemical, and genetic criteria.

2. Mutagenic agents used to induce polyploidy include primarily mitotic poisons, plant growth substances, and carcinogens.

3. The mechanism of action of such mutagens is known only for colchicine and similar spindle poisons. Treatment with colchicine leads to a disturbance of the spindle apparatus so that anaphase movement of the chromosomes fails to occur.

4. Polyploid nuclei are generally reduced to the haploid state after a time interval.

Literature

ALPER, T., and P. HOWARD-FLANDERS: Role of oxygen in modifying the radiosensitivity of *E. coli B*. Nature (Lond.) **178**, 978—979 (1956).

APIRION, D.: Formal and physiological genetics of ascospore colour in *Aspergillus nidulans*. Genet. Res. **4**, 276—283 (1963).

ARDITTI, R. R., and G. SERMONTI: Modification by manganous chloride of the frequency of mutation induced by nitrogen mustard. Genetics **47**, 761—768 (1962).

ATWOOD, K. C.: The role of lethal mutation in the killing of *Neurospora* conidia by ultraviolet light. Genetics **35**, 95—96 (1950).

—, and F. MUKAI: Indispensable gene functions in *Neurospora*. Proc. nat. Acad. Sci. (Wash.) **39**, 1027—1035 (1953a).

— — High spontaneous incidence of a mutant of *Neurospora crassa*. Genetics **38**, 654 (Abstr.) (1953b).

— — Survival and mutation in *Neurospora* exposed at nuclear detonations. Amer. Naturalist **88**, 295—314 (1954a).

— — Homology patterns of x-ray-induced lethal mutations in *Neurospora*. Radiat. Res. **1**, 125 (Abstr.) (1954b).

AUERBACH, C.: Spontaneous mutations in dry *Neurospora* conidia. Heredity **13**, 414 (Abstr.) (1959).

— Mutation. An introduction to research on mutagenesis, part I: Methods. Edinburgh 1962.

— Summary of the conference proceedings. I. Session on mutagenesis. Neurospora Newsletter **5**, 8—11 (1964).

— B. J. KILBEY, and G. KØLMARK: Differences in dose-effect curves for UV-induced reverse mutations at two different loci. Neurospora Newsletter **2**, 4 (1962a).

— — — Response of two loci to interaction treatment. Neurospora Newsletter **2**, 4 (1962b).

—, and M. WESTERGAARD: A discussion of mutagenic specificity. Abh. dtsch. Akad. Wiss. Berlin, Kl. Medizin **1**, 116—123 (1960).

BACQ, Z. M., and P. ALEXANDER: Fundamentals of radiobiology. Deutsche Übersetzung herausgeg. von H. J. MAURER. Stuttgart 1958.

BARNETT, W. E., and F. J. DE SERRES: Fixed genetic instability in *Neurospora crassa*. Genetics **48**, 717—723 (1963).

BAUCH, R.: Experimentelle Mutationsauslösung bei Hefe und anderen Pilzen durch Behandlung mit Campher, Acenaphthen und Colchicin. Naturwissenschaften **29**, 503—504 (1941a).

— Experimentell erzeugte Polyploidreihen bei der Hefe. Naturwissenschaften **29**, 687—688 (1941b).

— Experimentelle Auslösung von Gigas-Mutationen bei der Hefe durch carcinogene Kohlenwasserstoffe. Naturwissenschaften **30**, 263—264 (1942a).

BAUCH, R.: Über Beziehungen zwischen polyploidisierenden, carcinogenen und phytohormonalen Substanzen. Auslösung von Gigas-Mutationen bei der Hefe durch pflanzliche Wuchsstoffe. Naturwissenschaften **30**, 420—421 (1942b).
— Experimentelle Mutationsauslösung bei der Hefe durch chemische Stoffe. Wschr. Brauerei, H. 1 u. 2 (1942c).
— Die Erblichkeit der durch Radiumbestrahlung bei der Hefe ausgelösten Riesenzellbildung. Arch. Mikrobiol. **13**, 352—364 (1944).
— Die Konstanz der chemisch induzierten Gigas-Rassen der Hefe. Wiss. Z. Univ. Greifswald, math.-nat. Reihe **3**, 123—158 (1953).
BAUTZ, E., and E. FREESE: On the mutagenic effect of alkylating agents. Proc. nat. Acad. Sci. (Wash.) **46**, 1585—1594 (1960).
BAUTZ-FREESE, E.: Transitions and transversions induced by depurinating agents. Proc. nat. Acad. Sci. (Wash.) **47**, 540—545 (1961).
BEADLE, G. W., and E. L. TATUM: Neurospora. II. Methods of producing and detecting mutations concerned with nutritional requirements. Amer. J. Bot. **32**. 678—686 (1945).
BEAM, C. A.: The influence of ploidy and division stage on the anoxic protection of Saccharomyces cerevisiae against X-ray inactivation. Proc. nat. Acad. Sci. (Wash.) **41**, 857—861 (1955).
— R. K. MORTIMER, R. G. WOLFE, and C. A. TOBIAS: The relation of radioresistance to budding in Saccharomyces cerevisiae. Arch. Biochem. **49**, 110—122 (1954).
BENZER, S.: On the topology of the genetic fine structure. Proc. nat. Acad. Sci. (Wash.) **45**, 1607—1620 (1959).
— On the topography of the genetic fine structure. Proc. nat. Acad. Sci. (Wash.) **47**, 403—415 (1961).
—, and E. FREESE: Induction of specific mutations with 5-bromouracil. Proc. nat. Acad. Sci. (Wash.) **44**, 112—119 (1958).
BERAHA, L., E. D. GARBER, and O. STROMNAES: Genetics of phytopathogenic fungi. X. Virulence of color and nutritionally deficient mutants of Penicillium italicum and Penicillium digitatum. Canad. J. Bot. **42**, 429—436 (1964).
BEUKERS, R., and W. BERENDS: Isolation and identification of the irradiation product of thymine. Biochim. biophys. Acta (Amst.) **41**, 550—551 (1960).
— J. IJLSTRA, and W. BERENDS: The effect of ultraviolet light on some components of the nucleic acids. II. In rapidly frozen solutions. Rec. Trav. chim. Pays-Bas **77**, 729—732 (1958).
— — — The effect of ultraviolet light on some components of the nucleic acids. III. Apurinic acid. Rec. Trav. chim. Pays-Bas **78**, 247—251 (1959).
— — — The effect of ultraviolet light on some components of the nucleic acids. VI. The origin of the ultraviolet sensitivity of deoxyribonucleic acid. Rec. Trav. chim. Pays-Bas **79**, 101—104 (1960).
BLAU, M., u. K. ALTENBURGER: Über einige Wirkungen von Strahlen. II. Z. Physik **12**, 315—324 (1922).
BONNIER, G., and K. G. LÜNING: Studies on X-ray mutations in the white and forked loci of Drosophila melanogaster. I. A statistical analysis of mutation frequencies. Hereditas (Lund) **35**, 163—189 (1949).
BOONE, D. M., J. F. STAUFFER, M. A. STAHMANN, and G. W. KEITT: Venturia inaequalis (CKE.) WINT. VII. Induction of mutants for studies on genetics, nutrition and pathogenicity. Amer. J. Bot. **43**, 199—204 (1956).
BRENNER, S., L. BARNETT, F. H. C. CRICK, and A. ORGEL: The theory of mutagenesis. J. molec. Biol. **3**, 121—124 (1961).
— S. BENZER, and L. BARNETT: Distribution of proflavin-induced mutations in the genetic fine structure. Nature (Lond.) **182**, 983—985 (1958).
BRESCH, C.: Klassische und molekulare Genetik. Berlin-Göttingen-Heidelberg: Springer 1964.
BROCKMAN, H. E., and F. J. DE SERRES: Induction of ad-3 mutants of Neurospora crassa by 2-aminopurine. Genetics **48**, 597—604 (1963).

BUXTON, E. W., and A. C. HASTIE: Spontaneous and ultraviolet irradiation-induced mutants of *Verticillium albo-atrum*. J. gen. Microbiol. **28**, 625—632 (1962).

CALDAS, L. R., et T. CONSTANTIN: Courbes de survie de levures haploides et diploides soumises aux rayons ultraviolets. C. R. Acad. Sci. (Paris) **232**, 2356—2358 (1951).

CATCHESIDE, D. G.: Isolation of nutritional mutants of *Neurospora crassa* by filtration enrichment. J. gen. Microbiol. **11**, 34—36 (1954).

CLARKE, C. H.: A case of mutagen specificity attributable to a plating medium effect. Z. Vererbungsl. **93**, 435—440 (1962).

— Suppression by methionine of reversions to adenine independence in *Schizosaccharomyces pombe*. J. gen. Microbiol. **31**, 353—363 (1963).

COSTELLO, W. P., E. A. BEVAN, and M. W. MILLER: A comparison of ultraviolet and ethyl methane sulphonate induced mutations of adenine loci in *Saccharomyces cerevisiae*. Proc. XI. Inter. Congr. Genetics **1**, 60 (1963).

COX, B. S., and E. A. BEVAN: Aneuploidy in yeast. New Phytologist **61**, 342—355 (1962).

DAVIDSON, D.: Protection and recovery from ionizing radiation: Mechanisms in seeds and roots. In: A. HOLLAENDER (edit.), Radiation protection and recovery, p. 175—211. Oxford-London-New York-Paris 1960.

DAY, P. R., and G. E. ANDERSON: Two linkage groups in *Coprinus lagopus*. Genet. Res. **2**, 414—423 (1961).

DEMEREC, M.: What is a gene? — Twenty years later. Amer. Naturalist **89**, 5—20 (1955).

— Genetic structure of the *Salmonella* chromosome. Proc. X. Inter. Congr. Genetics **1**, 55—62 (1959).

— Frequency of deletions among spontaneous and induced mutations in *Salmonella*. Proc. nat. Acad. Sci. (Wash.) **46**, 1075—1079 (1960).

— E. L. LABRUM, I. GALINSKY, J. HEMMERLY, A. M. M. BERRIE, J. F. HANSON, I. BLOMSTRAND, and Z. DEMEREC: Bacterial genetics. Carnegie Inst. Wash. Year Book **52**, 210—221 (1953).

— E. L. LAHR, E. BALBINDER, T. MIYAKE, J. ISHIDSU, K. MIZOBUCHI, and B. MAHLER: Bacterial genetics. Carnegie Inst. Wash. Year Book **59**, 426—441 (1960).

— — — — C. MACK, D. MACKEY, and J. ISHIDSU: Bacterial genetics. Carnegie Inst. Wash. Year Book **58**, 433—439 (1959).

—, and J. SAMS: Induction of mutations in individual genes of *Escherichia coli* by low X-radiation. In: A. A. BUZZATI-TRAVERSO (edit.), Proc. Symposium on Immediate and Low-Level Effects of Ionizing Radiations, Venice, 1959, p. 238—291. Int. J. Radiat. Biol. Suppl. (1960).

— E. M. WITKIN, E. L. LABRUM, I. GALINSKY, J. F. HANSON, H. MONSEES, and T. H. FETHERSTON: Bacterial genetics. Carnegie Inst. Wash. Year Book **51**, 193—205 (1952).

DESSAUER, F. R.: Über einige Wirkungen von Strahlen. I. Z. Physik **12**, 38—47 (1922).

DICK, S., and J. R. RAPER: Origin of expressed mutations in *Schizophyllum commune*. Nature (Lond.) **189**, 81—82 (1961).

DOUDNEY, C. O.: Macromolecular synthesis in bacterial recovery from ultraviolet light. Nature (Lond.) **184**, 189—190 (1959).

—, and F. L. HAAS: Modification of ultraviolet-induced mutation frequency and survival in bacteria by post-irradiation treatment. Proc. nat. Acad. Sci. (Wash.) **44**, 390—401 (1958).

— — Mutation induction and macromolecular synthesis in bacteria. Proc. nat. Acad. Sci. (Wash.) **45**, 709—722 (1959).

— — Some biochemical aspects of the post-irradiation modification of ultraviolet-induced mutation frequency in bacteria. Genetics **45**, 1481—1502 (1960).

DRACULIC, M., and M. ERRERA: Chloramphenicol sensitive DNA synthesis in normal and irradiated bacteria. Biochim. biophys. Acta (Amst.) **31**, 459—463 (1959).

Mutation

DULBECCO, R.: Photoreactivation. In: A. HOLLAENDER (edit.), Radiation biology, vol. II: Ultraviolet and related radiations, p. 455—486. New York-Toronto-London 1955.
DUNN, C. G., W. L. CAMPBELL, H. FRAM, and A. HUTCHINS: Biological and photo-chemical effects of high energy, electrostatistically produced roentgen rays and cathode rays. J. appl. Physiol. **19**, 605—616 (1948).
DUPIN, M.: Mise en évidence d'une radiorésistance de *Saccharomyces cerevisiae* au course de la méiose. C. R. Soc. Biol. (Paris) **257**, 282—284 (1963).
EMERSON, R.: An experimental study of the life cycles and taxonomy of *Allomyces*. Lloydia **4**, 77—144 (1941).
— Current trends of experimental research on the aquatic Phycomycetes. Ann. Rev. Microbiol. **4**, 169—200 (1950).
—, and C. M. WILSON: The significance of meiosis in *Allomyces*. Science **110**, 86—88 (1949).
ERRERA, M.: Biochemical aspects of mutagenesis. R. C. Ist. Sci. Univ. Camerino **3**, 3—36 (1962).
ESSER, K., u. J. STRAUB: Genetische Untersuchungen an *Sordaria macrospora* AUERSW., Kompensation und Induktion bei genbedingten Entwicklungsdefekten. Z. Vererbungsl. **89**, 729—746 (1958).
EVANS, H. J.: Chromosome aberrations induced by ionizing radiations. Int. Rev. Cytol. **13**, 221—321 (1962).
FAHMY, O. G., and M. J. FAHMY: Cytogenetic analysis of the action of carcinogens and tumour inhibitors in *Drosophila melanogaster*. V. Differential genetic response to the alkylating mutagens and x-radiation. J. Genet. **54**, 146—164 (1956).
FINCHAM, J. R. S., and P. R. DAY: Fungal genetics. Oxford 1963.
FRANKE, G.: Versuche zur Genomverdoppelung des Ascomyceten *Podospora anserina* (CES.) REHM. Z. Vererbungsl. **93**, 109—117 (1962).
FREESE, E.: The difference between spontaneous and base-analogue induced mutations of phage *T4*. Proc. nat. Acad. Sci. (Wash.) **45**, 622—633 (1959a).
— On the molecular explanation of spontaneous and induced mutations. Brookhaven Symp. Biol. **12**, 63—75 (1959b).
— The specific mutagenic effect of base analogues on phage *T4*. J. molec. Biol. **1**, 87—105 (1959c).
— Molecular mechanism of mutations. In: J. H. TAYLOR, Molecular genetics, part I, p. 207—269. New York and London 1963.
— E. BAUTZ, and E. BAUTZ-FREESE: The chemical and mutagenic specificity of hydroxylamine. Proc. nat. Acad. Sci. (Wash.) **47**, 845—855 (1961).
— E. BAUTZ-FREESE, and E. BAUTZ: Hydroxylamine as a mutagenic and inactivating agent. J. molec. Biol. **3**, 133—143 (1961).
FRIES, N.: Experiments with different methods of isolating physiological mutations of filamentous fungi. Nature (Lond.) **159**, 199 (1947).
— The nutrition of fungi from the aspect of growth factor requirements. Trans. Brit. Mycol. Soc. **30**, 118—134 (1948a).
— Viability and resistance of spontaneous mutations in *Ophiostoma* representing different degree of heterotrophy. Physiol. plantarum (Kbh.) **1**, 330—341 (1948b).
— The production of mutations by caffeine. Hereditas (Lund) **36**, 134—149 (1950).
— Further studies on mutant strains of *Ophiostoma* which require guanine. J. biol. Chem. **200**, 325—333 (1953).
FRITZ-NIGGLI, H.: Strahlenbiologie, Grundlagen und Ergebnisse. Stuttgart 1959.
— Rückmutationen in Abhängigkeit vom Zellzustand und vom Agens der Vorwärtsinduktion. In: H. FRITZ-NIGGLI (edit.), Strahlenwirkung und Milieu, p. 129—139. München u. Berlin 1962.
— Induktion von Rückmutationen und Suppressoren mit Ribonukleinsäure bei *Schizosaccharomyces pombe*. Naturwissenschaften **50**, 530 (1963).

FUERST, R., and W. M. SKELLENGER: A *Neurospora* plate method for testing antimetabolites. Antibiot. and Chemother. **8**, 76—80 (1958).

GILES, N. H.: Studies on the mechanism of reversion in biochemical mutants of *Neurospora crassa*. Cold Spr. Harb. Symp. quant. Biol. **16**, 283—313 (1951).

— Forward and back mutation at specific loci in *Neurospora*. Brookhaven Symp. Biol. **8**, 103—125 (1955).

— Mutations at specific loci in *Neurospora*. Proc. 10th Int. Congr. Genet. (Montreal) **1**, 261—279 (1959).

GLASS, B., and L. E. METTLER: The oxygen effect in respect to point mutations in *Drosophila melanogaster*. Proc. 10th Int. Congr. Genet. (Montreal) **2**, 97—98 (1958).

GLOVER, S. W.: A comparative study of induced reversion in *Escherichia coli*. In: M. DEMEREC et al. (edits.), Genetic studies with bacteria, p. 121—136. Washington 1956.

GOLDSCHMIDT, R. B.: Theoretical genetics. Berkeley and Los Angeles 1955.

GORDON, W. W., and K. McKEHNIE: Colchicine induced autopolyploidy in *Penicillium notatum*. Lancet **1945**, 47—49.

GREER, S., and S. ZAMENHOF: Effect of 5-bromouracil in desoxyribonucleic acid of *E. coli* on sensitivity to ultraviolet irradiation. Amer. chem. Soc. Abstr. (131. meeting) 3C (1957).

GRIGG, G. W.: Competitive suppression and the detection of mutations in microbial populations. Aust. J. biol. Sci. **11**, 69—84 (1957).

—, and D. SERGEANT: Compound loci and coincident mutation in *Neurospora*. Z. Vererbungsl. **92**, 380—388 (1961).

GUTZ, H.: Distribution of X-ray and nitrous acid-induced mutations in the genetic fine structure of the *ad-7* locus of *Schizosaccharomyces pombe*. Nature (Lond.) **191**, 1125—1126 (1961).

— Untersuchungen zur Feinstruktur der Gene *ad-7* und *ad-6* von *Schizosaccharomyces pombe* LIND. Habil.-Schr. der Technischen Universität Berlin, 111 S., 1963.

HAAS, F. L., and C. O. DOUDNEY: A relation of nucleic acid and protein synthesis on ultraviolet induced mutation in bacteria. Proc. nat. Acad. Sci. (Wash.) **43**, 871—883 (1957).

— — Mutation induction and expression in bacteria. Proc. nat. Acad. Sci. (Wash.) **45**, 1620—1624 (1959).

HAEFNER, K.: Zur Ploidiegradabhängigkeit strahleninduzierter Mutationsraten in einem System weitgehend homozygoter und isogener *Saccharomyces*-Stämme. Z. Naturforsch. **19**b, 451—453 (1964a).

— Über die UV-Induktion prototropher Mutanten bei *Saccharomyces*. Biophysik **1**, 413—417 (1964b).

—, u. W. LASKOWSKI: Zur Induktion prototropher *Saccharomyces*-Mutanten durch ultraviolettes Licht in Abhängigkeit von Dosis und Nachbehandlung. Z. Naturforsch. **18**b, 301—309 (1963).

HAINZ, H., u. R. W. KAPLAN: Einfluß der Temperatur während der UV-Bestrahlung auf Inaktivierung und Mutation von *Serratia marcescens* sowie des Bakteriophagen *Kappa*. Z. allg. Mikrobiol. **3**, 113—125 (1963).

HARM, W., u. W. STEIN: Zur Deutung von Maxima und Sättigungs-Effekten bei Dosis-Effekt-Kurven für strahleninduzierte Mutationen. Z. Naturforsch. **11**b, 89—105 (1956).

HAROLD, F. M., and Z. Z. ZIPORIN: Synthesis of protein and of DNA in *Escherichia coli* irradiated with ultraviolet light. Biochim. biophys. Acta (Amst.) **29**, 439—440 (1958).

HARRIS, R. J. C.: The initial effects of ionizing radiations on cells. London and New York 1961.

HAWTHORNE, D. C.: A deletion in yeast and its bearing on the structure of the mating locus. Genetics **48**, 1727—1729 (1963).

HAYES, W.: The genetics of bacteria and their viruses. Studies in basic genetics and molecular biology. Oxford 1964.

21*

HEAGY, F. C., and J. A. ROPER: Desoxyribonucleic acid content of haploid. and diploid *Aspergillus* conidia. Nature (Lond.) **170**, 713—714 (1952).

HESLOT, H.: Contribution à l'étude cytogénétique et génétique des Sordariacées. Rev. Cytol. et Biol. végét. **19**, Suppl. 2, 1—235 (1958).

— *Schizosaccharomyces pombe:* un nouvel organisme pour l'étude de la mutagenèse chimique. Abh. dtsch. Akad. Wiss. Berlin, Kl. Medizin **1**, 98—105 (1960).

— Étude quantitative de réversions biochimiques induites chez la levure *Schizosaccharomyces pombe* par des radiations et des substances radiomimétiques. Abh. dtsch. Akad. Wiss. Berlin, Kl. Medizin **1**, 193—228 (1962).

HOLLAENDER, A. (edit.): Radiation biology. New York-Toronto-London 1954 (vol. 1), 1955 (vol. 2), 1956 (vol. 3).

—, and C. W. EMMONS: Wave length dependence of mutation production in the ultraviolet with special emphasis on fungi. Cold Spr. Harb. Symp. quant. Biol. **9**, 179—186 (1941).

HOLLIDAY, R.: A new method for the identification of biochemical mutants of micro-organisms. Nature (Lond.) **178**, 987 (1956).

— The genetics of *Ustilago maydis*. Genet. Res. **2**, 204—230 (1961).

— Mutation and replication in *Ustilago maydis*. Genet. Res. **3**, 472—486 (1962).

HOROWITZ, N. H., M. B. HOULAHAN, M. G. HUNGATE, and B. WRIGHT: Mustard gas mutations in *Neurospora*. Science **104**, 233—234 (1946).

HOWARD-FLANDERS, P.: Physical and chemical mechanisms in the injury of cells by ionizing radiations. Biol. med. Physics **6**, 544—603 (1958).

— Primary physical and chemical processes in radiobiology. In: Radiation Biology and Cancer, p. 29—40. Texas 1959.

HOWE, H. B., and C. E. TERRY: Genetic studies of resistance to chemical agents in *Neurospora crassa*. Canad. J. Genet. Cytol. **4**, 447—452 (1962).

ISHIKAWA, T.: Genetic studies of *ad-8* mutants in *Neurospora crassa*. I. Genetic fine structure of the *ad-8* locus. Genetics **47**, 1147—1161 (1962).

ISHITANI, C., Y. IKEDA, and K. SAKAGUCHI: Hereditary variation and genetic recombination in Koji-molds (*Aspergillus oryzae* und *A. sojae*). VI. Genetic recombination in heterozygous diploids. J. gen. appl. Microbiol. **2**, 401—430 (1956).

JAGGER, J.: Photoreactivation. Bact. Rev. **22**, 99—138 (1958).

— Photoreactivation. In: A. HOLLAENDER (edit.), Radiation protection and recovery, p. 352—377. Oxford-London-New York-Paris 1960.

JENSEN, K. A., J. KIRK, and M. WESTERGAARD: Biological action of "mustard gas" compounds. Nature (Lond.) **166**, 1019 (1950).

— — G. KØLMARK, and M. WESTERGAARD: Chemically induced mutations in *Neurospora*. Cold Spr. Harb. Symp. quant. Biol. **16**, 245—261 (1951).

— G. KØLMARK, and M. WESTERGAARD: Back-mutations in *Neurospora crassa* induced by diazomethane. Hereditas (Lund) **35**, 521—525 (1949).

KADA, T., C. O. DOUDNEY, and F. L. HAAS: Some biochemical factors in X-ray induced mutation in bacteria. Genetics **45**, 995 (Abstr.) (1960).

— — — Some biochemical factors in X-ray induced mutation in bacteria. Genetics **46**, 683—702 (1961).

KÄFER, E.: The processes of spontaneous recombination in vegetative nuclei of *Aspergillus nidulans*. Genetics **46**, 1581—1609 (1961).

— Translocations in stock strains of *Aspergillus nidulans*. Genetica **33**, 59—68 (1962).

— Origin and pedigree of a VI—VII translocation in *Aspergillus nidulans*. Microbial Genetics Bull. **19**, 12—13 (1963a).

— Radiation effects and mitotic recombination in *Aspergillus nidulans*. Genetics **48**, 27—45 (1963b).

—, and A. CHEN: UV-induced mutations and mitotic crossing-over in dormant and germinating conidia of *Aspergillus*. Microbial Genetics Bull. **20**, 8—9 (1964).

KAKAR, S. N., F. K. ZIMMERMANN, and R. P. WAGNER: Reversion behavior of isoleucine-valine mutants of yeast. Mutation Research **1**, 381—386 (1964).

KANAZIR, D., and M. ERRERA: Alterations of intracellular deoxyribonucleic acid and their biological consequence. Cold Spr. Harb. Symp. quant. Biol. **21**, 19—29 (1956).

KAPLAN, R. W.: Neuere Entwicklungen in der Mikrobengenetik. Zbl. Bakt., I. Abt. Orig. **160**, 181—193 (1953).

— Dose-effect curves of s-mutation and killing in *Serratia marcescens*. Arch. Mikrobiol. **24**, 60—79 (1956).

— Genetik der Mikroorganismen. Fortschr. Bot. **19**, 288—323 (1957).

— Strahlengenetik der Mikroorganismen. In: H. SCHINZ (Hrsg.), Strahlenbiologie, Nuklearmedizin und Krebsforschung, p. 109—156. Stuttgart: Georg Thieme 1959.

— Genetik der Mikroorganismen. Fortschr. Bot. **24**, 286—313 (1962a).

— Einfluß von Kälte oder Trockenheit während sowie von Behandlungen mit Biochemikalien nach der UV-Bestrahlung auf die Mutationsauslösung bei *Serratia*. Abh. dtsch. Akad. Wiss. Berlin, Kl. Medizin **1**, 167—170 (1962b).

— Photoreversion von vier Gruppen UV-induzierter Mutationen zur Giftresistenz in nichtphotoreaktivierbaren *E. coli*. Photochem. and Photobiol. **2**, 461—470 (1963).

—, H. BECKMANN, and W. RÜGER: Different "spectra" of mutant types by extracellular treatment of phage *Kappa* with differing mutagens. Nature (Lond.) **199**, 932—933 (1963).

—, u. W. GUNKEL: Reversion der Mutationen und Reaktivierung durch sichtbares Licht sowie verschiedene Salzlösungen nach UV-Bestrahlung von *Serratia*. Arch. Mikrobiol. **35**, 63—91 (1960).

—, and C. KAPLAN: Influence of water content on UV-induced s-mutation and killing in *Serratia*. Exp. Cell Res. **11**, 378—392 (1956).

KAUDEWITZ, F.: Inaktivierende und mutagene Wirkung salpetriger Säure auf Zellen von *Escherichia coli*. Z. Naturforsch. **14**b, 528—537 (1959a).

— Linear transfer of a potentially mutant state in bacteria. Nature (Lond.) **183**, 871—873 (1959b).

— Production of bacterial mutants with nitrous acid. Nature (Lond.) **183**, 1829—1830 (1959c).

KELNER, A.: Effect of visible light on the recovery of *Streptomyces* conidia from ultraviolet irradiation injury. Proc. nat. Acad. Sci. (Wash.) **35**, 73—79 (1949).

— Growth, respiration and nucleic acid synthesis in ultraviolet-irradiated and photoreactivated *Escherichia coli*. J. Bact. **65**, 252—262 (1953).

KIHLMAN, B. A.: Biochemical aspects of chromosome breakage. Advanc. Genet. **10**, 1—59 (1961).

KILBEY, B. J.: Mutagenic studies with *Aspergillus nidulans*. Microbial Genetics Bull. **19**, 14 (1963a).

— The influence of temperature of the ultraviolet induced revertant frequencies of two auxotrophs of *Neurospora crassa*. Z. Vererbungsl. **94**, 385—391 (1963b).

KIMBALL, R. F., N. GAITHER, and S. M. WILSON: Reduction of mutation by postirradiation treatment after ultraviolet and various kinds of ionizing radiations. Radiat. Res. **10**, 490—497 (1959).

KNAPP, E., A. REUSS, O. RISSE u. H. SCHREIBER: Quantitative Analyse der mutationsauslösenden Wirkung monochromatischen UV-Lichtes. Naturwissenschaften **27**, 304 (1939).

—, u. H. SCHREIBER: Quantitative Analyse der mutationsauslösenden Wirkung monochromatischen UV-Lichtes in Spermatozoiden von *Sphaerocarpus*. Proc. 7th Internat. Genet. Congr., Edinburgh. Suppl., J. Genet. 175—176 (1939).

KNIEP, H.: Über den Generationswechsel von *Allomyces*. Z. Bot. **22**, 433—441 (1930).

Mutation

Kølmark, G.: Differential response to mutagens as studied by the *Neurospora* reverse mutation test. Hereditas (Lund) **39**, 270—276 (1953).
— Mutagenic properties of certain esters of inorganic acids investigated by the *Neurospora* back-mutation test. C. R. Lab. Carlsberg, Sér. physiol. **26**, 205—220 (1956).
—, and N. H. Giles: Comparative studies of monoepoxides as inducers of reverse mutations in *Neurospora*. Genetics **40**, 890—902 (1955).
—, and B. J. Kilbey: An investigation into the mutagenic after-effect of butadiene diepoxide using *Neurospora crassa*. Z. Vererbungsl. **93**, 356—365 (1962).
—, and M. Westergaard: Induced back-mutations in a specific gene of *Neurospora crassa*. Hereditas (Lund) **35**, 490—506 (1949).
— — Validity of the *Neurospora* back-mutation test. Nature (Lond.) **169**, 626 (1952).
— — Further studies on chemically induced reversions at the adenine locus of *Neurospora*. Hereditas (Lund) **39**, 209—224 (1953).
Kostoff, D.: Gigantism in *Penicillium*, experimentally produced. Bull. chambre de culture nationale, Sér. Biologie, agriculture et silviculture Sofia **1**, 240 (1946).
Kubitschek, H. E., and H. E. Bendigkeit: Delay in the appearance of caffeine-induced *T5* resistance in *Escherichia coli*. Genetics **43**, 647—661 (1958).
— — Latent mutants in chemostats. Genetics **46**, 105—122 (1961).
Laskowski, W.: Inaktivierungsversuche mit homozygoten Hefestämmen verschiedenen Ploidiegrades. I. Aufbau homozygoter Stämme und Dosiseffektkurven für ionisierende Strahlen, UV und organische Peroxyde. Z. Naturforsch. **15b**, 495—506 (1960a).
— Die hefeartigen Pilze. V. Entwicklungscyclen und Erbverhalten der Hefen. Die Hefen **1**, 178—208 (1960b).
— Inaktivierungsversuche mit homozygoten Hefestämmen verschiedenen Ploidiegrades. VI. Über den Aufbau weitestgehend isogener, homozygoter penta- und hexaploider Stämme sowie den Einfluß bestimmter mutierter Allele auf die Strahlenresistenz. Z. Naturforsch. **17b**, 93—108 (1962a).
— Strahleninaktivierung von *Saccharomyces* in Abhängigkeit von Ploidiegrad und Genotyp. Abh. dtsch. Akad. Wiss. Berlin, Kl. Medizin **1**, 171—177 (1962b).
—, and K. Haefner: Determination of radiation-induced mutation rates of recessive lethal alleles in *Saccharomyces*. Nature (Lond.) **200**, 795—796 (1963).
—, u. W. Stein: Inaktivierungsversuche mit homozygoten Hefestämmen verschiedenen Ploidiegrades. II. Mikroskopische Beobachtungen nach Inaktivierung mit ionisierenden Strahlen, UV und organischen Peroxyden. Z. Naturforsch. **15b**, 604—612 (1960).
Latarjet, R., et B. Ephrussi: Courbes de survie de levures haploides et diploides soumises aux rayons X. C. Rend. Acad. Sci. (Paris) **229**, 306—308 (1949).
Lea, D. E.: Actions of radiations on living cells, 2. Aufl. Cambridge 1956.
—, and C. A. Coulson: The distribution of the numbers of mutants in bacterial populations. J. Genet. **49**, 264—285 (1949).
Leben, C., D. M. Boone, and G. W. Keitt: *Venturia inaequalis* (Cke.) Wint. IX. Search for mutants resistent to fungicides. Phytopathology **45**, 467—472 (1955).
Lederberg, J., and E. M. Lederberg: Replica plating and indirect selection of bacterial mutants. J. Bact. **63**, 399—406 (1952).
Lein, J., H. K. Mitchell, and M. B. Houlahan: A method for the selection of biochemical mutants of *Neurospora*. Proc. nat. Acad. Sci. (Wash.) **34**, 435—442 (1948).

LERMAN, L. S.: The structure of the DNA-acridine complex. Proc. nat. Acad. Sci. (Wash.) **49**, 94—102 (1963).

LESTER, H. E., and S. R. GROSS: Efficient method for selection of auxotrophic mutants of *Neurospora*. Science **129**, 572 (1959).

LEUPOLD, U.: Tetraploid inheritance in *Saccharomyces*. J. Genet. **54**, 411—426 (1956a).

— Tetrad analysis of segregation in autotetraploids. J. Genet. **54**, 427—439 (1956b).

— Some data on polyploid inheritance in *Schizosaccharomyces pombe*. C. R. Lab. Carlsberg, Sér. physiol. **26**, 221—251 (1956c).

— Studies on recombination in *Schizosaccharomyces pombe*. Cold Spr. Harb. Symp. quant. Biol. **23**, 161—170 (1958).

— Intragene Rekombination und allele Komplementierung. Arch. Klaus-Stift. Vererb.-Forsch. **26**, 89—117 (1961).

LIEB, M.: Deoxyribonucleic acid synthesis and ultraviolet-induced mutation. Biochim. biophys. Acta (Amst.) **37**, 155—157 (1960).

LINDEGREN, C. C., and G. LINDEGREN: X-ray and ultraviolet induced mutations in *Neurospora*, I. X-ray mutations; II. Ultraviolet mutations. J. Hered. **32**, 405—412 (1941).

— — Tetraploid *Saccharomyces*. J. gen. Bact. **5**, 885—893 (1951).

LISSOUBA, P.: Mise en evidence d'une unité génétique polarisée et essai d'analyse d'un cas d'interférence négative. Ann. Sci. nat. bot. **44**, 641—720 (1960).

— J. MOUSSEAU, G. RIZET, and J. L. ROSSIGNOL: Fine structure of genes in the Ascomycete *Ascobolus immersus*. Advanc. Genet. **11**, 343—380 (1962).

LOPRIENO, N., G. ZETTERBERG, R. GUGLIELMINETTI, and E. MICHEL: The lethal and mutagenic effects of N-nitroso-N-methylurethane and N-nitroso-N-ethylurethane in *Colletotrichum coccodes*. Mutation Research **1**, 37—44 (1964).

LORKIEWICZ, Z., and W. SZYBALSKI: Genetic effects of halogenated thymidine analogs incorporated during thymidylate synthetase inhibition. Biochem. biophys. Res. Commun. **2**, 413—418 (1960).

LU, B. C.: Polyploidy in the basidiomycete *Cyathus stercoreus*. Amer. J. Bot. **51**, 343—347 (1964).

—, and H. J. BRODIE: Chromosomes of the fungus *Cyathus*. Nature (Lond.) **194**, 606 (1962).

LUCKE, W. H., and A. SARACHEK: X-ray inactivation of polyploid *Saccharomyces*. Nature (Lond.) **171**, 1014—1015 (1953).

LURIA, S. E., and M. DELBRÜCK: Mutations of bacteria from virus sensitivity to virus resistance. Genetics **28**, 491—511 (1943).

MACKINTOSH, M. E., and R. H. PRITCHARD: The production and replica plating of micro-colonies of *Aspergillus nidulans*. Genet. Res. **4**, 320—322 (1963).

MAGNI, G. E.: Mutation rates during the meiotic process in yeasts. In: F. H. SOBELS (edit.), Repair from genetic radiation damage and differential radiosensitivity in germ cells. Proc. Internat. Symp. Leiden 1962, p. 77—85. Oxford-London-New York-Paris 1963.

— R. C. v. BORSEL, and S. SORA: Mutagenic action during meiosis and antimutagenic action during mitosis by 5-aminoacridine in yeast. Mutation Research **1**, 227—230 (1964).

MALING, B. D.: Replica plating and rapid ascus collection of *Neurospora*. J. gen. Microbiol. **23**, 257—259 (1960).

MALLING, H., H. MILTENBURGER, M. WESTERGAARD, and K. G. ZIMMER: Differential response of a double mutant — adenineless, inositolless — in *Neurospora crassa* to combined treatment by ultra-violet radiation and chemicals. Int. J. Radiat. Biol. **1**, 328—343 (1959).

MARCOU, D.: Sur l'influence du mode d'association des gènes sur les propriétés de certains hétérocaryotes du *Podospora anserina*. C. R. Acad. Sci. (Paris) **256**, 768—770 (1963).

Mutation

MARKERT, C. L.: Radiation-induced nutritional and morphological mutants of *Glomerella*. Genetics 37, 339—352 (1952).
— Lethal and mutagenic effects of ultraviolet radiation in *Glomerella* conidia. Exp. Cell Res. 5, 427—435 (1953).
— Response of *Glomerella* conidia to irradiation by X-rays and fast neutrons. Papers Mich. Acad. Sci. 91, 27—31 (1956).
MARQUARDT, H., R. SCHWAIER u. F. K. ZIMMERMANN: Nicht-Mutagenität von Nitrosaminen bei *Neurospora crassa*. Naturwissenschaften 50, 135—136 (1963).
— F. K. ZIMMERMANN u. R. SCHWAIER: Die Wirkung krebsauslösender Nitrosamine und Nitrosamide auf das Adenin-6—45-Rückmutationssystem von *Saccharomyces cerevisiae*. Z. Vererbungsl. 95, 82—96 (1964).
McCLINTOCK, B.: *Neurospora*. I. Preliminary observations of the chromosomes of *Neurospora crassa*. Amer. J. Bot. 32, 671—678 (1945).
McELROY, W. D., J. E. CUSHING, and H. MILLER: The induction of biochemical mutations in *Neurospora crassa* by nitrogen mustard. J. cell. comp. Physiol. 30, 331—346 (1947).
—, and B. GLASS (edits.): The chemical basis of heredity. Baltimore 1957.
METZGER, K.: Der Einfluß des Wassergehaltes auf Inaktivierung und Mutabilität von *Serratia marcescens* durch UV- und Röntgenbestrahlung. Z. allg. Mikrobiol. 1, 29—45 (1960).
MITCHELL, J. S., B. E. HOLMES, and C. L. SMITH (edits.): Progress in radiobiology. Proc. IV. Internat. Conf. radiobiology, Cambridge 1955. Edinburgh and London 1956.
MITCHELL, M. B., and H. K. MITCHELL: The selective advantage of an adenineless double mutant over one of the single mutants involved. Proc. nat. Acad. Sci. (Wash.) 36, 115—119 (1950).
— — Observations on the behaviour of suppressors in *Neurospora*. Proc. nat. Acad. Sci. (Wash.) 38, 205—214 (1952).
MORPURGO, G.: A new method of estimating forward mutation in fungi: resistance to 8-azaguanine and p-fluorophenylalanine. Sci. Rep. Ist. sup. Sanità (Roma) 2, 9—12 (1962).
MORROW, J.: Dispensable and indispensable genes in *Neurospora*. Science 144, 307—308 (1964).
MORTIMER, R. K.: The relative radiation-resistance of haploid, diploid, triploid and tetraploid yeast cells. Med. and Health Physics Quart. p. 39—44. 1952.
MULLER, H. J.: Artificial transmutation of the gene. Science 66, 84—87 (1927).
— The production of mutations by X-rays. Proc. nat. Acad. Sci. (Wash.) 14, 714—726 (1928).
MUNDRY, K. W., u. A. GIERER: Die Erzeugung von Mutationen des Tabakmosaikvirus durch chemische Behandlung seiner Nucleinsäure in vitro. Z. Vererbungsl. 89, 614—630 (1958).
NADSON, G. A., et G. S. FILLIPOV: Influence des rayons X sur la sexualité et la formation des mutantes chez les champignons inférieurs (Mucoracéae). C. R. Soc. Biol. (Paris) 93, 473—475 (1925).
— — De la formation de nouvelles races stables chez champignons inférieur sous l'influence des rayons x. C. R. Soc. Biol. (Paris) 186, 1566—1568 (1928).
NAGAI, S.: Interferences between some induces of the respiration-deficient mutation in yeast. Exp. Cell Res. 27, 19—24 (1962).
NAKADA, D., E. STRELZOFF, R. RUDNER, and F. J. RYAN: Is DNA replication a necessary condition for mutation? Z. Vererbungsl. 91, 210—213 (1960).
NEWCOMBE, H. B., and J. F. McGREGOR: Dose-response relationships in radiation induced mutations. Saturation effects in *Streptomyces*. Genetics 39, 619—627 (1954).

NORMAN, A.: The nuclear role in the ultraviolet inactivation of *Neurospora* conidia. J. cell. comp. Physiol. **44**, 1—10 (1954).

OGUR, M., S. MINCKLER, and D. O. MCCLARY: Desoxyribonucleic acids and the budding cycle in the yeasts. J. Bact. **66**, 642—645 (1953).

— — G. LINDEGREN, and C. C. LINDEGREN: The nucleic acids in a polyploid series of *Saccharomyces*. Arch. Biochem. **40**, 175—184 (1952).

OPPENOORTH, W. F. F.: Modification of the hereditary character of yeast by investigation of cell-free extracts. Eur. Brewery Convention 1960, p. 180—207.

ORGEL, A., and S. BRENNER: Mutagenesis of bacteriophage *T4* by acridines. J. molec. Biol. **3**, 762—768 (1961).

OSTER, R. H.: Results of irradiating *Saccharomyces* with monochromatic ultra-violet-light. I. Morphological and respiratory changes. J. gen. Physiol. **18**, 71—88 (1934a).

— Results of irradiating *Saccharomyces* with monochromatic ultra-violet-light. II. The influence of modifying factors. J. gen. Physiol. **18**, 243—250 (1934b).

— Results of irradiating *Saccharomyces* with monochromatic ultra-violet-light. III. The absorption of ultra-violet-light energy by yeast. J. gen. Physiol. **18**, 251—254 (1934c).

— and W. A. ARNOLD: Results of irradiating *Saccharomyces* with monochromatic ultra-violet-light. IV. Relation of energy to observed inhibitory effects. J. gen. Physiol. **18**, 351—355 (1934).

PITTMAN, D., E. SHULT, A. ROSHANMANESH, and C. C. LINDEGREN: The procurement of biochemical mutants of *Saccharomyces* by the synergistic effect of ultraviolet radiation and 2,6-diamino purine. Canad. J. Microbiol. **9**, 103—109 (1963).

POMPER, S., and K. C. ATWOOD: Radiation studies on fungi. In: A. HOLLAENDER (edit.), Radiation biology, vol. II: Ultraviolet and related radiations, p. 431—453. New York-Toronto-London 1955.

PONTECORVO, G.: Auxanographic techniques in biochemical genetics. J. gen. Microbiol. **3**, 122—126 (1949).

—, and J. A. ROPER: Genetic analysis without sexual reproduction by means of polyploidy in *Aspergillus nidulans*. J. gen. Microbiol.: Proceedings **6**, 7—8 (1952).

— — Resolving power of genetic analysis. Nature (Lond.) **178**, 83—84 (1956).

— — L. M. HEMMONS, K. D. MACDONALD, and A. W. J. BUFTON: The genetics of *Aspergillus nidulans*. Advanc. Genet. **5**, 141—238 (1953).

RAJEWSKY, B.: Strahlendosis und Strahlenwirkung, 2. Aufl. Stuttgart 1956.

RANGANATHAN, B., and M. K. SUBRAMANIAM: Studies on the mutagenic action of chemical and physical agencies on yeasts. I. Induction of polyploidy by diverse agencies. J. Indian. Inst. Sci., Sect. A **32**, pt. 4, 51—72 (1950).

REAUME, S. E., and E. L. TATUM: Spontaneous and nitrogen mustard induced nutritional deficiencies in *Saccharomyces cerevisiae*. Arch. Biochem. **22**, 331—338 (1949).

REISSIG, J. L.: Replica plating with *Neurospora crassa*. Microbial Genetics Bull. **14**, 31—32 (1956).

— Forward and back mutation in the *pyr-3* region of *Neurospora*. I. Mutations from arginine dependence to prototrophy. Genet. Res. **1**, 356—374 (1960).

— Induction of forward mutants in the *pyr-3* region of *Neurospora*. J. gen. Microbiol. **30**, 317—325 (1963a).

— Spectrum of forward mutants in the *pyr-3* region of *Neurospora*. J. gen. Microbiol. **30**, 327—337 (1963b).

REVELL, S.: The accurate estimation of chromatid breakage and its relevance to a new interpretation of chromatid aberrations induced by ionizing radiations. Proc. roy. Soc. B **150**, 563—589 (1959).

RIEGER, R.: Die Genommutationen (Ploidiemutationen). In: H. STUBBE (Hrsg.), Genetik: Grundlagen, Ergebnisse und Probleme in Einzeldarstellungen. Jena 1963.

—, u. H. BÖHME: Strahleninduzierte Mutagenese — Gesichtspunkte des Genetikers. Abh. dtsch. Akad. Wiss. Berlin, Kl. Medizin 1, 38—62 (1962).

—, u. A. MICHAELIS: Über die radiomimetische Wirkung von Äthylalkohol bei *Vicia faba*. Abh. dtsch. Akad. Wiss. Berlin, Kl. Medizin 1, 54—65 (1960).

— — Die Auslösung von Chromosomenaberrationen bei *Vicia faba* durch chemische Agentien. Kulturpflanze 10, 212—292 (1962).

RIZET, G., P. LISSOUBA et J. MOUSSEAU: Sur l'interférence négative au sein d'une série d'allèles chez *Ascobolus immersus*. C. R. Soc. Biol. (Paris) 11, 1967—1970 (1960a).

— — — Les mutations d'ascospore chez l'ascomycète *Ascobolus immersus* et l'analyse de la structure fine des gènes. Bull. Soc. franç. Physiol. végétale 6, 175—193 (1960b).

ROBERTS, C.: Methods in yeast genetics. Meth. med. Res. 3, 37—50 (1950).

— A replica plating technique for the isolation of nutritionally exacting mutants of a filamentous fungus *(Aspergillus nidulans)*. J. gen. Microbiol. 20, 540 (1959).

RÖBBELEN, G.: Cytogenetik. Fortschr. Bot. 22, 316—346 (1960).

— Cytogenetik. Fortschr. Bot. 24, 314—359 (1962).

— Cytogenetik. Fortschr. Bot. 25, 393—417 (1963).

ROMAN, H.: A system selective for mutations affecting the synthesis of adenine in yeast. C. R. Lab. Carlsberg, Sér. physiol. 26, 299—314 (1955).

— D. C. HAWTHORNE, and H. C. DOUGLAS: Polyploidy in yeast and its bearing on the occurrence of irregularic genetic ratios. Proc. nat. Acad. Sci. (Wash.) 37, 79—84 (1951).

— M. M. PHILLIPS, and S. M. SANDS: Studies of polyploid *Saccharomyces*. I. Tetraploid segregation. Genetics 40, 546—561 (1955).

ROPER, J. A.: Production of heterozygous diploids in filamentous fungi. Experientia (Basel) 8, 14 (1952).

—, and E. KÄFER: Acriflavine-resistant mutants of *Aspergillus nidulans*. J. gen. Microbiol. 16, 660—667 (1957).

ROSEN, R.: An hypothesis of FREESE and the DNA-protein coding problem. Bull. math. Biophys. 23, 305—318 (1961).

ROYES, J.: The production of mosaic mutations in *Neurospora crassa*. Neurospora Newsletter 1, 5—6 (1962).

RUDNER, R.: Mutation as an error in base pairing. Biochem. biophys. Res. Commun. 3, 275—280 (1960).

—, and E. BALBINDER: Reversions induced by base analogues in *Salmonella typhimurium*. Nature (Lond.) 186, 180 (1960).

RUPERT, C. S: Photoenzymatic repair of ultraviolet damage in DNA. I. Kinetics of the reaction. J. gen. Physiol. 45, 703—741 (1962).

RYAN, F. J.: Selected methods of *Neurospora* genetics. Meth. med. Res. 3, 51—75 (1950).

— Natural mutation in non dividing bacteria. Trans. N.Y. Acad. Sci., Ser. 2, 19, 515—517 (1957).

—, and K. KIRITANI: Effect of temperature on natural mutation in *Escherichia coli*. J. gen. Microbiol. 20, 644—653 (1959).

— D. NAKADA, and M. J. SCHNEIDER: Is DNA replication a necessary condition for spontaneous mutation? Z. Vererbungsl. 92, 38—41 (1961).

— R. RUDNER, T. NAGATA, and Y. KITANI: Bacterial mutation and the synthesis of macromolecules. Z. Vererbungsl. 90, 148—158 (1959).

— M. SCHWARTZ, and P. FRIED: The direct enumeration of spontaneous and induced mutations in bacteria. J. Bact. 69, 552—557 (1955).

SAGER, R., and F. J. RYAN: Cell heredity. An analysis of the mechanisms of heredity at the cellular level. New York and London 1961.

SANSOME, E. R.: Induction of gigas forms of *Penicillium notatum* by treatment with camphor vapours. Nature (Lond.) **157**, 843 (1946).
— Spontaneous mutation in standard and "gigas" forms of *Penicillium notatum* strain 1249 B 21. Transact. Brit. Mycol. Soc. **32**, 305—314 (1949).
— Camphor-induced gigas forms in *Neurospora*. Transact. Brit. Mycol. Soc. **39**, 67—78 (1956).
— M. DEMEREC, and A. HOLLAENDER: Quantitative irradiation experiments with *Neurospora crassa*. I. Experiments with X-rays. Amer. J. Bot. **32**, 218—226 (1945).
—, and B. J. HARRIS: The use of camphor-induced polyploidy to determine the place of meiosis in fungi. Microbial Genetics Bull. **19**, 20—21 (1963).
SARACHEK, A.: X-ray inactivation of *Saccharomyces* during the budding cycle. Experientia (Basel) **10**, 377—378 (1954a).
— A comparative study of the retardation of budding and cellular inactivation by ultraviolet radiation in polyploid *Saccharomyces* with special reference to photoreactivation. Cytologia (Tokyo) **19**, 77—85 (1954b).
—, and W. H. LUCKE: Ultraviolet inactivation of polyploid *Saccharomyces*. Arch. Biochem. **44**, 271—279 (1953).
SCHOLES, G., and J. WEISS: Organic hydroxy-hydroperoxides: a class of hydroperoxides formed under the influence of ionizing radiation. Nature (Lond.) **185**, 305—306 (1960).
SCHULL, W. J. (edit.): Mutations. Second conference on genetics. Ann. Arbor (Mich.): The University of Michigan Press 1962.
SCHUSTER, H.: Die Reaktionsweise der Desoxyribonucleinsäure mit salpetriger Säure. Z. Naturforsch. **15b**, 298—304 (1960).
SCOTT, W. M.: Pyrimidine analogs and the mutation of *Neurospora crassa*. Biochem. biophys. Res. Commun. **15**, 147—150 (1964).
SERRES, F. J. DE: Studies with purple adenine mutants in *Neurospora crassa*. III. Reversion of X-ray-induced mutants. Genetics **43**, 187—206 (1958).
— Genetic analysis of the structure of the *ad-3* region of *Neurospora crassa* by means of irreparable recessive lethal mutations. Genetics **50**, 21—30 (1964).
—, and H. G. KØLMARK: A direct method for determination of forward mutation rates in *Neurospora crassa*. Nature (Lond.) **182**, 1249—1250 (1958).
—, and R. S. OSTERBIND: Estimation of the relative frequencies of X-ray-induced viable and recessive lethal mutations in the *ad-3* region of *Neurospora crassa*. Genetics **47**, 793—796 (1962).
SHAMOIAN, C. A., A. CANZANELLI, and J. MELROSE: Back-mutation of a *Neurospora crassa* mutant by a nucleic acid complex from the wild strain. Biochim. biophys. Acta (Amst.) **47**, 208—211 (1961).
SHARMA, A. K., and A. SHARMA: Spontaneous and chemically induced chromosome breaks. Int. Rev. Cytol. **10**, 101—136 (1960).
SHOCKLEY, T., and E. L. TATUM: A search for genetic transformation in *Neurospora crassa*. Biochim. biophys. Acta (Amst.) **61**, 567—572 (1962).
SIDDIQI, O. H.: Mutagenic action of nitrous acid on *Aspergillus nidulans*. Genet. Res. **3**, 303—314 (1962).
SINGLETON, J. R.: A mechanism intrinsic to heterozygous inversions affecting observed recombination frequencies in adjacent regions. Genetics **49**, 541—560 (1964).
SKOVSTEDT, A.: Induced camphor mutations in yeast. C. R. Trav. Lab. Carlsberg, Sér. physiol. **24**, 249—262 (1948).
SMITH, H. H., and A. M. SRB: Induction of mutations with β-propiolactone. Science **114**, 490—492 (1951).
SÖRGEL, G.: Über heteroploide Mutanten bei *Allomyces Kniepii*. Nachr. Ges. Wiss. Göttingen, Fachgr. VI, **2**, 155—170 (1936).
— Untersuchungen über den Generationswechsel von *Allomyces*. Z. Bot. **31**, 401—446 (1937).
SOMMERMEYER, K.: Quantenphysik der Strahlenwirkung in Biologie und Medizin. Leipzig 1952.

Sost, H.: Über die Determination des Generationswechsels von *Allomyces arbuscula* (Butl.) (Polyploidieversuche). Arch. Protistenk. **100**, 541—564 (1955).

Sparrow, A. H., J. P. Binnington, and V. Pond: Bibliography on effects of ionizing radiations on plants: 1896—1955. New York 1958.

Sparrow jr., F. K.: Aquatic Phycomycetes exclusive of the Saprolegniaceae and *Pythium*. Ann Arbor (Mich.): The University of Michigan Press 1943.

Spencer, W. P., and C. Stern: Experiments to test the validity of the linear r-dose mutation frequency relation in *Drosophila* at low dosage. Genetics **33**, 43—74 (1948).

Stadler, L. J., and F. M. Uber: Genetic effects of UV radiation in maize. IV. Comparison of monochromatic radiations. Genetics **27**, 84—118 (1942).

Stahl, F. W.: The mechanics of inheritance. Englewood Cliffs, New Jersey: Prentice-Hall, Inc. 1964.

Stapleton, G. E., and A. Hollaender: Mechanism of lethal and mutagenic action of ionizing radiations on *Aspergillus terreus*. II. Use of modifying agents and conditions. J. cell. comp. Physiol. **39** (Suppl. 1), 101—113 (1952).

— — and F. L. Martin: Mechanism of lethal and mutagenic action of ionizing radiation on *Aspergillus terreus*. I. Relationship of relative biological efficiency to ion density. J. cell. comp. Physiol. **39** (Suppl. 1), 87—100 (1952).

Stein, W.: Inaktivierungsversuche mit homozygoten Hefestämmen verschiedenen Ploidiegrades. V. Treffertheoretische Betrachtungen. Z. Naturforsch. **17**b, 179—187 (1962).

—, u. W. Laskowski: Zur mathematischen Analyse der Strahleninaktivierung mikrobiologischer Objekte verschiedenen Ploidiegrades unter Berücksichtigung genetischer und nichtgenetischer Anteile. Z. Naturforsch. **13**b, 651—657 (1958).

— — Zur mathematischen Analyse der Strahleninaktivierung homozygoter Hefestämme verschiedenen Ploidiegrades. Naturwissenschaften **46**, 88—89 (1959).

— — Inaktivierungsversuche mit homozygoten Hefestämmen verschiedenen Ploidiegrades. IV. Quantitative Deutung unter Berücksichtigung genetischer und nichtgenetischer Anteile. Z. Naturforsch. **15**b, 734—743 (1960).

Stent, G. S.: Molecular biology of bacterial viruses. San Francisco and London 1963.

Stevens, C. M., and A. Mylroie: Production and reversion of biochemical mutants of *Neurospora crassa* with mustard compounds. Amer. J. Bot. **40**, 424—429 (1953).

Straub, J.: Wege zur Polyploidie. Berlin 1950.

— Cytogenetik. Fortschr. Bot. **20**, 236—256 (1958).

Strauss, B. S.: An outline of chemical genetics. Philadelphia and London 1960.

Strelzoff, E.: Identification of base pairs involved in mutations induced by base analogues. Biochem. biophys. Res. Commun. **5**, 384—388 (1961).

— DNA synthesis and induced mutations in the presence of 5-bromouracil. II. Induction of mutations. Z. Vererbungsl. **93**, 301—318 (1962).

Strigini, P., C. Rossi, and G. Sermonti: Effects of desintegration of incorporated ^{32}P in *Aspergillus nidulans*. J. molec. Biol. **7**, 683—699 (1963).

Stubbe, H. (Hrsg.): Chemische Mutagenese. Erwin-Baur-Gedächtnisvorlesungen I, 1959. Abh. dtsch. Akad. Wiss. Berlin, Kl. Medizin **1**, Berlin 1960.

— Strahleninduzierte Mutagenese. Erwin-Baur-Gedächtnisvorlesungen II, 1961. Abh. dtsch. Akad. Wiss. Berlin, Kl. Medizin **1**, Berlin 1962.

SUBRAMANIAM, M. K.: Induction of polyploidy in *Saccharomyces cerevisiae*. Curr. Sci. **14**, 234 (1945).

—, and B. RANGANATHAN: Induction of mutations in yeast by low temperatures. Sci. and Culture (Calcutta) **13**, 102—105 (1947).

— — Chromosome constitution and characteristics of giant colonies in yeasts. Proc. nat. Inst. Sci. India **14**, 279—283 (1948).

SUYAMA, Y., K. D. MUNKERS, and V. W. WOODWARD: Genetic analyses of the *pyr-3* locus of *Neurospora crassa:* the bearing of recombination and gene conversion upon intra-allelic linearity. Genetica **30**, 293—311 (1959).

SWANSON, C. P.: Cytology and cytogenetics. Prentice-Hall 1957.

SZYBALSKI, W., and Z. LORKIEWICZ: On the nature of the principal target of lethal mutagenic radiation effects. Abh. dtsch. Akad. Wiss. Berlin, Kl. Medizin **1**, 63—71 (1962).

TAKAHASHI, T.: Filtration methods for selecting auxotrophic mutants of flocculent type yeast. Rep. Kihara Inst. Biol. Res. **10**, 57—59 (1959).

TATUM, E. L., R. W. BARRATT, N. FRIES, and D. M. BONNER: Biochemical mutant strains of *Neurospora* produced by physical and chemical treatment. Amer. J. Bot. **37**, 38—46 (1950).

TECTOR, M. A., and E. KÄFER: Radiation-induced chromosomal aberrations and lethals in *Aspergillus nidulans*. Science **136**, 1056—1057 (1962).

TESSMAN, I.: The induction of large deletions by nitrous acid. J. molec. Biol. **5**, 442—445 (1962).

THAYSEN, A. C., and M. MORRIS: Preparation of a giant strain of *Torulopsis utilis*. Nature (Lond.) **152**, 526—528 (1943).

TIMOFEEFF-RESSOVSKY, N. W., u. K. G. ZIMMER: Das Trefferprinzip in der Biologie. In: Biophysik, Bd. 1. Leipzig 1947.

— — u. M. DELBRÜCK: Über die Natur der Genmutationen und der Genstruktur. Nachr. Ges. Wiss. Göttingen, Kl. Biol. **1**, 190—245 (1935).

VAHARU, T.: Modification in ultra-violet-induced mutation frequency in *Neurospora crassa*. Genetics **46**, 247—256 (1961).

VIELMETTER, W., u. H. SCHUSTER: Die Basenspezifität bei der Induktion von Mutationen durch salpetrige Säure im Phagen *T 2*. Z. Naturforsch. **15**b, 304—311 (1960).

WACKER, A., H. DELLWEG u. E. LODEMANN: Strahlengenetische Veränderungen der Nucleinsäuren. Angew. Chem. **73**, 64—65 (1960).

— — u. D. WEINBLUM: Strahlenchemische Veränderung der Bakterien-Desoxyribonucleinsäure in vivo. Naturwissenschaften **47**, 477 (1961).

WALLACE, B., and T. DOBZHANSKY: Radiation, genes, and man. New York 1959.

WARSHAW, S. D.: Effect of ploidy in photoreactivation. Proc. Soc. exp. Biol. (N.Y.) **79**, 268—271 (1952).

WEATHERWAX, R. S., and O. E. LANDMAN: Ultraviolet light-induced mutation and deoxyribonucleic acid synthesis in *Escherichia coli*. J. Bact. **80**, 528—535 (1960).

WEINFURTNER, F., u. G. A. VOERKELIUS: Das Absterben von Hefen unter der Einwirkung von Noxen in Abhängigkeit vom Ploidiegrad. Z. Naturforsch. **10**b, 257—267 (1955).

WESTERGAARD, M.: Chemical mutagenesis in relation to the concept of the gene. Experientia (Basel) **13**, 224—234 (1957).

— Chemical mutagenesis as a tool in macromolecular genetics. Abh. dtsch. Akad. Wiss. Berlin, Kl. Medizin **1**, 30—40 (1960).

WILKIE, D.: The induction by monochromatic UV light of respiratory-deficient mutants in aerobic and anaerobic cultures of yeast. J. molec. Biol. **7**, 527—533 (1963).

WILSON, C. M.: Meiosis in *Allomyces*. Bull. Torrey bot. Club **79**, 139—160 (1952).

WINGE, Ö.: The segregation in the ascus of *Saccharomyces Ludwigii*. C. R. Lab. Carlsberg, Sér. physiol. **24**, 223—231 (1947).

—, and O. LAUSTSEN: *Saccharomyces Ludwigii*, a balanced heterozygote. C. R. Lab. Carlsberg, Sér. physiol. **22**, 357—370 (1939).

Mutation

WINKLER, U.: „Hot spots" oder „Brennpunkte" von Mutationsereignissen. Umschau **11**, 342—345 (1963).
WITKIN, E. M.: Time, temperature and protein synthesis: A study of ultraviolet-induced mutation in bacteria. Cold Spr. Harb. Symp. quant. Biol. **21**, 123—140 (1956).
— Post-irradiation metabolism and the timing of ultraviolet-induced mutations in bacteria. Proc. 10th Int. Congr. Genet. (Montreal) **1**, 280—299 (1959).
— Modification of mutagenesis initiated by ultraviolet light through posttreatment of bacteria with basic dyes. J. cell. comp. Physiol. **58** (Suppl. 1), 135—144 (1961).
WOLFF, S.: Chromosome aberrations. In: A. HOLLAENDER (edit.), Radiation protection and recovery, p. 157—174. Oxford-London-New York-Paris 1960.
— (edit.): Radiation-induced chromosome aberrations. New York and London 1963.
WOODWARD, V. W.: Mutation rates of several gene loci in *Neurospora*. Proc. nat. Acad. Sci. (Wash.) **42**, 752—758 (1956).
— J. R. DE ZEEUW, and A. M. SRB: The separation and isolation of particular biochemical mutants of *Neurospora* by differential germination of conidia, followed by filtration and selective plating. Proc. nat. Acad. Sci. (Wash.) **40**, 192—200 (1954).
ZAMENHOF, S.: The chemistry of heredity. Springfield 1959.
— Mutations. Symp. Amer. J. Med. **34**, 609—626 (1963).
—, and S. GREER: Heat as an agent producing high frequency of mutations and unstable genes in *Escherichia coli*. Nature (Lond.) **182**, 611—613 (1958).
ZELLE, M. R., J. E. OGG, and A. HOLLAENDER: Photoreactivation of induced mutation and inactivation of *Escherichia coli* exposed to various wave lengths of monochromatic ultraviolet radiation. J. Bact. **75**, 190—198 (1958).
ZETTERBERG, G.: The mutagenic effect of 8-ethoxycaffein, caffein and dimethylsulfate in the *Ophiostoma* back-mutation test. Hereditas (Lund) **46**, 279—311 (1960a).
— The mutagenic effect of N-nitroso-N-methylurethan in *Ophiostoma multiannulatum*. Exp. Cell Res. **20**, 659—661 (1960b).
— A specific and strong mutagenic effect of N-nitroso-N-methylurethan in *Ophiostoma*. Hereditas (Lund) **47**, 295—303 (1961).
— Genetic influence on the back-mutation rate in biochemical mutant strains of *Ophiostoma*. Exp. Cell Res. **27**, 560—569 (1962).
—, and N. FRIES: Spontaneous back-mutations in *Ophiostoma multiannulatum*. Hereditas (Lund) **44**, 556—558 (1958).
ZIMMER, K. G.: The development of quantum biology during the last decade. Acta radiol. (Stockh.) **46**, 595—602 (1956).
ZIRKLE, R. E., and C. A. TOBIAS: Effects of ploidy and linear energy transfer on radiobiological survival curves. Arch. Biochem. **48**, 282—306 (1953).

References

which have come to the authors' attention after conclusion of the German manuscript

A I, 1

BERLINER, M. D., and P. W. NEURATH: The band forming rhythm of *Neurospora mutants*. J. cell. comp. Physiol. **65**, 183—193 (1965a).
— — The rhythms of three clock mutants of *Ascobolus immersus*. Mycologia (N.Y.) **57**, 809—817 (1965b).
GREEN, G. J.: A color mutation, its inheritance, and the inheritance of pathogenicity in *Puccinia graminis* pers. Canad. J. Bot. **42**, 1653—1664 (1964).

Sussman, A. S., R. J. Lowry, and T. Durkee: Morphology and genetics of a periodic colonial mutant of *Neurospora crassa*. Amer. J. Bot. **51**, 243—252 (1964).

Yu-Sun, C. C. C.: Biochemical and morphological mutants of *Ascobolus immersus*. Genetics **50**, 987—998 (1964).

A I, 2

Metzenberg, R. L., M. S. Kappy, and R. W. Parson: Irreparable mutations and ethionine resistance in *Neurospora*. Science **145**, 1434—1435 (1964).

A I, 3

Apirion, D.: The two-way selection of mutants and revertants in respect to acetate utilization and resistance to fluoro-acetate in *Aspergillus nidulans*. Genet. Res. **6**, 317—329 (1965).

Balázs, O., u. J. Roppert: Auxanographisches System zur Ermittlung des Auxotrophiegrades sowie des Nährstoffbedarfes bei polyauxotrophen Mutanten von Mikroorganismen. Arch. Mikrobiol. **50**, 298—320 (1965).

Dee, J.: Genetic analysis of actidione-resistant mutants in the myxomycete *Physarum polycephalum* Schw. Genet. Res. **8**, 101—110 (1966).

Lilly, L. J.: An investigation of the suitability of the suppressors of *meth-1* in *Aspergillus nidulans* for the study of induced and spontaneous mutation. Mutation Res. **2**, 192—195 (1965).

Lingens, F., u. O. Oltmanns: Erzeugung und Untersuchung biochemischer Mangelmutanten von *Saccharomyces cerevisiae*. Z. Naturforsch. **19**b, 1058—1065 (1964).

Megnet, R.: A method for the selection of auxotrophic mutants of the yeast *Schizosaccharomyces pombe*. Experientia (Basel) **20**, 320—321 (1964).

— Alkoholdehydrogenasemutanten von *Schizosaccharomyces pombe*. Path. Microbiol. **28**, 50—57 (1965a).

— Screening of auxotrophic mutants of *Schizosaccharomyces pombe* with 2-deoxyglucose. Mutation Res. **2**, 328—331 (1965b).

Oltmanns, O., u. F. Lingens: Versuche zur Anreicherung von Hefe-Mangelmutanten, insbesondere mit Penicillin. Z. Naturforsch. **21**b, 266—273 (1966).

Warr, J. R., and J. A. Roper: Resistance to various inhibitors in *Aspergillus nidulans*. J. gen. Microbiol. **40**, 273—281 (1965).

Yu-Sun, C. C. C.: Biochemical and morphological mutants of *Ascobolus immersus*. Genetics **50**, 987—998 (1964).

A II, 1

Haefner, K., u. W. Laskowski: Zur Induktion prototropher *Saccharomyces*-Mutanten durch ultraviolettes Licht in Abhängigkeit von Dosis und Nachbehandlung. Z. Naturforsch. **18**b, 301—309 (1962).

— Zum Inaktivierungskriterium für Einzelzellen unter besonderer Berücksichtigung der Teilungsfähigkeit Röntgen- und UV-bestrahlter *Saccharomyces*-Zellen verschiedenen Ploidiegrades. Int. J. Radiat. Biol. **9**, 545—558 (1965).

Holliday, R.: Radiation sensitive mutants of *Ustilago maydis*. Mutation Res. **2**, 557—559 (1965).

James, A. P., M. M. MacNutt, and P. M. Morse: The influence of dose on the spectrum of radiation-induced mutants affecting a quantitative character in yeast. Genetics **52**, 21—29 (1965).

Käfer, E., and T. L. Chen: Translocations and recessive lethals in *Aspergillus* by ultra-violet light and gamma rays. Canad. J. Genet. Cytol. **6**, 249—254 (1964).

Kilbey, B. J., and F. J. de Serres: Quantitative und qualitative aspects of photoreactivation of premutational ultraviolet damage at the *ad-3* loci of *Neurospora crassa*. Mutation Res. **4**, 21—29 (1967).

Mutation

KIVI, E. I., and A. P. JAMES: The influence of environment on radiation-induced mutations affecting growth. Hereditas **48**, 247—263 (1962).

LASER, H.: Production by X-rays of petite colonies in yeast and their radio-sensitivity. Nature (Lond.) **203**, 314—315 (1964).

LASKOWSKI, W.: Der aα-Effekt, eine Korrelation zwischen Paarungstyp-konstitution und Strahlenresistenz bei Hefen. Zbl. Bakt. **184**, 251—258 (1962).

MOUSTACCHI, E.: Induction by physical and chemical agents of mutations for radioresistance in *Saccharomyces cerevisiae*. Mutation Res. **2**, 403—412 (1965).

MÜLLER, I., and A. P. JAMES: The influence of genetic background on the frequency and the direction of radiation-induced mutations affecting a quantitative character. Genetics **46**, 1721—1733 (1961).

SCHWARTZ, L. J., and J. F. STAUFFER: Three methods of assessing the mutagenic action of ultraviolet radiation on the fungus *Emericellopsis glabra*. Appl. Microbiol. **14**, 105—109 (1966).

SERRES, F. J. DE: Impaired complementation between non-allelic mutations in *Neurospora*. Symp. on Genes and Chromosomes — Structure and Function, Buenos Aires 1964.

WEBBER, B. B., and F. J. DE SERRES: Induction kinetics and genetic analysis of X-ray-induced mutations in the *ad-3* region of *Neurospora crassa*. Proc. nat. Acad. Sci. (Wash.) **53**, 430—437 (1965).

WITKIN, E. M.: Radiation-induced mutations and their repair. Science **152**, 1345—1353 (1966).

YAMASAKI, T., T. ITO, and Y. MATSUDAIRA: The whole- and fractional-colony mutation induced by soft X-rays in yeast. Japan. J. Genet. **39**, 147—150 (1964).

A II, 2

HAEFNER, K.: Zum Mechanismus der Entstehung auxotropher Zellen in heterozygoten *Saccharomyces*-Stämmen nach UV- und Röntgen-Bestrahlung. Z. Vererbungsl. **98**, 82—90 (1966).

HESLOT, H.: Les mécanismes moléculaires de la mutagène et la nature des mutations. Ann. Amélior. Plantes **15**, 111—157 (1965).

A II, 3

BACCHETTI, S., R. ELLI, R. FALCHETTI, F. MAURO, and A. SACCHI: Recovery from X-ray-induced sub-lethal damage in *Saccharomyces cerevisiae* cells of different ploidy. Int. J. Radiat. Biol. **10**, 213—221 (1966).

HESLOT, H.: Les mécanismes moléculaires de la mutagenèse et la nature des mutations. Ann. Amélior. Plantes **15**, 111—157 (1965).

KLINGMÜLLER, W.: Der Einfluß des Wassergehaltes auf Inaktivierung und Mutationsrate von Röntgen- und gammabestrahlten *Neurospora crassa*-Conidien. Z. Vererbungsl. **96**, 116—127 (1965).

SUGIMURA, T., and H. TANOOKA: Photoreactivation of hemeprotein lacking yeast. Biochim. biophys. Acta (Amst.) **136**, 154—155 (1967).

A II, 4

CALVORI, C., and G. MORPURGO: Analysis of induced mutations in *Aspergillus nidulans*. I. UV- and HNO_2-induced mutations. Mutation Res. **2**, 145—151 (1966).

HESLOT, H.: Les mécanismes moléculaires de la mutagenèse et la nature des mutations. Ann. Amélior. Plantes **15**, 111—157 (1965).

TAKAMORI, Y., E. R. LOCHMANN u. W. LASKOWSKI: Inaktivierungsversuche mit homozygoten Hefestämmen verschiedenen Ploidiegrades. IX. Nucleinsäuregehalt und Nucleinsäuresynthese nach Röntgenbestrahlung. Z. Naturforsch. **21**b, 960—967 (1966).

WITKIN, E. M.: Radiation-induced mutations and their repair. Science **152**, 1345—1353 (1966).

A III, 1

ABBONDANDOLO, A., and N. LOPRIENO: Forward mutation studies with
N-nitroso-N-methyl-urethane and N-nitroso-N-ethylurethane in *Schizo-saccharomyces pombe*. Mutation Res. **4**, 31—36 (1967).

BALL, C., and J. A. ROPER: Studies on the inhibition and mutation of
Aspergillus nidulans by acridines. Genet. Res. Camb. **7**, 207—221 (1966).

BROCKMAN, H. E., and W. GOBEN: Mutagenicity of a monofunctional alkyl-ating agent derivative of acridine in *Neurospora*. Science **147**, 750—751
(1965).

GUGLIELMINETTI, R., S. BONATTI, and N. LOPRIENO: The mutagenic acti-vity of N-nitroso-N-methylurethane and N-nitroso-N-ethylurethane in
Schizosaccharomyces pombe. Mutation Res. **3**, 152—157 (1966).

LINDEGREN, G., Y. L. HWANG, Y. OSHIMA, and C. C. LINDEGREN: Genetical
mutants induced by ethyl methanesulfonate in *Saccharomyces*. Canad.
J. Genet. Cytol. **7**, 491—499 (1965).

LINGENS, F., u. O. OLTMANNS: Über die mutagene Wirkung von 1-Nitroso-3-nitro-1-methyl-guanidin (NNMG) auf *Saccharomyces cerevisiae*. Z.
Naturforsch. **21 b**, 660—663 (1966).

LOPRIENO, N.: Cysteine protection against reversion to methionine inde-pendence induced by N-nitroso-N-methylurethane in *Schizosaccharo-myces pombe*. Mutation Res. **1**, 469—472 (1964).

— G. ZETTERBERG, R. GUGLIELMINETTI, and E. MICHEL: The lethal and
mutagenic effects of N-nitroso-N-methylurethane and N-nitroso-N-ethylurethane in *Colletotrichum coccodes*. Mutation Res. **1**, 37—44 (1964).

MARQUARDT, H., F. K. ZIMMERMANN u. R. SCHWAIER: Nitrosamide als
mutagene Agentien. Naturwissenschaften **50**, 625 (1963).

— U. v. LAER u. F. K. ZIMMERMANN: Das spontane Nitrosamid- und Nitrit-induzierte Mutationsmuster von 6 Adenin-Genloci der Hefe. Z. Ver-erbungsl. **98**, 1—9 (1966).

MEGNET, R.: Screening of auxotrophic mutants of *Schizosaccharomyces
pombe* with 2-deoxyglucose. Mutation Res. **2**, 328—331 (1965 b).

MORITA, T., and I. MIFUCHI: Effect of methylene blue on the action of
4-nitroquinoline N-oxide and acriflavine in inducing respiration-deficient
mutants of *Saccharomyces cerevisiae*. Jap. J. Microbiol. **9**, 123—129
(1965).

NASHED, N., and G. JABBUR: A genetic and functional characterization of
adenine mutants induced in yeast by 1-nitroso-imidazolidone-2 and
nitrous acid. Z. Vererbungsl. **98**, 106—110 (1966).

NASIM, A., and C. H. CLARKE: Nitrous acid-induced mosaicism in *Schizo-saccharomyces pombe*. Mutation Res. **2**, 395—402 (1965).

SCHWAIER, R., F. K. ZIMMERMANN, and U. v. LAER: The effect of tempera-ture on the mutation induction in yeast by N-alkylnitrosamides and
nitrous acid. Z. Vererbungsl. **97**, 72—74 (1965).

ZETTERBERG, G., and B. A. KIHLMAN: Production of mutations by strepto-nigrin in the ascomycete *Ophiostoma multiannulatum*. Mutation Res. **2**,
470—471 (1965).

ZIMMERMANN, F. K., u. R. SCHWAIER: Teilweise Reversion einer histidin-adenin-bedürftigen Einfachmutante bei *Saccharomyces cerevisiae*. Z. Ver-erbungsl. **94**, 253—260 (1963 a).

— — Eine ungewöhnliche Dosiswirkungs-Beziehung der N-Nitroso-N-Methylacetamid induzierten Mutationsraten bei *Saccharomyces cerevisiae*.
Z. Vererbungsl. **94**, 261—268 (1963 b).

— R. SCHWAIER, and U. v. LAER: The influence of pH on the mutagenicity
in yeast of N-methylnitrosamides and nitrous acid. Z. Vererbungsl. **97**,
68—71 (1965).

— — — Nitrous acid and alkylating nitrosamides: Mutation fixation in
Saccharomyces cerevisiae. Z. Vererbungsl. **98**, 152—166 (1966a).

— — — The effect of residual growth on the frequency of reverse muta-tions induced with nitrous acid and 1-nitroso-imidazolidone-2 in yeast.
Mutation Res. **3**, 171—173 (1966b).

Mutation

ZIMMERMANN, F. K., R. SCHWAIER, and U. v. LAER: The effect of temperature on the mutation fixation in yeast. Mutation Res. **3**, 90—92 (1966c).

A III, 2

BROCKMAN, H. E., and W. GOBEN: Mutagenicity of a monofunctional alkylating agent derivative of acridine in *Neurospora*. Science **147**, 750—751 (1965).
CALVORI, C., and G. MORPURGO: Analysis of induced mutations in *Aspergillus nidulans*. I. UV- and HNO₂-induced mutations. Mutation Res. **2**, 145—151 (1966).
HESLOT, H.: Les mécanismes moléculaires de la mutagenèse et la nature des mutations. Ann. Amélior. Plantes **15**, 111—157 (1965).
LINGENS, F.: Wirkungsmechanismus einiger chemischer Mutagene. Naunyn-Schmiedebergs Arch. exp. Path. Pharmak. **253**, 116—131 (1966).
LOPRIENO, N.: Differential response of *Schizosaccharomyces pombe* to ethylmethanesulfonate and methylmethanesulfonate. Mutation Res. **3**, 486—493 (1966).
MALLING, H. V.: Identification of the genetic alterations in nitrous acid-induced *ad-3* mutants of *Neurospora crassa*. Mutation Res. **2**, 320—327 (1965).
NASIM, A., and C. H. CLARKE: Nitrous acid-induced mosaicism in *Schizosaccharomyces pombe*. Mutation Res. **2**, 395—402 (1965).
SARACHEK, A.: Mutagenicity of cystosine for an adenine and arginine requiring strain of *Candida albicans*. Mycopathologia (Den Haag) **26**, 72—78 (1965).

A IV, 1

CALVORI, C., and G. MORPURGO: Analysis of induced mutations in *Aspergillus nidulans*. I. UV- and HNO₂-induced mutations. Mutation Res. **2**, 145—151 (1966).
CLARKE, C. H.: Mutagen specificity among reversions of ultraviolet-induced *adenine-1* mutants of *Schizosaccharomyces pombe*. Genet. Res. Camb. **6**, 433—441 (1965).
—, and N. LOPRIENO: The influence of genetic background on the induction of methionine reversions by di-epoxybutane in *Schizosaccharomyces pombe*. Microbial Gen. Bull. **22**, 11—12 (1965).
LOPRIENO, N.: Differential response of *Schizosaccharomyces pombe* to ethyl methanesulfonate and methyl methanesulfonate. Mutation Res. **3**, 486—493 (1966).
—, and C. H. CLARKE: Investigations on reversions to methionine independence induced by mutagens in *Schizosaccharomyces pombe*. Mutation Res. **2**, 312—319 (1965).
MARQUARDT, H., U. v. LAER u. F. K. ZIMMERMANN: Das spontane Nitrosamid- und Nitrit-induzierte Mutationsmuster von 6 Adenin-Genloci der Hefe. Z. Vererbungsl. **98**, 1—9 (1966).
SCHWAIER, R.: Vergleichende Mutationsversuche mit sieben Nitrosamiden im Rückmutationstest an Hefen. Z. Vererbungsl. **97**, 55—67 (1965).

A IV, 2

LINDEGREN, G., Y. L. HWANG, Y. OSHIMA, and C. C. LINDEGREN: Genetical mutants induced by ethyl methanesulfonate in *Saccharomyces*. Canad. J. Genet. Cytol. **7**, 491—499 (1965).
GILMORE, R. A., and R. K. MORTIMER: Super-suppressor mutations in *Saccharomyces cerevisiae*. J. molec. Biol. **20**, 307—311 (1966).
ISHIKAWA, T.: Studies on the mechanism of forward and reverse mutations in *Ustilago maydis*. Japan. J. Bot. **18**, 1—17 (1962).
MAGNI, G. E.: The origin of spontaneous mutations during meiosis. Proc. nat. Acad. Sci. (Wash.) **50**, 975—980 (1963).

MAGNI, G. E.: Origin and nature of spontaneous mutations in meiotic organisms. J. cell. comp. Physiol. **64**, 165—171 (1964).
— R. C. v. BORSTEL, and C. M. STEINBERG: Super-suppressors as addition-deletion mutations. J. molec. Biol. **16**, 568—570 (1966).
MORPURGO, G., and C. CALVORI: Variable frequency of back-mutation in different genotypes. Ann. Ist. Super. Sanità **2**, 429—430 (1966).
PASZEWSKI, A., and S. SURZYCKI: "Selfers" and high mutation rate during meiosis in *Ascobolus immersus*. Nature (Lond.) **204**, 809 (1964).
UPSHALL, A.: Somatically unstable mutants of *Aspergillus nidulans*. Nature (Lond.) **209**, 1113—1115 (1966).

B I

BAINBRIDGE, B. W., and J. A. ROPER: Observations on the effects of a chromosome duplication in *Aspergillus nidulans*. J. gen. Microbiol. **42**, 417—424 (1966).
KÄFER, E.: Origins of translocations in *Aspergillus nidulans*. Genetics **52**, 217—232 (1965).
—, and T. L. CHEN: Translocations and recessive lethals in *Aspergillus* by ultra-violet light and gamma rays. Canad. J. Genet. Cytol. **6**, 249—254 (1964).

C I

JAMES, A. T., and H. P. PAPAZIAN: Enumeration of quad types in diploids and tetraploids. Genetics **46**, 817—829 (1961).
SCHEDA, R.: Untersuchungen über die Maltose- und Glucosevergärung bei homozygoten Hefestämmen mit verschiedenen Genomzahlen. Arch. Mikrobiol. **45**, 65—100 (1963).

C II

CASSELTON, L.: The production and behavior of diploids of *Coprinus lagopus*. Genet. Res. Camb. **6**, 190—208 (1965).
—, and D. LEWIS: Compatibility and stability of diploids in *Coprinus lagopus*. Genet. Res. Camb. **8**, 61—72 (1966).
PARAG, Y., and B. NACHMAN: Diploidy in the tetrapolar heterothallic basidiomycete *Schizophyllum commune*. Heredity **21**, 151—154 (1966).

Function

The hereditary material is able not only to duplicate itself (replication; *autocatalytic* function, p. 127 ff.), but also to transfer the information stored within it *(heterocatalytic* function). Both functions depend upon the fact that DNA serves as a template for the synthesis of macromolecular substances from smaller building blocks. *By function in the narrow sense we mean here the heterocatalytic action of DNA, i.e. its capacity to produce a phenotype.*

The genetic information is translated from the pattern of nucleotides in the DNA, which determines the sequence of amino acids in proteins. The process is not direct, however; RNA serves as an intermediary. Proteins, mainly enzymes, are able to catalyze metabolic reactions by virtue of their structural specificity, which lead to the formation of substances needed for anabolic and catabolic processes. Mutational alterations of the genetic information, i.e. changes in the nucleotide sequence, are passed on precisely during replication and are expressed as proteins with modified structures.

This currently accepted concept of gene action was developed in exactly the reverse sequence. Through the biochemical analysis of deficiency mutants (p. 268), the latter were shown to have lost the capacity to synthesize particular substances (amino acids, vitamins, etc.) as a result of gene mutation. It was then discovered that the auxotrophy resulted from the absence of activity of one or more specific enzymes. Analysis of the enzymes in question was responsible for a further advance by showing that an enzyme was either not produced at all or was structurally so modified through gene mutation as to be inactive. Only then was the link between gene and enzyme sought.

Fungi, particularly the ascomycete, *Neurospora crassa*, served as material for these investigations. Later, bacteria such as *Escherichia coli* were also used, especially for the elucidation of information transfer.

In line with the title of the present volume, results obtained with fungi are the central feature of our treatment. The work based on bacteria will be discussed only insofar as it is necessary for the coherence of this presentation.

Literature survey of a general nature: KARLSON (1954), YANOFSKY and ST. LAWRENCE (1960), LEVINTHAL and DAVIDSON (1961), RILEY and PARDEE (1962), FINCHAM (1962a), EGELHAAF (1962), KAUDEWITZ (1962), PERUTZ (1962), WINKLER and KAPLAN (1963), DELBRÜCK (1963), WAGNER and MITCHELL (1964), JOLY (1964).

Cold Spring Harbor Symposium on Quantitative Biology, volume **26** (1961) and **28** (1963).

A. Transfer of genetic information

In spite of innumerable investigations the following account of genetic information transfer remains speculative in many details. It can thus be considered as only a working hypothesis which will certainly be modified with further work.

Literature survey: NOVELLI (1960), SUTTON (1960), GIERER (1961), ROSEN (1961), CRICK et al. (1961), LINDEGREN (1961), BROWN (1962), CHANTRENNE (1962), JUKES (1962), LANNI (1962), YCAS (1962), WOESE (1962), MELCHERS (1962), CAVALLIERI and ROSENBERG (1963), MEDVEDEN (1962), SPIEGELMAN (1963), WITTMANN (1963), WINKLER and KAPLAN (1963), TSUGITA and FRAENKEL-CONRAT (1962), VOLKIN (1963), WITTMANN and WITTMANN-LIEBOLD (1963, 1964), WITTMANN-LIEBOLD and WITTMANN (1963, 1964), CRICK (1963), BRESCH (1964), HAYES (1964), HOROWITZ and METZENBERG (1965).

Additional references are found in the volumes of McELROY and GLASS (1957), KASHA and PULLMAN (1962), TAYLOR (1963), and STUBBE (1964).

I. The role of nucleic acids in protein synthesis

1. **DNA → messenger RNA.** The transfer of information stored in the DNA molecule begins with the formation of an RNA molecule adjacent to the DNA. A template mechanism similar to that of DNA synthesis is probably involved (p. 131 f.). As a result of a specific accumulation of free nucleotides along the DNA, a RNA molecule with a nucleotide sequence complementary to the DNA template is built up (Fig. VI-1, left). Cytosine pairs with guanine as in the replication of DNA; however, the base complementary to adenine in RNA is not thymine, but uracil (p. 129). During in vivo RNA synthesis, probably only one strand of the DNA double helix is copied, while the other serves for the synthesis of a second DNA strand (GEIDUSCHEK, 1961; GEIDUSCHEK et al., 1961, 1962). Both strands may act as templates in vitro (CHAMBERLIN and BERG, 1962; WOOD and BERG, 1962). Since the information copied from the DNA is not transferred directly to the protein but is carried to distinct particles (ribosomes) within the cell, it is called messenger RNA.

The demonstration that DNA must be present for the formation of messenger RNA in vitro is important for this hypothesis. Enzymes which catalyze the reaction were first isolated from liver (WEISS and GLADSTONE, 1959), later also from bacteria (for literature see WITTMANN, 1963) and fungi (HUANG et al., 1960; SCHULMAN and BONNER, 1962).

The template function of DNA in the synthesis of messenger RNA can be deduced from the RNA formed in vitro; the latter pairs fully with the DNA to yield hybrid DNA-RNA double strands (SPIEGELMAN, 1961). The synthesis of RNA complementary to DNA was demonstrated in vivo in *Neurospora* (SCHULMAN and BONNER, 1962; WAINWRIGHT and McFARLANE, 1962).

2. **Messenger RNA → transfer RNA.** After separating from the DNA template, the messenger RNA moves to the ribosomes; protein synthesis takes place on the surface of the ribosomes.

Ribosomes are organelles of the cell composed of RNA and protein; they are found both in the cytoplasm and in the nucleus. Ribosomal RNA makes up the bulk of the RNA of the cell (about 80%).

According to the so called adaptor hypothesis, the amino acids do not attach directly to the messenger RNA to form protein, but are first brought into the correct position by other RNA molecules. This intermediary role is assigned to another type of RNA called transfer RNA (Fig. VI-1). Transfer RNA differs from messenger RNA in both structure and function.

Analysis has shown that the molecules of transfer RNA are much smaller than those of other types of RNA, the former being approximately 70—90 nucleotides long. The base sequence is still unknown to a large extent. It has been shown, however, that the sequence for the last three bases, namely, cytosine-cytosine-adenine, is the same for all types of transfer RNA molecules. On the basis of other work the molecules are assumed to take the form of closed double helices (Fig. VI-1).

The molecules of transfer RNA take positions next to one another in such a way that a specific part is always attached to the messenger

RNA serving as a template (Fig. VI-1, right). Three consecutive nucleotides of messenger RNA are believed to be responsible for binding a specific transfer RNA molecule; thus an exact transfer of the genetic information from messenger to transfer RNA is possible (p. 345 f.).

The attachment of transfer RNA to messenger RNA occurs in part through hydrogen bonds between complementary bases. The attachment points are assumed to be groups (triplets) of three consecutive nucleotides, because with less than three an unequivocal translation of the genetic code (p. 344 ff.) from the four-base language of nucleic acids to the 20-element (amino acids) language of the polypeptide is not assured.

Nevertheless, it is difficult to understand why a transfer RNA molecule should consist of 80 nucleotides if coding for an amino acid only requires a sequence of three. MEDVEDEN (1962) has published a model to explain this phenomenon (for details see his paper).

3. **Transfer RNA → amino acids.** At least one specific kind of transfer RNA occurs for each of the 20 amino acids; these bring the amino acids to their particular positions on the messenger RNA. A certain site on each transfer RNA molecule is assumed to interact with an enzyme that determines which amino acid is attached to the transfer RNA molecule.

The binding of an amino acid to its transfer RNA occurs in two steps. The first step consists of an activation of amino acids in the presence of an enzyme. In the second step the activated amino acid is bound to the transfer RNA.

Because each of the different molecules of transfer RNA attaches to the messenger RNA by only one end, the amino acid which reflects the nucleotide sequence of the messenger RNA and therefore that of the DNA is found at the other end (Fig. VI-1). Each amino acid is determined by at least three bases, since the attachment of a transfer RNA molecule requires a nucleotide triplet of the messenger RNA.

4. **Amino acids → protein.** The final steps in protein synthesis remain largely unknown. When a transfer RNA molecule has brought an amino acid into the correct position, the latter is joined to those already positioned by a peptide bond. Subsequently the transfer RNA is released from the messenger RNA and "its" amino acid (Fig. VI-1, right). Such successive linkage to a polypeptide chain is catalyzed enzymatically. How many enzymes are involved, the mechanism of bond formation, and the processes by which the finished polypeptide chain separates from the ribosome all remain uncertain.

Analysis of protein structure has shown that the number and sequence of amino acids for any polypeptide is constant. The amino acid sequence of a protein is known as its *primary* structure.

A particular polypeptide generally consists of several hundred amino acids. The exact amino acid sequence has been determined in a number of cases, e.g. insulin (SANGER and SMITH, 1957), ribonuclease of mammals (HIRS et al., 1960), hemoglobin (BRAUNITZER et al., 1961) and tobacco mosaic protein (ANDERER et al., 1960).

Polypeptide chains join together to form proteins by a mechanism as yet unknown. The number and kinds of polypeptides are characteristic for each kind of protein. They also vary in different organisms.

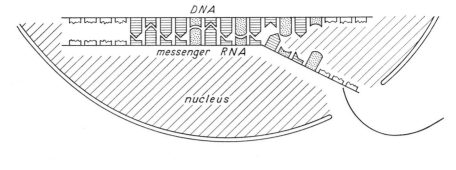

$\mathbb{P}$ *uracil*

$\mathbb{K}$ *thymine* $\text{IIIIII}\!\!\triangleright$ *adenine*

$\mathbb{X}$ *cytosine* $\text{E}\!\!\cdots\!\!\text{D}$ *guanine*

Fig. VI-1. Transfer of genetic information: through the mediation of messenger and transfer RNA, the nucleotide sequence of DNA (left) specifies the amino acid sequence in the protein (right). The symbols used for the bases of DNA and RNA correspond to those in Fig. III-2. The base uracil,

For example, in the ribonuclease of mammals and the tobacco mosaic protein there is only one polypeptide per molecule (HIRS et al., 1960; ANDERER et al., 1960), while in the phosphorylase of muscle probably a series of identical chains occur (MADSEN and GURD, 1956); in hemoglobin two kinds of polypeptides are found (RHINESMITH et al., 1958).

Native proteins do not consist of extended polypeptide chains, but each polypeptide assumes a helical configuration, to a greater or lesser extent. Hydrogen bonds stabilize the helical portions. This feature represents the *secondary* structure of the protein. The *tertiary* structure involves the folding of the polypeptide chain. In certain proteins, the proper folding is maintained through disulfide linkages which form between specific amino acid residues (cysteine) in different parts of the chain. Such monomeric tertiary structures may associate polymerically to give what is called a *quaternary* structure. The folded polypeptide chains join with each other in a specific way; how they are held together is not clearly understood.

II. The genetic code

As we have seen, the sequence of amino acids in a protein is determined by the sequence of the nucleotides in the DNA. We must now take up the question of how the information coded in the nucleotide sequence of DNA and messenger RNA is translated into the amino acid sequence

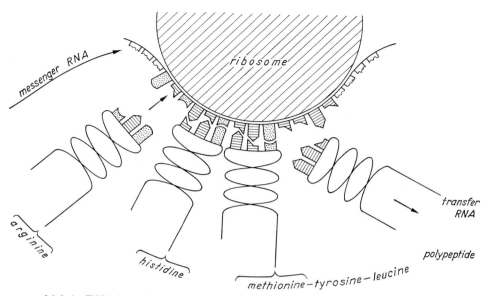

which in RNA is equivalent to thymine in DNA, is represented by a symbol similar to thymine (see also Fig. III-1, IV-24, and V-6). Further explanation in text. (Adapted from NULTSCH, 1964)

of the protein. The coding problem arises from the fact that the nucleic acid code of four letters (four bases) must be translated into a protein code of 20 elements (amino acids). *The correlation between the four nucleotides of DNA or messenger RNA and the 20 amino acids of proteins is designated the genetic code.* This is established through an adaptor which is probably identical with transfer RNA (see model in Fig. VI-1); the adaptor carries a particular amino acid on one end and a specific nucleotide grouping on the other. This is complementary to a nucleotide grouping in the messenger RNA, and the latter is in turn complementary to a segment of DNA. Thus there is an exact correspondence between a specific group of nucleotides of DNA and a specific amino acid.

"Breaking the code", i.e. identifying nucleotide sequences which correspond to amino acid sequences, has been studied intensively in numerous laboratories in recent years. Essentially two approaches have been used: 1. Determining the effect of chemically induced mutations on amino acid sequence of proteins (in phage *T 4*, tobacco mosaic virus, and tryptophan synthetase of *E. coli*). 2. Studying the synthesis of protein in a cell-free system with the aid of artificial messenger RNA, the base composition of which is known.

1. **The nucleotide triplet as the coding unit (codon).** Since 4 nucleotide "letters" are opposed to 20 amino acid "letters", it is unlikely that an amino acid is determined by a single nucleotide; two or more such elements must be involved in the coding of an amino acid. *A group of nucleotides which determine an amino acid is called a coding unit or*

345

codon. If one assumes that the codons are all of uniform length, then at least three nucleotides per codon are necessary. Two nucleotides are insufficient, since the four bases in combinations of two give only $4^2 = 16$ combinations, i.e. a maximum of 16 different amino acids can be coded by a doublet code. In contrast, with three nucleotides per codon (triplet) $4^3 = 64$ different combinations are possible, if different arrangements of the bases within a triplet are assumed to represent different codons (e.g. AGC differs from GAC). Experimental evidence for the triplet nature of the codon comes from investigations on proflavine-induced mutants of phage *T4* (CRICK et al., 1961) and nitrite-induced mutants of tobacco mosaic (WITTMANN, 1961, 1962).

In contrast to the fairly clear ideas on the nucleotide composition of the codons (MATTHAEI et al., 1962; SPEYER et al., 1962; GARDNER et al., 1962; WAHBA et al., 1962), the identification of sequence of nucleotides within the triplets has been achieved during the past years. Trinucleotides with defined sequences which affect the specific adhesion of amino acids to ribosomes, have been used in in-vitro experiments (NIRENBERG et al., 1966; MATTHAEI et al., 1966). Using this and other methods (i.e. OCHOA, 1963; KHORANA et al., 1966) it was possible to establish a *"code-dictionary"* for the 20 amino acids.

2. Reading of the code. The nucleotides of messenger RNA correspond to a continuous series of letters in the following way (compare Fig. VI-1):

.... G U C A U C U A G U U A U .....

Such a nucleotide sequence can be read theoretically by the transfer RNA (in triplets) in various ways:

1. Successively from a fixed starting point at one end of the RNA molecule (e.g. from left to right) without overlap of adjacent triplets and without spaces between each set of triplets ("comma-free") as in Fig. VI-1:

. ... GUC AUC UAG UUA U.. ... .

2. Beginning at any point on the RNA molecule (e.g. starting from the second nucleotide in one case, and from the third in the next), but otherwise as in (1):

. ... UCA UCU AGU UAU ... .

or　　　　　. ... ... CAU CUA GUU AU. ... .

In this case the information may have more than a single interpretation.

3. With overlapping of adjacent triplets, but otherwise as in (1): single overlap

. ... G̲UC C̲AU U̲CU U̲AG G̲UU U̲AU U̲.. ... ..

double overlap

. ... G̲UC U̲CA C̲AU A̲UC U̲CU C̲UA ... .

4. With spaces between adjacent triplets (i.e. "with commas"); e.g. each codon may be separated from its neighbor by an A as a comma, but otherwise as in (1):

. A ... **A** ... **A** GUC **A** UCU **A** GUU **A** U.. **A** ... **A** ..

Other, more complex codes are possible. Nevertheless, *all experimental evidence points to a relatively simple genetic code* (see case 1 in Fig. VI-1). Experiments on phage *T4* protein, on chemically-induced mutants of tobacco mosaic, and the study of polynucleotide-induced protein synthesis in cell-free systems warrant the following conclusions:

1. *The genetic information of the nucleic acids is read from a fixed point* (BISHOP et al., 1960; DINTZIS, 1961; CHAMPE and BENZER, 1962; WAHBA et al., 1962; OCHOA, 1963; KANOSUEOKA and SPIEGELMAN, 1962). This point is determined by the triplet AUG (CAPECCHI, 1966; MARCKER, 1965; see also *"degenerate nature of the code"*).

2. *Codons do not overlap.* A particular nucleotide belongs to only one codon (BRENNER, 1957; YANOFSKY, 1960; WITTMANN, 1961, 1962, 1963; TSUGITA and FRAENKEL-CONRAT, 1962; HENNING and YANOFSKY, 1962b; YANOFSKY et al., 1963; INGRAM, 1963).

3. *The code is "comma-free"*, i.e. adjacent codons are not separated by nucleotides which do not belong to a triplet (CRICK et al., 1961). The individual codons are identified by counting the nucleotides in groups of three from the beginning of the molecule.

3. Degenerate nature of the code. With a triplet code coding 20 amino acids 64 different triplets are available. This leads to two possible relationships between triplets and amino acids:

1. Each amino acid corresponds to a single triplet, i.e. only twenty of the 64 triplets determine amino acids, the remainder being "nonsense" triplets.

2. More than a single triplet may code for many or all of the amino acids; there are more than 20 triplets which "make sense".

The latter case is referred to as a degenerate, the former, a nondegenerate code. Numerous experimental results support the idea of a degenerate code (SUEOKA, 1961; CRICK et al., 1961; WITTMANN, 1961, 1962; MATTHAEI et al., 1962; SPEYER et al., 1962; SUEOKA and YAMANE, 1962; WEISBLUM et al., 1962; BENZER and CHAMPE, 1961, 1962; JONES and NIRENBERG, 1962; NIRENBERG and JONES, 1963; GARDNER et al., 1962, NIRENBERG et al., 1966; MATTHAEI et al., 1966; KHORANA et al., 1966). The experiments of the three groups mentioned last show further, that there are only two triplets (UAA, UAG) which do not code for any amino acid; they determine the end of the amino acid chain. The codon AUG is responsible for the beginning of the chain, but also codes for methionine. The degeneration of the code exhibits certain regularities; for instance, the four codons which differ in the last nucleotide code for the same amino acid.

4. **Universality of the code.** The question, "Is the genetic code universal, i.e. is it identical in all organisms?" is of particular interest. Identity of the code means that the same codons specify the same amino acids in all organisms. The results of experiments with phage in which nonsense codons in one bacterium can become "sense" codons in another

("ambivalent" mutants: BENZER and CHAMPE, 1961, 1962) argue against the universality of the code. We may broaden the definition of an universal code, however, to mean any code in which a codon in one organism does not code a different amino acid in another.

A number of results argues for such universality:

1. The relationship between triplets and amino acids found in the analysis of tobacco mosaic mutants are consistent with the results obtained with cell-free bacterial systems (WITTMANN, 1962; MATTHAEI et al., 1962; SPEYER et al., 1962).

2. By adding RNA from phage *f2* to bacterial extracts, *f2* protein is produced in vitro (ZINDER, 1963).

3. In vitro experiments with synthetic polynucleotides in cell-free bacterial and mammalian systems lead to essentially the same results (ARNSTEIN et al., 1962; GRIFFIN and O'NEAL, 1962; MAXWELL, 1962; WEINSTEIN and SCHECHTER, 1962).

4. In a cell-free system of rabbit reticulocytes a protein corresponding to normal rabbit hemoglobin can be produced with amino acid-charged transfer RNA from *E. coli* (EHRENSTEIN and LIPMAN, 1961).

5. Activating enzymes of an organism can load the transfer RNA of another organism with specific amino acids in certain cases (e.g. arginine to transfer RNA of *E. coli* with enzymes from *E. coli* and rabbit liver: BENZER and WEISBLUM, 1961).

6. If cells of *Bacillus subtilis* are infected with a DNA preparation of vaccinia virus and incubated, they will yield infectious viruses when lysed artificially. This result permits the conclusion that the DNA code of viruses can be read not only in the cells of higher animals (the normal case) but also in bacteria (ABEL and TRAUTNER, 1964).

Since the guanine-cytosine fraction of the DNA varies widely in different organisms (from about 35 to 75%) a universal code is possible only if certain amino acids have numerous synonymous codons. The different synonyms for the same amino acid must thus vary in frequency in different organisms (SUEOKA, 1961).

Summary

1. The nucleotide sequence of DNA determines the amino acid sequence of protein through the intermediary of RNA. Messenger RNA receives the genetic information from DNA by a template mechanism. The messenger carries the information to the ribosomes, which are the sites of protein synthesis. At the ribosomes, molecules of another type of RNA (transfer RNA) become attached in sequence to the messenger RNA; each molecule of transfer RNA is bound at one end to three nucleotides of the messenger RNA and at the other end to a specific amino acid. In this way a polypeptide chain is formed which reflects the base sequence of the messenger RNA and the DNA.

2. The genetic code, i.e. the correlation of the four nucleotides of the nucleic acid with the 20 amino acids of protein has already been determined to some extent, although approaches to the problem have only been worked out in recent years. The following assumptions can be made on the basis of results with viruses and bacteria: The genetic information, when transferred to the messenger RNA, is read continuously starting at one end of the molecule. Each three consecutive nucleotides of messenger RNA is read by a specific transfer RNA molecule. Thus the nucleotide triplet (coding unit or codon) specifies the amino acid which is attached to the other end of the corresponding transfer RNA molecule. The codons of messenger RNA lie adjacent to one another (without commas) and do not overlap. The code is degenerate, i.e. some or all amino acids may be coded by more than one triplet. The code is assumed to be universal; a given codon does not determine different amino acids in different organisms.

B. Genes and biochemical reactions

The preceding discussion has shown that the mechanism by which genetic information specifies protein molecules was completely unknown until a few years ago. On the other hand, the relation between specific genes and particular metabolic reactions had already been recognized at the beginning of the century (WHELDALE, 1903; GARROD, 1909).

WHELDALE utilized *Antirrhinum* as experimental material, because its flower pigments were easily analyzed and thus well suited for biochemical genetic studies. He found that the formation of anthocyanin depended upon individual genes. GARROD was concerned with the study of alkaptonuria in humans. He assumed that this disease resulted from the loss of activity of an enzyme which catalyzes a specific biochemical reaction in normal individuals. He attributed the absence of the enzyme to mutation of a single gene.

In spite of the simplicity of the explanation for alkaptonuria proposed by GARROD (1909, 1923) his conception of a gene-dependent loss of enzymatic activity with the resultant block of a reaction had relatively little impact on genetic thought of his period. The problem was not attacked again until years later when DANNEEL (1938) investigated melanin formation in different races of rabbits and found that the concentration of enzyme responsible for melanin production in hair-forming cells is gene-dependent. *The investigations carried out independently by the* KÜHN-BUTENANDT *and the* EPHRUSSI-BEADLE *groups provided the recent and profound insight into the manner in which the hereditary factors act.* Comparative genetic and biochemical studies on eye pigment formation in the flour moth *Ephestia* and in the fruit fly *Drosophila* were in agreement that *genes affect specific metabolic steps by determining enzymes* (BUTENANDT et al., 1940, 1942, 1943, 1949, 1951; KÜHN, 1941,

1948; BUTENANDT and HALLMANN, 1950; BEADLE and EPHRUSSI, 1936; EPHRUSSI, 1942). Furthermore, discovery of gene-determined reaction sequences revealed a pathway for tryptophan breakdown which had a significance beyond the mere study of insect pigmentation.

More precise ideas about the mechanism of gene action were not forthcoming, however, until TATUM and BEADLE (1942) recognized the advantages of *Neurospora crassa*, particularly in comparison with insects. They found that in one-gene mutants of the fungus, a single step in a biochemical synthetic pathway was blocked. Since there was no doubt about the enzymatic control of metabolic reactions, such results led to the *one gene-one enzyme hypothesis* (BEADLE, 1945a, b). The most lucid statement of this hypothesis was formulated by HOROWITZ (1950) after reviewing critically all pertinent experimental data from *N. crassa* as well as other fungi: *"a large class of genes exists in which each gene controls the synthesis or activity of but a single enzyme"*. The hypothesis, which can be considered the culmination of the investigations initiated by WHELDALE and GARROD, proved exceptionally valuable as a basis for further investigations of gene function because of its elegant simplicity. In spite of numerous objections which ultimately led to a more precise formulation of the hypothesis (p. 406f.), it can be considered today as essentially confirmed. It should be pointed out that at the time the hypothesis was conceived, the facts concerning the transfer of genetic information discussed in the preceding section were not yet known. The one gene-one enzyme hypothesis is thus to be considered as the first important step in the elucidation of gene action. The experimental data which led to its formulation will be discussed below.

Literature: TATUM and BEADLE (1942), BEADLE (1945a, b, c, 1948, 1955, 1956, 1957, 1959a, b, 1960a, b, 1961), HOROWITZ et al. (1945), BONNER (1946b, 1951, 1956), TATUM (1949, 1959), HOROWITZ (1950, 1951), CAMPBELL (1954), ADELBERG (1955a), AMES (1955), BLACK and WRIGHT (1955a), DAVIS (1955), EHRENSVÄRD (1955), MCELROY and GLASS (1955), RATNER (1955), SAKAMI (1955), VOGEL (1955), WORK (1955), YANOFSKY (1955), VOGEL and BONNER (1958), GREENBERG (1960, 1961), ABRAMS (1961), HELLMANN and LINGENS (1961).

I. Mutants with physiological defects (nutritional deficiency mutants)

The mutant strains of Neurospora crassa used for biochemical analysis of gene function have lost the capacity to synthesize certain substances essential to life (nutritional deficiency mutants). The survival of such auxotrophs becomes possible by adding to the minimal medium the substances which the mutant is unable to synthesize.

The methods which have been developed for the production, isolation, and characterization of nutritional deficiency mutants have already been described in detail in the chapter on mutation (p. 270ff.).

Mutants with physiological defects (nutritional deficiency mutants)

Table VI-1. *Fungi in which deficiency mutants have been produced (with the exception of N. crassa; see Table VI-2)*

object	reference
Ascobolus immersus	YU-SUN, 1964
Aspergillus nidulans	PONTECORVO, 1953
Cochliobolus sativus	TINLINE, 1962
Coprinus lagopus	LEWIS, 1961; DAY, 1960, 1963
Coprinus radiatus	CABET et al., 1962
Glomerella cingulata	MARKERT, 1952; WHEELER, 1956
Ophiostoma multiannulatum	FRIES, 1947; FRIES and KIHLMAN, 1948
Penicillium chrysogenum	BONNER, 1946a
Penicillium notatum	BONNER, 1946a
Penicillium expansum	BARRON, 1962
Podospora anserina	PERHAM, unpublished
Saccharomyces cerevisiae	LINDEGREN, 1949; POMPER and BURKHOLDER, 1949
Schizophyllum commune	RAPER and MILES, 1958; ELLINGBOE and RAPER, 1962
Schizosaccharomyces pombe	HESLOT, 1960, 1962
Sordaria fimicola	EL-ANI et al., 1961; EL-ANI, 1964
Torulopsis utilis	EHRENSVÄRD et al., 1947; 1951; STRASSMAN and WEINHOUSE, 1953; STRASSMAN et al., 1956
Ustilago maydis	PERKINS, 1949; HOLLIDAY, 1961
Venturia inaequalis	LAMEY et al., 1956

They involve primarily selection for particular kinds of auxotrophs. With such techniques, *Neurospora* mutants were found with nutritional defects for almost all amino acids, as well as vitamins, purines and pyrimidines (Table VI-2). Nutritional deficiency mutants have been produced in other fungi (Table VI-1) and bacteria in similar ways. By backcrossing the mutant with the wild type and analyzing the progeny genetically, most mutants were shown to differ from the wild type by a single gene.

The realization that many independently isolated mutants are not identical even though they are auxotrophic for the same end product of a synthetic pathway was a significant discovery based on the following observations:

1. Linkage studies showed that genes responsible for a particular kind of auxotrophy frequently occur at different loci in the genome, often in different chromosomes.

2. Complementation tests showed in many cases that defects of two mutants may be alleviated in a heterokaryon composed of the two mutants (p. 386f.).

3. Growth tests led to the discovery that intermediates of a synthetic pathway may or may not be metabolized by a given mutant in place of the end product in such a way as to insure survival.

Thus the synthesis of an amino acid or a vitamin is assumed to be controlled not by a single gene, but by a series of genes which affect different steps in the biosynthetic pathway through the formation of specific enzymes.

II. Blocks in Biosyntheses

1. Monoauxotrophic mutants

The discussion here will be introduced with a hypothetical example (Fig. VI-2; see also Fig. VI-5). Compound E (e.g. a specific amino acid) is synthesized from an unknown precursor through the intermediates A, B, C, and D. The enzymes α, β, γ, δ, ε are required for the individual synthetic steps as catalysts. Their formation results from the specific action of the genes a^+, b^+, c^+, d^+, and e^+. The gene c^+ is so altered through mutation $(c^+ \rightarrow c)$ that the enzyme γ is no longer produced, or its activity sharply decreased. As a result, B can no longer be transformed into C. The biosynthetic pathway is blocked

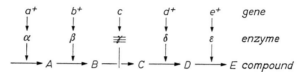

Fig. VI-2. Diagram of the action of genes in an unbranched biosynthetic pathway. (For explanation see text)

between B and C. Since the end product E can no longer be formed, the mutation is auxotrophic for E. Such a mutation is monoauxotrophic.

In the above example a sequence of events is inferred from the known mutation $c^+ \rightarrow c$. In an experiment, however, only the end result of the reaction sequence is known, namely the absence of compound E. It is relatively easy to determine through genetic analysis whether or not this defect involves a gene mutation. For the time being, nevertheless, the intermediate steps of the reaction sequence are unknown. Neither the enzyme, whose activity has been lost by mutation, nor the point where the synthetic pathway is interrupted, nor the intermediate products preceding or following the block are known. Therefore, we wish to indicate the experimental procedures which have been developed to elucidate such a gene-controlled biosynthetic pathway.

1. *Accumulation of intermediates before the block in the reaction sequence.* If a biosynthetic pathway is intact up to the point of the block caused by a mutation (between B and C in Fig. VI-2), an accumulation of the intermediate compound directly before the block occurs, since it continues to be produced but is not metabolized further. In the wild type the intermediates are generally not detectable because of their low concentration and thus cannot be easily isolated. On the other hand, in mutant strains which are provided with the end product of the blocked pathway (E) the accumulation may reach such a high level that the intermediates may be isolated from the cells or the culture medium and identified. *The accumulation of a particular*

substance indicates that the block in the reaction sequence occurs imme-diately following its synthesis. A chemical analysis allows identification of the intermediate.

Accumulated substances isolated from the culture media do not always prove to be true intermediates of a synthetic pathway. For example, in *N. crassa* the phosphate-free derivatives, rather than the phosphorylated precursors of histidine were found (AMES et al., 1953; AMES and MITCHELL, 1955; compare Fig. VI-5 and Table VI-2, No. 1 f).

On the other hand, complications may arise through the formation of labile intermediates prior to the block. These may be transformed into compounds which allow the completion of a synthetic pathway in spite of a genetic block. Phenylalanine synthesis is an example involving the precursor stages of shikimic and phenylpyruvic acid (Fig. VI-6 and Table VI-2, No. 1 g). If the synthetic pathway is blocked between the two intermedi-ates through mutation, the mutant is still capable of producing phenyl-pyruvic acid. However, the labile intermediate, prephenic acid could be demonstrated in such mutants; at pH ≤ 6 it is spontaneously converted into phenylpyruvic acid. Thus, these mutants are able to circumvent the genetic block.

2. *Growth on intermediates of the pathway subsequent to the block.* If a nutritional mutant is provided a substance which is normally produced as an intermediate or end product in the wild type subsequent to the block (C, D, or E in Fig. VI-2), the mutant can overcome the genetic block. The addition of the intermediates (A and B) which are formed in the pathway prior to the block does not permit growth. *The fact that a mutant grows after the addition of a specific compound to the medium shows that the synthetic pathway must be interrupted prior to the formation of this compound.* The sequence of intermediates as well as the effects of the genes involved can be determined when a series of different mutants which require the same end product of a reaction sequence are used in the growth experiments.

In the above example (Fig. VI-2) the substance D accumulating in mutant *e* supports the growth of the mutants. The intermediate C, B, or A isolated from mutants *d, c,* or *b* allow growth only of those organisms blocked in a step preceding the formation of the accumulated substance. Through the reactions of various mutants it becomes possible to determine the order of steps in the synthetic pathway without actually knowing the inter-mediates.

Experiments with biochemical mutants are often complicated by the fact that accumulated materials isolated from culture media do not restore the viability of the mutants as expected. The substances which have been isolated are not true intermediates in these cases, but rather chemically altered secondary products which, under experimental condi-tions, are not reconverted into the original compounds.

Synthetic compounds can be substituted in growth experiments for the accumulated substances from mutants, if the specific intermediates are known. Moreover the use of known compounds may facilitate the identification of intermediates of a pathway.

DAVIS (1951) used polyauxotrophic mutants (p. 354f.) of bacteria which grew only with the addition of phenylalanine, tyrosine, tryptophan, p-amino-benzoic acid and p-hydroxybenzoic acid. Of 55 substances tested, only one could replace the mixture of compounds listed above, namely, shikimic acid. This acid was later shown to accumulate in cultures of bacterial mutants (DAVIS and MINGIOLI, 1953) and a mutant of *Saccharomyces* (LINGENS and HELLMANN, 1958). It was also shown in *N. crassa* with growth tests that shikimic acid is a precursor of aromatic amino acids (Fig. VI-6 and Table VI-2, No. 1 g) (TATUM and PERKINS, 1950; TATUM, 1951; TATUM et al., 1954).

3. *Absence or inactivation of enzymes at the block*. As explained above, a block in a synthetic pathway occurs if an enzyme which catalyzes a specific reaction in the wild type is no longer produced or if its activity is greatly reduced by mutation. In certain cases it is possible to identify the particular intermediate and to determine where the block occurs through the correlation between enzyme and genetic block. The absence or inactivation of the wild type enzyme in the mutant indicates that *the substrate of the enzyme is the intermediate compound which is formed immediately before the block in the synthetic pathway*.

4. *Labelling of intermediates*. The methods described for elucidating biosynthetic pathways are supplemented by the use of radioactive compounds. *Labelling substances with specific isotopes enables determination of the fate of definite atoms or atomic groupings from step to step*, if the labelled compound is incorporated by the organism as an intermediate in the reaction sequence. In many cases the initial clue to the transformation of intermediates in the pathway was obtained in this way (description of methods and literature in WEYGAND, 1949; WEYGAND and SIMON, 1955; ARANOFF, 1957; BRODA, 1958).

For example, with C^{14} incorporated in the carboxyl group of anthranilic acid in *N. crassa*, the labelled carbon was shown to be split off prior to the formation of indole (NYC et al., 1949). Experiments with N^{15}-labelled anthranilic acid showed, on the other hand, that the nitrogen of the anthranilic acid becomes part of the indole ring (PARTRIDGE et al., 1952; Table VI-2, No. 1 g).

2. Polyauxotrophic mutants

Some nutritional mutants are viable only if the culture medium is supplemented with more than one substance. Such cases are often the result of single, rather than multiple gene mutations; these are known as simple polyauxotrophs and can generally be attributed to two causes:

1. *A genetic block prior to branching of the synthetic pathway* (between B and C in Fig. VI-3, see also Fig. VI-6). In such a case all steps in the pathway subsequent to the block are inoperative and formation of the end products (E and G) is prevented. The mutant will grow only if the medium is supplemented with either the substance in the pathway between the block and the point of branching (C in Fig. VI-3) or with an intermediate or the end product of each of the branches of the pathway (D or E as well as F or G).

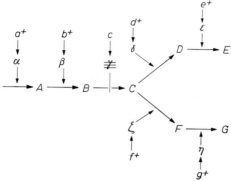

Fig. VI-3. Diagram of gene action in a branched synthetic pathway (explanation in text)

Branched pathways are known in the synthesis of amino acids. Glutamic acid is the precursor for both proline and arginine (Table VI-2, No. 1a). Similarly, homoserine is one of a number of precursors common to methionine and isoleucine (Table VI-2, No. 1b). Phenylalanine, tyrosine, tryptophan, and p-aminobenzoic acid likewise arise from a series of common precursors (among others, shikimic acid; Fig. VI-6, Table VI-2, No. 1g). Mutants with a genetic block in the synthetic pathway before the point of branching demand the addition of all four end products or the corresponding intermediates of the reaction series subsequent to the point of branching (TATUM, 1949, 1951; TATUM et al., 1950, 1954; TATUM and PERKINS, 1950).

2. *A simultaneous block in two separate synthetic pathways because of the absence or inactivation of a single enzyme* (between B_2 and C_2 as well as between B_1 and C_1 in Fig. VI-4; see also Fig. VI-7). Such a case always involves synthetic pathways in which some of the intermediates undergo the same chemical change. The formation of end products (E_1 and E_2) is prevented as a result of the multiple block. Such polyauxotrophic mutants will grow only if intermediates or end products of each of the pathways to the block are added to the medium (C_2, D_2, or E_2 as well as C_1, D_1, or E_1).

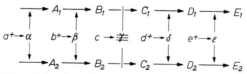

Fig. VI-4. Diagram of the simultaneous action of genes in separate synthetic pathways (explanation in text)

For example, in *Neurospora crassa* polyauxotrophic mutants have been found which require isoleucine as well as valine. Separate pathways which are so similar in their final reactions that the latter are catalyzed by the same enzymes are involved here (Fig. VI-7, Table VI-2, No. 1b and d; MEYERS and ADELBERG, 1954; RADHAKRISHNAN et al., 1960; WAGNER et al., 1960; BERNSTEIN and MILLER, 1961).

23*

III. Biosynthetic pathways

With the discovery that genes control the individual steps in metabolic sequences (p. 349 f.), an increasing number of syntheses of amino acids, vitamins, pyrimidines, purines, and other compounds were worked out. It has become apparent that the individual pathways are not only linear, but may be branched, or may converge to a greater or lesser degree. We have selected for illustration in Fig. VI-5 through VI-8 four typical

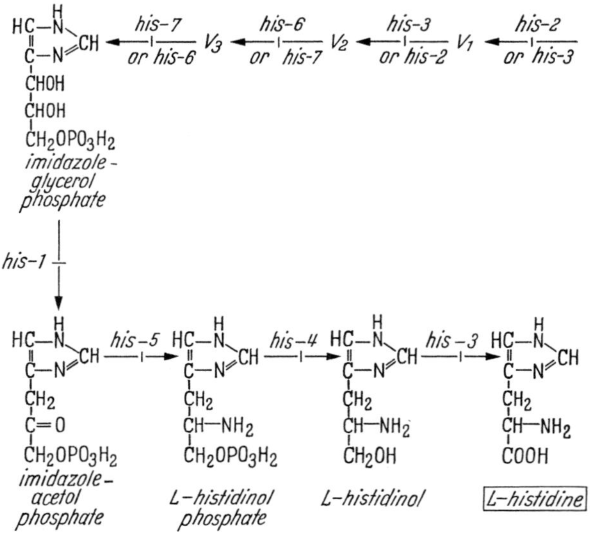

Fig. VI-5. Synthesis of histidine in *N. crassa* as an example of an unbranched synthetic pathway (see also Table VI-2, No. 1 f and Fig. VI-2). (Adapted from CATCHESIDE, 1960 b and WEBBER and CASE, 1960)

synthetic pathways from the large number of examples known for *N. crassa*. The following explanations are concerned primarily with the genetic basis of the pathways; the biochemical details are described in Table VI-2 along with additional examples.

Biosynthesis of histidine (Fig. VI-5): Up to now seven genes are known which are responsible for the formation of this amino acid. These are located in four linkage groups (see Fig. IV-14, p. 202). Three of the genes direct the formation of three precursors (V_1—V_3) of imidazoleglycerol phosphate, the details of which have not yet been fully clarified. It is noteworthy that the gene *his-3* acts not only in the formation of the precursors, but is also responsible for the last step in the synthesis. A model for interpreting this dual function will be discussed later (p. 378).

Biosynthesis of tryptophan, tyrosine, phenylalanine, p-aminobenzoic acid and p-hydroxybenzoic acid (Fig. VI-6): The four "allelic" *arom* mutants control the synthesis of the benzene ring. While *arom-1, -3,* and *-4* apparently involve point mutations which are each responsible for a step in the synthesis, *arom-2* appears to be a deletion with a pleiotropic effect (see also p. 409). The starting material for the formation of the benzene ring is shikimic acid. The synthesis proceeds via 5-phosphoshikimic acid, 3-enolpyruvic ether into chorismic acid (LINGENS, pers. communication). From this point the pathway branches to form five different aromatic compounds (Fig. VI-6 and p. 368).

In the mutants *arom-1, -3,* and *-4,* which are blocked prior to the formation of the benzene ring, two enzymes (dehydroshikimic acid reductase and protocatechuic acid oxidase) are produced constitutively. These enzymes, which in the mutant *arom-2* have been detected in only small quantities, and not at all in the wild type, prevent the accumulation of 5-dehydroshikimic acid by transforming this compound into protocatechuic acid; the latter is then converted into an aliphatic oxidation product.

According to recent investigations of DE MOSS and WEGMAN in *Neurospora* the pathway from chorismic acid to tryptophan consists of five steps (Fig. VI-6 and p. 368f.) (DEMOSS and WEGMAN, 1965; WEGMAN and DE MOSS, 1965). The anthranilate synthetase, which catalyses the transformation of chorismic acid into anthranilic acid is controlled by the genes *try-1* and *try-2* (see also AHMAD et al., 1964). Furthermore, the *try-1* gene is responsible for the formation of two other enzymes: Phosphoribosyl anthranilate isomerase and indole-3-glycerinphosphate isomerase. The first enzyme catalyses the step from phosphoribosylanthranilate (PRA) to 1-(o-carboxyphenylamino)-1-deoxyribulose-5-phosphate (CDRP); the second enzyme is responsible for the step from CDRP to indoleglycerol phosphate. Mutations in the *try-1* locus therefore lead to a loss of all three enzymes. In some mutants, however, the activity of one or the other enzyme is retained. Mutations in the *try-2* locus only affect the anthranilate synthetase. On the basis of complementation tests with *try-1* and *try-2* mutants CATCHESIDE (pers. communication) has assumed that both loci function in the formation of a polymeric enzyme protein which controls the synthesis of anthranilic acid. The functional characteristics of the gene *try-3* $(=td)$, which codes for the formation of tryptophan synthetase will be discussed later (p. 380 ff.).

Biosynthesis of valine and isoleucine (Fig. VI-7): While the initial steps of the synthesis of these amino acids are quite different, the latter reactions are almost the same. The only difference is that each of the final precursors of isoleucine has a methyl group which is lacking in the corresponding precursors of valine. This similarity is the basis for the same enzyme functioning in analogous reactions in the valine and isoleucine pathways. The activity of the isomerase and reductase, which catalyze two sequential steps, probably involves the dual function of a single enzyme. The fact that neither the isomerase nor the reductase is present in *iv-2* mutants is evidence for this interpretation.

Another enzyme, a dehydrase, is responsible for the transformation of the dihydroxy- to the keto-form of the compounds. It is not produced by the *iv-1* mutants. A transaminase serves as biocatalyst for the last step in the synthesis of valine and isoleucine. This enzyme can be demonstrated in the wild type. Mutants which lack the transaminase are not known in *Neurospora* (MEYERS and ADELBERG, 1954; RADHAKRISHNAN et al., 1960; WAGNER et al., 1960; BERNSTEIN and MILLER, 1961).

Recently it was shown that the synthesis of valine and isoleucine depends upon a particulate fraction (WAGNER and BERGQUIST, 1963)[1]. A mutant of *Neurospora* auxotrophic for both amino acids was found in which the enzymes necessary for valine and isoleucine synthesis are present, but the

---

[1] Additional genetic and biochemical data on valine and isoleucine mutants may be obtained from the study of WAGNER et al. (1964).

Function

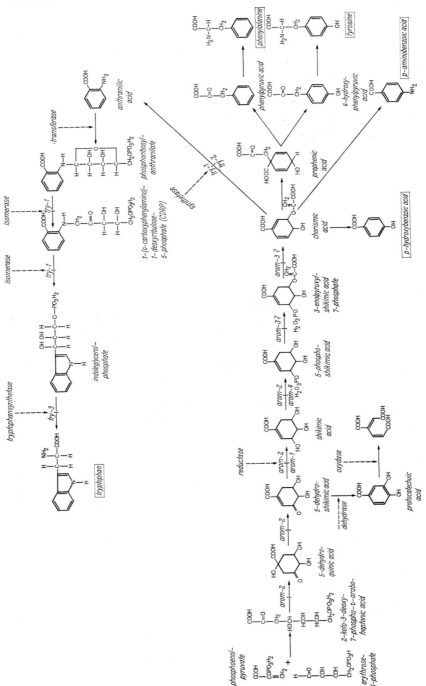

particulate fraction is inactive. As a result it was concluded that the enzymes must be bound or oriented in a specific way on the particles in order to catalyze the synthesis of both amino acids.

Biosynthesis of uridylic acid, arginine, and proline (Fig. VI-8): A double link exists between pyrimidine (uridylic acid) and arginine synthesis at the stage between ornithine and argininosuccinic acid on the one hand; on the

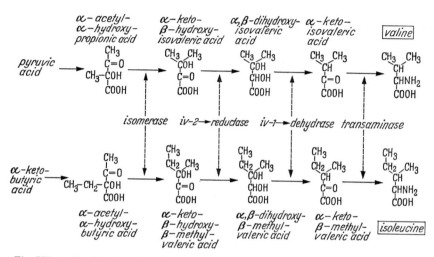

Fig. VI-7. Synthesis of valine and isoleucine in *Neurospora crassa* as an example of the catalytic role of a series of enzymes in two synthetic pathways (see also Table VI-2, No. 1 b and d as well as Fig. VI-4). (Adapted from FINCHAM and DAY, 1963)

other, proline and arginine synthesis are joined through a common precursor, glutamic acid.

Two cycles are noteworthy in the biosynthesis of arginine, namely, the ornithine cycle and the acetylglutamate cycle. Carbamyl phosphate, an intermediate for both arginine (CAP-A) and pyrimidine synthesis (CAP-P) arises from precursors which have not yet been identified and which serve as carbon and nitrogen donors. The genes responsible for the individual synthetic reactions are shown in the figure so far as they are known. Genes the locations of which remain uncertain are indicated by question marks (FINCHAM, 1953; VOGEL and KOPAC, 1960; NEWMEYER, 1962; DAVIS, 1962a, b, 1963, and personal communication; DAVIS and THWAITES, 1963; VOGEL and VOGEL, 1963).

Fig. VI-6. Synthesis of tryptophan, tyrosine, phenylalanine, p-amino-benzoic, and p-hydroxybenzoic acid in *Neurospora crassa* as an example of a branched synthetic pathway (see also Table VI-2, No. 1 g and Fig. VI-3). (Adapted from GROSS and FEIN, 1960)

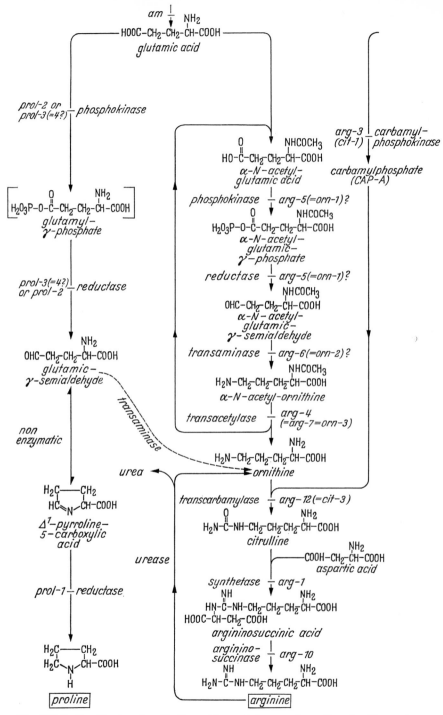

Fig. VI-8. Synthesis of uridylic acid, arginine, and proline in *Neurospora crassa* as an example of the interconnection of three synthetic pathways

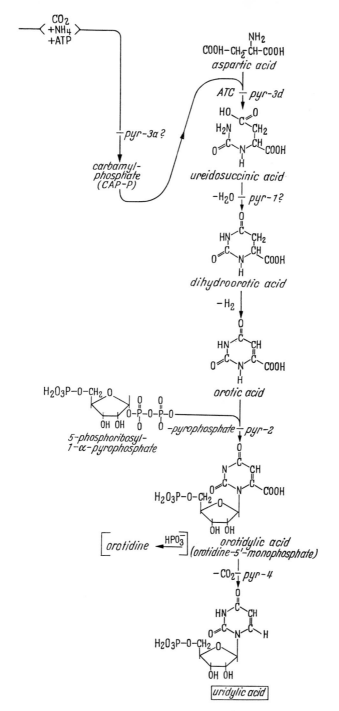

(see also Table VI-2, No. 1a and 3 as well as p. 398 ff.). (From DAVIS, 1962a, and personal communication)

Table VI-2. *Biosynthesis of certain amino acids, a vitamin, and a pyrimidine*

Unless otherwise noted all data are taken from investigations on *Neurospora crassa* (from the survey of VOGEL and BONNER, 1958).

In addition to the reaction sequences compiled in the table, the syntheses of certain other organic substances have been either partially or entirely elucidated using auxotrophic mutants of *N. crassa:* choline (HOROWITZ et al., 1945; JUKES and DORNBUSH, 1945; HOROWITZ, 1946), thiamine (TATUM and BELL, 1946; EBERHART and TATUM, 1963), riboflavin (MITCHELL and HOULAHAN, 1946b), pantothenic acid (TATUM and BEADLE, 1945; WAGNER and GUIRARD, 1948; WAGNER, 1949), succinic acid (LEWIS, 1948), purine (MITCHELL and HOULAHAN, 1946a; PIERCE and LORING, 1945; McELROY and MITCHELL, 1946; LORING and FAIRLEY, 1948).

metabolic product	evidence and reference

1. Biosynthesis of amino acids

a) proline, arginine (see also Fig. VI-8)

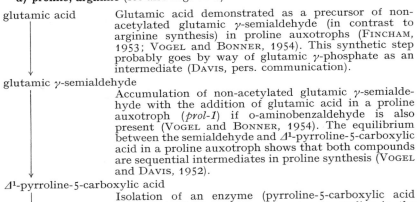

glutamic acid

    Glutamic acid demonstrated as a precursor of non-acetylated glutamic γ-semialdehyde (in contrast to arginine synthesis) in proline auxotrophs (FINCHAM, 1953; VOGEL and BONNER, 1954). This synthetic step probably goes by way of glutamic γ-phosphate as an intermediate (DAVIS, pers. communication).

glutamic γ-semialdehyde

    Accumulation of non-acetylated glutamic γ-semialdehyde with the addition of glutamic acid in a proline auxotroph (*prol-1*) if o-aminobenzaldehyde is also present (VOGEL and BONNER, 1954). The equilibrium between the semialdehyde and Δ^1-pyrroline-5-carboxylic acid in a proline auxotroph shows that both compounds are sequential intermediates in proline synthesis (VOGEL and DAVIS, 1952).

Δ^1-pyrroline-5-carboxylic acid

    Isolation of an enzyme (pyrroline-5-carboxylic acid reductase) which catalyzes the step to proline in the wild type (YURA and VOGEL, 1955). Proof from radioactive tracers that this carboxylic acid is the precursor of proline (ABELSON and VOGEL, 1955).

proline

glutamic acid

    Proof that glutamic acid is the precursor of ornithine through growth tests and from radioactive tracers (VOGEL and BONNER, 1954; ABELSON and VOGEL, 1955). Also demonstrated in mutants of *Penicillium notatum* and *P. chrysogenum* (BONNER, 1946a).

α-N-acetyl glutamic acid

    Arginine synthesis goes through acetylated intermediates, in contrast to proline synthesis (VOGEL and VOGEL, 1963). Also demonstrated in *Candida utilis* and *Saccharomyces cerevisiae* (MIDDELHOVEN, 1963, and DE DEKEN, 1962). Proof of an α-N-acetyl glutamic acid reductase.

α-N-acetyl glutamic γ-phosphate

    Postulated as the intermediate between α-N-acetyl glutamic acid and the corresponding semialdehyde by VOGEL and VOGEL (1963).

Table VI-2 (Continued)

metabolic product	evidence and reference

α-N-acetyl glutamic γ-semialdehyde

Identified as an intermediate by means of growth and enzymatic tests (VOGEL and BONNER, 1954; VOGEL and VOGEL, 1963). Also demonstrated in yeast (DE DEKEN, 1962). Isolation of an enzyme (ornithine-δ-transaminase) which catalyzes the transamination of the semialdehyde to form ornithine (FINCHAM, 1953).

α-N-acetyl ornithine

The α-N-acetyl ornithine is probably converted to ornithine by deacetylation by a transacetylase; at the same time glutamic acid is acetylated (VOGEL and VOGEL, 1963). Proof of an enzyme (α-N-acetylornithinase) which catalyzes the reaction forming ornithine; also shown in yeast (DE DEKEN, 1962).

ornithine + carbamyl phosphate

Growth of arginine auxotrophs following the addition of ornithine (SRB and HOROWITZ, 1944). Also demonstrated with radioactive tracers (ABELSON and VOGEL, 1955). Carbamyl phosphate (CAP-A) is required in addition to ornithine for the formation citrulline; the precursors of carbamyl phosphate are not yet known (block in *arg-3* mutants, DAVIS, 1962a, 1963). Proof of an enzyme (ornithine transcarbamylase) which catalyzes the step from ornithine to citrulline (MITCHELL and MITCHELL, 1952; DAVIS, 1962a, b, 1963; WOODWARD and SCHWARZ, 1964). Ornithine can be produced from α-N-acetyl ornithine, and also from the end product, arginine (ornithine cycle; Fig. VI-8).

citrulline + aspartic acid

Growth of arginine auxotrophs following addition of citrulline (SRB and HOROWITZ, 1944). Proof of enzymes which catalyze the conversion of citrulline and aspartic acid to argininosuccinic acid (FINCHAM and BOYLEN, 1955; see also Fig. VI-8).

argininosuccinic acid

Proof of enzymes which catalyze the conversion of argininosuccinic acid into arginine and fumaric acid (FINCHAM and BOYLEN, 1955; NEWMEYER, 1962).

arginine

b) methionine, isoleucine

aspartic acid

Proof through use of radioactive tracers that aspartic acid is the precursor of homoserine (ABELSON and VOGEL, 1955).

β-aspartyl phosphate

The discovery of enzymes in yeast which catalyze the conversion of β-aspartyl phosphate into the corresponding semialdehyde (BLACK and GRAY, 1953; BLACK and WRIGHT, 1955a, b).

aspartic β-semialdehyde

Enzymes demonstrated in yeast which control the step to homoserine (BLACK and WRIGHT, 1955c, d).

Function

Table VI-2 (Continued)

metabolic product	evidence and reference
homoserine ↓	Growth of the methionine-isoleucine double auxotroph following addition of methionine and threonine or of homoserine. Proof of branching of the pathway after homoserine. Accumulation of homoserine (TEAS et al., 1948; FLING and HOROWITZ, 1951).
cystathionine ↓	Growth of methionine auxotrophs after addition of cystathionine; accumulation of cystathionine (HOROWITZ, 1947). Demonstration of an enzyme which catalyzes the step to homocysteine (FISCHER, 1954).
homocysteine ↓	Growth of methionine auxotrophs following addition of homocysteine (HOROWITZ, 1947).

methionine

aspartic acid
β-aspartyl phosphate } see methionine synthesis
aspartic-β-semialdehyde

homoserine ↓	Growth of a threonine (and methionine) auxotroph after the addition of homoserine (TEAS et al., 1948). Evidence from radioactive tracers that homoserine is the precursor of threonine and isoleucine (ABELSON and VOGEL, 1955; KAPLAN and FLAVIN, 1965).
threonine ↓	Evidence from radioactive tracers in mutants (ADELBERG, 1954, 1955 a, b) and in the wild type (ABELSON and VOGEL, 1955) that threonine is an intermediate in isoleucine synthesis. Conversion to α-ketobutyric acid probably catalyzed by threonine dehydrase (YANOFSKY and REISSIG, 1953).
α-ketobutyric acid (see also Fig. VI-7) ↓	Proof from radioactive tracers in mutants that α-ketobutyric acid is the precursor of isoleucine (ADELBERG, 1954, 1955 a, b; ADELBERG et al., 1955). Similar data from *Torulopsis utilis* (STRASSMAN et al., 1954).
α-acetyl-α-hydroxy-butyric acid ↓	Evidence of an isomerase which catalyzes the conversion of this butyric acid derivative into methyl valeric acid (RADHAKRISHNAN et al., 1960; WAGNER et al., 1960; KIRITANI et al., 1966 b).
α-keto-β-hydroxy-β-methyl valeric acid ↓	Evidence of a reductase as catalyst for the conversion into the subsequent dihydroxy-derivative (RADHAKRISHNAN et al., 1960; WAGNER et al., 1960; KIRITANI et al., 1966 b).
α,β-dihydroxy-β-methyl valeric acid ↓	Accumulation of this valeric acid and its isolation from culture filtrates of an isoleucine auxotroph (ADELBERG, et al., 1951). Use of radioactive tracers (ADELBERG, 1955 b). Demonstration of an enzyme (dehydrase) that catalyzes the conversion of the dihydroxy- into the keto-derivative (MEYERS and ADELBERG, 1954; KIRITANI et al., 1966 a).

Table VI-2 (Continued)

metabolic product	evidence and reference

α-keto-β-methyl valeric acid

| |
↓

isoleucine

Evidence of a transaminase which controls the transamination of isoleucine (FINCHAM and BOULTER, 1956; WAGNER et al., 1960).

c) lysine

α-ketoglutaric acid

Demonstration by radioactive tracers that α-ketoglutaric acid is an intermediate in lysine synthesis (ANDERSSON-KOTTÖ et al., 1954; ABELSON and VOGEL, 1955). Similar data from *Torulopsis utilis* (STRASSMAN and WEINHOUSE, 1953; SAGISAKA and SHIMURA, 1959).

α-amino adipic acid

Growth of lysine auxotrophs following addition of α-amino adipic acid (MITCHELL and HOULAHAN, 1948) and ε-hydroxynorleucine (GOOD et al., 1950). Each of these compounds is probably converted to α-amino-adipic-δ-semialdehyde through transamination of the aldehyde group.

α-amino-adipic-δ-semialdehyde

Shown to be the precursor of lysine with radioactive tracers (ABELSON and VOGEL, 1955; TRUPIN and BROQUIST, 1965). In *S. cerevisiae* shown by JONES and BROQUIST (1965).

saccharopine

Evidence by analyses of accumulates and the enzyme saccharopine reductase in *N. crassa* and *S. cerevisiae* (TRUPIN and BROQUIST, 1965; JONES and BROQUIST, 1965).

lysine

d) α-alanine, valine, leucine

pyruvic acid

Evidence that pyruvic acid is a common precursor of α-alanine, valine, and leucine from tracer studies (ABELSON and VOGEL, 1955). Conversion of pyruvic acid to α-alanine probably by transamination (FINCHAM, 1951a; FINCHAM and BOULTER, 1956).

α-alanine

pyruvic acid (see also Fig. VI-7)

Demonstrated as an intermediate in valine synthesis with radioactive tracers (ABELSON and VOGEL, 1955).

α-acetyl-α-hydroxypropionic acid

Evidence of an enzyme (identical with the isomerase in isoleucine synthesis) which catalyzes the conversion of aceto-lactate into hydroxyisovaleric acid (RADHAKRISHNAN et al., 1960; WAGNER et al., 1960).

α-keto-β-hydroxyisovaleric acid

Proof of an enzyme (identical with the reductase in isoleucine synthesis) which is responsible for the conversion to the subsequent dihydroxy compound (RADHAKRISHNAN et al., 1960; WAGNER et al., 1960).

Function

Table VI-2 (Continued)

metabolic product	evidence and reference

α, β-hydroxyisovaleric acid

Accumulation of this compound in culture filtrates of a valine auxotroph (ADELBERG and TATUM, 1950; ADELBERG et al., 1951). Demonstration of a dehydrase which, as in isoleucine synthesis, catalyzes the conversion of the dihydroxy- to the keto-form (MEYERS and ADELBERG, 1954; WAGNER et al., 1960). Shown to be a precursor of valine with radioactive tracers (ADELBERG, 1955a). Similar observations in yeast (STRASSMAN et al., 1953; MCMANUS, 1954; STRASSMAN and WEINHOUSE, 1955).

α-ketoisovaleric acid

Growth of a valine auxotroph with the addition of this compound (BONNER et al., 1943). Demonstration of a transaminase which, as in the synthesis of isoleucine, catalyzes the transamination of the ketovaleric acid (FINCHAM and BOULTER, 1956; WAGNER et al., 1960).

valine

pyruvic acid

Shown to be an intermediate in leucine synthesis with the use of radioactive tracers (ABELSON and VOGEL, 1955).

α-acetyl-α-hydroxypropionic acid
α-keto-β-hydroxyisovaleric acid } see valine synthesis (and Fig. VI-7)
α, β-dihydroxyisovaleric acid

α-keto-isovaleric acid

Demonstrated as a precursor of leucine with radioactive tracers (ABELSON and VOGEL, 1955). Similar result in *Torulopsis utilis* (EHRENSVÄRD et al., 1951; STRASSMAN et al., 1956).

α-carboxy-β-hydroxyisocaproic acid

Accumulation of this compound in leucine auxotrophs. Indication of an enzyme (isomerase) which catalyzes the isomerization of β-hydroxy to α-hydroxyisocaproic acid. The polymeric protein appears to be composed of non-identical monomers which are determined by the genes *leu-2* and *leu-3* (GROSS, 1962). This step may possibly go by way of isopropyl maleic acid as an intermediate (JUNGWIRTH et al., 1963).

α-hydroxy-β-carboxycaproic acid

Accumulation in leucine auxotrophs. Indication of an enzyme which catalyzes the oxidation and decarboxylation to α-ketoisocaproic acid (JUNGWIRTH et al., 1963).

α-ketoisocaproic acid

Growth of a leucine auxotroph following addition of α-ketoisocaproic acid (REGNERY, 1944). Demonstration of an enzyme which catalyzes the reaction to leucine (FINCHAM and BOULTER, 1956). α-ketoisocaproic acid was also shown to be the precursor of leucine with radioactive tracers (ABELSON and VOGEL, 1955).

leucine

366

Table VI-2 (Continued)

metabolic product	evidence and reference

e) glycine, cysteine

serine

Several lines of evidence indicate that serine is the common precursor of glycine and cysteine (SAKAMI, 1955). Growth of a glycine auxotroph following addition of serine (TATUM, 1949). Proof from radioactive tracers, in the yeast, *Torulopsis utilis*, among others (EHRENSVÄRD et al., 1947; ABELSON and VOGEL, 1955; COMBÉPINE and TURIAN, 1965).

glycine

serine

Serine shown to be the precursor of cysteine with radioactive tracers, also in *Torulopsis utilis* (ABELSON and VOGEL, 1955).

cysteine

f) L-histidine (see also Fig. VI-5)

imidazoleglycerol phosphate

Accumulation of this substance in the mycelium of histidine auxotrophs (AMES, 1955; AMES and MITCHELL, 1955). This and the following two intermediates have only been found in the dephosphorylated form (p. 353). Proof of enzymes which catalyze the dehydration of imidazoleglycerol phosphate to imidazoleacetol phosphate (AMES, 1955). The accumulation of imidazoleglycerol phosphate in yeast as a result of inhibition of histidine synthesis by 3-amino-1,2,4-triazole (3-AT). The inhibition of 3-AT probably involves imidazoleglycerol phosphate dehydrogenase, which catalyzes the conversion of imidazoleglycerol phosphate to imidazoleacetol phosphate (KLOPOTOWSKI and HULANICKA, 1963; FINK, 1964).

imidazoleacetol phosphate

Accumulation of this compound in the mycelium of a histidine auxotroph (AMES, 1955; AMES and MITCHELL, 1955). Demonstration of an enzyme (glutamic acid phosphohistidinol transaminase) which catalyzes the synthetic step to L-histidinol phosphate (AMES and HORECKER, 1956). In *S. cerevisiae* by identification of accumulates and enzymes (FINK, 1964).

L-histidinol phosphate

Accumulation of this substance in the mycelium of a histidine auxotroph (AMES, 1955; AMES and MITCHELL, 1955). Proof of a phosphatase which catalyzes the conversion to L-histidinol (AMES, 1955). In *S. cerevisiae* (FINK, 1964).

L-histidinol

Accumulation in the culture filtrate of a histidine auxotroph (*his-3*) (AMES, 1955). Proof of an enzyme (L-histidinol dehydrogenase) which catalyzes the conversion of L-histidinol into L-histidine (CATCHESIDE, 1960b). In *S. cerevisiae* (FINK, 1964).

L-histidine

Table VI-2 (Continued)

metabolic product	evidence and reference

g) tryptophan, tyrosine, phenylalanine, p-aminobenzoic acid, p-hydroxy-benzoic acid (see also Fig. VI-6)

5-dehydroquinic acid

$\downarrow$

Growth of a mutant which is auxotrophic for all four aromatic compounds following the addition of quinic acid. The quinic acid is probably transformed into 5-dehydroquinic acid (GORDON et al., 1950). This compound is a common precursor for all five aromatic compounds. Demonstrated as the precursor of shikimic acid by using radioactive tracers (TATUM and GROSS, 1956).

5-dehydroshikimic acid

$\downarrow$

Accumulation of this acid in a shikimic acid auxotroph (TATUM et al., 1954). Proof of an enzyme (reductase) which catalyzes the conversion of the 5-dehydroshikimic acid into shikimic acid (GROSS, 1958; GROSS and FEIN, 1960).

shikimic acid

$\downarrow$

Growth of mutants which are simultaneously auxotrophic for tryptophan, tyrosine, phenylalanine, p-aminobenzoic acid and p-hydroxybenzoic acid following addition of shikimic acid (TATUM, 1949; TATUM et al., 1950). Thus shikimic acid is a common precursor for all five aromatic compounds (TATUM and PERKINS, 1950; TATUM, 1951; TATUM et al., 1954). Recent investigations indicate that shikimic acid is phosphorylated to 5-phosphoshikimic acid (DAVIS, pers. commun.; LINGENS, pers. commun.; DEMOSS, 1965a, b; EDWARDS and JACKMAN, 1965).

5-phosphoshikimic acid

$\downarrow$

Analyses of accumulates and enzymes have shown that the conversion of 5-phosphoshikimic acid to chorismic acid includes the step via 3-enolpyruvylshikimic acid-7-phosphate (LINGENS, pers. commun.; DEMOSS, 1965a).

3-enolpyruvylshikimic acid-7-phosphate

$\downarrow$

Found in bacteria (EDWARDS and JACKMAN, 1965), and probably in fungi (LINGENS, pers. commun.; DEMOSS, 1965a, b).

chorismic acid

$\downarrow$

Is the branching point for the further pathways of the five aromatic compounds. Identification of anthranilate synthetase, which catalyses the step to anthranilic acid (DEMOSS, 1965a, b; DEMOSS and WEGMAN, 1965; WEGMAN and DEMOSS, 1965). Chorismic acid is also found in S. cerevisiae (LINGENS et al., 1966a, b, c).

anthranilic acid

$\downarrow$

Growth of tryptophan auxotrophs following addition of anthranilic acid; accumulation of the acid in culture filtrates (TATUM et al., 1944, 1954); accumulation also in yeast (LINGENS and LÜCK, 1963). Analysis of the conversion of anthranilic acid into indoleglycerol phosphate by using radioactive tracers (NYC et al., 1949; PARTRIDGE et al., 1952). This conversion proceeds in two steps. Phosphoribosyl transferase controls the conversion of anthranilic acid to phosphoribosyl-anthranilate (DEMOSS and WEGMAN, 1965; WEGMAN and DEMOSS, 1965).

Table VI-2 (Continued)

metabolic product	evidence and reference

phosphoribosylanthranilate (PRA)

    Isolation of PRA-Isomerase which catalyses the step to CDRP (Lester, 1963; Wegman and DeMoss, 1965; DeMoss and Wegman, 1965).

1-(o-carboxyphenylamino)-1-deoxyribulose-5-phosphate (CDRP)

    Found as accumulate and by enzyme analyses. Identification of indole-3-glycerol phosphate isomerase which controls the step to indoleglycerol phosphate (Wegman and DeMoss, 1965; DeMoss and Wegman, 1965).

indoleglycerol phosphate

    Growth of tryptophan auxotrophs following addition of indole (Tatum and Bonner, 1943, 1944). Tryptophan can also be synthesized from indole and serine in cell free extracts (Umbreit et al., 1946). Demonstration of an enzyme (tryptophan synthetase) which catalyzes this reaction (Yanofsky, 1952a; Tatum and Shemin, 1954; Wegman and DeMoss, 1965; DeMoss and Wegman, 1965) (p. 380 ff.). Mutants which cannot produce tryptophan synthetase are auxotrophic for tryptophan and accumulate anthranilic acid, and to some extent, indole in their culture filtrates (Mitchell and Lein, 1948; Yanofsky, 1952b, 1955). Later investigations on tryptophan synthesis have shown, however, that in vivo indoleglycerol phosphate, and not indole, is the precursor of tryptophan.

tryptophan

5-dehydroquinic acid to chorismic acid } see tryptophan synthesis

    Evidence that the pathway follows the above steps from growth and radioactive tracer studies (Tatum et al., 1954; DeMoss and Wegman, 1965; Edwards and Jackman, 1965).

p-aminobenzoic acid + p-hydroxybenzoic acid

5-dehydroquinic acid to chorismic acid } see tryptophan synthesis

    Branching off from the pathways of tryptophan and p-aminobenzoic acid. Evidence for chorismic acid as intermediate (Colburn and Tatum, 1965; Tatum et al., 1954). Conversion of chorismic acid to prephenic acid is probably catalysed like in bacteria by two isozymes (Lingens and Goebel, 1965; Lingens et al., 1966b).

prephenic acid   Isolation of prephenic acid in *N. crassa* (Metzenberg and Mitchell, 1956). The step from prephenic acid to tyrosine via 4-hydroxyphenylpyruvic acid was found in bacteria (Cotton and Gibson, 1965; Edwards and Jackman, 1965), and also in *Neurospora crassa* and *S. cerevisiae* (Colburn and Tatum, 1965; Lingens and Goebel, 1965; Lingens et al., 1966b, c).

Table VI-2 (Continued)

metabolic product	evidence and reference

4-hydroxyphenylpyruvic acid

|

↓

Probably an intermediate in fungi (like in bacteria) (COL-BURN and TATUM, 1965; LINGENS and GOEBEL, 1965; LINGENS et al., 1966b, c).

| tyrosine |

5-dehydroquinic acid ⎫
to prephenic acid   ⎬ see tyrosine synthesis
                     ⎭

|

↓

Evidence from bacteria and fungi (ref. see tyrosine synthesis).

phenylpyruvic acid   Evidence from bacteria and fungi (ref. see tyrosine
↓   synthesis).

| phenylalanine |

2. Biosynthesis of a vitamin
Nicotinic acid

tryptophan

|

↓

Growth of certain nicotinic acid auxotrophs following the addition of tryptophan (BONNER and BEADLE, 1946; BEADLE et al., 1947). Proof that tryptophan is the precursor of nicotinic acid from radioactive tracers (PARTRIDGE et al., 1952; BONNER et al., 1952). Demonstration of an enzyme (kynurenine-formamidase), which catalyzes the reaction from tryptophan to kynurenine (JAKOBY, 1954). In *S. cerevisiae* detected as precursor of nicotinic acid (LINGENS and VOLLPRECHT, 1964).

kynurenine

|

↓

Growth of nicotinic acid auxotrophs after addition of kynurenine (BEADLE et al., 1947). Accumulation of α-N-acetyl kynurenine, a substance which is probably formed from kynurenine (YANOFSKY and BONNER, 1950, 1951). Demonstration through growth and accumulation experiments that kynurenine is converted to 3-hydroxy-anthranilic acid by way of 3-hydroxykynurenine (BONNER, 1948; YANOFSKY and BONNER, 1950).

3-hydroxykynurenine

|

↓

Growth of nicotinic acid auxotrophs following the addition of 3-hydroxykynurenine (HASKINS and MIT-CHELL, 1949). Accumulation of this compound (YA-NOFSKY and BONNER, 1950). Demonstration of an enzyme (kynureninase) which catalyzes the formation of 3-hydroxy-anthranilic acid from 3-hydroxykynure-nine (JAKOBY and BONNER, 1953a, b; JAKOBY, 1955).

3-hydroxy-anthranilic acid

|

↓

Growth of nicotinic acid auxotrophs after addition of 3-hydroxy-anthranilic acid (MITCHELL and NYC, 1948). Accumulation and isolation of this acid from culture filtrates (BONNER and BEADLE, 1946; BONNER, 1948). Evidence from radioactive tracers that 3-hydroxy-anthranilic acid is converted into nicotinic acid by way of quinolinic acid (YANOFSKY and BONNER, 1951).

quinolinic acid

|

↓

Restitution of the vitality of nicotinic acid auxotrophs after addition of quinolinic acid (YANOFSKY and BONNER, 1951). Accumulation of quinolinic acid in culture filtrates (HENDERSON, 1949; BONNER and YANOFSKY, 1949).

| nicotinic acid |

Table VI-2 (Continued)

metabolic product	evidence and reference

3. Biosynthesis of a pyrimidine
Uridylic acid (see also Fig. VI-8)

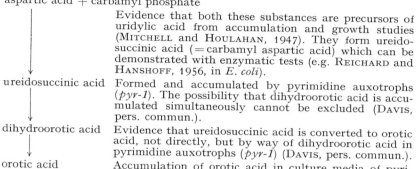

aspartic acid + carbamyl phosphate

Evidence that both these substances are precursors of uridylic acid from accumulation and growth studies (MITCHELL and HOULAHAN, 1947). They form ureidosuccinic acid (= carbamyl aspartic acid) which can be demonstrated with enzymatic tests (e.g. REICHARD and HANSHOFF, 1956, in *E. coli*).

ureidosuccinic acid — Formed and accumulated by pyrimidine auxotrophs (*pyr-1*). The possibility that dihydroorotic acid is accumulated simultaneously cannot be excluded (DAVIS, pers. commun.).

dihydroorotic acid — Evidence that ureidosuccinic acid is converted to orotic acid, not directly, but by way of dihydroorotic acid in pyrimidine auxotrophs (*pyr-1*) (DAVIS, pers. commun.).

orotic acid — Accumulation of orotic acid in culture media of pyrimidine auxotrophs (*pyr-2*) (MITCHELL et al., 1948). Demonstration of the conversion of orotic acid into orotidylic acid through enzyme tests with pure substrates; 5-phosphoribosyl-1-α-pyrophosphate is simultaneously involved in the formation of orotidylic acid with the liberation of pyrophosphate (DAVIS, pers. commun.). Proof in yeast using radioactive-labelled orotic acid that the latter is the precursor of uridylic acid (EDMONDS et al., 1952).

orotidylic acid — Accumulation of orotidine which is formed by the dephosphorylation of orotidylic acid (MICHELSON et al., 1951). Proof of a similar synthetic step in *Schizosaccharomyces pombe* (MEGNET, 1959). Proof in pyrimidine auxotrophs (*pyr-4*) that orotidylic acid is converted into uridylic acid through decarboxylation (TUTTLE, unpublished). Isolation of a decarboxylase which catalyzes this step (LIEBERMANN et al., 1955).

uridylic acid

Summary

1. Monogenic nutritional mutants of fungi have proved particularly suitable for elucidating biosynthetic pathways of amino acids, vitamins, purines, and pyrimidines. Certain of these pathways are described in Fig. VI-5 through VI-8 and in Table VI-2.

2. Auxotrophy in a mutant can generally be attributed to the loss of activity of a single enzyme. Mono- and polyauxotrophic mutants occur. A single gene mutation may be involved in both instances. Mono-auxotrophs have lost the capacity to synthesize the end product of a particular synthetic pathway. Polyauxotrophy, on the other hand, may be induced by a genetic block prior to the branching of a synthetic pathway. It may also result from the loss or inactivation of a single

24*

enzyme through which two separate but related reaction sequences are blocked.

3. The results of investigations with auxotrophic mutants led to the one gene-one enzyme hypothesis (see also p. 405 ff.).

C. Genes and enzymes

The fact that single genes control specific metabolic steps through enzymes raises the question of the relation between genes and enzymes: Is there always a direct correlation between the structure of the genetic material and enzyme specificity, or do the genes function indirectly in the individual steps of an enzymatically controlled synthesis? In the former case the blocking of a particular synthesis must result from the loss of an enzyme or a change in its specificity. In the latter case, although the enzyme is produced in its unaltered form, its catalytic activity is inhibited by other substances. With such considerations in mind a number of research teams have sought since the 1940's to do the following:

1. *Isolate the enzyme responsible for a particular synthetic step.*
2. *Establish that the mutational change affects the enzyme concerned.*
3. *Find a correlation between gene structure and enzyme specificity.*

These investigations have shown that both of the alternatives, the direct ces well as the indirect correlation between gene and enzyme, occur. The experiments, particularly those with bacteria, have led to the development of a model of genetic control of enzyme formation.

Literature: BONNER (1951, 1952, 1955, 1956, 1959, 1964a, b), HORO-WITZ (1951), BEADLE (1956, 1957, 1960a, b, 1961), FINCHAM (1959a, 1960), CATCHESIDE (1960a), YANOFSKY (1960), YANOFSKY and ST. LAWRENCE (1960), CATCHESIDE (1964).

I. Gene mutation and enzyme specificity

After MITCHELL and LEIN (1948) demonstrated that a tryptophan-deficient mutant of *Neurospora crassa* is not able to produce an enzyme which catalyzes the synthesis of the amino acid in this species, innumerable enzymes responsible for many different metabolic steps have been identified and analyzed biochemically. In many cases the way in which the gene acts in enzyme formation has also been elucidated.

1. Direct effects

We have brought together in Table VI-3 examples of the direct effect of mutation on enzymes. The table indicates whether the gene mutation influences the enzyme quantitatively or qualitatively.

a) Quantitative effects

The most frequently observed quantitative effect of a gene mutation is the absence of enzyme formation in a particular mutant. In other cases the mutation causes a change in the amount of enzyme produced. The relative activity is mostly decreased; an increase has been observed in only one case (TAVLITZKI, 1954) (Table VI-3).

Quantitative changes in enzyme activity can only be demonstrated unequivocally if the enzymatic comparison of wild type and mutant is based on the following criteria: 1. Identical culture conditions. 2. The same reference system for measuring the different enzyme activities, e.g. activity/protein content = specific activity. 3. Proof that a qualitative change has not occurred. Unfortunately not all three conditions have been satisfied in all cases. An intensive analysis of the enzyme for qualitative change is lacking in most examples of a quantitative change in enzyme activity cited in Table VI-3. A qualitative change in the enzyme is indicated if an auxotrophic mutant produces a protein related to the enzyme, but inactive instead of the active enzyme (details on p. 381f.).

Tyrosinase

The investigations of HOROWITZ et al. (1960) on *N. crassa* showed that *single gene mutations can influence enzyme production quantitatively without altering the structure of the enzyme.*

Tyrosinase, which has recently been prepared in crystalline form by FLING et al. (1963), is involved in the synthesis of melanin pigments. It controls the oxidation of mono- and dihydroxyphenols. Phenol oxidases are not essential for growth in fungi. The wild type of *N. crassa* produces tyrosinase constitutively only under culture conditions which are unfavorable for the proliferation of mycelium and protein synthesis, for example, after the nutrient supply in the culture medium is exhausted (Fox et al., 1963). Tyrosinase is likewise constitutively synthesized in a synthetic medium with a low sulfate concentration (HOROWITZ and SHEN, 1952). Moreover, the enzyme production is inhibited by an increase in sulfate concentration. Tyrosinase formation is induced if an aromatic amino acid (e.g. dihydroxyphenylalanine) which can serve as a substrate is added to a culture medium which has enough sulfate for the inhibition of constitutive enzyme synthesis.

The non-allelic mutants *ty-1* and *ty-2* do not form tyrosinase under culture conditions which are optimal for constitutive enzyme formation. However, they can be induced to form enzymes in two ways: 1. through the effect of the wild type gene in heterokaryons and 2. inductively, i.e. following addition of aromatic amino acids, as mentioned above. The tyrosinase produced in each case cannot be differentiated from that of the wild type. This can be demonstrated by marking *ty-1* and *ty-2* mutants with genes which produce structural changes in the tyrosinase (p.378 f.); the tyrosinase characteristic of the marker gene is always produced in the heterokaryon as well as through induction. As experiments with C¹⁴-labelled valine have shown, tyrosinase is formed "de novo" from amino acids and not out of pre-existing protein.

Experiments which indicate quantitative alteration of tyrosinase production have been carried out on *N. crassa* by SCHAEFFER (1953), FOX and BURNETT (1962), FOX et al. (1963), and on *Glomerella cingulata* by MARKERT (1950), and SUSSMAN et al. (1955).

Function

Table VI-3. *Examples of enzymatic changes in mutants of various fungi*
With the exception of the prototrophic tyrosinase and laccase mutants, all mutants are auxotrophs. Mutant enzymes for which structural changes have been demonstrated are designated with a +. All other enzymes are either changed only quantitatively or not yet sufficiently tested for qualitative changes (data taken in part from FINCHAM, 1959a, and CATCHESIDE, 1960a).

object	locus	enzyme	enzymatic activity of the mutants	qualitative change	reference
Saccharomyces cerevisiae	?	galactokinase	loss		DE ROBICHON-SZULMAJSTER, 1958
	ga-4	galactokinase, transferase, epimerase	loss		DOUGLAS and HAWTHORNE, 1964
	ga-7	transferase	loss		
	ga-10	epimerase	loss		
	?	diphosphothiamine-phosphatase	increase		TAVLITZKI, 1954
	MZ	α-glucomelizitase	loss		PALLERONI and LINDEGREN, 1953
	M-1 to M-6	α-glucosidase	loss		HALVORSON et al. 1963
Aspergillus nidulans	?	nitrate reductase	loss		COVE and PATEMAN, 1963; PATEMAN et al., 1964
Aspergillus oryzae	?	amylase	loss		SEARASHI, 1962
Glomerella cingulata	several loci	tyrosinase	decrease		MARKERT, 1950; MARKERT and OWEN, 1954
Podospora anserina	several loci	laccase	decrease	+	ESSER, 1963, 1966
Neurospora crassa	ad-4	adenylosuccinase	loss	+	GILES et al., 1957a, b
	ad-8	adenylosuccinate synthetase	loss		ISHIKAWA, 1960, 1962a, b
	am	glutamic acid dehydrogenase	loss	+	FINCHAM, 1954, 1957; FINCHAM and PATEMAN, 1957a, b; PATEMAN and FINCHAM, 1958
	arg-1	argininosuccinate synthetase	loss		NEWMEYER,, 1957
	arg-10	argininosuccinase	loss		FINCHAM and BOYLEN, 1955
	arom-1	dehydroshikimic acid reductase	loss		GROSS and FEIN, 1960
	car	pyruvate carboxylase	decrease		STRAUSS, 1953
	his-1	imidazoleglycerol phosphate dehydrase	loss		AMES, 1957a
	his-3	L-histidinol dehydrogenase	decrease		AMES, 1957b WEBBER, 1960
	his-4	L-histidinol phosphate phosphatase	loss		AMES, 1957b

Table VI-3 (Continued)

object	locus	enzyme	enzymatic activity of the mutants	quali-tative change	reference
Neurospora crassa	*his-5*	glutamic acid-phospho-histidinol trans-aminase	loss		AMES and HOR-ECKER, 1956
	iv-1	α,β-dihydroxy-iso-valeric acid dehydrase	decrease		MEYERS and ADELBERG, 1954
	iv-2	α-acetyl-α-hydroxy-propionic acid iso-merase, α-keto-β-hydroxyisovaleric acid reductase	loss		WAGNER et al., 1960
	leu-2	β-carboxy-β-hydroxy-isocaproic acid-isomerase	loss		GROSS, 1962
	?	lactase	decrease		LANDMAN, 1950
	me-2	cystathionase II	loss and		FISCHER, 1957
	me-7	cystathionase I	decrease		
	?	nitrate reductase	loss		SILVER and McELROY, 1954; McELROY and SPENCER, 1956
	oxD	D-amino acid oxidase	loss		OHNISHI et al., 1962
	prol-1	pyrroline-5-carb-oxylate-reductase	decrease	+	YURA, 1959
	pyr-3	aspartic transcarb-amylase	loss		DAVIS, 1960; DA-VIS and WOOD-WARD, 1962
	su pyr-3	ornithine trans-carbamylase	decrease	+	DAVIS, 1962a, b
	suc	oxalacetic acid carboxylase	decrease		STRAUSS, 1957
	td=tryp-3	tryptophan synthetase (syn. tryptophan desmolase)	decrease loss	+	YANOFSKY, 1952a; YANOFSKY and BONNER, 1955a; SUSKIND et al., 1955; SUSKIND and KUREK, 1959; MOHLER and SUSKIND, 1960; DEMOSS and BONNER, 1959; ESSER et al., 1960
	thr-2	threnonine synthetase	loss		FLAVIN and SLAUGHTER 1960; KAPLAN and FLAVIN, 1965
	T *ty-1* *ty-2*	tyrosinase	altered decrease and loss	+	HOROWITZ and FLING, 1953, 1956; HOROWITZ et al., 1960, 1961 a, b; FOX et al., 1963
	?	tryosinase	loss		
	?	β-glucosidase	decrease		EBERHART et al., 1964

SCHAEFFER studied a single gene mutant which produced more melanin pigment in comparison with the wild type. It was not possible to determine, however, whether the increased enzyme activity resulted from a greater production of the enzyme or from the absence of a tyrosinase inhibitor.

Fox and his coworkers investigated a race of *Neurospora* in which tyrosinase activity was absent. This strain also failed to produce a tyrosinase-related protein; however, it did contain a non-dialyzable factor able to activate the wild type tyrosinase. The genetic basis of this situation is not yet clear.

Following UV-irradiation of *G. cingulata* MARKERT obtained a number of unifactorial mutants which were macroscopically recognizable by a more or less weak mycelial pigmentation. The mutants showed a great variation in tyrosinase activity. Genetic analysis indicated that possibly six different loci were involved.

In our opinion it is questionable that the changes in tyrosinase activity observed by MARKERT were in all cases quantitative in nature. 1. Enzyme activity was based on the weight of a dialyzed and lyophilized crude extract and not on the protein content of the solution. 2. Tyrosine formation is subject to wide variation during mycelial growth and is dependent upon environmental conditions (MARKERT, 1950; SUSSMANN and MARKERT, 1953). The optimum is reached after about 200 hours, if a petri dish is inoculated with a conidial suspension and incubated at 25° C. Since all mycelia were harvested after the same period of growth, it is perfectly possible that optimal enzyme formation was not attained because of the delay in growth of individual mutants and thus the reduction in enzyme activity was simulated.

That the tyrosinase produced by the individual mutants is qualitatively changed could very likely be excluded through immunological experiments (MARKERT and OWEN, 1954; OWEN and MARKERT, 1955). These authors obtained an antiserum after injection of tyrosinase preparations of the wild type into rabbits that quickly inactivated and slowly precipitated the tyrosinase. A test of enzyme preparations from different mutants against the antityrosinase showed the bound anti-tyrosinase from this reaction mixture was proportional to the amounts of tyrosinase added. Cross-reacting material (p. 382) could not be demonstrated.

α-glucosidase

Six non-allelic genes ($M1$—$M6$) control the fermentation of maltose in yeast. These hereditary factors have their origin in different species (*S. cerevisiae:* WINGE and ROBERTS, 1948, 1950; *S. distaticus:* GILLILAND, 1954; *S. carlsbergensis:* WINGE and ROBERTS, 1950). The M genes are responsible for the formation of α-glucosidase. HALVORSON et al. (1963) investigated the problem of whether these genes control the formation of structurally different enzymes or whether they all direct the synthesis of the same enzyme.

It was shown that the recessive forms of all the M genes investigated ($M5$ was not included in the experiment) prevented the formation of α-glucosidase or a serologically related protein. One dominant allele was sufficient for constitutive formation of the enzyme, the amount of which could be increased inductively by the addition of maltose. The α-glucosidases formed by the different M strains were partially purified. They could not be differentiated on the basis of the following criteria: heat inactivation, electrophoretic migration, chromatography in cellulose or DEAE-cellulose, substrate specificity, and neutralization with specific antigens. HALVORSON and his coworkers concluded from their results that the individual enzyme proteins were possibly identical in tertiary structure.

376

The M genes often have a purely quantitative influence on enzyme synthesis. This idea was confirmed by the fact that the formation of α-glucosidase depends upon the dosage of M genes. It increases proportionally to the number of dominant M factors in a nucleus (RUDERT and HALVORSON, 1963).

α-glucomelizitase

Gene mutations have a quantitative effect not only on the constitutive, but also the induced enzymes in yeast. LINDEGREN and his coworkers found that yeast strains form an adaptive (induced) enzyme system which permits the hydrolysis of melizitose, furanose, maltose, sucrose, or α-methyl glucoside following culture on these substances (PALLERONI and LINDEGREN, 1953). Because the capacity to split melizitose is always correlated with a hydrolytic activity for the other four sugars, PALLERONI and LINDEGREN assumed that a single enzyme, α-glucomelizitase, is responsible for the hydrolysis of the five sugars. A series of monogenic mutants were found which no longer reacted to all of the sugars with adaptive enzyme formation, but only toward single substrates (LINDEGREN and LINDEGREN, 1953). The LINDEGRENS (1956) explained this phenomenon with the assumption that the mutants represented a series of multiple alleles at the MZ locus which are responsible for the induced enzyme synthesis. ROBERTSON and HALVORSON (1957) disagreed with this interpretation. On the one hand, genes other than the MZ locus are known which influence induced enzyme formation; on the other, "new alleles" of the MZ locus arise in all crosses between the individual mutants (LINDEGREN et al., 1956). ROBERTSON and HALVORSON attributed the differences between the mutants to the permease-controlled differential permeability of the cell membrane to the specific substrates. Such permeases were demonstrated a short time later by ROBICHON-SZULMAJSTER (1958) for galactose mutants of yeast. Rebuttals by LINDEGREN (1957) have not clarified the situation.

Enzymes of histidine biosynthesis
(compare Fig. VI-5 and Table VI-2, No. 1f)

About a hundred allelic mutants for the *his-3* locus of *Neurospora crassa* are known which can be classified into different physiological groups by complementation tests (Fig. VI-11) (CATCHESIDE, 1960b); the complementation groups can be arranged into different enzymatic classes (AMES et al., 1953; AMES, 1955, 1957a, b; WEBBER, 1960).

Class I: Mutants of complementation groups C, D, E, and F (Fig. VI-11) form large amounts of L-histidinol after addition of L-histidine to the culture medium. The last step of histidine synthesis appears to be blocked. Biochemical investigation of certain mutants showed that the enzyme histidinol dehydrogenase was absent.

Class II: Mutants of complementation groups A and B (Fig. VI-11) accumulate neither L-histidinol nor a detectable imidazole. These mutants belong to three categories: 1. strains with an L-histidinol dehydrogenase defect, 2. strains with an enzyme deficiency in an early stage of histidine synthesis, 3. strains in which both enzyme systems are absent.

The various enzymatic activities of *his-3* mutants can also be attributed to qualitative enzyme alterations. This suggests that the *his-3* locus is responsible for the formation of an enzyme with different catalytic functions, like, for example, the *td* locus (p. 380 ff.). Apart from the fact that no biochemical data are available in support of this interpretation, the genetic investigations of WEBBER (1960) also argue against this explanation. His results support the assumption that this locus is not only structurally, but functionally complex.

b) Qualitative effects

A much deeper insight into the functional relationship between gene and enzyme has come from instances in which qualitative changes in enzymes are produced through mutation (Table VI-3).

Proof of a qualitative change of an enzyme requires a knowledge of the characteristics of the wild type enzyme which is as precise as possible. For example, the following criteria can be used to characterize the wild type enzyme: substrate specificity, sensitivity to inhibitors or to specific pH ranges, heat stability, enzyme kinetics, behavior in an electrical and in a centrifugal field, molecular weight, sequence analysis of the enzyme protein, and immunological behavior. In most cases, thorough purification of the specific enzyme is necessary for determining these characteristics. The three criteria for establishing quantitative enzyme alterations previously discussed (p. 372) must also be considered in comparing enzyme preparations which come from strains of different genetic constitution.

Tyrosinase

In addition to the genes mentioned in the previous section which influence tyrosinase formation in *Neurospora* quantitatively, HOROWITZ and his coworkers identified a gene which is responsible for the structure of tyrosinase (HOROWITZ and FLING, 1953, 1956; HOROWITZ et al., 1960, 1961 a, b). Four alleles, which were isolated from different wild type strains, are known for this locus. The enzymes produced under the influence of the individual alleles differ qualitatively in their heat stability and electrophoretic mobility (Table VI-4).

The existence of structural differences between the four different tyrosinases was confirmed by an enzyme analysis (heat inactivation, electrophoresis) of tyrosinase in heterokaryons (e. g. $T_L + T_{Sing-2}$ and $T_L + T_{PR-15}$).

Table VI-4. *Characteristics of tyrosinase from N. crassa the formation of which is controlled by different alleles of the T locus* (from HOROWITZ et al., 1961 a)

genotype	thermostability (50% activity in minutes at 59° C)	electrophoretic migration mm/hour on paper at pH 6 1.25 mA per strip
T_S	70	2
T_L	5	2.25
T_{PR-15}	20	1.5
T_{Sing-2}	70	1.5

Heat inactivation of the individual tyrosinases is always a first order reaction. This indicates the homogeneity of each of the types of enzyme. The heat inactivation of an enzyme preparation obtained from the heterokaryon, $T_S + T_L$ is not a first order reaction; it follows a course which is expected if both forms of the enzyme are present in equal quantities. Similar investigations with the heterokaryon, $T_L + T_{Sing-2}$, gave the same results. The tyrosinases produced by this heterokaryon can also be separated electrophoretically. They are represented by two distinct bands on the paper strip.

A further characterization of the T_S and T_L tyrosinases was undertaken by SUSSMAN (1961). He failed to find a difference between the two enzymes with respect to substrate specificity, pH optimum, Michaelis constants, and specificity of inhibitors. However, the thermolabile enzyme loses its activity more rapidly than the thermostable form after incubation with inhibitors.

Pyrroline-5-carboxylic acid reductase
(Fig. VI-8 and Table VI-2, No. 1a)

Pyrroline-5-carboxylic acid reductase (PC reductase) catalyzes the last step in proline biosynthesis: the reduction of pyrroline-5-carboxylic acid which follows from the presence of di- and tripyridine nucleotide (DPN and TPN) (YURA and VOGEL, 1955). The mutant *prol-1* cannot carry out this synthetic step and needs proline in order to grow (BEADLE and TATUM, 1945). MEISTER et al. (1958) showed that this strain possesses only a small PC reductase activity. YURA (1959) demonstrated that the reduced enzyme activity in the mutant is not a quantitative effect, nor is it brought about by an inhibitor, but that the mutation at the *proline* locus causes a qualitative alteration in PC reductase.

The investigation of partially purified extracts has shown that the PC reductase prepared from the mutant requires a 3 to 4 times higher activation energy and is less thermostable than that of wild type. When the two extracts are mixed, each type of enzyme retains its own characteristics.

Glutamic acid dehydrogenase
(Fig. VI-8)

Neurospora crassa produces two different glutamic acid dehydrogenases (GAD) that require either diphosphopyridine nucleotide (DPN) or triphosphopyridine nucleotide (TPN) as a coenzyme (FINCHAM, 1951b; SANWAL and LATA, 1961). FINCHAM and his coworkers have concerned themselves with genetic studies of TPN-specific GAD for the last several years (FINCHAM, 1954, 1957, 1959a, 1962b; FINCHAM and PATEMAN, 1957a, b; FINCHAM and BOND, 1960; PATEMAN, 1957; FINCHAM and CODDINGTON, 1963a, b). The preparation of pure TPN-GAD was undertaken by BARRATT and STRICKLAND (1963). This enzyme, the production of which the *am* locus is responsible, catalyzes the formation of glutamic acid from α-ketoglutaric acid and ammonium ion. In the meantime thirteen allelic mutants of the *am* locus have become known, the nutritional deficiencies of which can be alleviated by the addition of α-amino nitrogen.

379

The formation of DPN-dependent GAD is not influenced by any of the mutational alterations of the *am* locus. Because of the presence of this enzyme, the *am* mutants merely show a delayed growth on an unsupplemented medium.

Under normal conditions the *am* mutants show less than 0.2% of the GAD activity of the wild type. Nevertheless, thoroughly purified protein preparations obtained from the strains, *am-2*, *am-3*, and *am-12*, show an enzyme activity if TPN and glutamate is added to the system and the reaction initiated at pH values above 8. The mutant *am-1* produces a protein which behaves electrophoretically like GAD.

Since the suppression of the normal GAD reaction by inhibitors could be excluded, these experiments led to the assumption that the mutants produce a qualitatively altered enzyme. Comprehensive investigations of the Fincham group on the activity and characterstics of GAD from heterokaryons of the *am* mutants and reverse mutants with partially restored enzyme activity (p. 401 ff.) have unequivocally demonstrated the existence of structurally altered GAD. It should be pointed out that the different types of enzymes possess largely the same physical and chemical properties; they can not be differentiated either by electrophoretic criteria nor by the amino acid constitution (fingerprinting) of the peptides obtained from trypsin digestion.

Adenylosuccinase

Adenylosuccinase can catalyze two different reactions: (1) the conversion of succinyl adenosine monophosphate (AMPS) into adenosine monophosphate through the liberation of succinic acid, and (2) the separation of the succinic acid from 5 amino-4 imidazole (N succinylocarboxy-amide)-riboside (SAICAR) (GOTS and GOLLUB, 1957; PARTRIDGE and GILES, 1957; BUCHANAN et al., 1957). The *ad-4* locus of *Neurospora crassa* is responsible for the formation of this enzyme; *ad-4* mutants require adenine. GILES and his coworkers have studied both the genetic and biochemical features of *ad-4* mutants (GILES et al., 1957a, b). They have found 35, in part heteroallelic mutants at this locus. All these mutants show only traces of adenylosuccinase activity. According to preliminary results one strain of the *ad-4* series appears to form a qualitatively altered enzyme (GILES, 1958) which differs from the wild type enzyme in its thermostability.

Tryptophan synthetase
(compare Fig. VI-6 and Table VI-2, No. 1g)

Tryptophan synthetase (previously called tryptophan desmolase; abbreviated to T-ase) catalyzes the last step in tryptophan synthesis in *Neurospora crassa* and *Escherichia coli* (Fig. VI-9): the glyceraldehyde phosphate moeity of the indoleglycerol phosphate (the latter derived from anthranilriboside) is replaced by L-serine. In addition to this conversion (reaction 1), the same enzyme can catalyse two other reactions: the synthesis of tryptophan from indole and L-serine (reaction 2), and the splitting of indoleglycerol phosphate into indole and glycerol-aldehyde phosphate (reaction 3). The last reaction is reversible (YANOF-SKY, 1952a, b, 1956, 1957, 1958, 1960; CRAWFORD and YANOFSKY, 1958; DEMOSS and BONNER, 1959; DEMOSS, 1962). Pyridoxal phosphate is required as a cofactor for reactions 1 and 2 (UMBREIT et al., 1946). Since free indole has never been found as an intermediate in wild type strains (YANOFSKY and RACHMELER, 1958; DEMOSS et al., 1958; DE MOSS, 1962), reaction 1 is assumed to be the normal pathway of tryptophan synthesis. This assumption has recently been confirmed in tryptophan-deficient mutants of *Coprinus radiatus* by CABET et al. (1962).

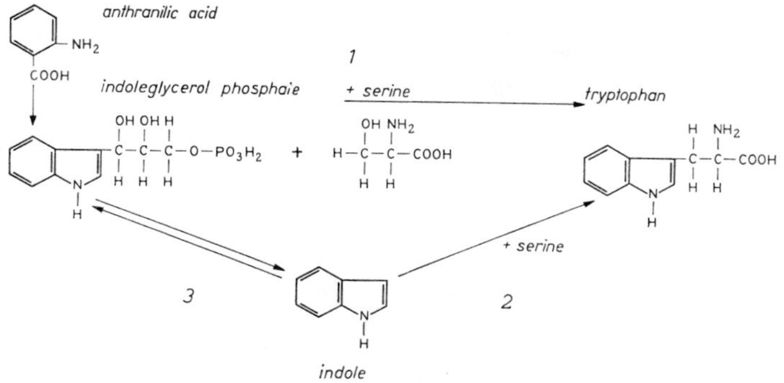

Fig. VI-9. The reactions in *Neurospora crassa* and *Escherichia coli* that are controlled by tryptophan synthetase. (Adapted from BONNER, 1959)

Tryptophan synthetase has been prepared in a highly purified form (MOHLER and SUSKIND, 1960). Its molecular weight has been estimated at 140,000. A further characterization of T-ase by GARRICK and SUSKIND (1964a, b) has recently appeared. Investigations on the mechanism of T-ase formation in vitro and in vivo have been carried out by WAINWRIGHT (1959, 1963) and LESTER (1961a). WAINWRIGHT found that particles separable from a cell-free extract from conidia or mycelia by ultracentrifugation are capable of synthesis of the enzyme after suitable supplementation. LESTER demonstrated that germinating conidia from the wild type increase their enzyme production with the addition of indole acetic acid. Tryptophan does not have this effect. An inhibition of the enzyme system responsible for anthranilic acid synthesis was observed in conidia of one tryptophan mutant which were germinated in the presence of tryptophan.

In recent years BONNER and his coworkers have isolated more than 200 monogenic tryptophan auxotrophs which have arisen both spontaneously and from UV irradiation. All these mutants were alleles of the *td* locus (also designated *try-3*). They do not produce a T-ase that is capable of catalyzing reaction 1 (YANOFSKY and BONNER, 1955a; SUSKIND et al., 1955; BONNER et al., 1960). However, a biochemical analysis of a series of these mutants grown on minimal medium supplemented with tryptophan showed differences that allowed a classification of the strains into different categories (Table VI-5). The CRM test, which has already been mentioned (e.g. p. 373), was used for the initial subdivision of the mutants.

Function

This test is based on a method worked out by COHN and TORRIANI (1952, 1953). An antibody is produced after injection of partially purified enzyme into rabbits. T-ase becomes inactivated a few minutes following the addition of anti-T-ase (SUSKIND et al., 1955). The anti-T-ase content of the serum can be quantitatively determined by titration of antiserum with wild-type T-ase.

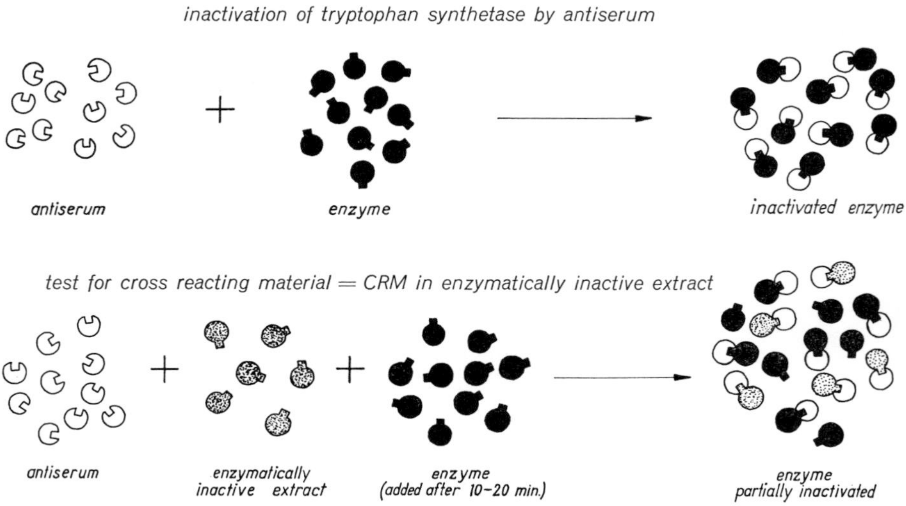

inactivation of tryptophan synthetase by antiserum

antiserum   enzyme   inactivated enzyme

test for cross reacting material = CRM in enzymatically inactive extract

antiserum   enzymatically   enzyme   enzyme
                   inactive extract   (added after 10–20 min.)   partially inactivated

Fig. VI-10. Diagram of the CRM test for tryptophan synthetase (for further details see text)

After the mutant extract and anti-T-ase are mixed, whether or not antibodies are bound by the enzymatically inactive extract can be determined by titrating the mixture against active T-ase. If the mutant extract reacts with the antibodies, one can conclude that the mutant used has produced a protein which is serologically related to T-ase. This is designated cross-reacting material (CRM). A negative CRM test leads to the conclusion that such a protein is absent. The principle of the CRM test is diagrammed in Fig. VI-10.

The *td* mutants can be classified into two groups on the basis of the CRM test:

Group I, CRM +: The mutants do not produce an active T-ase for reaction 1, but a protein serologically related to the enzyme (= CRM). The gene mutation has produced a qualitative alteration in the enzyme (Table VI-5).

A more exact biochemical comparison of the physical characteristics of CRM from a *td* mutant with the wild type T-ase has shown that the proteins are almost identical (SUSKIND, 1957; MOHLER and SUSKIND, 1960).

Group II, CRM −: Neutralization tests are negative with these mutants. Thus, BONNER and his coworkers first assumed that the CRM mutants do not produce a protein serologically related to T-ase and that the mutation produces a quantitative effect. However, reexamination

Table VI-5. *Functional differences between mutants of the td locus of Neurospora crassa.* (Adapted from BONNER et al., 1960)

Group I: CRM +, Production of a protein serologically related to T-ase; shown to be cross-reacting material in a neutralization test.

Class 1: Inactive for reaction 1, 2, and 3
Class 2: Inactive for reaction 1 and 3, active for reaction 2
Class 3: Inactive for reaction 1 and 2, active for reaction 3

Cofactor requirements for mutants of class 3

	pyridoxal phosphate	L-serine
Subclass a	—	—
Subclass b	+	—
Subclass c	+	+

Group II: CRM —, Formation of a protein which is serologically related to T-ase, but that is not identical with the cross-reacting material; demonstrated by a complement fixation test.

with another serological method (complement fixation test) showed that these mutants also form a characteristic product which is related to the T-ase protein (BONNER, 1964a; KAPLAN et al., 1964c).

The mutants of group I could be divided into three classes (Table VI-5) on the basis of other criteria, namely their capacity to carry out reaction 2 and 3. Most of the CRM + mutants are unable to do this. However, one strain was found which could catalyze reaction 2; it grows following the addition of indole (RACHMELER and YANOFSKY, 1959, 1961). On the other hand, a whole series of mutants are able to carry out reaction 3. They require tryptophan for growth, but form considerable amounts of indole which cannot be metabolized further. This class can be subdivided into three subclasses on the basis of the cofactor requirements of the enzyme (DEMOSS and BONNER, 1959; SUSKIND and JORDAN, 1959).

An additional classification of *td* mutants can be made, if other criteria such as temperature sensitivity, suppressor effects (p. 397ff.), sensitivity toward zinc ions, and behavior of the enzyme after dialysis are used (YANOFSKY and BONNER, 1955a, b; SUSKIND and KUREK, 1959; WUST, 1961).

These investigations show that the *td* locus, which forms a multiplicity of different enzymes following mutation, must possess a complex genetic structure. Supplementation of the biochemical data with genetic experiments on the fine structure of the *td* locus has led to a clear idea of the correlation between gene structure and enzyme specificity (p. 395ff.).

Parallel results have been obtained in the biochemical analyses of *td* mutants in *E. coli*. (Review in YANOFSKY, 1960; YANOFSKY and ST. LAWRENCE, 1960.)

2. Indirect effects

The most obvious indirect effect of a gene mutation on an enzyme is the formation of an inhibitor which influences the synthesis or the activity of the particular enzyme. Thus in mutants which show a reduced enzyme activity with respect to the wild type, one first looks for the presence of an inhibitor. A test is made to determine whether after the mutant and wild type extracts are mixed, the enzyme from the latter is inhibited or not. Such an effect has been demonstrated in only a few cases.

SILVER and McELROY (1954) reported a mutation in *N. crassa* which leads to the formation of a thermo-labile protein-like product that inhibits the production of nitrate reductase, controlled by another locus.

HOGNESS and MITCHELL (1954) found that different loci of *N. crassa* have a quantitative inhibitory effect on the formation of tryptophan synthetase.

In *Neurospora* a series of examples of gene-dependent nutritional defects are known which cannot be clearly related to enzyme deficiencies: FINCHAM, in his literature survey (1960), has assembled a number of such unpublished observations from different investigators which allow one to assume an indirect effect on a particular enzyme.

1. A mutant which is auxotrophic for pantothenic acid can be shown to be capable of forming the enzyme responsible for the synthesis of this peptide (WAGNER and GUIRARD, 1948; WAGNER, 1949; WAGNER and HADDOX, 1951).

2. A series of allelic mutants in which arginine synthesis (compare Fig. VI-8 and Table VI-2, No. 1a) is blocked prior to citrulline formation grows only after addition of citrulline even though they possess the enzyme necessary for the conversion of ornithine to citrulline, namely, ornithine transcarbamylase; they also have the enzyme system necessary for the formation of ornithine (VOGEL and KOPAC, 1959; FINCHAM, 1960, see also p. 400f.).

As explanation for such abnormal behavior of auxotrophic mutants FINCHAM discusses several possibilities which may serve as bases for further experiments. At the moment we can only say that a few examples of indirect effects on enzymes are known, but that these do not permit any conclusions as to the mechanism by which such genes produce their effects.

In conclusion, if we review the experimental data on the effects of individual genes on enzyme formation, the question inevitably arises whether the variable nature of this influence (quantitative, qualitative, or indirect) is determined by different categories of genes which have different functions. Two such categories of genes have been demonstrated in extensive investigations with bacteria (literature in MONOD and COHN, 1952; JACOB and MONOD, 1961a, b). These are: *structural genes*, which carry the genetic information controlling enzyme protein structure, and *regulatory genes*, which have no effect on enzyme structure, but control the activity of the structural genes. The interaction of the two kinds of genes in enzyme synthesis will be discussed in the last section of this chapter (p. 405ff.). We wish to consider first the data obtained from fungi and determine if the genes responsible for enzyme synthesis can be found in both of these categories.

If *qualitative enzyme alterations* occur following a mutation, we can assume that a *structural gene* has been affected. According to the hypothesis the genetic information of this gene is altered by the mutation only to the extent that the original information is transferred in altered form (= missense mutation).

It should be made clear, however, that a gene can be classified with certainty as structural, only if it is shown that the amino acid sequence (primary structure) (p. 343) of the enzyme protein is changed in a mutant. This has been demonstrated only for hemoglobin (INGRAM, 1961), for RNA-dependent proteins of tobacco mosaic (WITTMANN, 1960; TSUGITA and FRAENKEL-CONRAT, 1962), and for tryptophan synthetase of *E. coli* (literature in YANOFSKY et al., 1963). All other changes in physical or chemical properties of mutant enzyme allow at best the conclusion that the tertiary or secondary structure has been altered. Since the secondary and tertiary structure of proteins is in all likelihood determined by the primary structure, such changes can be considered as indications of alterations in amino acid sequences.

It is much more difficult to classify mutant *hereditary factors* which influence *enzyme synthesis only quantitatively*. The rate of enzyme formation is either altered or completely prevented in such mutants. However, a quantitative effect can be attributed to the first group of genes only with reservation, since the experimental data in any of the cases described until now are insufficient to exclude completely a structural alteration of the gene in question. These genes are accordingly eliminated from our consideration. It is more reasonable to ascribe a regulating function to the genes of the second group which are incapable of enzyme synthesis following mutation. This has been certain in only one case in fungi.

The *ty* genes of *N. crassa*, depending upon the physiological environment, determine whether tyrosinase, controlled by the structural gene *T*, is produced or not (p. 373).

According to the recently published data of MATCHETT and DeMOSS (1962), a mutational site (*td* 201) within the *td* locus of *N. crassa* exists which is responsible for the regulation of tryptophan synthesis. The function of this mutational site has not yet been fully clarified.

Regulator genes have been demonstrated in bacteria in a number of cases (literature survey: MONOD and JACOB, 1961; JACOB and MONOD, 1961a, b; STARLINGER, 1963; WINKLER and KAPLAN, 1963).

All other genes which cause an irreparable loss of enzyme following mutation and which have not been investigated further *can also be considered structural genes*. One can assume that the genetic code of these hereditary units has been altered so extensively through mutation that information transfer is no longer possible in the changed form (nonsense mutation; p. 347).

As previously mentioned, the cases of indirect effects of genes on enzyme synthesis occasionally observed have not been thoroughly enough investigated in order to be evaluated in the present context. A more precise analysis of mutants in which enzyme formation (as far as is known) is quantitatively or indirectly controlled by genes will lead to a clearer conception of gene function in fungi if such experiments are carried out with the existence of structural and regulator genes in mind.

Summary

1. Biochemical analysis of auxotrophic mutants has shown that individual genes directly or indirectly determine the formation and characteristics of enzymes.

2. Genes can influence particular enzymes directly either quantitatively or qualitatively. Many examples of such cases are known.

3. The quantitative effect of a mutation can be a complete absence of enzyme synthesis or a decreased amount of enzyme produced, in the case of both constitutive and inducible enzymes.

4. Qualitative effects lead to the formation of structurally altered enzymes. Such changes in enzyme structure are generally correlated with a change in enzyme activity which, in extreme cases, is characterized by a complete loss of activity.

5. An indirect effect of genes may occur either by the formation of an inhibitor or by depression of enzyme activity in an unknown way. Only a few examples of this kind of gene effect are known and do not permit a generalized explanation.

6. More recent investigations on bacteria dealing with the relation between gene and enzyme have shown that two categories of genes, structural and regulatory, exist; these control enzyme formation directly or indirectly. In fungi the assignment of a mutational enzyme alteration to the latter of these categories of genes has been possible in only one case. In all other examples of direct or indirect effects on an enzyme it has been possible only on the basis of circumstantial evidence to determine that a structural or regulatory gene has been mutatively altered.

II. Gene structure and enzyme specificity

The preceding discussion has shown that specific genes, called structural genes, serve as templates for the synthesis of enzyme protein. Mutational changes within such genes determine structural alterations of the respective enzymes; in the simplest cases these are expressed as qualitative changes. This has led to the consideration of the correlation between the structural elements of a gene and the functional characteristics of an enzyme. In the fungi this problem has been approached through a series of different conceptions.

1. Intragenic complementation

By *complementation* we mean that the *functional defects of different mutants are not expressed phenotypically if their genomes occur together in a single cell*. Complementation is possible only if both mutants have undergone mutation at different sites in the genetic material. The loss of function as a result of mutation in one genome is compensated by the corresponding normal site in the partner genome.

Since the discovery of this phenomenon in deficiency mutants of *Neurospora crassa* (BEADLE and COONRADT, 1944) complementation has frequently been observed and analyzed not only for other auxotrophs of this and other species, but also for morphological mutants (p. 44f.). It occurs in heterokaryons (haplonts), in heterozygotes (diplonts, e.g. *Saccharomyces cerevisiae*), or following abortive transduction (bacteria; HARTMAN et al., 1958). As a result of such mutal compensation between mutants, prototrophic and morphologically normal phenotypes are produced. For a long time the existence of complementation was the criterion for the non-allelic nature of the respective mutants. For this reason the test for allelism was a complementation test (p. 211).

The investigations of MITCHELL and MITCHELL (1956) on the complementation behavior of alleles of the *pyr-3* locus of *N. crassa* showed that allelic mutants could also complement each other in certain combinations. Since it had become clear that genes consist of a number of linearly arranged mutational sites (p. 211 ff.), an analysis of *intragenic or interallelic complementation* appeared to be promising for revealing the functional characteristics of the gene. An insight into the correlation between gene structure and function should come from an analysis of loci which can be related to the absence of a specific enzyme (Literature survey: LEUPOLD, 1961; CATCHESIDE, 1962).

A compilation of loci of different species which display intragenic complementation is found in Table VI-6. The results of these experiments may be summarized under the following points:

1. *Occurrence*. In addition to the loci showing inter-allelic complementation, a series of complex loci are known which fail to do so (e.g. 3 loci of *N. crassa*, CATCHESIDE, 1962; the *ad-7* locus of *Schizosaccharomyces pombe*, LEUPOLD, 1957, 1961). Generally complementation can be demonstrated only between certain allelic mutants while the remainder do not show the phenomenon, e.g. of 97 mutants of the *his-3* locus, 53 fail to complement each other or any other mutants of the gene (Fig. VI-11). An exception is the *pyr* locus, the mutants of which all complement in at least one combination.

As DE SERRES (1962) has recently demonstrated, the absence of intragenic complementation may be determined by other genes which prevent heterokaryon formation between two auxotrophic strains (see also *vegetative incompatibility*, p. 98ff.). To demonstrate the failure of the complementation reaction unequivocally the use of isogenic stocks is required.

2. *Characteristics of heterokaryons and diplonts arising through complementation*. Most of these strains do not exhibit the normal phenotype of the wild type. They show great variability with respect to their growth characteristics and mycelial habit. In so far as the enzymes controlled by the loci concerned were tested, those from heterokaryotic and diploid strains showed quantitative as well as qualitative differences from the wild type enzymes.

WOODWARD et al. (1958) found that amounts of adenylosuccinase produced by the complementing mutants of the *ad-4* locus of *N. crassa* vary widely. Mutants of sites which are close together show a very low enzyme activity in heterokaryons while those farther apart reach 25% of the activity of the wild type.

Function

Table VI-6. *Compilation of auxotrophic loci of different objects for which intragenic complementation has been demonstrated*
In so far as known, the table includes the enzymes under the control of these loci.

object	locus	type of auxotrophy	enzyme	reference
Saccharo- myces cerevisiae	*ad-2* *ad-5* *ad-7* *ad-5/7*	adenine		BEVAN and WOODS, 1962 ROMAN, 1956, 1958 COSTELLO and BEVAN, 1964; DORFMAN, 1964
Schizo- saccharo- myces pombe	*ad-6*			LEUPOLD, 1961
Neurospora crassa	*ad-3 B* *ad-4*	adenine	adenylosuccinase	DE SERRES, 1956, 1963 GILES et al., 1957a, b; GILES, 1958; WOOD- WARD et al., 1958; WOODWARD, 1959, 1960
	ad-8		adenylosuccinate synthetase	ISHIKAWA, 1960, 1962a, b; KAPULER and BERNSTEIN, 1963
	am	α-amino- nitrogen	glutamic acid dehydrogenase	FINCHAM and PATEMAN, 1957a, b; PATEMAN and FINCHAM, 1958; FINCHAM, 1957, 1959b, c; PATEMAN, 1960; FINCHAM, 1962b
	arg-1	arginino- succinate	argininosuccinate synthetase	CATECHSIDE and OVER- TON, 1958
	arom-1 to 4	tryptophan, tyrosine, phenyl- alanine, p-amino- benzoic acid	dehydroshikimic acid reductase, dehydroshikimic acid dehydrase, protocatechuic acid oxidase	GROSS, 1958; GROSS and FEIN, 1960
	his-1 *his-2* *his-3* *his-5*	histidine	L-histidinol dehydrogenase glutamic phospho- histidinol- transaminase	CATCHESIDE, 1960b CATCHESIDE, 1960b WEBBER, 1960; CATCHE- SIDE, 1960b CATCHESIDE, 1960b
	iv-2 *iv-3* *leu-2*	isoleucine, valine leucine	β-carboxy-β-hydroxy- isocaproic isomerase	BERNSTEIN and MILLER, 1961 GROSS, 1962
	meth-2 *pan-2*	methionine pantothenic acid		MURRAY, 1960 CASE and GILES, 1958, 1960
	pyr-3	pyrimidine	aspartate trans- carbamylase	SUYAMA et al., 1959; WOODWARD, 1962; WOODWARD and DA- VIS, 1963

388

Table VI-6 (Continued)

object	locus	type of auxotrophy	enzyme	reference
	td	tryptophan	tryptophan synthetase	LACY and BONNER, 1958, 1961; SUYAMA and BONNER, 1964
	tryp-1		indoleglycerol phosphate synthetase/ anthranilic acid synthetase	CATCHESIDE, pers. commun.; AHMAD et al., 1964
	tryp-3 = td		tryptophan synthetase	AHMAD and CATCHESIDE, 1960

Qualitative enzyme differences were demonstrated by FINCHAM (1959b, 1962b) for different complementing combinations of mutants of the *am* locus in *N. crassa*. These involve differences in heat stability and substrate affinity of the enzymes among the various heterokaryons and as compared to the wild type [PARTRIDGE (1960) made similar observations on the *ad-4* heterokaryons]. A more precise analysis of the glutamic acid dehydrogenase from heterokaryons formed from the complementing nuclear types, *am-1* and *am-3*, showed that beside the enzyme proteins of the mutants a protein is produced which in its physico-chemical characteristics is similar to the glutamic acid dehydrogenase of the wild type (FINCHAM and CODDINGTON, 1963a).

As SUYAMA and BONNER (1964) have recently shown, the enzymes produced by *td* mutants following intragenic complementation differ from the wild type enzyme with respect to substrate affinity, cofactor requirements, and temperature stability.

3. *Complementation in vitro.* The investigations of WOODWARD (1959, 1960) are particularly significant in this connection. He succeeded in demonstrating a low enzyme activity in a mixture of extracts from complementing *ad-4* mutants. The activity reached 20% of that of the corresponding heterokaryons. Since the extracts showed no activity prior to mixing and no protein synthesis occurs normally under the experimental conditions, this result could not be attributed to a de novo formation of enzyme but resulted from an in vitro complementation of inactive enzyme material.

Similar results were obtained by FINCHAM and CODDINGTON (1963 a, b). In a mixture of purified enzyme proteins isolated from the complementing *am-1* and *am-3* mutants, the same enzyme spectrum was observed as that formed in the heterokaryon in vivo (see above), namely different glutamic acid dehydrogenases with the characteristics of both mutants as well as the characteristics of the wild type enzyme.

In vitro complementation is also possible between allelic *td* mutants (SUYAMA, 1963; SUYAMA and BONNER, 1964). The tryptophan synthetases formed in this way are qualitatively different from the enzymes arising after in vivo complementation and from the T-ase of the wild type.

4. *Complementation maps.* In all cases in which a large enough number of auxotrophic mutants of a locus have been tested for complementation

Function

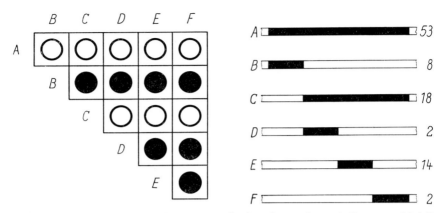

Fig. VI-11. Complementation pattern (left) and complementation map (right) for the *his-3* locus of *N. crassa*. The classification of the 97 mutants investigated into individual complementation groups can be noted from the numbers next to the map. Solid circle = complementation; open circle = failure to complement. The individual groups are characterized in the map by defects in the region represented in black. Non-complementing mutants have overlapping defects while the defects of those which complement do not overlap, e.g. in the non-complementing group A mutants the entire segment is black while in those of group B which complements all the other groups, only a short segment is black. (From CATCHESIDE, 1960a)

the mutants can be subdivided into groups depending upon whether or not they complement. Complementation does not occur within each group; it is possible only between mutants of different groups. However, complementation between specific mutants of different groups also fails to occur in certain cases. A characteristic complementation pattern is obtained for each group of a locus. In Fig. VI-11 (left) the complementation pattern of the *his-3* locus of *N. crassa* is shown as an example. The largest group (A) includes all the mutants which do not complement. The same is true for the mutants of group C which only complement with those of group B. The mutants of group B, D, E, and F, on the other hand, complement with all the groups. The complementation pattern of a locus can be represented in the form of a complementation map (Fig. VI-11 right).

Individual groups overlap within the complementation map as shown in Fig. VI-11. A satisfactory explanation for this has not been presented. Because the phenomenon is important for a concept of the structure and function of the gene, the problem will be discussed further in connection with the mechanism of intragenic complementation.

Similar linear complementation maps have been worked out for a series of loci which have been more intensively investigated (e.g. the *td, pan-2, ad-4, ad-8* loci of *N. crassa*; the *ad-6* locus of *S. pombe* and the *ad-2* locus of *S. cerevisiae*; reference: Table VI-6).

The almost complete agreement of the complementation maps of LACY and BONNER (1958, 1961) and AHMAD and CATCHESIDE (1960), which were worked out independently, is evidence of the accuracy of this method of mapping.

The complementation groups of a locus cannot always be represented by a linear map. The mutants of the *his-1*, *his-5*, and *lys-5* loci may be cited as examples (CATCHESIDE, 1962). This difficulty can be circumvented in the *iv-3* mutants if the complementation groups are arranged in a circle (BERNSTEIN and MILLER, 1961). Such circular complementation maps have also been devised for the *ad-8* (KAPULER and BERNSTEIN, 1963), and the *leu-2* locus (GROSS, 1962)[1]. The remainder of the loci cited above have not yet been mapped by this method.

What conclusions can be drawn about gene physiology from the phenomenon of intragenic complementation? The most obvious is that the complementation map reflects a subdivision of the particular locus into different functional regions, each of which includes numerous mutational sites and serves as a template for one of many polypeptide chains of the enzyme molecule. This idea, originally developed by WOODWARD et al. (1958) is no longer valid for the following reasons:

1. The complementation groups of thoroughly investigated loci number between 10 and 20 [e.g. WOODWARD (1962) found 18 for the *pyr-3* locus]. It is unlikely that this corresponds to the number of polypeptide chains in an enzyme protein.

2. The overlapping of the individual groups cannot be reconciled with the linear arrangement of mutational sites in a gene.

However, one can assume that the mutants of an overlapping group (e.g. A in Fig. VI-11) are not point mutations, but defects involving a larger portion of the gene (e.g. block mutations, p. 311; multiple point mutations). The following data (CATCHESIDE, 1960a) argue against this idea. In the first place mutants of overlapping groups show a normal rate of reversion to wild type and secondly produce prototrophs at a normal frequency in crosses with one another. A restitution of the wild type character is not probable through either recombination or point or chromosomal mutation. Another possible explanation for overlapping involves the assumption that the individual functional regions have segments in common (ROMAN, 1958). A mutation in such a zone of overlap gives a defect in both functional regions. For the example which is used here, the *his-3* locus, this would mean that all mutants of group A (see Fig. VI-11) lie in a region which overlaps all other regions. This is irreconcilable with a linear gene structure.

3. The enzymes produced in complementary heterokaryons show in part structural alterations.

These arguments, which led to a rejection of the idea described above, cannot be raised against the second hypothesis (BRENNER, 1959; FINCHAM, 1960) discussed in detail by CATCHESIDE (1962) and ISHIKAWA (1962b).

The first question to be raised is at what point in the pathway from gene to enzyme complementation exercises its effect. Theoretically it could occur at every stage from DNA to messenger RNA through the formation of polypeptide chains to their aggregation and folding into protein. Surely interaction does not occur between the DNA of the two complementing

---

[1] *Acknowledgement in proof:* Circular complementation maps were independently established for the *ad-5/7* locus of *Saccharomyces cerevisiae* (COSTELLO and BEVAN, 1964; DORFMAN, 1964).

nuclei in the heterokaryon, since the two kinds (and only two kinds) of nuclei can be isolated from the heterokaryon in an unaltered state at any time. Complementation between RNA or polypeptides appears possible, but does not explain the fact that certain heteroallelic mutants complement and others do not.

Therefore, *complementation is assumed to occur between enzyme proteins which have already been folded into their tertiary structure. This means that in the ultimately active enzyme molecule two or more molecules, each of which is not functional as a monomeric mutant protein, join to form the quaternary structure of a hybrid protein.* However, this assumes that intragenic complementation can only occur with polymeric enzymes.

A polymeric structure for enzyme proteins has recently been demonstrated in several cases: e.g. CRAWFORD and YANOFSKY (1958) for tryptophan synthetase of *E. coli* (p. 396); KÄGI and VALLEE (1960) for alcohol dehydrogenase of yeast; GROSS (1962) for an isomerase acting in the leucine synthesis in *N. crassa* (for additional literature see ISHIKAWA, 1962b).

According to this idea, the mutational alteration of DNA leads by way of RNA to a change in the amino acid sequence of the monomers and is expressed in alterations in the folding of the chains. *The aggregation of the individual protein components to form an active enzyme is only possible if the tertiary structure fits.* The protein interaction required in such a hypothesis is represented schematically in Fig. VI-12.

On the basis of this theory, which has the great advantage of leaving the concept of the gene as a functional unit unchallenged, the following experimental data can be explained:

1. Loci which show no intragenic complementation may be responsible for the formation of monomeric enzyme proteins; the alteration of their tertiary structure cannot be repaired by the formation of a hybrid protein.

2. Enzyme formation occurs in vitro in a mixture of extracts from complementary mutants (p. 388).

3. All mutants of the *td* locus which complement form a protein which is immunologically related to the enzyme (CRM, p. 382) (LACY and BONNER, 1961).

4. The structurally altered enzymes which have been repeatedly demonstrated are reconcilable with the hybrid nature of proteins obtained following complementation.

It is essential to point out in this connection that the qualitative alterations described by FINCHAM (1962b; see also p. 387f.) in the enzymes obtained through complementation can be attributed to changes in tertiary structure.

5. However, the probability that two mutant proteins have become so altered in their tertiary structure that they fit together is slight. This explains the fact that a large portion of the mutants of a locus do not complement and thus accounts for overlaps within the complementation map.

6. According to LEUPOLD (1961) the variable extent and overlapping of complementation groups may be a matter of the effect of the folding of the polypeptide chain the amino acid sequence of which has been altered by mutation. Correction of the deformation occurs only in those

regions which in the final protein are in contact with the non-deformed homologous regions of the partner chain. In those regions in which both polypeptide chains show a deformation of their 3 dimensional structure, there will be no correction.

7. The investigations of SCHLESINGER and LEVINTHAL (1963) on the complementation behavior of *E. coli* mutants which fail to form active alkaline phosphatase also support this theory. This enzyme possesses a

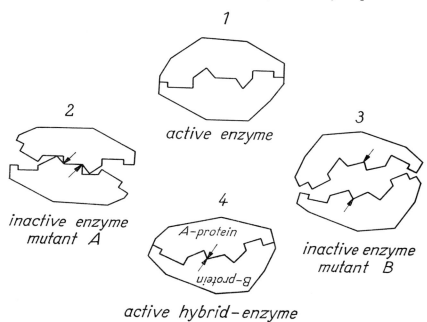

Fig. VI-12. Model of intragenic complementation. A dimeric enzyme protein is postulated. The normal enzyme consists of two identical proteins (1). The two mutants A (2) and B (3) form proteins which differ in their tertiary structure from that of the wild type protein (arrows). The two homologous proteins do not fit together and no active enzyme is formed. The A and B proteins are able to aggregate, however, and produce a dimeric hybrid (4) which resembles the wild type protein in form and which is active. (From CATCHESIDE, 1962)

dimeric structure and consists of two identical subunits (ROTHMAN and BYRNE, 1963). SCHLESINGER and LEVINTHAL succeeded in purifying and separating the enzymatically inactive, but serologically related protein (= CRM, p. 382) of the mutants. The monomers of different mutants aggregated in vitro to form a hybrid molecule with a slight enzymatic activity.

This raises the question of whether or not *the sequence of the mutational sites of a gene that are brought together in a complementation group coincides with the linear arrangement obtained through recombination mapping.*

Function

Such comparisons have been made for a number of loci. For *N. crassa:* *pan-2* locus (CASE and GILES, 1960); *me-2* locus (MURRAY, 1960); *pyr* locus (WOODWARD, 1962); *ad-8* locus (ISHIKAWA, 1962b); for *S. pombe:* *ad-6* locus (LEUPOLD, 1961).

These investigations have shown that:

1. A correlation generally exists between the sequence of the complementation groups and the mutational sites in both maps.

Of a total of 23 allelic mutants of the *pan-2* locus which could be arranged into 12 partially overlapping groups, only 3 failed to agree with the "recombinational sequence" (CASE and GILES, 1960). A distinctly lower degree of colinearity in the sequence was found for the *pyr* locus; 7 out of 27 mutants did not correspond in arrangement in this case (WOODWARD, 1962).

2. The non-complementing mutants are distributed at random over the entire recombination region.

3. The relative distance between individual complementation groups and corresponding mutational sites of both maps differs.

These results can likewise be explained with the hybrid-molecule theory of intragenic complementation. If one assumes that the relative position of the damage induced in a protein monomer by mutation coincides with the sequence of the mutational sites in the gene, it is entirely possible that the original linear arrangement is completely altered through the folding of the protein molecule. The colinearity of gene and primary structure of protein originally present is thus no longer recognizable. The random distribution of the non-complementing mutational sites may be explained through the assumption that either the genetic information of these sites is so disturbed that no protein synthesis can occur or that the proteins concerned are so altered in their tertiary structure that they no longer can form hybrid molecules.

Another model for explaining intragenic complementation has been proposed by KAPULER and BERNSTEIN (1963). It is based on a comparison of a circular complementation map for the *ad-8* locus of *N. crassa* (p. 389) with a novel, spirally-arranged recombination map. The two authors assume that the *ad-8* locus specifies the formation of a polypeptide which takes the form of a spiral with two turns in its tertiary structure. Further, these individual polypeptide helices are so stacked to form the enzyme molecule that the homologous sites of adjacent spirals are matched. In a mutant each polypeptide spiral has a defect at a specific site; thus the defects form a line through the entire protein. The polypeptide spirals produced by two different nuclei in a heterokaryon may both participate in the formation of the stacked protein. With such a structure the same explanation of complementation and non-complementation holds on the molecular level as indicated by the complementation map, namely, complementation occurs with non-overlapping, and non-complementation with overlapping of defects. Unfortunately experimental evidence for this model is lacking.

The investigations on intragenic complementation have strengthened the view that the gene is the functional unit which determines structure; individual mutational sites within a gene show either no or incomplete complementation in contrast to the complementation behavior of non-allelic genes. The reconstitution of the normal wild type character through intragenic complementation could not be demonstrated in the more carefully investigated cases. Further, these experiments have also shown that the individual mutational sites of a gene participate in different ways in forming the gene product. A colinearity between location and function of the mutational sites appears possible. The investigations of the *td* mutants of *N. crassa* discussed in the following section also point in this direction.

2. Genetic fine structure and function

The problem of the gene map corresponding to the function of the mutational sites raised by the investigations on intragenic complementation was attacked in a different way by BONNER and his group (BONNER et al., 1960; SUYAMA et al., 1964; KAPLAN et al., 1964a). These workers have mapped the mutants of the *td* locus of *N. crassa* which are characterized by different enzyme specificities (Table VI-5, p. 383). They used the method presented in detail in the chapter on *recombination* (p. 211 ff.). Surprisingly, the functionally different mutants are not distributed at random over the entire region of the *td* locus, and a correlation between position and function does exist (Fig. VI-13).

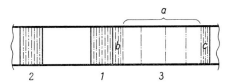

Fig. VI-13. Map of the *td* locus of *N. crassa*. The relative positions of the different mutational sites of the *td* mutants of group I (Table VI-5) are designated by the interrupted vertical lines. The subdivision of the total region corresponds to the classification of the mutants into the functional groups shown in Table VI-5. Explanation in text. (Adapted from BONNER et al., 1960)

The diagram in Fig. VI-13 and the data of Table VI-5 show that the region of the *td* locus which controls the formation of tryptophan synthetase (T-ase) can be subdivided into a number of units which correspond to the individual specificities of the CRM-positive mutant classes (Table VI-5, Group I). The region begins with mutants of class 2, that only form tryptophan following the addition of indole (Fig. VI-9). A genetically inactive segment follows in which no mutants have been found. Mutants of class 1, which only form CRM and do not exhibit any enzyme activity, are found adjacent to it. The mutants of class 3, which are able to convert indoleglycerol phosphate to indole, fall into the second half of the region.

A correlation between mutational site and function can be recognized for this segment with respect to cofactor requirements. Mutants of the subclasses b and c which require a cofactor for the formation of indole, fall at each end of the region. The mutants without any cofactor requirements (subclass a) are scattered between the ends.

The CRM-negative mutants of group II (Table VI-5) cannot be mapped without contradiction (BONNER, 1964a, b; KAPLAN et al., 1964b).

Function

The existence of a genetically inactive region requires a brief comment. Since a relatively small number of mutants have been mapped, the existence of such a region may be only apparent. On the other hand, it is also possible that within the region of the gene sites exist whose alteration has no significant influence on the function of the enzyme formed. The fact that the mutational sites which are responsible for the enzyme specificity occur in groups suggests that sites having an influence on the amino acid constitution without affecting the catalytic characteristics of the enzyme may likewise be grouped into definite regions.

These results can only be discussed in light of the data on the T-ase of *E. coli*.

Literature: YANOFSKY (1960), more recent studies: HELINSKI und YANOFSKY (1962a, b), HENNING und YANOFSKY (1962a, b), HATANAKA et al. (1962), HENNING et al. (1962), CARLTON und YANOFSKY (1962), YANOFSKY et al. (1963).

The T-ase of this bacterium, which has also been isolated in pure form, has the same catalytic characteristics as that of *N. crassa*. The pure enzyme, the molecular weight of which is about 140,000, can be separated into two protein components, A and B, with molecular weights of 29,500 and 114,000 respectively (WILSON and CRAWFORD, personal communication). Although the A and B proteins catalyze the reactions 2 and 3 to a small extent (Fig. VI-9), the essential biological function of the T-ase (reaction 1) can only be controlled by the complex of the two proteins. The mutational sites which code the two proteins lie in two non-overlapping regions. In contrast to the fine structure of the *td* locus of *N. crassa* (Fig. VI-13), the mutants of *E. coli* which have lost the A as well as the B function do not fall into a third region, but are distributed throughout only two regions.

It has further been shown through comparative biochemical analyses of wild and mutant T-ase that the *qualitative enzyme alteration of individual point mutations is determined by alteration of the amino acid sequence*. Such clear proof that a point mutation leads to an alteration of the primary structure of an enzyme has not been demonstrated in gene-enzyme systems of fungi.

In addition to the constitutively formed T-ase studied by the Yanofsky group, NEWTON and SNELL (1962) found an inducible T-ase in tryptophan-deficient mutants. Its production is induced by high concentrations of tryptophan. It catalyzes the synthesis of tryptophan only from indole and serine (Fig. VI-9, reaction 2). This T-ase was shown to be neither identical with the A nor the B protein of the constitutively produced T-ase. Since a correlation between the biochemical and genetic data is incomplete, these experiments are not definitive.

The investigations of T-ase in *N. crassa* and *E. coli* indicate that *although a gene consists of innumerable mutational sites, it is composed of only a few functional regions*. The *td* locus of *E. coli* encompasses the two non-overlapping regions A and B; the *Neurospora* data can also be interpreted to indicate two functional regions which correspond to the A and B regions of the bacterium (BONNER, 1963).

The mutants of the INGP region (Fig. VI-13, region 2) form a T-ase which is unable to catalyze the breakdown of indoleglycerol phosphate. This enzyme, which corresponds to the B protein of *E. coli* in function, can only control the conversion of indole and serine to tryptophane (reaction 2, Fig. VI-9). Similarly an enzyme is formed by the mutants of the indole region (Fig. VI-13, region 3) which corresponds to the A protein in its functional characteristics. This enzyme cannot catalyze the indole-serine reaction, but can accelerate the splitting of the indole from indoleglycerol phosphate (reaction 3, Fig. VI-9). The mutants of region 1 (Fig. VI-13) form a

CRM which lacks catalytic activity for INGP as well as indole. BONNER assumes that in this part of the INGP and indole domain regions 2 and 3 overlap.

The essential difference between the two functional regions which are responsible for the characteristics of the fungal and bacterial T-ases is that in the fungus they overlap while in the bacterium they apparently do not. One can assume that the genetic information for both functions is coded in the zone of overlap. Evidently this is reflected in the protein structure of *Neurospora* T-ase, since it has not been possible to split this enzyme into two protein components, A and B, characteristic of *E. coli*.

In discussing intragenic complementation (p. 394) the results of which have led to the idea that a gene consists of a number of functional regions, we came to the conclusion that *the gene as a whole can nevertheless be viewed as a structure-determining functional unit*. This also holds for the *td* locus, since every point mutation within the gene leads to tryptophan auxotrophy. The mutants concerned have lost the capacity to form an enzyme which catalyzes the essential reaction 1 in vivo (Fig. VI-9).

The investigations of DAVIS (1960) and DAVIS and WOODWARD (1962) with *pyr-3* mutants of *N. crassa* allow generalization of the correlation between structure and function shown by the *td* locus.

30 pyrimidine auxotrophs of the *pyr-3* locus could be subdivided into two classes through enzymatic analysis: 7 strains produced the wild-type form of the enzyme, aspartate transcarbamylase (ATC), on a supplemented minimal medium; the remaining 23 did not. The two categories of mutants could also be differentiated by other criteria. [Discussed in greater detail in connection with suppressor genes (p. 400f.).] While the ATC-positive mutants, which correspond to the CRM-positive mutants of the *td* locus (Group I, Table VI-5), are localized in a single genetic region with one exception, the ATC-negative mutants are not (compare with CRM-negative mutants). Their mutational sites are scattered over the entire region.

3. Suppressor genes

The effect of a mutation may be partially or completely nullified by a non-allelic suppressor gene. Suppressor genes may be linked to the loci which they affect, or they may occur on other chromosomes. Suppressor mutations can be distinguished from reversions (p. 267, 401) by back-crossing with the wild type.

A dihybrid segregation is obtained among the progeny from a cross between a suppressor mutant and the wild type ($su\ a \times su^+a^+$). Both the parental and recombinant type ($su\ a^+$) exhibit the wild phenotype. The recombinant type, su^+a, shows the mutant phenotype. The frequency of the recombinant types depends upon whether or not the suppressor and mutant locus are linked. For example, with random assortment 25% of the recombinant, su^+a, occurs.

Since a reversion takes place within the same locus in which the forward mutation, occurred, the progeny from a backcross ($a^+ \times a^+$) is uniformly wild type.

Such crossing results are only valid, of course, in the case of complete restoration of the wild type character by suppression or reversion. If the restoration of the wild type character is incomplete, a 1:1 segregation for reverse mutants and wild type or a 2:1:1 segregation for wild (su^+a^+, $su\ a^+$): suppressed mutant ($su\ a$): original mutant (su^+a) is obtained in the F_1 if there is no linkage.

If tetrad analysis is possible, only a few asci are necessary to differentiate between the two phenomena, provided the suppressor is not closely linked to the mutated gene.

The occurrence of suppressor mutations is of particular interest in connection with a discussion of the structure and function of the genetic material, because a superficial consideration gives the impression that several loci are responsible for the structure of an enzyme. This contradiction can be resolved if an auxotrophic-prototrophic system which can be enzymatically analyzed is used for the study of suppressor effects.

Suppressor genes for auxotrophic mutants have been described repeatedly (GILES and PARTRIDGE, 1953; STRAUSS and PIEROG, 1954; LEWIS, 1961). Biochemical data on the action of suppressor genes on enzymes are available for only the *td* locus of *N. crassa* (and *E. coli*) (YANOFSKY, 1952a, 1958; YANOFSKY and BONNER, 1955b; YANOFSKY and CRAWFORD, 1959; CRAWFORD and YANOFSKY, 1959; SUSKIND and KUREK, 1959; SUSKIND and JORDAN, 1959) and the *pyr-3* locus of *N. crassa* (DAVIS, 1961, 1962a, b; DAVIS and WOODWARD, 1962; WOODWARD and DAVIS, 1963; WOODWARD and SCHWARZ, 1964).

25 allelic mutants of the *td* locus of *N. crassa* (*td-1* to *td-25*) were tested for suppressor genes which arose either spontaneously or after UV irradiation. 4 suppressor genes were isolated from about 50 strains which became more or less strongly prototrophic. Two of the suppressors were not linked either to each other or to the others. The remaining two appeared to be either closely linked or allelic. The mechanism of action of the four suppressors on the various *td* mutants was investigated genetically and biochemically.

The results of the experiments with *td* mutants are summarized under the following points:

1. *Incomplete restoration of the wild type phenotype.* In no case do the suppressed mutants show the same growth characteristics on minimal medium as the wild type.

2. *Specificity of suppressor genes.* By introducing suppressor genes into the *td* mutants, it could demonstrated that a suppressor operates on a specific *td* allele (Table VI-7). No suppressor gene is generally able to alleviate the tryptophan deficiency in a number of different mutants. Since *td-3* and *td-24* possess the same functional defect, both mutants are suppressed by the same suppressor. The gene, *su-6*, which suppresses *td-6* as well as *td-2*, is an exception.

Table VI-7 also shows that suppressors were not found in CRM-negative mutants (p. 383) nor did such genes have an effect on them. (However, in *E. coli* CRM-negative mutants do react to suppressor genes.) On the other hand, CRM-positive mutants are not necessarily subject to suppression (e.g. *td-7*).

Table VI-7. *Effect of suppressor genes on td mutants of N. crassa*

The index shows the *td* mutants from which the particular suppressor was isolated. The column "CRM" records the presence or absence of enzymatically inactive, but "serologically related to enzyme" protein. Under the heading "suppressor gene" + = suppression; — = no suppression; 0 = not tested. (From YANOFSKY, 1960, supplemented)

td strain	CRM	suppressor gene			
		su-2	*su-6*	*su-3*	*su-24*
td-1	—	—	—	—	—
td-16	—	—	—	0	0
td-7	+	—	—	—	—
td-2	+	+	+	—	—
td-6	+	—	+	—	—
td-3	+	—	—	+	+
td-24	+	—	—	+	+
remaining 18 *td* strains	+ or —	—	—	—	—

3. *Characteristics of tryptophan synthetase formed as a result of suppressor action.* All suppressible strains produce quantitatively less tryptophan synthetase (T-ase) than the wild type. It was necessary, of course, to determine whether suppressor genes exert a qualitative as well as quantitative effect on enzyme formation. As the investigations described below show, this seems to be true in one case, but not in others.

The mutant *td-2* forms a structurally altered tryptophan synthetase which can catalyze only the hydrolysis of indoleglycerol phosphate (Fig. VI-9, reaction 3). Under the influence of the gene *su-2* this strain forms a small amount of normal T-ase in addition to the defective enzyme. Similar observations were made in *E. coli*, in which the same type of enzyme damage was noted.

A biochemical comparison of T-ase formed by the suppressed mutant, *td-6 su-6*, with that of the wild type showed, nevertheless, no qualitative differences with respect to substrate affinity, activity, or stability.

Unequivocal proof of the absence of a structural effect on the T-ase through a suppressor was obtained in the experiment with the mutant *td-24*. The mutant fails to show any enzyme activity if it is cultured at a temperature of 25° C. However, if the temperature is raised to 35° C, small amounts of T-ase are produced. In contrast to the wild type T-ase, this enzyme is sensitive to zinc ions. Biochemical analysis of partially purified enzyme preparations showed that the inhibition of enzyme formation at 25° C is induced by a metallic ion-containing inhibitor the effect of which can be doubled by the addition of zinc ions. Following suppression of this mutant (through *su-24* as well as *su-3*) it retains its zinc sensitivity, although the enzyme is produced at 25° C. Thus one can conclude that the suppressor makes the inhibitor ineffective in this case.

What conclusions about the *mechanism of suppressor action* can be drawn from these experiments? It is *not possible to develop a general theory of suppressor action* on the basis of the currently available information. In light of the specificity of suppressor genes and the investigations on the mutant *td-24* one can conclude that suppressor genes cannot take over the function of the *td* locus, i.e. they have no influence on enzyme structure. This view agrees with the current dogma that *a single* structural gene exists for every enzyme. Let us now consider the investigations with *td-2* mutants; here through the action of the suppressor an enzyme is apparently produced which is qualitatively different from the mutant protein. Is this the first experimental evidence for the existence of more than a single structural gene per enzyme? This is not the case, as BRODY and YANOFSKY (1963) have shown through their analysis of suppressor gene effects on the *td* locus of *E. coli*. From their investigations one can conclude that the suppressor gene influences the amino acid sequence during formation of the enzyme protein. As a result of "incorporation errors" it is possible that the amino acid sequence (primary structure) of the mutated structural gene is by chance so altered at one or another position that a functional enzyme molecule is produced. Therefore, it would be appropriate to attack this problem in *N. crassa* again. One possibility is offered by a biochemical analysis of pure T-ase from the wild type and from suppressed mutants, the preparation of which has recently become possible (p. 381).

An entirely different mechanism of action from that of the suppressors of the *td* locus is shown by a suppressor which allows the mutants of the *pyr-3* locus to become prototrophic for pyrimidine.

pyr mutants are blocked in one of the first steps of pyrimidine synthesis. They are unable to convert aspartic acid into ureidosuccinic acid (see also Fig. VI-8). The enzyme, aspartic transcarbamylase (ATC), is responsible for this step.

As previously mentioned (p. 397), the *pyr-3* mutants can be subdivided into two classes: those which are able to produce ATC and those which fail to produce ATC when they are grown on supplemented minimal medium. The suppressor gene, *s*, which was first discovered by HOULAHAN and MITCHELL (1947, 1948) and analyzed by MITCHELL and MITCHELL (1952) can only compensate for the auxotrophy of the ATC-positive mutant class (DAVIS and WOODWARD, 1962). It has no effect on the mutants of the ATC-negative class.

The MITCHELLS had previously discovered that the suppressor, *s*, could also alleviate the proline auxotrophy of the *prol-2* and *prol-3* mutants. This is understandable since pyrimidine and proline synthesis are related to arginine synthesis (Fig. VI-8).

REISSIG (1960) had pointed out earlier the close relationship between pyrimidine and arginine synthesis. He found that the *pyr-3* locus suppresses the arginine auxotrophy produced by the *arg-2* locus, i.e. *pyr-3 arg-2* double mutants are only auxotrophic for pyrimidine (p. 276).

The physiological effect of the suppressor gene, *s*, became clear through the extensive genetic and biochemical investigations of DAVIS and WOODWARD cited above:

1. The ATC-positive *pyr-3* mutants which are suppressed by *s* produce a qualitatively altered ornithine transcarbamylase (OTC) the specific activity of which reaches only 2% of the wild type enzyme. (In *pyr-3 s⁺* strains the OTC activity is normal.)

2. The suppressor effect can be alleviated by addition of arginine to the culture medium.

3. Neither the synthesis not the activity of ATC is inhibited by arginine.

DAVIS concluded that the effect of the gene s, apart from its influence on the structure of OTC (DAVIS and THWAITES, 1963) consists mainly *of diverting precursors of arginine synthesis to different synthetic pathways.*

By reducing the OTC activity s makes it possible for ornithine to be converted to glutamic γ-semialdehyde (Fig. VI-8) which becomes available for proline synthesis. In this way the proline block can be overcome, since the *prol* mutants are blocked in the synthesis of this substance for proline synthesis (VOGEL and KOPAC, 1959, 1960).

The alleviation of the *pyr* defect can be explained in the same way, if it is assumed that carbamyl phosphate (CAP) can be produced in two different pools, one for arginine synthesis (CAP-A) and one for pyrimidine synthesis (CAP-P). The ATC-positive *pyr* mutants lack CAP-P. CAP-A is available for pyrimidine synthesis from the reduction in OTC activity as a result of the effect of s (for the physiological significance of CAP see JONES, 1963).

The inhibitory effect of arginine must then be traced back to a suppression of CAP-A synthesis, since this substance is no longer needed for arginine synthesis when arginine is present in the culture medium. Experimental evidence which appears to confirm this hypothesis has been published recently (DAVIS, 1963).

A further contribution to the clarification of the function of suppressor genes is found in the enzymatic analysis of suppressors of *Saccharomyces* carried out by HAWTHORNE and MORTIMER (1963). These investigators discovered that two non-allelic suppressors have a specific effect on different auxotrophies which have no physiological relationship with one another. However, only individual alleles of the 14 loci investigated respond to the effect of the suppressor genes.

It is already apparent from the relatively limited biochemical investigations on suppressor genes that although the *suppressor effect* (restoration of prototrophy from auxotrophy) is *uniform, the physiological processes leading to the alleviation of the defect are quite diverse.* With T-ase the suppressor effect leads to *repair of an enzyme defect.* Auxotrophy of the *pyr* and *prol* mutants is suppressed through the *induction of an enzyme defect* allowing the utilization of substances which normally cannot be used by the mutants for pyrimidine and proline synthesis. Accordingly, the *suppressors* appear *to act primarily as regulator genes.* A more precise biochemical and genetic analysis of suppressed mutants (on the basis of the JACOB and MONOD model, p. 407 ff.) could lead to a confirmation of this idea and the discarding of the concept of "suppressor." However, the function of suppressors as structural genes in individual instances cannot be excluded with certainty. This can be reconciled by assuming that the particular enzymes have a polymeric structure and are composed of non-identical monomers.

4. Reverse mutations

Reverse mutations, i.e. those which restore the original phenotype, have been frequently described in microorganisms (e.g. GILES and LEDERBERG, 1948; GILES, 1951, 1953, 1958; KØLMARK and WESTERGAARD,

1953). They are readily recovered from deficiency mutants (p. 276 f.). In contrast to suppressor mutants (p. 397 ff.), which produce the same phenotypic effect, a reversion always takes place within the same gene in which the original mutation occurred. These two mutational events can be readily differentiated by backcrossing the recovered normal strain to the wild type.

As already pointed out in the chapter on mutation, *one must ask whether or not a reverse mutation restores the original phenotype on the molecular level.* Since a gene consists of numerous mutational sites, it seems unlikely that a mutation, the direction of which cannot be controlled, should occur at exactly the point in the DNA at which a previous mutation took place. This problem can be attacked by using auxotrophy-prototrophy in the same way as in the analysis of suppressor genes, comparing the enzymes formed by the reverse mutants with those of the wild type. The results of such experiments should not only lead to a better understanding of the mutation process, but also yield valuable information on the relation between structure and function of the genetic material.

GILES and coworkers analyzed enzymatically revertants of the *ad-4* locus of *N. crassa* (GILES, 1958; WOODWARD et al., 1960). The *ad-4* locus is responsible for the formation of the enzyme, adenylosuccinase (p. 380, Table VI-5).

The prototrophic revertants obtained from different heteroallelic mutants showed in part quantitative differences in enzyme production. They could be grouped into several classes, the enzyme content of which lay significantly below the level of the wild type.

For example, among the revertants of the *F 12* mutant numerous strains were found which produced 3, 10, or 25% of the adenylosuccinase formed by the wild type.

Such quantitative enzyme differences may be traced in part to qualitative changes. Deviations from the wild type enzyme in heat stability, substrate affinity, stability in different buffers, and sensitivity to inhibitors argue for structural alterations in the adenylosuccinase of certain revertants.

These results agree with the assumption of STADLER and YANOFSKY (1959) that the quantitative enzyme differences demonstrated among the revertants of the *td* locus of *E. coli* could be explained partially through structural differences in the newly formed enzymes. This assumption was confirmed in later investigations. YANOFSKY et al. (1963) were able to show that the amino acids which were altered in the enzyme protein through forward mutation were replaced by other amino acids in the revertants. This led to a partial restoration of enzyme activity. Other investigations on bacteria point in the same direction (MAAS and DAVIS, 1952; GAREN, 1960).

More precise knowledge of the biochemical characteristics of revertants is available for the *td* locus of *Neurospora crassa* (ESSER et al., 1960). A large number of revertants were isolated from a UV irradiated, CRM-producing mutant (*td-2*) belonging to group I, class 3 (Table VI-5). 25 strains were selected which could not be differentiated from the wild type with respect to their growth rate and mycelial habit. The enzymatic activity of the revertants was determined for the three reactions of

T-ase and correlated with the amount of enzyme protein (CRM) (measured immunologically). It was possible to divide the revertants into four significantly different classes according to the ratio of enzyme activity to CRM. The T-ase produced by the revertants of class 1 (13 strains) could not be differentiated qualitatively from the wild type enzyme. In contrast, the representatives of class 2 (3 strains), class 3 (8 strains), and class 4 (1 strain) possessed T-ases with structural characters differing from those of the wild type enzyme.

How can such enzymatic differences be explained as the result of reverse mutation? If the existence of these categories actually involves enzymatic differences, at least 4, or perhaps 6 alternative enzyme configurations must exist among the 25 revertants and they must all be able to carry out the 3 reactions of the wild type enzyme. Thus it follows that at least the same number of alternative gene configurations which are responsible for the restoration of the wild type character must exist. In light of the current theory of the genetic code (p. 344 ff.) it seems unlikely that so many gene configurations could arise at a single mutational site (td-2) as a result of reverse mutation. Therefore, *reversions must be produced by a mutational alteration at another site in the td locus, at least in those strains with an altered enzymatic behavior.* The term *"intragenic suppression"* was introduced for this phenomenon. This hypothesis can also be used to explain similar experiments carried out on td mutants of E. coli (HELINSKI and YANOFSKY, 1963; ALLEN and YANOFSKY, 1963).

WOODWARD (1962) explained the change in the complementation pattern of a secondary mutant obtained from a UV-irradiated mutant of the *pyr-3* locus by intragenic suppression. As genetic experiments showed, the second mutational event apparently did not involve the original mutational site, but a closely adjacent site within the *pyr* gene.

Extensive enzymatic investigations have been carried out on the revertants of the *am* locus of *N. crassa* (PATEMAN, 1957; FINCHAM, 1957; FINCHAM and BOND, 1960; FINCHAM, 1962b). The *am* locus is responsible for the synthesis of glutamic acid dehydrogenase (GAD) (p. 379f.; Fig. VI-8, Table VI-3). Revertants of the allelic mutants, *am-1*, *am-2*, *am-3*, were obtained by UV-irradiation. While none of the *am-1* revertants could be distinguished enzymatically from the wild type, the remaining revertants produced highly variable amounts of enzyme. Purified enzyme from three of these revertants, which under standard conditions produced considerably less enzyme than the wild type, were analyzed biochemically.

All three enzyme varieties could be brought to an activity level of 50% or more of the GAD activity of the wild type either by short term heat treatment (35—40° C) or through addition of substrate (ketoglutaric acid, ammonium ion) and cofactor (TPN). The difference between the enzymes of individual revertants consisted of their reaction to these activation criteria. Moreover, all three species of enzymes showed a characteristically lower heat stability than the wild enzyme.

The defect in the enzyme protein appears in this case to consist not of an absence of catalytically active groups but of a deformation of the tertiary structure which is responsible for the function of the active sites. While this structure is very stable for normal GAD, in the enzymes of revertants it appears to be dependent upon external conditions to a large extent. It is not yet known whether or not such structural characteristics are the result of differences in primary structure (amino acid sequence).

The enzymatic characteristics of one of the three strains the GAD of which is normally inactive argue against an alteration in the tertiary structure of the revertant enzymes. This enzyme can be activated after heat treatment. After cooling, the activity of the GAD drops, but with reheating it increases again.

Genetic and biochemical investigations on revertants indicate that the *restoration of deficiency mutants to prototrophy by reverse mutation may differ in genetic basis as well as in physiological manifestation.* Thus the concept *"reverse mutation"* like that of *"suppression"* (p. 401) should be used *only for characterizing a specific phenotypic effect.* In the genetic sense reverse mutation means: restoration of the genetic information through mutational change within a gene. The mutational event, in contrast to suppression, always occurs within the gene affected by the original mutation; however, it does not have to occur at the same site at which the original mutation took place, as shown by the investigations on the *td* locus (p. 403). The physiological effect of a reverse mutation, like that of a suppressor, can lead to a complete restoration of the catalytic function of an enzyme the structure of which nevertheless need not be identical to that of the wild type.

Summary

Clues to the correlation between structural elements of a gene and the functional characteristics of an enzyme have been derived from a series of different experiments:

1. Intragenic complementation: Allelic auxotrophic mutants can mutually compensate for their defects to a limited extent in the heterokaryotic or diploid state. Prototrophic strains arise, however, which are not always completely identical with the wild type in habit and physiological characters; a large portion of the mutants of a particular gene may not complement at all. Allelic mutants can be arranged into groups the members of which show the same complementation behavior; these may form a linear or circular complementation map. Complementation and recombination maps are not completely colinear. As a possible explanation for intragenic complementation defective proteins formed by individual mutants are believed to aggregate into a hybrid protein which approximates the wild type enzyme in structure.

2. Genetic fine structure and function: The mutant classes of the *td* locus of *N. crassa*, which are characterized by differences in the structure of the tryptophan synthetase show a correlation between position and function, i.e. the mutants of each class are localized in a narrow part of the *td* gene. The individual segments along with a genetically quiet region in which no mutant has yet been found can be linearly arranged in the *td* region.

3. Suppressor genes: The effect of a non-allelic mutation may be partially or completely neutralized through a suppressor mutation (e.g. restoration of prototrophy in deficiency mutants). The physiological effect of suppressor genes may differ widely. It may lead to the repair of an enzyme defect or, by inducing an enzyme defect, so shift biochemical reactions as to compensate for the auxotrophy.

4. Reverse mutations have the same phenotypic effect as suppressor mutations. In contrast, reversions always occur within the locus of the original mutation. However, through a reversion the catalytic function of an enzyme is restored even though the structure of the enzyme may differ from that of the wild type enzyme.

III. Genetic control of enzyme formation

The experiments discussed in the preceding sections of this chapter have shown that the genetic information stored in the DNA controls the production of enzymes. The exhaustive biochemical analyses of nutritional mutants indicate that a single complex genetic functional unit, the gene, is responsible for the structure of an enzyme.

For example, 200 different tryptophan auxotrophic mutants of *N. crassa* could be assigned to a single locus (BONNER et al., 1960) (p. 381 ff.). Similar relationships were found for 35 mutants with an adenylosuccinase defect (GILES et al., 1957a, b; GILES, 1958) as well as for different mutants with an arginino-succinase defect (NEWMEYER, 1957). Studies of histidine synthesis (CATCHESIDE, 1960b; WEBBER and CASE, 1960) and tyrosinase formation in *N. crassa* (HOROWITZ et al., 1960, 1961a, b) yield additional evidence for this concept.

These results are in complete agreement with the one gene-one enzyme hypothesis. However, we cannot overlook a series of experimentally-based inconsistencies in seeking universality of this hypothesis.

1. Inconsistency: *Selection.* This exception is based on the fact that a high percentage of mutations lead to irreparable defects (lethal mutations; p. 270f.) which can only be alleviated in heterokaryotic or heterozygotic cells by the action of the wild type allele (ATWOOD and MUKAI, 1953). Among these mutations may be those with simple as well as multiple primary effects. Experimental evidence on this question is difficult to obtain because mutants with irreparable defects will not grow on any medium and therefore are not detected as auxotrophs. Because of this selective disadvantage it is possible that the only difference between mutants with irreparable and those with reparable defects may be that in the former case the supplementary growth substances cannot pass through the cell membrane. However, this phenomenon can just as well be attributed to multiple primary effects.

Investigations of HOROWITZ and LEUPOLD (1951) on temperature mutants of *N. crassa* and *E. coli* do not support the idea of a selective advantage of mutants with a simple primary effect. Nevertheless, these experiments do not conclusively indicate that the relatively simple gene function generally observed in auxotrophs is representative of all gene effects occurring in nature. An unequivocal refutation of this objection is thus not possible at the moment (see also MORROW, 1964).

405

2. Inconsistency: *one gene → several enzymes.* The pleiotropic effect of genes which can be observed in higher as well as lower forms of life is in certain cases considered irreconcilable with the one gene-one enzyme hypothesis. Nevertheless, the analysis of polyauxotrophs shows that multiple end effects of a gene mutation can be determined by the loss of *a single* enzyme (Fig. VI-6 and p. 357). Other examples point to the same conclusion.

BÜRK and PATEMAN (1962) failed to find either a glutamic acid dehydrogenase or an alanine dehydrogenase activity in the crude extracts from *am* mutants of *N. crassa* (p. 380). Attempts to isolate both enzymes led to the discovery that a single protein molecule with two catalytic effects rather than two different enzymes was involved. According to investigations of WOODWARD and DAVIS (1963), the *pyr-3* locus of *N. crassa* likewise appears responsible for the synthesis of a protein with two catalytic functions (see also Fig. VI-8 and Table VI-2, No. 3).

3. Inconsistency: *several genes → a single enzyme.* Already shortly after formulation of the one to one hypothesis, it was shown that several genes may be responsible for the formation of a single enzyme (LANDMAN, 1950 in *N. crassa*; MARKERT (1950) in *Glomerella cingulata*, LEDERBERG et al., 1951 in *E. coli*). This phenomenon has been observed repeatedly (e.g. HOROWITZ et al., 1961 b in *N. crassa*, KURAHASHI, 1957 in *E. coli*; GLASSMAN and MITCHELL, 1959 in *Drosophila melanogaster*; ESSER, 1963 in *Podospora anserina*).

This contradiction of the one gene-one enzyme hypothesis was reconciled primarily through the work of JACOB and MONOD (1961 a, b) on *E. coli*. These two investigators succeeded in proving that two categories of genes are necessary for the synthesis of an enzyme, namely structural and regulatory genes (p. 384). In the sense of the hypothesis only the structural gene can be considered as the "one gene," i.e. a structural gene is responsible for the formation of an enzyme. Thus this exception has not detracted from the principal significance of the one gene-one enzyme hypothesis for the explanation of gene function, but has only led to a more precise characterization of the genes responsible for enzyme formation.

Since the main interest of the investigators which has concerned itself with the study of gene-enzyme systems in fungi has been directed toward clarification of the function of structural genes, a systematic search for regulatory genes has not been carried out. Mutants which did not exhibit any qualitative alteration of their enzymes were generally not further analyzed. Among the innumerable genes which quantitatively or indirectly influence the synthesis of an enzyme (p. 372 ff., 383) only a few have been characterized without question as regulatory genes (p. 384).

Not only the concept of the gene, but also that of the enzyme needs more precise formulation in the context of more recent knowledge about the transfer of genetic information and the structure of protein (p. 341 ff.). We know that proteins may be polymeric and built up of several identical or non-identical monomeric polypeptides. Just recently an increasing number of enzymes has been discovered which exhibit a quaternary structure (p. 344). It is entirely conceivable that in a polymeric enzyme composed of non-identical monomers, different genes may code for particular monomers. For this reason it is better to speak of a "one structural gene-one polypeptide" than a one gene-one enzyme relationsship.

One must ask in what way structural and regulatory genes interact in enzyme synthesis. An idea of the interaction of these two kinds of genes cannot be developed on the basis of results with fungi. The extensive genetic and biochemical experiments which have been carried out on bacteria have been more productive in this regard (literature in PAIGEN, 1962; FISHER, 1962; STARLINGER, 1963; WINKLER and KAPLAN, 1963). They enabled JACOB and MONOD (1961 a) to formulate a theory of control of enzyme formation in bacteria.

JACOB and MONOD assume that in the bacterial cell *regulatory genes are responsible for two regulatory mechanisms, namely induction and repression*. Induction (formerly called adaptive enzyme formation, p. 377) is the formation of an enzyme only when its substrate is present in the culture medium. The substrates of such inducible enzymes are mostly energy-rich substances which cannot be utilized by the organism without particular enzymes. Repression is the inhibition of enzyme synthesis by the product of its catalytic activity.

The formation of tryptophan synthetase in specific strains of *E. coli* and *N. crassa* following addition of tryptophan to the culture medium can be inhibited (MONOD and COHN, 1952, and LESTER, 1961 a, b). The same holds for tyrosine of *N. crassa* (HOROWITZ et al., 1960).

Another repressor gene was discovered in *N. crassa* by METZENBERG (1962). This was a mutant in which the synthesis of invertase and trehalase is repressed by mannose.

With the aid of these two mechanisms the bacterial cell is able to adapt the synthesis of its enzymes to the current environmental conditions; synthesis of superfluous enzymes is prevented.

Experiments on enzyme synthesis in *E. coli* suggest that *induction and repression are different manifestations* of the same regulatory mechanism. Accordingly, induction of enzyme formation involves inhibition of a repression which is induced by the enzyme substrate. Contrarily it is believed that substances which repress enzyme synthesis activate a repressor in some unknown way. Such a substance, e.g. tryptophan (see above) is called a corepressor. The repressor, which seems to be a protein (GILBERT and MÜLLER-HILL, 1966), is thought to be a product of the regulatory gene. Further, it is believed to prevent the transfer of genetic information from the structural gene to the messenger RNA and thereby to inhibit enzyme synthesis. The repressor is considered to be highly specific and able to influence only a single enzyme or the enzymes of a single reaction system which are controlled by a structural gene or a series of consecutive structural genes.

The best known example of the correlation between position and function is offered by the structural gene responsible for histidine synthesis in *E. coli*. All the mutational sites controlling the different steps of the biosynthesis of this amino acid lie in a single genetic region (DEMEREC and HARTMAN, 1959).

A group of structural genes which are under the influence of a single repressor is called an "operon" by JACOB and MONOD. It represents a "supra" functional unit in which individual structural genes may lose their function through mutation without impairing the function of the others. A short genetic region at one end of the operon is an exception; the functional capacity of the latter, designated as the *operator gene*, is a requirement for the function of the operon. A mutation of the operator can stop the synthesis of all enzymes controlled by it. Thus the operator gene is assumed in some way to initiate the reading of the genetic information of the operon by the messenger RNA and that a specific repressor can impair the function of the operator. Mutations within the operator gene can make it insensitive to the repressor or interfere with its function even when the repressor is not present.

The operon does not have to consist of a series of structural genes. It may include only one or two structural genes joined to an operator.

According to this conception three different genetic elements are responsible for the formation of an enzyme or a group of functionally related enzymes: structural, operator, and regulatory genes. While the structural and operator genes adjoin one another, the regulatory gene does not necessarily have to be linked with this region. A diagram of the interaction between these genes during enzyme synthesis is shown in Fig. VI-14. Depending upon whether the repressor produced by the regulatory gene is activated or inhibited by a metabolic product, either repression or induction of enzyme synthesis occurs. The model has the advantage of invoking the same genetic basis for the formation of constitutive and induced enzymes.

If one wishes to determine whether or not the JACOB and MONOD model can be utilized as a working hypothesis for further analysis of the genetic basis of enzyme synthesis in the fungi, one must first clarify to what extent the genetic prerequisites for such a mechanism exist in this group of plants.

The investigations of HOROWITZ and coworkers previously cited (p. 373, 378) conclusively establish that regulatory genes which induce enzyme synthesis also occur in *Neurospora*. However, in contrast to the results with bacteria, functionally related genes only rarely appear to be closely linked with one another and localized in the unit which can be designated as an operon. Numerous examples of genes (of a synthetic pathway) which are randomly distributed among different chromosomes (literature in PONTECORVO, 1958) stand against only three cases of a correlation between position and function.

Two genes (*iv-2* and *iv-1*) which control the formation of two enzymes responsible for sequential steps in the synthetic pathway of isoleucine and valine lie four map units apart in linkage group V of *N. crassa* (see Fig. IV-14, p. 202, Fig. VI-7) (WAGNER et al., 1960; BERNSTEIN and MILLER, 1961). The complementation pattern of a series of allelic mutants of these two loci suggests that they are not completely independent in their function. DOUGLAS and HAWTHORNE (1964) found 3 tightly linked genes in *Saccharomyces cerevisiae* that are responsible for three enzymes functioning sequentially in galactose degradation (see Table VI-3).

The results obtained by GROSS and FEIN (1960) in a genetic and biochemical analysis of 4 linked *arom* mutants which control the synthesis of the benzene ring in aromatic substances are more informative (Fig. VI-6 and Table VI-2, No. 1 g). Three of these loci (*arom-1, arom-2, arom-4*) are

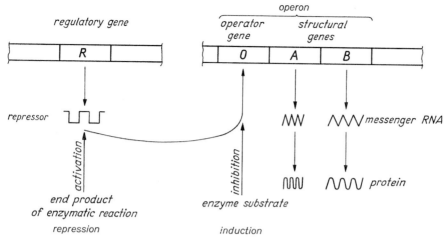

Fig. VI-14. Model of the genetic control of enzyme synthesis; explanation in text. (Adapted from Jacob and Monod, 1961)

close together, while the fourth (*arom-3*) is some distance from the others (Fig. IV-14). *arom-1* and *arom-4* are 0.3 map units apart and are responsible for the formation of two enzymes acting sequentially in the formation of the benzene ring. In contrast, *arom-2* appears to be a deletion encompassing several genes. This mutant fails to form four enzymes which function in successive steps of a synthetic pathway. Another possible explanation for the behavior of the *arom-2* gene is that it acts as an operator and forms an operon together with the other two genes. The regulator belonging with this unit is unknown, however.

The presence of an operon in the *his-3* region of *N. crassa* has been postulated by Giles (1964). Three adjacent structural genes appear to be involved; these control the synthesis of three enzymes acting in L-histidine synthesis. The function of the three genes are thought to be controlled by other mutational sites which are located in the same region.

These few data do not exclude the presence of an operator-regulator mechanism in the fungi, but they are by no means sufficient to take such a mechanism for granted. Above all, data on the existence of regulatory genes are lacking; these would allow adoption of the bacterial model. Nevertheless, this catagory of hereditary factors is likely to be found after a more detailed analysis of induced enzyme systems in *Neurospora* and yeasts, and further study of the phenomena described in the sections on *quantitative* (p. 373 ff.) and *indirect genetic effects* (p. 383 ff.). Moreover, ultimate clarification of the function of suppressor genes (p. 397 ff.) may lead to the identification of some of these as regulatory genes.

The discovery of an apparent correlation between regulatory and structural genes in enzyme synthesis has revealed a way by which the organism can adapt metabolic processes to internal and external conditions with the help of its genetic information.

Function

The investigations concerned with gene function described and discussed in this chapter along with those presented in the chapters on *replication, recombination,* and *mutation* have contributed to a more precise concept of the gene. The current state of knowledge allows us *to define the gene as follows: the gene is a complex linear structure consisting of numerous mutational sites which may recombine; nevertheless, it is still to be considered a single unit in a functional sense.*

Summary

1. Recent investigations on the relation between genes and enzymes have led to a modification and more precise formulation of the BEADLE-TATUM one gene-one enzyme hypothesis. This has involved the discovery that (a) formation of an enzyme may be controlled by two different categories of genes, namely structural and regulatory genes, and (b) enzymes frequently have a polymeric structure. Currently the expression, one structural gene-one polypeptide, is to be preferred.

2. On the basis of genetic and biochemical investigation of enzyme systems of bacteria a theory of the cooperative action of structural and regulatory genes has been developed. Although extensive information is available on the function of structural genes in fungi, clear cut examples of regulatory genes are few. At present it is not possible to determine whether or not this model is valid for the genetic control of enzyme synthesis in fungi.

Literature

ABEL, P., and T. A. TRAUTNER: Formation of an animal virus within a bacterium. Z. Vererbungsl. **95**, 66—72 (1964).
ABELSON, P. H., and H. J. VOGEL: Amino acid biosynthesis in *Torulopsis utilis* and *Neurospora crassa.* Biol. Chem. **213**, 355—364 (1955).
ABRAMS, R.: Nucleic acid metabolism and biosynthesis. Ann. Rev. Biochem. **30**, 165—188 (1961).
ADELBERG, E. A.: Isoleucine biosynthesis from threonine. J. Amer. chem. Soc. **76**, 4241 (1954).
— The biosynthesis of isoleucine, valine and leucine. In: W. D. McELROY and B. GLASS (edits.), Amino acid metabolism, p. 419—429. Baltimore 1955a.
— The biosynthesis of isoleucine and valine. III. Tracer experiments with L-threonine. J. biol. Chem. **216**, 431—437 (1955b).
— D. BONNER, and E. L. TATUM: A precursor of isoleucine obtained from a mutant strain of *Neurospora crassa.* J. biol. Chem. **190**, 837—841 (1951).
— C. A. COUGHLIN, and R. W. BARRATT: The biosynthesis of isoleucine and valine. II. Independence of the biosynthetic pathways in *Neurospora.* J. biol. Chem. **216**, 425—433 (1955).
—, and E. L. TATUM: Characterization of a valine analog accumulated by a mutant strain of *Neurospora crassa.* Arch. Biochem. **29**, 235—236 (1950).
AHMAD, M., and D. G. CATCHESIDE: Physiological diversity amongst tryptophan mutants in *Neurospora crassa.* Heredity **15**, 55—64 (1960).
— MD. KHALIL, N. A. KHAN, and A. MOZMADAR: Structural and functional complexity at the tryptophan-1 locus in *Neurospora crassa.* Genetics **49**, 925—933 (1964).
ALLEN, K. M., and C. YANOFSKY: A biochemical and genetic study of reversion with the *A*-gene *A*-protein system of *Escherichia coli* tryptophan synthetase. Genetics **48**, 1065—1083 (1963).

AMES, B. N.: The biosynthesis of histidine. In: W. D. McELROY and B. GLASS (edits.), McCollum Pratt Symp. on amino acid metabolism. p. 357—372. Baltimore 1955.

AMES, B. N.: The biosynthesis of histidine: L-histidinol phosphate phosphatase. J. biol. Chem. **226**, 583—593 (1957a).

— The biosynthesis of histidine: D-erythro-imidazole-glycerol phosphate dehydrase. J. biol. Chem. **228**, 131—143 (1957b).

—, and B. L. HORECKER: The biosynthesis of histidine: Imidazole-acetol phosphate-transaminase. J. biol. Chem. **220**, 113—128 (1956).

—, and H. K. MITCHELL: The biosynthesis of histidine: Imidazole-glycerol phosphate, imidazole-acetol phosphate and histidinol phosphate. J. biol. Chem. **212**, 687—696 (1955).

— — and M. B. MITCHELL: Some new naturally occurring imidazoles related to the biosynthesis of histidine. J. Amer. chem. Soc. **75**, 1015—1018 (1953).

ANDERER, F. A., H. UHLIG, E. WEBER, and G. SCHRAMM: Primary structure of the protein of tobacco mosaic virus. Nature (Lond.) **186**, 922—925 (1960).

ANDERSSON-KOTTÖ, J., G. EHRENSVÄRD, G. HÖGSTRÖM, L. REIO, and E. SALUSTE: Amino acid formation and utilization in *Neurospora*. J. biol. Chem. **210**, 455—463 (1954).

ARANOFF, S.: Technics of radiobiochemistry. Ames (Iowa): Iowa State College Press 1957.

ARNSTEIN, H. R. V., R. A. COX, and J. A. HUNT: Function of polyuridylic acid and ribonucleic acid in protein biosynthesis by ribosomes from mammalian reticulocytes. Nature (Lond.) **194**, 1042—1044 (1962).

ATWOOD, K. C., and F. MUKAI: Indispensable gene functions in *Neurospora*. Proc. nat. Acad. Sci. (Wash.) **39**, 1027—1035 (1953).

BARRATT, R. W., and W. N. STRICKLAND: Purification and characterization of a TPN-specific glutamic-acid dehydrogenase from *Neurospora crassa*. Arch. Biochem. **102**, 66—76 (1963).

BARRON, G. L.: The parasexual cycle and linkage relationship in the storage root fungus *Penicillium expansum*. Canad. J. Bot. **40**, 1603—1613 (1962).

BEADLE, G. W.: Biochemical genetics. Chem. Rev. **37**, 15—96 (1945a).

— Genetics and metabolism in *Neurospora*. Physiol. Rev. **25**, 643—663 (1945b).

— Genes and the chemistry of the organism. Amer. Scientist **34**, 31—75 (1945c).

— Physiological aspects of genetics. Ann. Rev. Physiol. **10**, 17—42 (1948).

— Gene structure and gene action. Fortschr. Chem. organ. Naturstoffe **12**, 366—384 (1955).

— Some recent advances in *Neurospora* genetics. Proc. Internat. Genet. Symp. Cytologia (Tokyo), Suppl. Vol., 142—145 (1956).

— The role of nucleus in heredity. In: W. D. McELROY and B. GLASS (edits.), The chemical basis of heredity, p. 3—22. Baltimore 1957.

— Genes and chemical reactions in *Neurospora*. Stockholm: Nobel Lecture 1959a.

— Genes and chemical reactions in *Neurospora*. The concepts of biochemical genetics with Garrod's "inborn errors" and have evolved gradually. Science **129**, 1715—1719 (1959b).

— Physiological aspects of genetics. Ann. Rev. Physiol. **22**, 45—74 (1960a).

— Evolution in microorganisms with special reference to the fungi. Proc. Internat. Colloq. Evoluzione e Genet., Acad. Nazl. Linnei, Rome **47**, 301—319 (1960b).

— The language of the genes. Advanc. Sci. **17**, 511—521 (1961).

—, and V. L. COONRADT: Heterocaryosis in *Neurospora crassa*. Genetics **29**, 291 —308 (1944).

—, and B. EPHRUSSI: The differentiation of eye pigments in *Drosophila* as studied by transplantation. Genetics **21**, 225—247 (1936).

— H. K. MITCHELL, and J. F. NYC: Kynurenine as an intermediate in the formation of nicotinic acid from tryptophan by *Neurospora*. Proc. nat. Acad. Sci. (Wash.) **33**, 155—158 (1947).

BEADLE, G. W., and E. L. TATUM: *Neurospora*. II. Methods of producing and detecting mutations concerned with nutritional requirements. Amer. J. Bot. **32**, 678—686 (1945).

BENZER, S., and S. P. CHAMPE: Ambivalent *rII* mutants of phage *T4*. Proc. nat. Acad. Sci. (Wash.) **47**, 1025—1038 (1961).

— — A change form nonsense to sense in the genetic code. Proc. nat. Acad. Sci. (Wash.) **48**, 1114—1121 (1962).

—, and B. WEISBLUM: On the species specificity of acceptor RNA and attachment enzymes. Proc. nat. Acad. Sci. (Wash.) **47**, 1149—1154 (1961).

BERNSTEIN, H., and A. MILLER: Complementation studies with isoleucine-valine mutants of *Neurospora crassa*. Genetics **46**, 1039—1052 (1961).

BEVAN, E. A., and R. A. WOODS: Complementation between adenin requiring mutants in yeast. Heredity **17**, 141 (1962).

BISHOP, J., J. LEAHY, and R. SCHWEET: Formation of the peptide chain of hemoglobin. Proc. nat. Acad. Sci. (Wash.) **46**, 1030—1038 (1960).

BLACK, S., and N. M. GRAY: Enzymatic phosphorylation of L-aspartate. J. Amer. chem. Soc. **75**, 2271—2272 (1953).

—, and N. G. WRIGHT: Intermediate steps in the biosynthesis of threonine. In: W. D. MCELROY and B. GLASS (edits.), Amino acid metabolism, p. 591—600. Baltimore 1955a.

— — β-aspartokinase and β-aspartyl phosphate. J. biol. Chem. **213**, 27—38 (1955b).

— — Aspartic β-semialdehyde dehydrogenase and aspartic β-semialdehyde. J. biol. Chem. **213**, 39—50 (1955c).

— — Homoserine dehydrogenase. J. biol. Chem. **213**, 51—60 (1955d).

BONNER, D. M.: Production of biochemical mutations in *Penicillium*. Amer. J. Bot. **33**, 788—791 (1946a).

— Further studies of mutant strains of *Neurospora* requiring isoleucine and valine. J. biol. Chem. **166**, 545—554 (1946b).

— The identification of a natural precursor of nicotinic acid. Proc. nat. Acad. Sci. (Wash.) **34**, 5—9 (1948).

— Gene-enzyme relationship in *Neurospora*. Cold Spr. Harb. Symp. quant. Biol. **16**, 143—157 (1951).

— The genetic control of enzyme formation. In: W. D. MCELROY and B. GLASS (edits.), Phosphorus metabolism, vol. II, p. 153—163. Baltimore 1952.

— Aspects of enzyme formation. In: W. D. MCELROY and B. GLASS (edits.), Amino acid metabolism, p. 193—197. Baltimore 1955.

— The genetic unit. Cold Spr. Harb. Symp. quant. Biol. **21**, 163—170 (1956).

— Gene action. In: Genetics and Cancer, p. 207—225. Univ. of Texas 1959.

— Gene-enzyme relationship in micro-organisms. Proc. XI. Internat. Congr. of Genetics, vol. 2, p. 141—149. The Hague 1964a.

— Correlation of the gene and protein structure. J. exp. Zool. **157**, 9—20 (1964b).

—, and G. W. BEADLE: Mutant strains of *Neurospora* requiring nicotine amide or related compounds for growth. Arch. Biochem. **11**, 319—328 (1946).

— Y. SUYAMA, and J. A. DEMOSS: Genetic fine structure and enzyme formation. Fed. Proc. **19**, 926—930 (1960).

— E. L. TATUM, and G. W. BEADLE: The genetic control of biochemical reactions in *Neurospora*: A mutant strain requiring isoleucine and valine. Arch. Biochem. **3**, 71—91 (1943).

—, and C. YANOFSKY: Quinolinic acid accumulation in the conversion of 3-hydroxyanthranilic acid to niacin in *Neurospora*. Proc. nat. Acad. Sci. (Wash.) **35**, 576—581 (1949).

— — and C. W. H. PARTRIDGE: Incomplete genetic blocks in biochemical mutants of *Neurospora*. Proc. nat. Acad. Sci. (Wash.) **38**, 25—34 (1952).

BRAUNITZER, G., R. GEHRING-MÜLLER, N. HILLSCHMANN, K. HILSE, G. HOBAM, V. RUDLOFF u. B. WITTMANN-LIEBOLD: Die Konstitution des normalen adulten Humanhämoglobins. Hoppe-Seylers Z. physiol. Chem. **325**, 283—286 (1961).

BRENNER, S.: On the impossibility of all overlapping triplet codes in information transfer from nucleic acids to proteins. Proc. nat. Acad. Sci. (Wash.) **43**, 687—694 (1957).
— The mechanism of gene action. In: G. E. W. WOLSTENHOLME and C. M. O'CONNOR (edits.), Symp. on Biochemistry of human genetics. Ciba Found. and internat. Union of Biol. Sci., p. 304—317. London 1959.
BRESCH, C.: Klassische und molekulare Genetik. Berlin-Göttingen-Heidelberg: Springer 1964.
BRODA, E.: Radioaktive Isotope in der Biochemie. Wien: Franz Deuticke 1958.
BRODY, S., and C. YANOFSKY: Suppressor gene alteration of protein primary structure. Proc. nat. Acad. Sci. (Wash.) **50**, 9—16 (1963).
BROWN, G. L.: Ribonucleic acid and bacterial genetics. Brit. med. Bull. **18**, 10—13 (1962).
BUCHANAN, J. M., J. G. FLAKS, S. C. HARTMAN, B. LEVENBERG, L. N. LUKENS, and L. WARREN: The enzymatic synthesis of inosinic acid de novo. Ciba Found. Symp. on Chem. and Biol. of Purines, G. E. W. WOLSTENHOLME and C. M. O'CONNOR (edits.), p. 233—255. London 1957.
BÜRK, R. R., and J. A. PATEMAN: Glutamic and alanine dehydrogenase determined by one gene in *Neurospora crassa*. Nature (Lond.) **196**, 450—451 (1962).
BUTENANDT, A., u. G. HALLMANN: Neue Synthesen des d,l-Kynurenins und d,l-3-Oxy-kynurenins. Z. Naturforsch. **5b**, 444—446 (1950).
— P. KARLSON u. W. ZILLIG: Über das Vorkommen von Kynurenin in Seidenspinnerpuppen. Hoppe-Seylers Z. physiol. Chem. **288**, 125—132 (1951).
— W. WEIDEL u. E. BECKER: Kynurenin als Augenpigmentbildung auslösendes Agens bei Insekten. Naturwissenschaften **28**, 63—64 (1940).
— — u. W. v. DERJUGIN: Zur Konstitution des Kynurenins. Naturwissenschaften **30**, 51 (1942).
— — u. H. SCHLOSSBERGER: 3-Oxy-kynurenin als CN$^+$-Gen, abhängiges Glied im intermediären Tryptophan-Stoffwechsel. Naturforsch. Z. **4b**, 242—244 (1949).
— R. WEICHERT u. W. v. DERJUGIN: Über Kynurenin. Physiologie, Konstitutionsermittlung und Synthese. Hoppe-Seylers Z. physiol. Chem. **279**, 27—43 (1943).
CABET, D., C. ANAGNOSTOPOULOS et M. GANS: Contribution à l'étude de la biosynthèse du tryptophane chez le *Coprinus radiatus*. C. R. Acad. Sci. (Paris) **255**, 1007—1009 (1962).
CAMPBELL, J. J. R.: Metabolism of microorganisms. Ann. Rev. Microbiol. **8**, 71—104 (1954).
CAPECCHI, M. R.: Initiation of *E. coli* proteins. Proc. nat. Acad. Sci. (Wash.) **55**, 1517—1524 (1966).
CARLTON, B. C., and C. YANOFSKY: The amino terminal sequence of the A protein of tryptophan synthetase of *E. coli*. J. biol. Chem. **237**, 1531—1534 (1962).
CASE, M. E., and N. H. GILES: Recombination mechanism at the *pan-2* locus in *Neurospora crassa*. Cold Spr. Harb. Symp. quant. Biol. **23**, 119—135 (1958).
— — Comparative complementation and genetic maps of the *pan-2* locus in *Neurospora crassa*. Proc. nat. Acad. Sci. (Wash.) **46**, 659—676 (1960).
CATCHESIDE, D. G.: Relation of genotype to enzyme content. Microbiol. Genetics, vol. 10, Symp. Soc. gen. Microbiol. Cambridge 1960a, p. 181—207.
— Complementation among histidine mutants of *Neurospora crassa*. Proc. roy. Soc. B **153**, 179—194 (1960b).
— Functional structure of genes. The scientific basis of Medicine annual Rev. 1962, p. 140—151.
— Gene action and interaction. Biol. J. **2**, 35—47 (1964).
—, and A. OVERTON: Complementation between alleles in heterocaryons. Cold Spr. Harb. Symp. quant. Biol. **23**, 137—140 (1958).

Function

CAVALLIERI, L. F., and B. H. ROSENBERG: Nucleic acids: Molecular biology of DNA. Ann. Rev. Biochem. **31**, 247—270 (1963).
CHAMBERLIN, M., and P. BERG: Deoxyribonucleic acid-directed synthesis ribonucleic acid by an enzyme from *Escherichia coli*. Proc. nat. Acad. Sci. (Wash.) **48**, 81—94 (1962).
CHAMPE, S. P., and S. BENZER: An active cistron fragment. J. molec. Biol. **4**, 288—292 (1962).
CHANTRENNE, H.: Aspects of the biosynthesis of enzymes. In: F. F. NORD (edit.), Advances in enzymology and related subjects of biochemistry, vol. 24, p. 1—34. New York and London 1962.
COHN, M., and A. M. TORRIANI: Immunological studies with the β-galacto-sidase and structurally related proteins of *Escherichia coli*. J. Immunol. **69**, 471—491 (1952).
— — The relationship in biosynthesis of the β-galactosidase- and Pz-proteins in *Escherichia coli*. Biochim. biophys. Acta (Amst.) **10**, 280—289 (1953).
COLBURN, R. W., and E. L. TATUM: Studies of a phenylalanine-tyrosine requiring mutant of *Neurospora crassa*. Biochim. biophys. Acta (Amst.) **97**, 442—448 (1965).
COMBÉPINE, G., et G. TURIAN: Recherches sur la biosynthèse de la glycine chez *Neurospora crassa*, type sauvage et mutants. Path. Microbiol. **28**, 1018—1030 (1965).
COSTELLO, W. P., and E. A. BEVAN: Complementation between *ad 5/7* alleles in yeast. Genetics **50**, 1219—1230 (1964).
COTTON, R. G. H., and F. GIBSON: The biosynthesis of phenylalanine and tyrosine; enzymes converting chorismic acid into prephenic acid and their relationships to prephenate dehydratase and prephenate dehydrogenase. Biochim. biophys. Acta (Amst.) **100**, 76—88 (1965).
COVE, D. J., and J. A. PATEMAN: Independently segregating genetic loci concerned with nitrate reductase activity in *Aspergillus nidulans*. Nature (Lond.) **198**, 262—263 (1963).
CRAWFORD, I. P., and C. YANOFSKY: On the separation of tryptophan synthetase of *Escherichia coli* into two protein components. Proc. nat. Acad. Sci. (Wash.) **44**, 1161—1170 (1958).
— — The formation of a new enzymatically active protein as a result of suppression. Proc. nat. Acad. Sci. (Wash.) **45**, 1280—1288 (1959).
CRICK, F. H. C.: The recent excitement in the coding problem. Progr. in Nucl. Ac. Res. **1**, 163—217 (1963).
— L. BARNETT, S. BRENNER, and R. J. WATTS-TOBIN: General nature of the genetic code for proteins. Nature (Lond.) **192**, 1227—1232 (1961).
DANNEEL, R.: Die Wirkungsweise der Grundfaktoren für Haarfärbung beim Kaninchen. Naturwissenschaften **26**, 505 (1938).
DAVIS, B. D.: Aromatic biosynthesis. I. The role of shikimic acid. J. biol. Chem. **191**, 315—325 (1951).
— Biosynthesis of the aromatic acids. In: W. D. MCELROY and B. GLASS (edits.), Amino acid metabolism, p. 799—811. Baltimore 1955.
—, and E. S. MINGIOLI: Aromatic biosynthesis. VII. Accumulation of two derivatives of shikimic acid by bacterial mutants. J. Bact. **66**, 129 (1953).
DAVIS, R. H.: An enzymatic difference among *pyr-3* mutants of *Neurospora crassa*. Proc. nat. Acad. Sci. (Wash.) **46**, 677—682 (1960).
— Suppressor of pyrimidine-3 mutants of *Neurospora* and it relation to arginine synthesis. Science **134**, 470—471 (1961).
— Consequences of a suppressor gene effect with pyrimidine and proline mutants of *Neurospora*. Genetics **47**, 351—360 (1962a).
— A mutant form of ornithine transcarbamylase found in a strain of *Neurospora* carrying a pyrimidine-proline suppressor gene. Arch. Biochem. **97**, 185—191 (1962b).

DAVIS, R. H.: *Neurospora* mutant lacking an arginine-specific carbamyl phosphokinase. Science **142**, 1652—1654 (1963).

—, and W. M. THWAITES: Structural gene for ornithine transcarbamylase in *Neurospora*. Genetics **48**, 1551—1558 (1963).

—, and V. W. WOODWARD: The relationship between gene suppression and aspartate transcarbamylase activity in *pyr-3* mutants of *Neurospora*. Genetics **47**, 1075—1083 (1962).

DAY, P. R.: The structure of the *A* mating type locus in *Coprinus lagopus*. Genetics **45**, 641—650 (1960).

— The structure of the *A* mating type factor in *Coprinus lagopus* wild alleles. Genet. Res. **4**, 323—325 (1963).

DE DEKEN, R. H.: Pathway of arginine biosynthesis in yeast. Biochem. biophys. Res. Commun. **8**, 462—466 (1962).

DELBRÜCK, M.: Die Vererbungschemie. Naturwiss. Rdsch. **16**, 85—89 (1963).

DEMEREC, M., and P. E. HARTMAN: Complex loci in microorganisms. Ann. Rev. Microbiol. **13**, 377—406 (1959).

DE MOSS, J. A.: Studies on the mechanism of the tryptophan synthetase reaction. Biochim. biophys. Acta (Amst.) **62**, 279—293 (1962).

— The conversion of shikimic acid to anthranillic acid by extracts of *Neurospora crassa*. J. biol. Chem. **240**, 1231—1235 (1965a).

— Biochemical diversity in the tryptophan pathway. Biochem. and Biophys. Res. Comm. **18**, 850 (1965b).

—, and D. M. BONNER: Studies on normal and genetically altered tryptophan synthetase from *Neurospora crassa*. Proc. nat. Acad. Sci. (Wash.) **45**, 1405—1412 (1959).

— M. IMAI, and D. M. BONNER: Studies on tryptophan biosynthesis in *Neurospora crassa*. Bact. Proc. **112** (1958).

—, and J. WEGMAN: An enzyme aggregate in the tryptophan pathway of *Neurospora crassa*. Proc. nat. Acad. Sci. (Wash.) **54**, 241—247 (1965).

DINTZIS, H. M.: Assembly of the peptide chains of hemoglobin. Proc. nat. Acad. Sci. (Wash.) **47**, 247—261 (1961).

DORFMAN, B.: Allelic complementation at the *ad 5/7* locus in yeast. Genetics **50**, 1231—1243 (1964).

DOUGLAS, H. C., and D. C. HAWTHORNE: Enzymatic expression and genetic linkage of genes controlling galactose utilization in *Saccharomyces*. Genetics **49**, 837—844 (1964).

EBERHART, B., D. F. CROSS, and L. R. CHASE: β-Glucosidase system of *Neurospora crassa*. I. β-Glucosidase and cellulase activities of mutant and wild type strains. J. Bact. **87**, 761—770 (1964).

EBERHART, B. M., and E. L. TATUM: Thiamine metabolism in wild-type and mutant strains of *Neurospora crassa*. Arch. Biochem. **101**, 378—387 (1963).

EDMONDS, M., A. M. DELLUVA, and D. W. WILSON: The metabolism of purines and pyrimidines by growing yeast. J. biol. Chem. **197**, 251—259 (1952).

EDWARDS, J. M., and L. M. JACKMAN: Chorismic acid. A branch point intermediate in aromatic biosynthesis. Aust. J. Chem. **18**, 1227—1239 (1965).

EGELHAAF, A.: Genphysiologie: Biochemische Genwirkungen. Fortschr. Zool. **15**, 378—423 (1962).

EHRENSTEIN, G. v., and F. LIPMAN: Experiments on hemoglobin biosynthesis. Proc. nat. Acad. Sci. (Wash.) **47**, 941—950 (1961).

EHRENSVÄRD, G.: Metabolism of amino acids and proteins. Ann. Rev. Biochem. **24**, 275—310 (1955).

— L. REIO, E. SALUSTE, and R. STJERNHOLM: Acetic acid metabolism in *Torulopsis utilis*. III. Metabolic connection between acetic acid and various amino acids. J. biol. Chem. **189**, 93—108 (1951).

— E. SPERBER, E. SALUSTE, L. REIO, and R. STJERNHOLM: Metabolic connection between proline and glycine in the amino acid utilization of *Torulopsis utilis*. J. biol. Chem. **169**, 759—760 (1947).

EL-ANI, A. S.: Self-sterile auxotrophs and their relation to heterothallism in *Sordaria fimicola*. Science **145**, 1067—1068 (1964).
— L. S. OLIVE, and Y. KITANI: Genetics of *Sordaria fimicola*. IV. Linkage group I. Amer. J. Bot. **48**, 716—723 (1961).
ELLINGBOE, A. H., and J. R. RAPER: Somatic recombination in *Schizophyllum commune*. Genetics **47**, 85—98 (1962).
EPHRUSSI, B.: Chemistry of "eye color hormones" of *Drosophila*. Quart. Rev. Biol. **17**, 327—338 (1942).
ESSER, K.: Quantitatively and qualitatively altered phenoloxidases in *Podospora anserina*, due to mutations at non-linked loci. Proc. of the XI. internat. Congr. of Genetics, vol. 1, p. 51—52. The Hague 1963.
— Die Phenoloxydasen des Ascomyceten *Podospora anserina*. III. Quantitative und qualitative Enzymunterschiede nach Mutation an nicht gekoppelten Loci. Z.Vererbungsl. **97**, 327—344 (1966).
— J. A. DeMOSS, and D. M. BONNER: Reverse mutations and enzyme heterogeneity. Z. Vererbungsl. **91**, 291—299 (1960).
FINCHAM, J. R. S.: Transaminases in *Neurospora crassa*. Nature (Lond.) **168**, 957—958 (1951a).
— The occurrence of glutamic dehydrogenase in *Neurospora* and its apparent absence in certain mutant strains. J. gen. Microbiol. **5**, 793—806 (1951b).
— Ornithine transaminase in *Neurospora* and its relation to the biosynthesis of proline. Biochem. J. **53**, 313—320 (1953).
— Effects of gene mutation in *Neurospora crassa* relating to glutamic dehydrogenase formation. J. gen. Microbiol. **11**, 236—246 (1954).
— A modified glutamic acid dehydrogenase as a result of gene mutation in *Neurospora crassa*. Biochem. J. **65**, 721—728 (1957).
— The biochemistry of genetic factors. Ann. Rev. Biochem. **28**, 343—364 (1959a).
— On the nature of glutamic dehydrogenase produced by interallele complementation at the *am* locus of *Neurospora crassa*. J. gen. Microbiol. **21**, 600—611 (1959b).
— The role of chromosomal loci in enzyme formation. Proc. X. Internat. Congr. of Genetics, vol. I, p. 335—363, Montreal 1958. University of Toronto Press 1959c.
— Genetically controlled differences in enzyme activity. Advanc. Enzymol. **22**, 1—43 (1960).
— Genes and enzymes in micro-organisms. Brit. med. Bull. **18**, 14—18 (1962a).
— Genetically determined multiple forms of glutamic dehydrogenase in *Neurospora crassa*. J. molec. Biol. **4**, 257—274 (1962b).
—, and P. A. BOND: A further genetic variety of glutamic acid dehydrogenase in *Neurospora crassa*. Biochem. J. **77**, 96—105 (1960).
—, and A. B. BOULTER: Effects of amino acids on transaminase production in *Neurospora crassa*: Evidence for four different enzymes. Biochem. J. **62**, 72—77 (1956).
—, and J. B. BOYLEN: A block in arginine synthesis in *Neurospora crassa*, due to gene mutation. Biochem. J. **61** (Proc. Biochem. Soc.) XXIII—XXIV (1955).
—, and A. CODDINGTON: Complementation at the *am* locus of *Neurospora crassa*: A reaction between different mutant forms of glutamate dehydrogenase. J. molec. Biol. **6**, 361—373 (1963a).
— — The mechanism of complementation between *am* mutants of *Neurospora crassa*. Cold Spr. Harb. Symp. quant. Biol. **28**, 517—527 (1963b).
—, and P. R. DAY: Fungal genetics. Oxford: Blackwell 1963.
—, and J. A. PATEMAN: A new allele at the *am* locus in *Neurospora crassa*. J. Genet. **55**, 456—466 (1957a).
— — Formation of an enzyme through complementary action of mutant "alleles" in separate nuclei in a heterocaryon. Nature (Lond.) **179**, 741—742 (1957b).

FINK, G. R.: Gene-enzyme in histidine biosynthesis in yeast. Science **146**, 525—527 (1964).

FISCHER, G. A.: Genetic and biochemical studies of the cysteine-methionine series of mutants of *Neurospora crassa*. Thesis. Pasadena: California Institute of Technology 1954.

— The cleavage and synthesis of cystathionine in wild type and mutant strains of *Neurospora crassa*. Biochim. biophys. Acta (Amst.) **25**, 50—55 (1957).

FISHER, K. W.: Regulation of bacterial metabolism. Brit. med. Bull. **18**, 19—23 (1962).

FLAVIN, M., and C. SLAUGHTER: Purification and properties of threonine synthetase of *Neurospora*. J. biol. Chem. **235**, 1103—1108 (1960).

FLING, M., and N. H. HOROWITZ: Threonine and homoserine in extracts of a methionineless mutant of *Neurospora*. J. biol. Chem. **190**, 277—285 (1951).

— — and S. F. HEINEMANN: The isolation and properties of crystalline tyrosinase from *Neurospora*. J. biol. Chem. **238**, 2045—2053 (1963).

FOX, A. S., and J. B. BURNETT: Tyrosinases of diverse thermostabilities and their interconversion in *Neurospora crassa*. Biochim. biophys. Acta (Amst.) **61**, 108—120 (1962).

— — and M. S. FUCHS: Tyrosinase as a model for genetic control of protein synthesis. Ann. N.Y. Acad. Sci. **100**, 840—856 (1963).

FRIES, N.: Experiments with different methods of isolating physiological mutations of filamentous fungi. Nature (Lond.) **159**, 199 (1947).

—, and B. KIHLMAN: Fungal mutations obtained with methyl xanthines. Nature (Lond.) **162**, 573—574 (1948).

GARDNER, R. S., A. J. WAHBA, C. BASILIO, R. S. MILLER, P. LENGYEL, and J. F. SPEYER: Synthetic polynucleotides and the amino acid code. VII. Proc. nat. Acad. Sci. (Wash.) **48**, 2091—2094 (1962).

GAREN, A.: Genetic control of the specificity of the bacterial enzyme, alkaline phosphatase. In: W. HAYES and R. C. CLOWES (edits.), Microbiol Genetics, p. 239—247. London 1960.

GARRICK, M. D., and S. R. SUSKIND: Trypsin treated *Neurospora* tryptophan synthetase. I. Enzymic properties. J. molec. Biol. **9**, 70—82 (1964a).

— — Trypsin treated *Neurospora* tryptophan synthetase. II. Antigenic properties. J. molec. Biol. **9**, 83—99 (1964b).

GARROD, A. E.: Inborn errors of metabolism, 1st edit. London: Oxford University Press 1909.

— Inborn errors of metabolism, 2nd edit., London: Oxford University Press 1923.

GEIDUSCHEK, E. P.: "Reversible" DNA. Proc. nat. Acad. Sci. (Wash.) **47**, 950—955 (1961).

— J. W. MOOHR, and S. B. WEISS: The secondary structure of complementary RNA. Proc. nat. Acad. Sci. (Wash.) **48**, 1078—1086 (1962).

— T. NAKAMOTO, and S. B. WEISS: The enzymatic synthesis of RNA: complementary interaction with DNA. Proc. nat. Acad. Sci. (Wash.) **47**, 1405—1415 (1961).

GIERER, A.: Molekulare Grundlagen der Vererbung. Naturwissenschaften **48**, 283—289 (1961).

GILBERT, W., and B. MÜLLER-HILL: Isolation of the *lac* repressor. Proc. nat. Acad. Sci. (Wash.) **56**, 1891—1898 (1966).

GILES, N. H.: Studies on the mechanism of reversion in biochemical mutants of *Neurospora crassa*. Cold Spr. Harb. Symp. quant. Biol. **16**, 283—313 (1951).

— Studies on reverse mutation in *Neurospora crassa*. Trans. N.Y. Acad. Sci., Ser. II, **15**, 251—253 (1953).

— Mutations at specific loci in *Neurospora*. Proc. X. Internat. Congr. of Genetics, vol. 1, p. 261—279. Montreal 1958.

— Genetic fine structure in relation to function in *Neurospora*. Proc. XI. Intern. Congr. of Genetics, vol. 2, p. 17—30. The Hague 1964.

GILES, N. H., and E. Z. LEDERBERG: Induced reversion of biochemical mutants in *Neurospora crassa*. Amer. J. Bot. **35**, 150—157 (1948).
—, and C. W. H. PARTRIDGE: The effect of a suppressor on allelic inositolless mutants in *Neurospora crassa*. Proc. nat. Acad. Sci. (Wash.) **39**, 479—488 (1953).
— C. W. H. PARTRIDGE, and N. J. NELSON: The genetic control of adenylosuccinase in *Neurospora crassa*. Proc. nat. Acad. Sci. (Wash.) **43**, 305—317 (1957a).
— — — Genetic control of adenylosuccinase in *Neurospora crassa*. Proc. Internat. Genetics Symp. Cytologia (Tokyo), Suppl. Vol., 543—546 (1957b).
GILLILAND, R. B.: Identification of the genes for maltose fermentation in *Saccharomyces distaticus*. Nature (Lond.) **173**, 409 (1954).
GLASSMAN, E., and H. K. MITCHELL: Mutants of *Drosophila melanogaster* deficient in xanthine dehydrogenase. Genetics **44**, 153—162 (1959).
GOOD, N., R. HEILBRONNER, and H. K. MITCHELL: ε-Hydroxynorleucine as a substitute for lysine for *Neurospora*. Arch. Biochem. **28**, 264—265 (1950).
GORDON, M., F. A. HASKINS, and H. K. MITCHELL: The growth-promoting properties of quinic acid. Proc. nat. Acad. Sci. (Wash.) **36**, 427—430 (1950).
GOTS, J. S., and E. G. GOLLUB: Sequential blockade in adenine biosynthesis by genetic loss of an apparent bifunctional deacylase. Proc. nat. Acad. Sci. (Wash.) **43**, 826—834 (1957).
GREENBERG, D. M.: Metabolic pathways. New York and London: Academic Press, vol. I 1960, vol. II 1961.
GRIFFIN, A. C., and M. A. O'NEAL: Effect of polyuridylic acid upon incorporation in vitro of (^{14}C) phenylalanine by ascites tumor components. Biochim. biophys. Acta (Amst.) **61**, 469—471 (1962).
GROSS, S. R.: The enzymatic conversion of 5-dehydroshikimic acid to protocatechuic acid. J. biol. Chem. **233**, 1146—1151 (1958).
— On the mechanism of complementation at the *leu-2* locus of *Neurospora*. Proc. nat. Acad. Sci. (Wash.) **48**, 922—930 (1962).
—, and A. FEIN: Linkage and function in *Neurospora*. Genetics **45**, 885—904 (1960).
HALVORSON, H. O., S. WINDERMAN, and J. GORMAN: Comparison of the glucosidases of *Saccharomyces* produced in response to five non-allelic maltose genes. Biochim. biophys. Acta (Amst.) **67**, 42—53 (1963).
HARTMAN, P. E., Z. HARTMAN, D. SERMAN, and J. C. LOPER: Genetic complementarity in histidineless *Salmonella typhimurium*. Proc. X. internat. Congr. of Genetics, vol. 2, p. 115. Montral 1958.
HASKINS, F. A., and H. K. MITCHELL: Evidence for a tryptophane cycle in *Neurospora*. Proc. nat. Acad. Sci. (Wash.) **35**, 500—506 (1949).
HATANAKA, M., E. A. WHITE, K. HORIBATA, and I. P. CRAWFORD: A study of catalytic properties of *Escherichia coli* tryptophan synthetase, a two component enzyme. Arch. Biochem. **97**, 596—606 (1962).
HAWTHORNE, D. C., and R. K. MORTIMER: Supersuppressors in yeast. Genetics **48**, 716—620 (1963).
HAYES, W.: The genetics of bacteria and their viruses. Oxford 1964.
HELINSKI, D. R., and C. YANOFSKY: Correspondence between genetic data on the position of amino acid alteration in a protein. Proc. nat. Acad. Sci. (Wash.) **48**, 173—182 (1962a).
— — Peptide pattern studies on the wild protein of the tryptophan synthetase of *Escherichia coli*. Biochim. biophys. Acta (Amst.) **63**, 10—19 (1962b).
— — A genetic and biochemical analysis of second site reversion. J. biol. Chem. **238**, 1043—1048 (1963).
HELLMANN, H., u. F. LINGENS: Aufklärung biologischer Syntheseketten an Mikroorganismen. Angew. Chem. **73**, 107—113 (1961).

HENDERSON, L. M.: Quinolinic acid excretion by the rat receiving trypto-phan. J. biol. Chem. **178**, 1005—1006 (1949).

HENNING, U., D. R. HELINSKI, F. C. CHAO, and C. YANOFSKY: The A protein of the tryptophan synthetase in *E. coli.* J. biol. Chem. **237**, 1523—1530 (1962).

—, and C. YANOFSKY: An alteration in the primary structure of a protein predicted on the basis of genetic recombination data. Proc. nat. Acad. Sci. (Wash.) **48**, 183—190 (1962a).

— — Amino acid replacements associated with reversion and recombina-tion within the *A* gene. Proc. nat. Acad. Sci. (Wash.) **48**, 1497—1504 (1962b).

HESLOT, H.: *Schizosaccharomyces pombe:* un nouvel organisme pour l'étude de la mutagénèse chimique. Abh. dtsch. Akad. Wiss., Berlin, Kl. Medizin **1**, 98—105 (1960).

— Étude quantitative de réversions biochemiques induites chez la levure *Schizosaccharomyces pombe* par des radiations et des substances radio-métriques. Abh. dtsch. Akad. Wiss., Berlin, Kl. Medizin **1**, 192—228 (1962).

HIRS, C. H. W., S. MOORE, and W. H. STEIN: The sequence of amino acid residues in performic acid-oxidized ribonuclease. J. biol. Chem. **235**, 633—647 (1960).

HOGNESS, D. S., and H. K. MITCHELL: Genetic factors influencing the acti-vity of tryptophan desmolase in *Neurospora crassa.* J. gen. Microbiol. **11**, 401—411 (1954).

HOLLIDAY, R.: The genetics of *Ustilago maydis.* Genet. Res. **2**, 204—230 (1961).

HOROWITZ, N. H.: The isolation and identification of a natural precursor of choline. J. biol. Chem. **162**, 413—419 (1946).

— Methionine synthesis in *Neurospora.* The isolation of cystathionine. J. biol. Chem. **171**, 255—264 (1947).

— Biochemical genetics of *Neurospora.* Advanc. Genet. **3**, 33—71 (1950).

— Genetic and non-genetic factors in the production of enzymes by *Neurospora.* Growth Symp. **10**, 47—62 (1951).

— D. M. BONNER, and M. B. HOULAHAN: The utilization of choline analogs by cholineless mutants of *Neurospora.* J. biol. Chem. **159**, 145—151 (1945).

—, and M. FLING: Genetic determination of tyrosinase thermostability in *Neurospora.* Genetics **4**, 360—374 (1953).

— — Studies of tyrosinase production by a heterocaryon of *Neurospora.* Proc. nat. Acad. Sci. (Wash.) **42**, 498—501 (1956).

—, M. FLING, H. L. MACLEOD, and N. SUEOKA: Genetic determination and enzymatic induction of tyrosinase in *Neurospora.* J. molec. Biol. **2**, 96—104 (1960).

— — — — A genetic study of two new structural forms of tyrosinase in *Neurospora.* Genetics **46**, 1015—1024 (1961a).

— — — and Y. WATANABE: Structural and regulative genes controlling tyrosinase synthesis in *Neurospora.* Cold Spr. Harb. Symp. quant. Biol. **26**, 233—238 (1961b).

—, and U. LEUPOLD: Some recent studies bearing on the one gene one enzyme hypothesis. Cold Spr. Harb. Symp. quant. Biol. **16**, 65—74 (1951).

—, and R. L. METZENBERG: Biochemical aspects of genetics. Ann. Rev. Biochem. **34**, 527—564 (1965).

—, and S. C. SHEN: *Neurospora* tyrosinase. J. biol. Chem. **197**, 513—520 (1952).

HOULAHAN, M. B., and H. K. MITCHELL: A suppressor in *Neurospora* and its use as evidence for allelism. Proc. nat. Acad. Sci. (Wash.) **33**, 223—229 (1947).

— — Evidence for an interrelation in the metabolism of lysine, arginine and pyrimidine in *Neurospora.* Proc. nat. Acad. Sci. (Wash.) **34**, 465—470 (1948).

HUANG, P. C., N. MAHESHWARI, and J. BONNER: Enzymatic synthesis of RNA. Biochem. biophys. Res. Commun. **3**, 689—694 (1960).

INGRAM, V.: Hemoglobin and its abnormalities. Springfield (Ill.): Ch. C. Thomas 1961.

INGRAM, V. M.: The hemoglobins in genetics and evolution. New York and London 1963.

ISHIKAWA, T.: Complementation and genetic maps of the *ad-8* locus in *Neurospora crassa*. Genetics **45**, 993 (1960).

— Genetic studies of *ad-8* mutants in *Neurospora crassa*. I. Genetic fine structure of the *ad-8* locus. Genetics **47**, 1147—1161 (1962a).

— Genetic studies of *ad-8* mutants in *Neurospora crassa*. II. Interallelic complementation at the *ad-8* locus. Genetics **47**, 1755—1770 (1962b).

JACOB, F., and J. MONOD: Genetic regulatory mechanism in the synthesis of proteins. J. molec. Biol. **3**, 318—356 (1961a).

— — On the regulation of gene action. Cold Spr. Harb. Symp. quant. Biol. **26**, 193—211 (1961b).

JAKOBY, W. B.: Kynurenine formamidase from *Neurospora*. J. biol. Chem. **207**, 657—663 (1954).

— An interrelationship between tryptophan, tyrosine and phenylalanine in *Neurospora*. In: W. D. McELROY and B. GLASS (edits.), Amino acid metabolism, p. 909—913. Baltimore 1955.

—, and D. M. BONNER: Kynureninase from *Neurospora:* Purification and properties. J. biol. Chem. **205**, 699—707 (1953a).

— — Kynureninase from *Neurospora:* Interaction of enzyme with substrates, coenzyme, and amines. J. biol. Chem. **205**, 709—715 (1953b).

JOLY, P.: Données récentes sur la génétique des champignons supérieurs (Ascomycètes et Basidiomycètes). Rev. Mycol. (Paris) **29**, 115—186 (1964).

JONES, E. E., and H. P. BROQUIST: Saccharopine, an intermediate of the aminoadipic acid pathway of lysine biosynthesis. J. biol. Chem. **240**, 2531—2536 (1965).

JONES, M. E.: Carbamyl phosphate. Many forms of life use this molecule to synthesize arginine, uracil, and adenosine triphosphate. Science **140**, 1373—1379 (1963).

JONES jr., O. W., and M. W. NIRENBERG: Qualitative survey of RNA codewords. Proc. nat. Acad. Sci. (Wash.) **48**, 2115—2123 (1962).

JUKES, T. H.: Possible base sequences in the amino acid code. Biochem. biophys. Res. Commun. **7**, 497—502 (1962).

—, and A. C. DORNBUSH: Growth stimulatioᴜ of *Neurospora* cholineless mutant by dimethylaminoethanol. Proc. Soc. exp. Biol. (N.Y.) **58**, 142—143 (1945).

JUNGWIRTH, C., S. R. GROSS, P. MARGOLIN, and H. E. UMBARGER: The biosynthesis of leucine. I. The accumulation of β-carboxy-β-hydroxy-isocaproate by leucine auxotrophs of *Salmonella typhimurium* and *Neurospora crassa*. Biochemistry **2**, 1—6 (1963).

KÄGI, J. H. R., and B. L. VALLEE: The role of zinc in alcohol dehydrogenase. V. The effect of metal binding agents on the structure of yeast alcohol dehydrogenase molecule. J. biol. Chem. **235**, 3188—3192 (1960).

KANO-SUEOKA, T., and S. SPIEGELMAN: Evidence for a nonrandom reading of the genome. Proc. nat. Acad. Sci. (Wash.) **48**, 1942—1949 (1962).

KAPLAN, M. M., and M. FLAVIN: Threonine biosynthesis. On the pathway in fungi and bacteria and the mechanism of the isomerization reaction. J. biol. Chem. **240**, 3928—3933 (1965).

KAPLAN, S., S. ENSIGN, D. M. BONNER, and S. E. MILLS: Gene products of CRM-mutants at the *td* locus. Proc. nat. Acad. Sci. (Wash.) **51**, 372—378 (1964a).

— ST. E. MILLS, ST. ENSIGN, and D. M. BONNER: Genetic determination of the antigenic specificity of tryptophan synthetase. J. molec. Biol. **8**, 801—813 (1964b).

— Y. SUYAMA, and D. M. BONNER: Fine structure analysis at the *td* locus of *Neurospora crassa*. Genetics **49**, 145—158 (1964c).

KAPULER, A. M., and H. BERNSTEIN: A molecular model for an enzyme based on a correlation between genetic and complementation maps of the locus specifying enzyme. J. molec. Biol. **6**, 443—451 (1963).

KARLSON, P.: Biochemische Wirkungen der Gene. Ergebn. Enzymforsch. **13**, 85—206 (1954).

KASHA, M., and B. PULLMAN (edits.): Horizons in biochemistry. New York and London 1962.

KAUDEWITZ, F.: Ausgewählte Beispiele biochemisch genetischer Forschung. Z. menschl. Vererb.- u. Konstit.-Lehre **36**, 242—257 (1962).

KHORANA, H. G., H. BÜCHI, H. GHOSH, N. GUPTA, T. M. JACOB, H. KÖSSEL, R. MORGAN, S. A. NARANG, E. OHTSUKA and R. D. WELLS: Polynucleotide synthesis and the genetic code. Cold Spr. Harb. Symp. quant. Biol. **31**, 39—49 (1966).

KIRITANI, K., S. NARISE, and R. P. WAGNER: The dihydroxy dehydratase of *Neurospora crassa*. J. biol. Chem. **241**, 2042—2046 (1966a).

— — — The reductoisomerase of *Neurospora crassa*. J. biol. Chem. **241**, 2047—2051 (1966b).

KLOPOTOWSKI, T., and D. HULANICKA: Imidazol-glycerol accumulation by yeast resulting from inhibition of histidine biosynthesis by 3-amino-1,2,4-triacole. Acta biochim. pol. **10**, 209—218 (1963).

KØLMARK, G., and M. WESTERGAARD: Further studies on chemically induced reversions at the adenine locus of *Neurospora*. Hereditas (Lund) **39**, 209—224 (1953).

KÜHN, A.: Über eine Gen-Wirkkette der Pigmentbildung bei Insekten. Nachr. Akad. Wiss. Göttingen, Math.-physik. Kl. **1941** 231—261.

— Neue Mutationen und Phänogenetik bei Tieren. In: Naturforschung und Medizin in Deutschland 1939—1946, E. BÜNNING u. A. KÜHN (Hrsg.), Bd. 53, S. 77—93. Wiesbaden 1948.

KURAHASHI, K.: Enzyme formation in galactose negative mutants of *Escherichia coli*. Science **125**, 114—116 (1957).

LACY, A. M., and D. M. BONNER: Complementarity between alleles at the *td* locus in *Neurospora crassa*. Proc. X. Internat. Congr. of Genet., vol. 2, p. 157. Montreal 1958.

— — Complementation between alleles of the *td* locus in *Neurospora crassa*. Proc. nat. Acad. Sci. (Wash.) **47**, 72—77 (1961).

LAMEY, H. A., D. M. BOONE, and G. W. KEITT: *Venturia inaequalis* (CKE.) WINT. Growth responses of biochemical mutants. Amer. J. Bot. **43**, 828—834 (1956).

LANDMAN, O. E.: Formation of lactose in mutants and parental strains of *Neurospora*. Genetics **35**, 673—674 (1950).

LANNI, F.: Biological validity of amino acid codes deduced with synthetic ribonucleotide polymers. Proc. nat. Acad. Sci. (Wash.) **48**, 1623—1630 (1962).

LEDERBERG, J., E. M. LEDERBERG, N. ZINDER, and E. LIVELY: Recombination analysis of bacterial heredity. Cold Spr. Harb. Symp. quant. Biol. **16**, 413—443 (1951).

LESTER, G.: Some aspects of tryptophan synthetase formation in *Neurospora crassa*. J. Bact. **81**, 964—973 (1961a).

— Repression and inhibition of indole synthesizing activity in *Neurospora crassa*. J. Bact. **82**, 215—223 (1961b).

— Regulation of early reactions in the biosynthesis of tryptophan in *Neurospora crassa*. J. Bact. **85**, 468—475 (1963).

LEUPOLD, U.: Physiologisch-genetische Studien an adenin-abhängigen Mutanten von *Schizosaccharomyces pombe*. Schweiz. Z. Path. Bakt. **20**, 535—544 (1957).

— Intragene Rekombination und allele Komplementierung. Arch. Klaus-Stift. Vererb.-Forsch. **36**, 89—117 (1961).

LEVINTHAL, C., and P. F. DAVIDSON: Biochemistry of genetic factors. Amer. Rev. Biochem. **30**, 641—668 (1961).

Lewis, D.: Genetical analysis of methionine suppressors in *Coprinus*. Genet. Res. **2**, 141—155 (1961).

Lewis, R. W.: Mutants of *Neurospora* requiring succinic acid or a bio-chemically related acid for growth. Amer. J. Bot. **35**, 292—295 (1948).

Liebermann, I., A. Kornberg, and E. S. Simms: Enzymatic synthesis of pyrimidine nucleotides. Orotidine-5-phosphate and uridine-5-phosphate. J. biol. Chem. **215**, 403—415 (1955).

Lindegren, C. C.: The yeast cell, its genetics and cytology. St. Louis (Missouri): Educational publishers 1949.

— Gene control of fermentation in *Saccharomyces* without control of permeability. J. Bact. **74**, 689—690 (1957).

— The biological function of deoxyribonucleic acid. J. theor. Biol. **1**, 107—119 (1961).

—, and G. Lindegren: Asci in *Saccharomyces* with more than four spores. Genetics **38**, 73—78 (1953).

— — Eight genes controlling the presence or absence of carbohydrate fermentation in *Saccharomyces*. J. gen. Microbiol. **15**, 19—28 (1956).

— M. A. Williams, and D. O. McClary: The distribution of chromatin in budding yeast cells. Antonie v. Leeuwenhoek **22**, 1—20 (1956).

Lingens, F., u. W. Goebel: Untersuchungen an biochemischen Mangel-mutanten von *Saccharomyces cerevisiae* mit genetischem Block hinter einer Verzweigungsstelle in der Biosynthese der aromatischen Amino-säuren. Hoppe-Seylers Z. physiol. Chem. **342**, 1—12 (1965).

— — u. H. Uesseler: Regulation der Biosynthese der aromatischen Ami-nosäuren in *Saccharomyces cerevisiae*. I. Hemmung der Enzymaktivitäten (Feedback-Wirkung). Biochem. Z. **346**, 357—367 (1966a).

—, u. H. Hellmann: Isolierung von Shikimisäure aus dem Medium einer *Saccharomyces cerevisiae*-Mutante. Z. Naturforsch. **13b**, 462—463 (1958).

—, u. W. Lück: Über die Biosynthese des Tryptophans in *Saccharomyces cerevisiae*. Hoppe-Seylers Z. physiol. Chem. **333**, 190—198 (1963).

— — u. G. Müller: Über die Wirkung von 5-Oxo-6-diazonorleucin und Albizziin auf die Biosynthese der Anthranilsäure in *Saccharomyces cere-visiae*. Hoppe-Seylers Z. physiol. Chem. **343**, 282—289 (1966b).

— B. Sprössler u. W. Goebel: Zur Biosynthese der Anthranilsäure in *Sac-charomyces cerevisiae*. Biochim. biophys. Acta (Amst.) **121**, 164—166 (1966c).

—, u. P. Vollprecht: Zur Biosynthese der Nicotinsäure in Streptomyceten, Algen, Phycomyceten und Hefe. Hoppe-Seylers Z. physiol. Chem. **339**, 64—74 (1964).

Loring, H. S., and J. L. Fairley: Growth-promoting activity of guanine for the purine-deficient *Neurospora* 28610. J. biol. Chem. **172**, 843—844 (1948).

Maas, W. K., and B. D. Davis: Production of an altered panthothenate-synthesizing enzyme by a temperature sensitive mutant of *Escherichia coli*. Proc. nat. Acad. Sci. (Wash.) **38**, 785—797 (1952).

Madsen, N. B., and F. R. N. Gurd: The interaction of muscle phosphorylase with p-chloromercuribenzoate. III. The reversible dissociation of phos-phorylase. J. biol. Chem. **223**, 1055—1065 (1956)

Marcker, K.: The formation of N-formyl-methionyl-sRNA. J. mol. Biol. **14**, 63—70 (1965).

Markert, C. L.: The effects of genetic changes on tyrosinase activity in *Glomerella*. Genetics **35**, 60—75 (1950).

— Radiation-induced nutritional and morphological mutants of *Glomerella*. Genetics **37**, 339—352 (1952).

—, and R. D. Owen: Immunogenetic studies of tyrosinase specificity. Genetics **39**, 818—835 (1954).

Matchett, W. H., and J. A. DeMoss: Factors affecting increased produc-tion of tryptophan synthetase by a *td* mutant of *Neurospora crassa*. J. Bact. **83**, 1294—1300 (1962).

Matthaei, J. H., O. W. Jones, R. G. Martin, and M. W. Nirenberg: Characteristics and composition of RNA coding units. Proc. nat. Acad. Sci. (Wash.) **48**, 666—677 (1962).

MATTHAEI, J. H., H. P. VOIGT, G. HELLER, R. NETH, G. SCHÖCH, H. KÜBLER, F. AMELUNXEN, G. SANDER, and A. PARMEGGIANI: Specific interactions of ribosomes in decoding. Cold Spr. Harb. Symp. quant. Biol. **31**, 25—38 (1966).

MAXWELL, E. S.: Stimulation of amino acid incorporation into protein by natural and synthetic polyribonucleotides in a mammalian cell-free system. Proc. nat. Acad. Sci. (Wash.) **48**, 1639—1643 (1962).

McELROY, W. D., and B. GLASS (edits.): Amino acid metabolism. Baltimore 1955.

— — The chemical basis of heredity. Baltimore 1957.

— —, and H. K. MITCHELL: Enzyme studies on a temperature sensitive mutant of *Neurospora*. Fed. Proc. **5**, 376—379 (1946).

—, and D. SPENCER: Normal pathways of assimilation of nitrate and nitrite. In: W. D. McELROY and B. GLASS (edits.), Inorganic nitrogen metabolism, p. 137—152. Baltimore 1956.

McMANUS, I. R.: The biosynthesis of valine by *Saccharomyces cerevisiae*. J. biol. Chem. **208**, 639—644 (1954).

MEDVEDEN, Z. A.: A hypothesis concerning the way of coding interaction between transfer RNA and messenger RNA at the later stages of protein synthesis. Nature (Lond.) **195**, 38—39 (1962).

MEGNET, R.: Untersuchungen über die Biosynthese von Uracil bei *Schizosaccharomyces pombe*. Arch. Klaus-Stift. Vererb.-Forsch. **33**, 299—334 (1959).

MEISTER, A., A. N. RADHAKRISHNAN, and S. D. BUCKLEY: Enzymatic synthesis of L-pipecolic acid and l-proline. J. biol. Chem. **229**, 789—800 (1958).

MELCHERS, G.: Viruses and genetics. Plant Virology. Proc. 5th Conf. Czech. Plant Virologists, p. 101—109. Prague 1962.

METZENBERG, R. L.: A gene affecting the repression of invertase and trehalase in *Neurospora*. Arch. Biochem. **96**, 468—474 (1962).

—, and H. K. MITCHELL: Isolation of prephenic acid from *Neurospora*. Arch. Biochem. **64**, 51—56 (1956).

MEYERS, J. W., and E. A. ADELBERG: The biosynthesis of isoleucine and valine. I. Enzymatic transformation of the dihydroxy acid precursors to the keto acid precursors. Proc. nat. Acad. Sci. (Wash.) **40**, 493—499 (1954).

MICHELSON, M., W. DRELL, and H. K. MITCHELL: A new ribose nucleoside from *Neurospora:* "Orotidine". Proc. nat. Acad. Sci. (Wash.) **37**, 396—399 (1951).

MIDDELHOVEN, W. J.: The ornithine pathway in the yeast *Candida utilis*. Biochim. biophys. Acta (Amst.) **77**, 152—154 (1963).

MITCHELL, H. K., and M. B. HOULAHAN: Adenine requiring mutants of *Neurospora crassa*. Fed. Proc. **5**, 370—375 (1946a).

— — *Neurospora*. IV. A temperature-sensitive riboflavinless mutant. Amer. J. Bot. **33**, 31—35 (1946b).

— — Investigations on the biosynthesis of pyrimidine nucleosides in *Neurospora*. Fed. Proc. **6**, 506—509 (1947).

— — An intermediate in the biosynthesis of lysine in *Neurospora*. J. biol. Chem. **174**, 883—887 (1948).

— — and J. F. NYC: The accumulation of orotic acid by a pyrimidineless mutant of *Neurospora*. J. biol. Chem. **172**, 525—529 (1948).

—, and J. LEIN: A *Neurospora* mutant deficient in the enzymatic synthesis of tryptophan. J. biol. Chem. **175**, 481—482 (1948).

—, and J. F. NYC: Hydroxyanthranilic acid as a precursor of nicotinic acid in *Neurospora*. Proc. nat. Acad. Sci. (Wash.) **34**, 1—5 (1948).

MITCHELL, M. B., and H. K. MITCHELL: Observations on the behavior of suppressors in *Neurospora*. Proc. nat. Acad. Sci. (Wash.) **38**, 205—214 (1952).

— — Test for non-allelism at the pyrimidine-3 locus of *Neurospora*. Genetics **41**, 319—326 (1956).

Function

MOHLER, W. C., and S. R. SUSKIND: The similar properties of tryptophan synthetase and a mutationally altered enzyme in *Neurospora crassa*. Biochim. biophys. Acta (Amst.) **43**, 288—299 (1960).

MONOD, J., and M. COHN: La biosynthèse induite des enzymes (adaption enzymatique). Advanc. Enzymol. **13**, 67—119 (1952).

—, and F. JACOB: Telenomic mechanism in cellular metabolism, growth and differentiation. Cold Spr. Harb. Symp. quant. Biol. **26**, 389—411 (1961).

MORROW, J.: Dispensable and indispensable genes in *Neurospora*. Science **144**, 307—308 (1964).

MURRAY, N. E.: Complementation and recombination between methionine-2 alleles in *Neurospora crassa*. Heredity **15**, 207—217 (1960).

NEWMEYER, D.: Arginine synthesis in *Neurospora crassa:* Genetic studies. J. gen. Microbiol. **16**, 449—462 (1957).

— Genes influencing the conversion of citrulline to arginino-succinate in *Neurospora crassa*. J. gen. Microbiol. **28**, 215—230 (1962).

NEWTON, W. A., and E. E. SNELL: An inducible tryptophan synthetase in tryptophan auxotrophs of *Escherichia coli*. Proc. nat. Acad. Sci. (Wash.) **48**, 1431—1439 (1962).

NIRENBERG, M. W., and O. W. JONES jr.: The current status of the RNA code. In: H. J. VOGEL, V. BRYSON and J. O. LAMPEN (edits.), Informational Macromolecules, p. 451—465. New York and London 1963.

—, T. CASKEY, R. MARSHALL, R. BRIMACOMBE, D. KELLOGG, B. DOCTOR, D. HATFIELD, J. LEVIN, F. ROTTMAN, S. PESTKA, M. WILCOX and F. ANDERSON: The RNA code and protein synthesis. Cold Spr. Harb. Symp. quant. Biol. **31**, 11—24 (1966).

NOVELLI, G. D.: Protein synthesis in microorganisms. Ann. Rev. Microbiol. **14**, 65—82 (1960).

NULTSCH, W.: Allgemeine Botanik. Stuttgart 1964.

NYC, J. F., H. K. MITCHELL, E. LEIFER, and W. H. LANGHAM: Use of isotopic carbon in a study of the metabolism of anthranilic acid in *Neurospora*. J. biol. Chem. **179**, 783—787 (1949).

OCHOA, S.: Synthetic polynucleotides and the genetic code. In: H. J. VOGEL, V. BRYSON and J. O. LAMPEN (edits.), Informational Macromolecules, p. 437—449. New York and London 1963.

OHNISHI, E., H. MACLEOD, and N. H. HOROWITZ: Mutants of *Neurospora crassa* deficient in D-amino acid oxidase. J. biol. Chem. **237**, 138—142 (1962).

OWEN, R. D., and C. L. MARKERT: Effects of antisera on tyrosinase in *Glomerella* extracts. J. Immunol. **74**, 257—269 (1955).

PAIGEN, K.: On the regulation of DNA transcription. J. theor. Biol. **3**, 268—282 (1962).

PALLERONI, N. J., and C. C. LINDEGREN: A single adaptive enzyme in *Saccharomyces* elicited by several related substrates. J. Bact. **65**, 122—130 (1953).

PARTRIDGE, C. W. H.: Altered properties of the enzyme, adenylosuccinase, produced by interallelic complementation at the *ad-4* locus in *Neurospora crassa*. Biochem. biophys. Res. Commun. **3**, 613—619 (1960).

— D. M. BONNER, and C. YANOFSKY: A quantitative study of the relationship between tryptophan and niacin in *Neurospora*. J. biol. Chem. **194**, 269—278 (1952).

—, and N. H. GILES: Identification of major accumulation products of adenine-specific mutants of *Neurospora*. Arch. Biochem. **67**, 237—258 (1957).

PATEMAN, J. A.: Back-mutation studies at the *am*-locus in *Neurospora crassa*. J. Genet. **55**, 444—455 (1957).

— Inter-relationship of alleles at the *am* locus in *Neurospora crassa*. J. gen. Microbiol. **23**, 393—399 (1960).

— D. J. COVE, B. M. REVER, and D. B. ROBERTS: A common co-factor for nitrate reductase and xanthine dehydrogenase which also regulates the synthesis of nitrate reductase. Nature (Lond.) **201**, 58—60 (1964).

PATEMAN, J. A., and J. R. S. FINCHAM: Gene-enzyme relationship at the *am* locus in *Neurospora crassa*. Heredity **12**, 317—332 (1958).

PERKINS, D. D.: Biochemical mutants in the smut fungus *Ustilago maydis*. Genetics **34**, 607—626 (1949).

PERUTZ, M. F.: Proteins and nucleic acids. Structure and function. Amsterdam-London-New York 1962.

PIERCE, J. G., and H. S. LORING: Growth requirements of a purine deficient strain of *Neurospora*. J. biol. Chem. **160**, 409—415 (1945).

POMPER, S., and P. R. BURKHOLDER: Studies on the biochemical genetics of yeast. Proc. nat. Acad. Sci. (Wash.) **35**, 456—464 (1949).

PONTECORVO, G.: The genetics of *Aspergillus nidulans*. Advanc. Genet. **5**, 142—239 (1953).

— Trends in genetic analysis. New York 1958.

RACHMELER, M., and C. YANOFSKY: Biochemical and genetic studies with a new *td* mutant type in *Neurospora crassa*. Bact. Proc. **30** (1959).

— — Biochemical, immunological and genetic studies with a new type tryptophan synthetase mutant of *Neurospora crassa*. J. Bact. **81**, 955—963 (1961).

RADHAKRISHNAN, A. N., R. P. WAGNER, and E. E. SNELL: Biosynthesis of valine and isoleucine. III. α-Keto-β-hydroxy acid reductase and α-hydroxy-β-keto- acid reductoisomerase. J. biol. Chem. **235**, 2322—2331 (1960).

RAPER, J. R., and P. G. MILES: The genetics of *Schizophyllum commune*. Genetics **43**, 530—546 (1958).

RATNER, S.: Arginine metabolism and interrelationships between the citric acid and urea cycles. In: W. D. McELROY and B. GLASS (edits.), Amino acid metabolism, p. 231—257. Baltimore 1955.

REGNERY, D. C.: A leucineless mutant strain of *Neurospora crassa*. J. biol. Chem. **154**, 151—160 (1944).

REICHARD, P., and G. HANSHOFF: Aspartate carbamyl transferase from *Escherichia coli*. Acta chem. Scand. **10**, 548—566 (1956).

REISSIG, J. L.: Forward and back mutation in the *pyr-3* region of *Neurospora*. I. Mutations from arginine dependence to prototrophy. Genet. Res. **1**, 356—374 (1960).

RHINESMITH, H. S., W. A. SCHROEDER, and N. J. MARTIN: The N-terminal sequence of the β chain of normal adult human hemoglobin. J. Amer. chem. Soc. **80**, 3358—3361 (1958).

RILEY, M., and A. B. PARDEE: Gene expression: its specificity and regulation. Ann. Rev. Microbiol. **16**, 1—34 (1962).

ROBERTSON, J. J., and H. O. HALVORSON: The components of maltozymase in yeast and their behavior during deadaption. J. Bact. **73**, 186—198 (1957).

ROBICHON-SZULMAJSTER, H. DE: Induction of enzymes of the galactose pathway in mutants of *Saccharomyces cerevisiae*. Science (Lancaster) **127**, 28—29 (1958).

ROMAN, H.: Studies of gene mutation in *Saccharomyces*. Cold Spr. Harb. Symp. quant. Biol. **21**, 175—185 (1956).

— Sur les récombinaisons nonréciproques chez *Saccharomyces cerevisiae* et sur les problèmes posés par ces phénomènes. Ann. Génét. **1**, 11—17(1958).

ROSEN, R.: An hypothesis of FREESE and the DNA-protein coding problem. Bull. math. Biophys. **23**, 305—318 (1961).

ROTHMAN, F., and R. BYRNE: Fingerprint analysis of alkaline phosphatase of *Escherichia coli K 12*. J. molec. Biol. **6**, 330—340 (1963).

RUDERT, F., and H. O. HALVORSON: The effect of gene dosage on the level of α-glucosidase in yeast. Bull. Res. Coun. Israel A 4, **11**, 337—344 (1963).

SAGISAKA, S., and K. SHIMURA: Enzymic reduction of α-aminoadipic acid by yeast enzyme. Nature (Lond.) **184**, 1709—1710 (1959).

SAKAMI, W.: The biochemical relationship between glycine and serine. In: W. D. McELROY and B. GLASS (edits.), Amino acid metabolism, p. 658—683. Baltimore 1955.

Function

SANGER, F., and L. F. SMITH: The structure of insulin. Endeavour **16**, 48—53 (1957).

SANWAL, B. D., and M. LATA: Glutamic dehydrogenase in single-gene mutants of *Neurospora crassa* deficient in amination. Nature (Lond.) **190**, 286—287 (1961).

SCHAEFFER, P.: A black mutant of *Neurospora crassa*. Mode of action of the mutant allele and action of light on melanogenesis. Arch. Biochem. **47**, 359—379 (1953).

SCHLESINGER, M. J., and C. LEVINTHAL: Hybrid protein formation of *E. coli* alkaline phosphatase leading to in vitro complementation. J. molec. Biol. **7**, 1—12 (1963).

SCHULMAN, H. M., and D. M. BONNER: A naturally occurring DNA-RNA complex from *Neurospora crassa*. Proc. nat. Acad. Sci. (Wash.) **48**, 53—63 (1962).

SEARASHI, T.: Genetical and biochemical studies on amylase in *Aspergillus oryzae*. Jap. J. Genet. **37**, 10—23 (1962).

SERRES, F. J. DE: Studies with purple adenin mutants in *Neurospora crassa*. I. Structural and functional complexity in the *ad-3* region. Genetics **41**, 668—676 (1956).

— Heterokaryon-incompatibility factor interaction tests between *Neurospora* mutants. Science **138**, 1342—1343 (1962).

— Studies with purple adenine mutants in *Neurospora crassa*. V. Evidence for allelic complementation among *ad-3 B* mutants. Genetics **48**, 351—360 (1963).

SILVER, W. S., and W. D. McELROY: Enzyme studies on nitrate and nitrite mutants of *Neurospora*. Arch. Biochem. **51**, 379—394 (1954).

SPEYER, J. F., P. LENGYEL, C. BASILIO, and S. OCHOA: Synthetic poly-nucleotides and the amino acid code. IV. Proc. nat. Acad. Sci. (Wash.) **48**, 441—448 (1962).

SPIEGELMAN, S.: The relation of information RNA to DNA. Cold Spr. Harb. Symp. quant. Biol. **26**, 75—90 (1961).

— Information transfer from the genome. Fed. Proc. **22**, 36—54 (1963).

SRB, A. M.:, and N. H. HOROWITZ The ornithine cycle in *Neurospora* and its genetic control. J. biol. Chem. **154**, 129—139 (1944).

STADLER, J., and C. YANOFSKY: Studies on a series of tryptophan-independent strains derived from a tryptophan requiring mutant of *Escherichia coli*. Genetics **44**, 105—123 (1959).

STARLINGER, P.: Die genetische Regulation der Enzymsynthese. Angew. Chem. **75**, 71—77 (1963).

STRASSMAN, M., A. J. THOMAS, L. A. LOCKE, and S. WEINHOUSE: Intra-molecular migration and isoleucine biosynthesis. J. Amer. chem. Soc. **76**, 4241—4242 (1954).

— — — — A study of leucine biosynthesis in *Torulopsis utilis*. J. Amer. chem. Soc. **78**, 1599—1602 (1956).

— — and S. WEINHOUSE: Valine biosynthesis in *Torulopsis utilis*. J. Amer. chem. Soc. **75**, 5135 (1953).

—, and S. WEINHOUSE: Biosynthetic pathways. III. The biosynthesis of lysine by *Torulopsis utilis*. J. Amer. chem. Soc. **75**, 1680—1684 (1953).

— — Isotope studies on biosynthesis of valine and isoleucine. In: W. D. McELROY and B. GLASS (edits.), Amino acid metabolism, p. 452—457. Baltimore 1955.

STRAUSS, B. S.: Properties of mutants of *Neurospora crassa* with low pyruvic carboxylase activity. Arch. Biochem. **44**, 200—210 (1953).

— Oxalacetic carboxylase deficiency of the succinate-requiring mutants of *Neurospora crassa*. J. biol. Chem. **225**, 535—544 (1957).

—, and S. PIEROG: Gene interaction: The mode of action of the suppressor of acetate requiring mutants of *Neurospora crassa*. J. gen. Microbiol. **10**, 221—235 (1954).

STUBBE, H. (Hrsg.): Struktur und Funktion des genetischen Materials. Erwin-Baur-Gedächtnisvorlesungen III, 1963. Abh. dtsch. Akad. Wiss. Berlin, Kl. Medizin 4, Berlin 1964.

SUEOKA, N.: Compositional correlation between deoxyribonucleic acid and protein. Cold Spr. Harb. Symp. quant. Biol. 26, 35—43 (1961).

—, and T. YAMANE: Fractionation of amino acyl-acceptor RNA on a methylated albumin column. Proc. nat. Acad. Sci. (Wash.) 48, 1454—1461 (1962).

SUSKIND, S. R.: Properties of a protein antigenically related to tryptophan synthetase in Neurospora crassa. J. Bact. 74, 308—318 (1957).

—, and E. JORDAN: Enzymatic activity of a genetically altered tryptophan synthetase in Neurospora crassa. Science 129, 1614—1615 (1959).

—, and L. I. KUREK: On a mechanism of suppressor gene regulation of tryptophan synthetase activity in Neurospora crassa. Proc. nat. Acad. Sci. (Wash.) 45, 193—196 (1959).

— C. YANOFSKY, and D. M. BONNER: Allelic strains of Neurospora lacking tryptophan synthetase: A preliminary immuno-chemical characterization. Proc. nat. Acad. Sci. (Wash.) 41, 577—582 (1955).

SUSSMAN, A. S.: A comparison of the properties of two forms of tyrosinase from Neurospora crassa. Arch. Biochem. 95, 407—415 (1961).

— P. COUGHEY, and J. C. STRAIN: Effect of environmental conditions upon tyrosinase activity in Glomerella cingulata. Amer. J. Bot. 42, 810—815 (1955).

—, and C. L. MARKERT: The development of tyrosinase and cytochrome oxidase activity in mutants of Glomerella cingulata. Arch. Biochem. 45, 31—40 (1953).

SUTTON, H. E.: Genetics. Genetic information and the control of protein structure and function. New York 1960.

SUYAMA, Y.: In vitro complementation in the tryptophan synthetase system of Neurospora. Biophys. biochem. Res. Commun. 10, 144—149 (1963).

—, and D. M. BONNER: Complementation between tryptophan synthetase mutants of Neurospora crassa. Biochim. biophys. Acta (Amst.) 81, 565—575 (1964).

— A. M. LACY, and D. M. BONNER: A genetic map of the td locus in Neurospora crassa. Genetics 49, 135—144 (1964).

— K. D. MUNKRES, and V. W. WOODWARD: Genetic analysis of the pyr-3 locus of Neurospora crassa. Genetica 30, 293—311 (1959).

TATUM, E. L.: Amino acid metabolism in mutant strains of microorganisms. Fed. Proc. 8, 511—517 (1949).

— Genetic aspects of growth responses in fungi. In: F. SKOOG (edit.), Plant growth substances, p. 447—461. Madison: Univ. Wisconsin Press 1951.

— A case history in biological research. Chance and the exchange of ideas played roles in the discovery that genes control biochemical events. Science 129, 1711—1715 (1959).

— R. W. BARRATT, N. FRIES, and D. M. BONNER: Biochemical mutant strains of Neurospora produced by physical and chemical treatment. Amer. J. Bot. 37, 38—46 (1950).

— —, and G. W. BEADLE: Genetic control of biochemical reactions in Neurospora. An "aminobenzoicless" mutant. Proc. nat. Acad. Sci. (Wash.) 28, 234—243 (1942).

— — Biochemical genetics of Neurospora. Ann. Missouri Botan. Garden 32, 125—129 (1945).

—, and T. T. BELL: Neurospora. III. Biosynthesis of thiamin. Amer. J. Bot. 33, 15—20 (1946).

—, and D. M. BONNER: Synthesis of tryptophan from indole and serine by Neurospora. J. biol. Chem. 151, 349 (1943).

— — Indole and serine in the biosynthesis and breakdown of tryptophan. Proc. nat. Acad. Sci. (Wash.) 30, 30—37 (1944).

— — and G. W. BEADLE: Anthranilic acid and the biosynthesis of indole and tryptophan by Neurospora. Arch. Biochem. 3, 477—478 (1944).

Function

TATUM, E. L., and S. R. GROSS: Incorporation of carbon atoms 1 and 6 of glucose into protocatechuic acid by *Neurospora*. J. biol. Chem. **219**, 797—807 (1956).
— — G. EHRENSVÄRD, and L. GARNJOBST: Synthesis of aromatic compounds by *Neurospora*. Proc. nat. Acad. Sci. (Wash.) **40**, 271—276 (1954).
—, and D. D. PERKINS: Genetics of microorganisms. Ann. Rev. Microbiol. **4**, 129—150 (1950).
—, and D. SHEMIN: Mechanism of tryptophan synthesis in *Neurospora*. J. biol. Chem. **209**, 671—675 (1954).
TAVLITZKI, J.: Sur la réalisation, chez une souche de *Saccharomyces cerevisiae*, du caractère «besoin en thiamine». C. R. Acad. Sci. (Paris) **238**, 2016—2018 (1954).
TAYLOR, J. H. (edit.): Molecular genetics. Part 1. New York and London 1963.
TEAS, H. J., N. H. HOROWITZ, and M. FLING: Homoserine as a precursor of threonine and methionine in *Neurospora*. J. biol. Chem. **172**, 651—658 (1948).
TINLINE, R. D.: *Cochliobolus sativus*. V. Heterokaryosis and parasexuality. Canad. J. Bot. **40**, 425—437 (1962).
TRUPIN, J. S., and H. P. BROQUIST: Saccharopine, an Intermediate of the aminoadipic acid pathway of lysine biosynthesis. J. biol. Chem. **240**, 2524—2530 (1965).
TSUGITA, A., and H. FRAENKEL-CONRAT: The composition of proteins of chemically evoked mutants of TMV RNA. J. molec. Biol. **4**, 73—82 (1962).
UMBREIT, W. W., W. A. WOODWARD, and I. C. GUNSALUS: The activity of pyridoxal phosphate in tryptophan formation by cell-free enzyme preparations. J. biol. Chem. **165**, 731—732 (1946).
VOGEL, H. J.: On the glutamate-proline-ornithine interrelation in various microorganisms. In: W. D. McELROY and B. GLASS (edits.), Amino acid metabolism, p. 335—346. Baltimore 1955.
—, and D. M. BONNER: On the glutamate-proline-ornithine interrelation in *Neurospora crassa*. Proc. nat. Acad. Sci. (Wash.) **40**, 688—694 (1954).
— — The use of mutants in the study of metabolism. In: W. RUHLAND (Hrsg.), Handbuch der Pflanzenphysiologie, vol. XI, p. 1—32. Berlin-Göttingen-Heidelberg: Springer 1958.
—, and B. D. DAVIS: Glutamatic gamma-semialdehyde and delta-1-pyrroline-5-carboxylic acid, intermediates in the biosynthesis of proline. J. Amer. chem. Soc. **74**, 109—112 (1952).
VOGEL, R. H., and M. J. KOPAC: Glutamic-γ-semialdehyde in arginine and proline synthesis in *Neurospora*. A mutant-tracer analysis. Biochim. biophys. Acta (Amst.) **36**, 505—510 (1959).
— — Some properties of ornithine-transaminase from *Neurospora*. Biochim. biophys. Acta (Amst.) **37**, 539—540 (1960).
—, and H. J. VOGEL: Evidence for acetylated intermediates of arginine synthesis in *Neurospora crassa*. Genetics **48**, 914 (1963).
VOLKIN, E.: Biosynthesis of RNA in relation to genetic coding problems. In: J. H. TAYLOR, Molecular Genetics, part I, p. 271—289. New York and London 1963.
WAGNER, R. P.: The in vitro synthesis of pantothenic acid by pantothenicless wild type *Neurospora*. Proc. nat. Acad. Sci. (Wash.) **35**, 185—189 (1949).
—, and A. BERGQUIST: Synthesis of valine and isoleucine in the presence of a particulate cell fraction of *Neurospora*. Proc. nat. Acad. Sci. (Wash.) **49**, 892—897 (1963).
— — T. BARBEE, and K. KIRITANI: Genetic blocks in the isoleucine-valine pathway of *Neurospora crassa*. Genetics **49**, 865—882 (1964).
—, and B. M. GUIRARD: A gene-controlled reaction in *Neurospora* involving the synthesis of pantothenic acid. Proc. nat. Acad. Sci. (Wash.) **34**, 398—402 (1948).

428

WAGNER, R. P., and C. H. HADDOX: A further analysis of the pantothenicless mutants of *Neurospora*. Amer. Naturalist **85**, 319—330 (1951).

—, and H. K. MITCHELL: Genetics and metabolism, 2. edit. New York 1964.

— C. E. SOMERS, and A. BERGQUIST: Gene structure and function in *Neurospora*. Proc. nat. Acad. Sci. (Wash.) **46**, 708—717 (1960).

WAHBA, A. J., C. BASILIO, J. F. SPEYER, P. LENGYEL, R. S. MILLER, and S. OCHOA: Synthetic polynucleotides and the amino acid code. VI. Proc. nat. Acad. Sci. (Wash.) **48**, 1683—1686 (1962).

WAINWRIGHT, S. D.: On the development of increased tryptophan synthetase enzyme activity by cell-free extracts of *Neurospora crassa*. Canad. J. Biochem. **37**, 1417—1430 (1959).

— On the formation of tryptophan synthetase enzyme by cell-free extracts of mycelium of *Neurospora crassa*. Canad. J. Biochem. **41**, 1327—1329 (1963).

—, and E. S. McFARLANE: Partial purification of the "messenger RNA" of *Neurospora crassa* controlling formation of tryptophan synthetase enzyme. Biophys. biochem. Res. Commun. **9**, 529—533 (1962).

WEBBER, B. B.: Genetical and biochemical studies of histidine-requiring mutants of *Neurospora crassa*. II. Evidence concerning heterogeneity among *hist-3* mutants. Genetics **45**, 1617—1625 (1960).

—, and M. E. CASE: Genetical and biochemical studies of histidine-requiring mutants of *Neurospora crassa*. I. Classification of mutants and characterization of mutant groups. Genetics **45**, 1605—1615 (1960).

WEGMAN, J., and J. A. DeMOSS: The enzymatic conversion of anthranilate to indolylglycerol phosphate in Neurospora crassa. J. biol. Chem. **240**, 3781—3788 (1965).

WEINSTEIN, I. B., and A. N. SCHECHTER: Polyuridylic acid stimulation of phenylalanine incorporation in animal cell extracts. Proc. nat. Acad. Sci. (Wash.) **48**, 1686—1691 (1962).

WEISBLUM, B., S. BENZER, and R. W. HOLLEY: A physical basis for degeneracy in the amino acid code. Proc. nat. Acad. Sci. (Wash.) **48**, 1449—1454 (1962).

WEISS, S. B., and L. GLADSTONE: A mammalian system for the incorporation of cytidine triphosphate into ribonucleic acid. J. Amer. Chem. Soc. **81**, 4118—4119 (1959).

WEYGAND, F.: Anwendungen der stabilen und radioaktiven Isotope in der Biochemie. Angew. Chem. **61**, 285—296 (1949).

—, u. H. SIMON: Herstellung isotopenhaltiger organischer Verbindungen. In: HOUBEN-WEYL (E. MÜLLER Hrsg.), Methoden der organischen Chemie, Bd. 4/2, S. 539—727. Stuttgart 1955.

WHEELER, H. E.: Linkage groups in *Glomerella*. Amer. J. Bot. **43**, 1—6 (1956).

WHELDALE 1903: Zit. bei BEADLE 1959a.

WINGE, Ö., and C. ROBERTS: Inheritance of enzymatic characters in yeast and the phenomenon of longterm adaption. C. R. Lab. Carlsberg, Sér. physiol. **24**, 263—315 (1948).

— — The polymeric genes for maltose fermentation in yeasts and their mutability. C. R. Lab. Carlsberg, Sér. Physiol. **25**, 35—83 (1950).

WINKLER, U., u. R. W. KAPLAN: Genetik der Mikroorganismen: Phänogenetik. Fortschr. Bot. **25**, 341—363 (1963).

WITTMANN, H. G.: Comparison of the tryptic peptides of chemically induced and spontaneous mutants of tobacco mosaic virus. Virology **12**, 609—612 (1960).

— Ansätze zur Entschlüsselung des genetischen Codes. Naturwissenschaften **48**, 729—734 (1961).

— Proteinuntersuchungen an Mutanten des Tabakmosaikvirus als Beitrag zum Problem des genetischen Codes. Z. Vererbungsl. **93**, 491—530 (1962).

— Übertragung der genetischen Information. Naturwissenschaften **50**, 76—88 (1963).

WITTMANN, H. G., and B. WITTMANN-LIEBOLD: Tobacco mosaic virus mutants ɐnd the genetic coding problem. Cold Spr. Harb. Symp. quant. Biol. **28**, 589—595 (1963).
— — Untersuchungen über Mutanten und Stämme des Tabakmosaikvirus. Abh. dtsch. Akad. Wiss. Berlin, Kl. Med. **4**, 141—146 (1964).
WITTMANN-LIEBOLD, B., u. H. G. WITTMANN: Die primäre Proteinstruktur von Stämmen des Tabakmosaikvirus. Aminosäuresequenzen des Proteins des Tabakmosaikvirusstammes *dahlemense*. Teil III. Z. Vererbungsl. **94**, 427—435 (1963).
— — Die primäre Proteinstruktur von Stämmen des Tabakmosaikvirus. Aminosäuresequenzen des Proteins des Tabakmosaikvirusstammes *dahlemense*. Teil I. Hoppe-Seylers Z. physiol. Chem. **335**, 69—116 (1964).
WOESE, C. R.: Nature of the biological code. Nature (Lond.) **194**, 1114—1115 (1962).
WOOD, B. W., and P. BERG: The effect of enzymatically synthesized ribonucleic acid on amino acid incorporation by a soluble protein-ribosome system from *Escherichia coli*. Proc. nat. Acad. Sci. (Wash.) **48**, 94—104 (1962).
WOODWARD, D. O.: Enzyme complementation in vitro between adenylosuccinaseless mutants of *Neurospora crassa*. Proc. nat. Acad. Sci. (Wash.) **45**, 846—850 (1959).
— A gene concept based on genetic and chemical studies in *Neurospora*. Quart. Rev. Biol. **35**, 313—323 (1960).
— C. W. H. PARTRIDGE, and N. H. GILES: Complementation at the *ad-4* locus in *Neurospora crassa*. Proc. nat. Acad. Sci. (Wash.) **44**, 1237—1244 (1958).
— — — Studies of adenylosuccinase in mutants and revertants of *Neurospora crassa*. Genetics **45**, 535—554 (1960).
WOODWARD, V. W.: Complementation and recombination among *pyr-3* hetero-alleles of *Neurospora crassa*. Proc. nat. Acad. Sci. (Wash.) **48**, 348—356 (1962).
—, and R. H. DAVIS: Co-ordinate changes in complementation, suppression and enzyme phenotypes of a *pyr-3* mutant of *Neurospora crassa*. Heredity **18**, 21—25 (1963).
—, and P. SCHWARZ: *Neurospora* mutants lacking ornithine transcarbamylase. Genetics **49**, 845—853 (1964).
WORK, E.: Some comparative aspects of lysine metabolism. In: W. D. McELROY and B. GLASS (edits.), Amino acid metabolism, p. 462—492. Baltimore 1955.
WUST, C. J.: Inactivation of tryptophan synthetase from *Neurospora crassa* during dialysis. Biochim. biophys. Res. Commun. **5**, 35—39 (1961).
YANOFSKY, C.: The effect of gene change on tryptophan desmolase formation. Proc. nat. Acad. Sci. (Wash.) **38**, 215—226 (1952a).
— Tryptophan desmolase of *Neurospora*. Partial purification and properties J. biol. Chem. **194**, 279—286 (1952b).
— Tryptophan and niacin synthesis in various organisms. In: W. D. McELROY and B. GLASS (edits.), Amino acid metabolism, p. 930—939. Baltimore 1955.
— The enzymatic conversion of anthranilic acid to indole. J. biol. Chem. **223**, 171—184 (1956).
— Enzymatic studies with a series of tryptophan auxotrophs of *Escherichia coli*. J. biol. Chem. **224**, 783—792 (1957).
— Restoration of tryptophan synthetase activity in *Escherichia coli* by suppressor mutations. Science **128**, 843 (1958).
— The tryptophan synthetase system. Bact. Rev. **24**, 221—245 (1960).
—, and D. M. BONNER: Evidence for the participation of kynurenine as a normal intermediate in the biosynthesis of niacin in *Neurospora*. Proc. nat. Acad. Sci. (Wash.) **36**, 167—176 (1950).

References

Yanofsky, C., and D. M. Bonner: Studies on the conversion of 3-hydroxy-anthranilic acid to niacin in *Neurospora*. J. biol. Chem. **190**, 211—218 (1951).
— — Gene interaction in tryptophan synthetase formation. Genetics **40**, 761—769 (1955a).
— — Non-allelic suppressor genes affecting a single *td*-allele. Genetics **40**, 602 (1955b).
—, and I. P. Crawford: The effect of deletions, point mutations, reversions and suppressor mutations on the two components of tryptophan synthetase in *Escherichia coli*. Proc. nat. Acad. Sci. (Wash.) **45**, 1016—1026 (1959).
—, and M. Rachmeler: The exclusion of free indole as an intermediate in biosynthesis of tryptophan in *Neurospora crassa*. Biochim. biophys. Acta (Amst.) **28**, 640—641 (1958).
—, and J. L. Reissig: L-Serine dehydrase of *Neurospora*. J. biol. Chem. **202**, 567—577 (1953).
—, and P. St. Lawrence: Gene action. Ann. Rev. Microbiol. **14**, 311—340 (1960).
— U. Henning, D. Helinski, and B. Carlton: Mutational alteration of protein structure. Fed. Proc. **22**, 75—79 (1963).
Ycas, M.: The coding hypothesis. Int. Rev. Cytol. **13**, 1—34 (1962).
Yura, T.: Genetic alteration of pyrroline-5-carboxylate reductase in *Neurospora crassa*. Proc. nat. Acad. Sci. (Wash.) **45**, 197—204 (1959).
—, and H. J. Vogel: On the biosynthesis of proline in *Neurospora crassa:* enzymatic reduction of Δ^1-pyrroline-5-carboxylate. Biochim. biophys. Acta (Amst.) **17**, 582 (1955).
Yu-Sun, C.: Nutritional studies of *Ascobolus immersus*. *A*mer. J. Bot. **51**, 231—237 (1964).
Zinder, N. D.: The information content of an RNA-containing bacteriophage. In: H. J. Vogel, V. Bryson and J. O. Lampen (edits.), Informational Macromolecules, p. 229—237. New York and London 1963.

References

which have come to the authors' attention after conclusion of the German manuscript

A I

Haidle, C. W., J. M. Kornfeld, and S. G. Knight: Incorporation of amino acids into protein by cell-free extracts of *Penicillium chrysogenum*. Arch. Mikrobiol. **53**, 41—49 (1966).
Halvorson, H., J. Gorman, P. Tauro, R. Epstein, and M. la Berge: Control of enzyme synthesis in synchronous cultures of yeast. Fed. Proc. **23**, 1002—1008 (1964).
Helinski, D. R., and C. Yanofsky: Genetic control of protein structure. In: H. Neurath (edit.), The proteins, vol. IV, 2nd ed., p. 1—93. New York and London 1966.
Lester, G.: Genetic control of amino acid permeability in *Neurospora crassa*. J. Bacter. **91**, 677—684 (1966).
Printz, D. B., and S. R. Gross: An apparent relationship between mistranslation and an altered leucyl-TRNA synthetase in a conditional lethal mutant of *Neurospora crassa*. Genetics **55**, 451—467 (1967).
Schweiger, H. G.: Proteinsynthese und Ribonucleinsäure in kernlosen Reticulocyten. Naturwissenschaften **51**, 521—533 (1964).
Stadler, D. R.: Genetic control of the uptake of amino acids in *Neurospora*. Genetics **54**, 677—685 (1966).
St. Lawrence, P., B. D. Maling, L. Altwasser, and M. Rachmeler: Mutational alteration of permeability in *Neurospora:* Effects on growth and the uptake of certain amino acids and related compounds. Genetics **50**, 1383—1402 (1964).

Function

SURDIN, Y., W. SLY, J. SIRE, A. M. BORDES et H. DE ROBICHON-SZUL-MAJSTER: Propriétés et contrôle génétique du système d'accumulation des acides amines chez *Saccharomyces cerevisiae*. Biochim. biophys. Acta (Amst.) **107**, 546—566 (1965).
WINTERSBERGER, E., u. H. TUPPY: DNA-abhängige RNA-Synthese in isolierten Hefe-Mitochondrien. Biochem. Z. **341**, 399—408 (1965).

A II

VERSTEEG, D. H. G., and J. F. G. VLIEGENTHART: A spatial depiction for the systematically degenerate genetic code. Experientia (Basel) **21**, 615 (1965).
YANOFSKY, C.: Possible RNA codewords for the eight amino acids that can occupy one position in the tryptophan synthetase A protein. Biochem. biophys. Res. Commun. **18**, 898—909 (1965).

B I

OGUR, M., L. COKER, and S. OGUR: Glutamate auxotrophs in *Saccharomyces*. I. The biochemical lesion in the *glt-1* mutants. Biochem. biophys. Res. Commun. **14**, 193—197 (1964).
THRELKELD, S. F. H.: Pantothenic acid requirement for spore color in *Neurospora crassa*. Canad. J. Genet. Cytol. **7**, 171—173 (1965).
WIEBERS, J. L., and H. R. GARNER: Use of S-methylcysteine and cystathionine by methionineless *Neurospora* mutants. J. Bact. **88**, 1798—1804 (1964).
YU-SUN, C. C. C.: Biochemical and morphological mutants of *Ascobolus immersus*. Genetics **50**, 987—998 (1964).

B II, 1

BHATTACHARJAC, I. K., and G. LINDEGREN: Gene control of pigmentation associated with a specific lysine requirement of *Saccharomyces*. Biochem. biophys. Res. Commun. **17**, 554—558 (1964).

B II, 2

LACROUTE, F.: Un cas de double rétrocontrôle: la chaîne de biosynthèse de l'uracile chez le levure. C. R. Acad. Sci. (Paris) **259**, 1357—1359 (1964).
ZIMMERMANN, F. K., u. R. SCHWAIER: Teilweise Reversion einer Histidin-Adenin-bedürftigen Einfachmutante bei *Saccharomyces cerevisiae*. Z. Vererbungsl. **94**, 253—260 (1963a).

B III

CABET, D., M. GANS, M. L. HIRSCH et G. PRÉVOST: Mode d'action de deux génes suppresseurs de l'effect des génes «*ur 1a*» chez *Coprinus radiatus*. C. R. Acad. Sci. (Paris) **261**, 5191—5194 (1965).
CARSIOTIS, M., and S. R. SUSKIND: The role of pyridoxal phosphate in the aldolytic activity of tryptophan synthetase from *Neurospora crassa*. J. biol. Chem. **239**, 4227—4231 (1964).
—, and A. M. LACY: Increased activity of tryptophan biosynthetic enzymes in histidine mutants of *Neurospora crassa*. J. Bact. **89**, 1472—1477 (1965).
CREASER, E. H., D. J. BENNETT, and R. B. DRYSDALE: Studies on biosynthetic enzymes. I. Mutant forms of histidinol dehydrogenase from *Neurospora crassa*. Canad. J. Biochem. **43**, 993—1000 (1965).
DEMAIN, A. L.: Nutrition of "adenineless" auxotrophs of yeasts. J. Bact. **88**, 339—345 (1964).
DONACHIE, W. D.: The regulation of pyrimidine biosynthesis in *Neurospora crassa*. I. End-product inhibition and repression of aspartate carbamyltransferase. Biochim. biophys. Acta (Amst.) **82**, 284—292 (1964).
— The regulation of pyrimidine biosynthesis in *Neurospora crassa*. II. Heterokaryons and the role of the "regulatory mechanisms". Biochim. biophys. Acta (Amst.) **82**, 293—302 (1964).

432

Finck, D., Y. Suyama, and R. H. Davis: Metabolic role of the pyrimidine-3 locus of *Neurospora*. Genetics **52**, 829—834 (1965).

Hütter, R., and J. A. de Moss: Enzyme analysis of the tryptophan pathway in *Aspergillus nidulans*. Genetics **55**, 241—247 (1967).

Lingens, F., u. W. Goebel: Nachweis von 3,3-Diindolyl-(3)-propandiol-(1,2) und 2-Oxo-3-indolyl-(3)-propanol-(1) und ihrer Phosphorsäureester bei Tryptophanmangelmutanten von *Saccharomyces cerevisiae*. Biochim. biophys. Acta (Amst.) **107**, 183—184 (1965).

—, u. H. D. Heilmann: Zur Biosynthese des Lysins in *Xanthomonas pruni*, *Streptomyces antibioticus* und *Cyanidium caldarium*. Hoppe-Seylers Z. physiol. Chem. **345**, 249—256 (1966).

— P. Vollprecht u. V. Gildemeister: Zur Biosynthese der Nicotinsäure in *Xanthomonas*- und *Pseudomonas*-Arten, *Mycobacterium phlei* und Rotalgen. Biochem. Z. **344**, 462—477 (1966).

— W. Goebel u. H. Uesseler: Regulation der Biosynthese der aromatischen Aminosäuren in *Saccharomyces cerevisiae*. 2. Repression, Induktion und Aktivierung. Europ. J. Biochem. **1**, 363—374 (1967).

— — — Nachweis von Enzymen zur Biosynthese der aromatischen Aminosäuren in *Claviceps paspali*. Naturwissenschaften **54**, 141 (1967).

Matchett, W. H., and J. A. DeMoss: Physiological channeling of tryptophan in *Neurospora crassa*. Biochim. biophys. Acta (Amst.) **86**, 91—99 (1964).

Murray, N. E.: Cysteine mutant strains of *Neurospora*. Genetics **52**, 801—808 (1965).

Piña, E., and E. L. Tatum: Inositol biosynthesis in *Neurospora crassa*. Biochim. biophys. Acta (Amst.) **136**, 265—271 (1967).

Roberts, C. F.: Complementation analysis of the tryptophan pathway in *Aspergillus nidulans*. Genetics **55**, 233—239 (1967).

Robichon-Szulmajster, H. de, et M. Somlo: Contrôle exercé par le tryptophane sur la biosynthèse de deux systèmes inductibles chez la levure. C. R. Acad. Sci. (Paris) **249**, 1583—1585 (1959).

Scarborough, G. A., and J. F. Nyc: Methylation of ethanolamine phosphatides by microsomes from normal and mutant strains of *Neurospora crassa*. J. biol. Chem. **242**, 238—242 (1967).

Staron, T., C. Allard et N. D. Xuong: Métabolisme de l'acide anthranilique et de l'acide p-aminobenzoique chez *Aspergillus ochraceus*. C. R. Acad. Sci. (Paris) D **263**, 81—84 (1966).

C I

Ahmad, S. I., and B. D. Sanwal: A structural gene for the DPN-specific glutamate dehydrogenase in *Neurospora*. Genetics **55**, 359—364 (1967).

Carsiotis, M., and S. R. Suskind: The role of pyridoxal phosphate in the aldolytic activity of tryptophan synthetase from *Neurospora crassa*. J. biol. Chem. **239**, 4227—4231 (1964).

— E. Appella, P. Provost, J. Germershausen, and S. R. Suskind: Chemical and physical studies of the structure of tryptophan synthetase from *Neurospora crassa*. Biochem. biophys. Res. Commun. **18**, 877—888 (1965).

Cove, D., and A. Coddington: Purification of nitrate reductase and cytochrome *c* reductase from *Aspergillus nidulans*. Biochim. biophys. Acta (Amst.) **110**, 312—318 (1965).

Dave, P. J., R. W. Kaplan u. G. Pfleiderer: Wirkung von Mutationen auf die Isoenzyme der Enolase aus Hefe und andere glykolytische Enzyme. Biochem. Z. **345**, 440—453 (1966).

Davis, R. H.: Carbamyl phosphate synthesis in *Neurospora crassa*. I. Preliminary characterization of arginine-specific carbamyl phosphokinase. Biochim. biophys. Acta (Amst.) **107**, 44—53 (1965).

DAVIS, R. H.: Carbamyl phosphate synthesis in *Neurospora crassa*. II. Genetics, metabolic position, and regulation of arginine-specific carbamyl phosphokinase. Biochim. biophys. Acta (Amst.) **107**, 54—58 (1965).

DORN, G.: Genetic analysis of the phosphatases in *Aspergillus nidulans*. Genet. Res. Camb. **6**, 13—20 (1965).

— Phosphatase mutants in *Aspergillus nidulans*. Science **150**, 1183—1184 (1965).

DOY, C. H.: Anthranilate synthetase and the allosteric protein model. Biochim. biophys. Acta (Amst.) **118**, 173—188 (1966).

FLAVIN, M., and C. SLAUGHTER: The depression and function of enzymes of reverse transsulfuration in *Neurospora*. Biochim. biophys. Acta (Amst.) **132**, 406—411 (1967).

— — Trypsin-treated *Neurospora* tryptophan synthetase. II. Antigenic properties. J. molec. Biol. **9**, 83—99 (1964).

GREGORY, K. F., and J. C. C. HUANG: Tyrosinase inheritance in *Streptomyces scabies*. I. Genetic recombination. J. Bact. **87**, 1281—1286 (1964).

— — Tyrosinase inheritance in *Streptomyces scabies*. II. Induction of tyrosinase deficiency by acridine dyes. J. Bact. **87**, 1287—1294 (1964).

GUEST, J. R., and C. YANOFSKY: Amino acid sequences surrounding the sulfhydryl groups of the A protein subunit of the *Escherichia coli* tryptophan synthetase. J. biol. Chem. **241**, 1—16 (1966).

HOROWITZ, N. H.: Evidence for common control of tyrosinase and L-amino acid oxidase in *Neurospora*. Biochem. biophys. Res. Commun. **18**, 686—692 (1965).

HWANG, D. S., and C. C. LINDEGREN: Palatinose element of the receptor of the melezitose locus in *Saccharomyces*. Nature (Lond.) **203**, 791—792 (1964).

— — Induction of an alpha-glucosidase by glucose. Nature (Lond.) **205**, 880—883 (1965).

KAPLAN, N. O.: Symposium on multiple forms of enzymes and control mechanisms. I. Multiple forms of enzymes. Bact. Rev. **27**, 155—169 (1963).

KLINGMÜLLER, W.: Die Aufnahme der Zucker Sorbose, Fructose und Glucose durch Sorbose-resistente Mutanten von *Neurospora crassa*. Z. Naturforsch. **22**b, 327—335 (1967).

LINDEGREN, C. C.: The receptor-hypothesis of induction of gene-controlled adaptive enzymes. J. theoret. Biol. **5**, 192—210 (1963).

METZENBERG, R. L.: Enzymatically active subunits of *Neurospora* invertase. Biochim. biophys. Acta (Amst.) **89**, 291—302 (1964).

MEYER, J. A., E. D. GARBER, and S. G. SHAEFFER: Genetics of phytopathogenic fungi. XII. Detection of esterases and phosphatases in culture filtrates of *Fusarium oxysporum* and *F. xylarioides* by starch-gel zone electrophoresis. Bot. Gaz. **125**, 298—300 (1964).

MUNKRES, K. D.: Structure of *Neurospora* malate dehydrogenase. I. Reconstitution from acid and urea. Biochemistry **4**, 2180—2185 (1965).

— Structure of *Neurospora* malate dehydrogenase. II. Isolation and partial characterization of polypeptide subunits. Biochemistry **4**, 2186—2196 (1965).

— Simultaneous genetic alteration of *Neurospora* malate dehydrogenase and aspartate aminotransferase. Arch. Biochem. **112**, 340—346 (1965).

— Physiochemical identity of *Neurospora* malate dehydrogenase and aspartate aminotransferase. Arch. Biochem. **112**, 347—354 (1965).

—, and R. M. RICHARDS: Genetic alteration of *Neurospora* malate dehydrogenase. Arch. Biochem. **109**, 457—465 (1965).

— — The purification and properties of *Neurospora* malate dehydrogenase. Arch. Biochem. **109**, 466—479 (1965).

Ouchi, S., and C. C. Lindegren: Genic interaction in *Saccharomyces*. Canad. J. Genet. Cytol. **5**, 257—267 (1963).

Pynadath, T. I., and R. M. Fink: Studies of orotidine 5′-phosphate decarboxylase in *Neurospora crassa*. Arch. Biochem. **118**, 185—189 (1967).

Roberts, D. B., and J. A. Pateman: Immunological studies of amination deficient strains of *Neurospora crassa*. J. gen. Microbiol. **34**, 295—305 (1964).

Sherman, F.: Respiration-deficient mutants of yeast. I. Genetics. Genetics **48**, 375—385 (1963).

— H. Taber, and W. Campbell: Genetic determination of isocytochromes *c* in yeast. J. molec. Biol. **13**, 21—39 (1965).

Sorger, G. J.: Nitrate reductase electron transport systems in mutant and in wildtype strains of *Neurospora*. Biochim. biophys. Acta (Amst.) **118**, 484—494 (1966).

— TPNH-cytochrome c reductase and nitrate reductase in mutant and wild type *Neurospora* and *Aspergillus*. Biochem. biophys. Res. Commun. **12**, 395—401 (1963).

— Simultaneous induction and repression of nitrate reductase and TPNH cytochrome c reductase in *Neurospora crassa*. Biochim. biophys. Acta (Amst.) **99**, 234—245 (1965).

—, and N. H. Giles: Genetic control of nitrate reductase in *Neurospora crassa*. Genetics **52**, 777—788 (1965).

St. Lawrence, P., R. Naish, and B. Burr: The action of suppressors of a tryptophan synthetase mutant of *Neurospora* in heterocaryons. Biochem. biophys. Res. Commun. **18**, 868—876 (1965).

Sundaram, T. K., and J. R. S. Fincham: A mutant enzyme in *Neurospora crassa* interconvertible between electrophoretically distinct active and inactive forms. J. molec. Biol. **10**, 423—437 (1964).

Suskind, S. R., M. L. Wickham, and M. Carsiotis: Antienzymes in immunogenetic studies. Ann. N. Y. Acad. Sci. **103**, 1106—1127 (1963).

Tsoi, A., and H. C. Douglas: The effect of mutation in two forms of phosphoglucomutase in *Saccharomyces*. Biochim. biophys. Acta (Amst.) **92**, 513—520 (1964).

Webster, R. E., and S. R. Gross: The α-isopropylmalate synthetase of *Neurospora*. I. The kinetics and end product control of α-isopropylmalate synthetase function. Biochemistry (Wash.) **4**, 2309—2318 (1965).

Wiebers, J. L., and H. R. Garner: Homocysteine and cysteine synthetases of *Neurospora crassa*. J. biol. Chem. **242**, 12—23 (1967).

C II, 1

Catcheside, D. G.: Interallelic complementation. Brookhaven Symp. in Biol. **17**, 1—14 (1964).

Coddington, A.: Hybridization in vitro between wild-type and mutant forms of glutamate dehydrogenase from *Neurospora crassa*. Biochem. J. **99**, 9c—11c (1966).

—, and J. R. S. Fincham: Proof of hybrid enzyme formation in a case of inter-allelic complementation in *Neurospora crassa*. J. molec. Biol. **12**, 152—161 (1965).

— J. R. S. Fincham, and T. K. Sundaram: Multiple active varieties of *Neurospora* glutamate dehydrogenase formed by hybridization between two inactive mutant proteins in vivo and in vitro. J. molec. Biol. **17**, 503—512 (1966).

Cohen, B. B., and J. O. Bishop: Purification of arginosuccinase from *Neurospora* and comparison of some properties of the wild-type enzyme and an enzyme formed by inter-allelic complementation. Genet. Res. Camb. **8**, 243—252 (1966).

Function

DUTTA, S. K., and V. W. WOODWARD: Complementation pattern changes at the pyr-3 locus of *Neurospora crassa*. Genetics **52**, 391—396 (1965).

FINCHAM, J. R. S., and D. R. STADLER: Complementation relationships of *Neurospora am* mutants in relation to their formation of abnormal varieties of glutamate dehydrogenase. Genet. Res. Camb. **6**, 121—129 (1965).

FOLEY, J. M., N. H. GILES, and C. F. ROBERTS: Complementation at the adenylosuccinase locus in *Aspergillus nidulans*. Genetics **52**, 1247—1263 (1965).

HUANG, P. C.: Recombination and complementation of albino mutants in *Neurospora*. Genetics **49**, 453—469 (1964).

INGE-VECHTOMOV, S. G., B. V. SIMAROV, T. R. SOIDLA, and S. A. KOZIN: Super-suppressor induced interallelic complementation. Z. Vererbungsl. **98**, 375—384 (1966).

ISHIKAWA, T.: A molecular model for an enzyme based on the genetic and complementation analyses at the *ad-8* locus in *Neurospora*. J. molec. Biol. **13**, 586—591 (1965).

JESSOP, A. P., and D. G. CATCHESIDE: Interallelic recombination at the *his-1* locus in *Neurospora crassa* and its genetic control. Heredity **20**, 237—256 (1965).

MANNEY, T. R.: Action of a super-suppressor in yeast in relation to allelic mapping and complementation. Genetics **50**, 109—121 (1964).

MEGNET, R., and N. H. GILES: Allelic complementation at the adenylosuccinase locus in *Schizosaccharomyces pombe*. Genetics **50**, 967—971 (1964).

PATEMAN, J. A., and J. R. S. FINCHAM: Complementation and enzyme studies of revertants induced in an *am* mutant of *N. crassa*. Genet. Res. Camb. **6**, 419—432 (1965).

RADFORD, A.: Further studies on complementation at the PDX-1 locus of *Neurospora crassa*. Canad. J. Genet. Cytol. **8**, 672—676 (1966).

SERRES, F. J. DE: Impaired complementation between nonallelic mutation in *Neurospora*. Nat. Cancer Inst. Monogr. **18**, 33—52 (1964).
— The utilization of leaky *ad-3* mutants of *Neurospora crassa* in heterokaryon tests for allelic complementation. Mutation Res. **3**, 3—12 (1966).
— Carbon dioxide stimulation of the *ad-3* mutants of *Neurospora crassa*. Mutation Res. **3**, 420—425 (1966).

SMITH, B. R.: Interallelic recombination at the *his-5* locus in *Neurospora crassa*. Heredity **20**, 257—276 (1965).
— Genetic controls of recombination. I. The recombination-2 gene of *Neurospora crassa*. Heredity **21**, 481—498 (1966).

WEBBER, B. B.: Genetical and biochemical studies of histidine-requiring mutants of *Neurospora crassa*. III. Correspondence between biochemical characteristics and complementation map position of *hist-3* mutants. Genetics **51**, 263—273 (1965).
— Genetical and biochemical studies of histidine-requiring mutants of *Neurospora crassa*. IV. Linkage relationships of *hist-3* mutants. Genetics **51**, 275—283 (1965).

WOODS, R. A., and E. A. BEVAN: Interallelic complementation at the *ad-2* locus of *Saccharomyces cerevisiae*. Heredity **21**, 121—130 (1966).

C II, 2

BRODY, S., and C. YANOFSKY: Mechanism studies of suppressor-gene action. J. Bact. **90**, 687—695 (1965).

GOLDBERG, M. E., T. E. CREIGHTON, R. L. BALDWIN, and C. YANOFSKY: Subunit structure of the tryptophan synthetase of *Escherichia coli*. J. molec. Biol. **21**, 71—82 (1966).

HARDMAN, J. K., and C. YANOFSKY: Studies on the active site of the A protein subunit of the *Escherichia coli* tryptophan synthetase. J. biol. Chem. **240**, 725—732 (1965).

KAPLAN, S., Y. SUYAMA, and D. M. BONNER: A genetic analysis of CRM negative mutants at the *td* locus of *Neurospora crassa*. Genet. Res. Camb. **4**, 470—479 (1963).

LACY, A. M.: Structural and physiological relationships within the *td* locus in *Neurospora crassa*. Biochem. biophys. Res. Commun. **18**, 812—823 (1965).

MURPHY, R. M., and S. E. MILLS: An antigenic comparison of tryptophan synthetase derived from various strains of *Neurospora crassa*. Biochem. biophys. Res. Commun. **18**, 843—849 (1965).

C II, 3

BARBEN, H.: Allelspezifische Suppressormutationen von *Schizosaccharomyces pombe*. Genetica **37**, 109—148 (1966).

BRODY, S., and C. YANOFSKY: Mechanism studies of suppressor-gene action. J. Bact. **90**, 687—695 (1965).

—, and E. L. TATUM: The primary biochemical effect of a morphological mutation in *Neurospora crassa*. Proc. nat. Acad. Sci. (Wash.) **56**, 1290—1297 (1966).

McDOUGALL, K. J., and V. W. WOODWARD: Suppression of arginine and pyrimidine-requiring mutants of *Neurospora crassa*. Genetics **52**, 397—406 (1965).

OSHIMA, Y., and I. OSHIMA: A super-suppressor on the thirteenth linkage group in *Saccharomyces*. Genet. Res. **8**, 1—7 (1966).

TERRY, K.: Suppression of a *Neurospora td* mutant that lacks cross-reacting material. Genetics **54**, 105—114 (1966).

WEGLENSKI, P.: Genetical analysis of proline mutants and their suppressors in *Aspergillus nidulans*. Genet. Res. Camb. **8**, 311—321 (1966).

— The mechanism of action of proline suppressors in *Aspergillus nidulans*. J. gen. Microbiol. **47**, 77—85 (1967).

YOURNO, J. D., and S. R. SUSKIND: Suppressor gene action in the tryptophan synthetase system of *Neurospora crassa*. I. Genetic studies. Genetics **50**, 803—816 (1964).

— — Suppressor gene action in the tryptophan synthetase system of *Neurospora crassa*. II. Biochemical studies. Genetics **50**, 817—828 (1964).

C II, 4

GILMORE, R. A., and R. K. MORTIMER: Super-suppressor mutations in *Saccharomyces cerevisiae*. J. molec. Biol. **20**, 307—311 (1966).

STADLER, D. R.: Glutamic dehydrogenase in revertants of *am* mutants in *Neurospora*. Genet. Res. **7**, 18—31 (1966).

C III

BECHET, J., J. M. WIAME et M. GRENSON: Mutation affectant la régulation de la synthèse de l'ornithine transcarbamylase chez *Saccharomyces cerevisiae*. Arch. Int. Physiol. **73**, 137—139 (1965).

—, and J. M. WIAME: Indication of a specific regulatory binding protein for ornithine transcarbamylase in *Saccharomyces cerevisiae*. Biochem. biophys. Res. Commun. **21**, 226—234 (1965).

DOUGLAS, H. C., and D. C. HAWTHORNE: Regulation of genes controlling synthesis of the galactose pathway enzymes in yeast. Genetics **54**, 911—916 (1966).

FINK, G. R.: A cluster of genes controlling three enzymes in histidine biosynthesis in *Saccharomyces cerevisiae*. Genetics **53**, 445—459 (1966).

GRUBER, M., and R. N. CAMPAGNE: Regulation of protein synthesis: An alternative to the repressor-operator hypothesis. Proceedings C **68**, 1—7 (1965).

JACOB, F., and J. MONOD: Méchanismes biochimiques et génétiques de la régulation dans la cellule bactérienne. Bull. Soc. Chim. biol. (Paris) **46**, 1499—1532 (1965).

Function

LESTER, G., and A. BYERS: Properties of two B-galactosidases of *Neurospora crassa*. Biochem. biophys. Res. Commun. **18**, 725—734 (1965).

LOOMIS jr., W. F., and B. MAGASANIK: The catabolite repression gene of the *lac* operon in *Escherichia coli*. J. molec. Biol. **23**, 487—494 (1967).

MUNKRES, K. D., and D. O. WOODWARD: On the genetics of enzyme locational specificity. Proc. nat. Acad. Sci. (Wash.) **55**, 1217—1224 (1966).

— — Interaction of *Neurospora* mitochondrial structural protein with other proteins and coenzyme nucleotides. Biochim. biophys. Acta (Amst.) **133**, 143—150 (1967).

NASHED, N., G. JABBUR, and F. K. ZIMMERMANN: Negative complementation among *ad-2* mutants of yeast. Molec. Gen. Genetics **99**, 69—75 (1967).

SLONIMSKI, P. P., R. ACHER, G. PÉRÉ, A. SELS et M. SOMLO: Éléments du système respiratoire et leur régulation: Cytochromes et iso-cytochromes. Coll. Intern. C.N.R.S. Marseille 1963 (Paris) 435—461 (1965).

SOMERVILLE, R. L., and C. YANOFSKY: Studies on the regulation of tryptophan biosynthesis in *Escherichia coli*. J. molec. Biol. **11**, 747—759 (1965).

YANOFSKY, C., and J. ITO: Nonsense codons and polarity in the tryptophan operon. J. molec. Biol. **21**, 313—334 (1966).

Chapter VII

Extrachromosomal inheritance

A. Introductory remarks

As early as 1909 CARL CORRENS, one of the rediscoverers of the Mendelian laws, recognized that the nucleus did not have a monopoly on heredity. He was able to show that *the cytoplasm also carries hereditary determinants*. In the years following, many cases of extrachromosomal inheritance were described in plants and animals. It became clear that hereditary determinants occurring outside of the chromosomes have the capacity of *self-replication* and can be transmitted *sexually or asexually*. Nevertheless, in some cases the phenotypic manifestation of extrachromosomal hereditary determinants depends upon the genotype. The extrachromosomal determinants may reside in the plastids or other cytoplasmic components. While the term genome includes all of the chromosomal hereditary determinants, the extrachromosomal hereditary factors are subdivided into the *plastome* (factors in the plastids) and the *plasmone* (factors in the remainder of the cytoplasm). Since the fungi lack plastids our concern here is with the plasmone.

The expressions "cytoplasmic" and "extra nuclear inheritance" as well as "extra chromosomal inheritance" have all been used in the literature. We prefer "extrachromosomal inheritance" in order to emphasize that the determinants of this kind of inheritance are not located in the chromosome, but may be found in any other cell component (for example, even in the karyoplasm).

Extrachromosomal, in contrast to chromosomal inheritance never results in a typical Mendelian segregation. This is explained by the fact that the cytoplasmic contribution of the egg cell in fertilization exceeds that of the sperm in most plants and animals (*oogamy*). The male parent delivers little or no cytoplasm. This is also the case in Ascomycetes (with the exception of yeasts) and in the Uredinales. To what extent cytoplasmic exchange is correlated with nuclear exchange in somatogamy in the Hymenomycetes and the Ustilaginales is uncertain. Such cases are also referred to as "maternal inheritance". The manner of inheritance leads to *reciprocal differences in the progeny* of crosses between hermaphrodites, which can be recognized as early as in the F_1 generation in haploids as well as in diploids (Fig. VII-1, left).

In *isogamy* the proportion of cytoplasm contributed to the zygote by the male and female germ cells is the same (e.g. in yeast). No reciprocal differences occur in contrast to oogamy. The different plasmones of the two parents become mixed during zygote formation and are passed on to the progeny as a mixture. The latter all receive the same cytoplasm and are uniform in this respect (Fig. VII-1, right). In haplonts the absence of segregation can be detected in the F_1; in diplonts it is not apparent until the F_2.

The criterion of reciprocal differences is only adequate for revealing extrachromosomal inheritance if the parents are isogenic. To avoid misinterpretations the following points must be considered:

1. The nuclei of reciprocal hybrids must be identical as well as equivalent. The researches of RENNER on the complex heterozygotes of *Oenothera* have shown that specific types of gametes may be transmitted only through the pollen and others only through the egg.

2. The cytoplasm may become temporarily altered through "Dauermodifikationen" or predetermination.

3. Selective fertilization, differential viability of gametes, recessive lethals, chromosome elimination, and chromosome anomalies can simulate extrachromosomal inheritance.

Thus the Mendelian principles are not valid for hereditary characters determined extrachromosomally. Because of the absence of segregation, for a long time the plasmone was considered an indivisible unit. Detectable components of the plasmone corresponding to the genes of the genome could not be demonstrated. Therefore, it is understandable that the geneticists focused their research primarily upon the segregating chromosomal factors. Although extrachromosomal inheritance was known and now and then new examples were added, its determinants remained obscure.

It is also true that the results of many crosses which could not be interpreted in clearcut Mendelian terms were not pursued further and tested for their possibly extrachromosomal basis. Moreover, extrachromosomal hereditary determinants may be lost as a result of selection while all characters inherited through the chromosomes are reproduced with great precision and regularity.

Extensive researches which have been carried out particularly with fungi during the last 20 years have shown that *specific structures in the cytoplasm can be considered as determiners of extrachromosomal heredity.* In the first place, *mitochondria,* which are microscopically recognizable, are involved. Further, *self-replicating, infectious cytoplasmic particles* come into consideration as hereditary determiners (p. 444f.). Only indirect, suggestive genetic and cytological evidence exists for such particles, the nature of which is still unknown. In other cases the determinants of extrachromosomal inheritance remain completely unidentified.

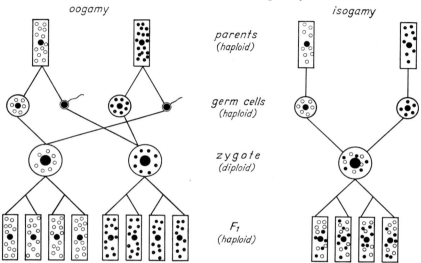

Fig. VII-1. Diagram of extrachromosomal inheritance in haplonts. The plasmone difference in the two parents is shown by differential stippling. See further in text

Reviews of extrachromosomal inheritance:
General: RENNER (1934), CORRENS (1937), V. WETTSTEIN (1937), CASPARI (1948), EPHRUSSI (1951, 1953), OEHLKERS (1952, 1953), RHOADES (1955), LINDEGREN (1957), NANNEY (1957), MATHER (1958), CATCHESIDE (1958, 1959), L'HÉRITIER (1962), HAGEMANN (1964, 1966)
Special (excluding fungi): L'HÉRITIER (1951), MICHAELIS (1951, 1961), BEALE (1954), SONNEBORN (1959).
Special (in fungi): EPHRUSSI and HOTTINGUER (1951), EPHRUSSI (1952, 1956), WINDISCH (1958), RIZET et al. (1958), JINKS (1958), JOLY (1964).

Summary

1. Extrachromosomal inheritance leads (in the case of oogamy) to reciprocal differences which are apparent in the F_1 in both haplonts and diplonts.

2. The F_1 is uniform in isogamy. One can conclude that extrachromosomal inheritance is involved in this case, if segregation fails to occur in the F_1 in haplonts and in the F_2 in diplonts.

3. It must be shown in every case that other genetic and phenotypic changes are not simulating extrachromosomal inheritance.

B. Mitochondria as hereditary determinants

I. „Petite"-colony mutants of yeast

One of the best-known examples of extrachromosomal inheritance is the case of *respiratory defects* in *Saccharomyces cerevisiae*, which has been studied by EPHRUSSI and his coworkers (reviews: EPHRUSSI, 1951, 1952, 1953, 1956; EPHRUSSI and HOTTINGUER, 1951).

If haploid and diploid cells of *S. cerevisiae* are placed on solid nutrient medium in which glucose is a limiting growth factor, colonies which are distinctly smaller than most arise at a frequency of about one percent. All the cells from such small colonies (petites) yield exclusively small colonies in asexual reproduction. In contrast, individual cells of a normal colony produce 99% normal and 1% small colonies under the same conditions. Cells from petite and from normal colonies are not morphologically different. The small size of the petite colonies results from a slower division rate of the cells, which compose the colonies. The decrease in growth rate stems from the fact that the cells cannot undergo respiration, but only fermentation (EPHRUSSI et al., 1949a, b; SLONIMSKI, 1949; SLONIMSKI and EPHRUSSI, 1949; TAVLITZKI, 1949).

Yeasts are generally able to satisfy their energy requirements either by respiration or fermentation of glucose. If respiration is inhibited by endogenous or exogenous factors (e.g. anaerobic culture) the viability of the yeast cells remains unaffected; they are able to grow vegetatively, but their division rate is significantly reduced because of the low energy yield of fermentation.

The *respiratory defect* of petite cells, which are also called *vegetative mutants, results from a change in the complement of respiratory enzymes* so that only about 3—5% of normal respiration can occur. Thus respiration is practically eliminated as a source of energy. The capacity to ferment is not affected by these changes (SLONIMSKI, 1949,a 1950, 1952; EPHRUSSI and SLONIMSKI, 1950).

The vegetative mutants lack at least four enzymes: cytochrome oxidase, succinate-cytochrome-c-reductase, coenzyme-I-cytochrome-c-reductase, and α-glycerol-phosphate dehydrogenase. The mutational changes were shown to involve the apoenzymes, precursors common to the latter, or their cytoplasmic carriers, and not the prosthetic groups of the enzymes (SLONIMSKI and HIRSCH, 1952). As a result of these alterations cytochromes a and b are absent in the mutated cells. Several other less critical changes are associated with these defects, e.g. a decreased level of malate dehydrogenase associated with coenzyme I and a higher content of alcohol dehydrogenase and cytochrome c. They can be considered as secondary effects of the enzyme defects.

If vegetative mutants are crossed with the wild type, the petite character (= respiratory defect) disappears. It does not reappear either in the diploid generation nor among the haploid progeny of the asci, nor following various backcrosses to the vegetative mutants (except for the usual 1%). This clearly indicates that the respiratory defect is not determined by nuclear genes. Its *control mechanism* must be *extrachromosomal*.

The frequency of the vegetative mutation can be increased considerably by treatment of cells with chemicals [acriflavines (e.g. BULDER, 1964), tetrazolium chloride] or a change in physical conditions (UV, heat) (see WINDISCH, 1958, for literature).

Vegetative mutants may be produced artificially by other means. GRENSON (1963) found that after crossing a glutamate auxotroph with the wild type, minute colony mutants regularly appeared among the offspring.

The spontaneous and induced vegetative mutants have to a large extent the same biochemical characteristics. However, they do show differences in their genetic behavior; this has led to the classification of the mutants into neutral (type so far discussed) and suppressive types (EPHRUSSI et al., 1954, 1955). The suppressive mutants behave just oppositely from the neutrals in crosses with the wild type, because their characteristic is found in all of the offspring; all cells have the respiratory defect. However, the suppressives mutate to the neutral form after a short period of vegetative reproduction. A mutation from neutral to suppressive has not been observed.

The following results are pertinent to the question whether specific particles in the cytoplasm are responsible for the respiratory defects of the vegetative mutants, thus representing extrachromosomal hereditary determiners:

1. The four respiratory enzyme systems which are absent in both types of vegetative mutants are those which are normally bound to grana which can be centrifuged from the cytoplasm (SLONIMSKI and HIRSCH, 1952).

2. The mitochondria contain the cytochrome oxidase.

3. Although no cytochrome oxidase is formed in the respiration-defective mutants, mitochondria are present. They exhibit an altered ultrastructure (EPHRUSSI and SLONIMSKI, 1955; YOTSUYANAGI, 1955, 1962; SCHATZ et al., 1963).

On the basis of these and other observations not detailed here, EPHRUSSI and his coworkers concluded that the *mitochondria are probably the extrachromosomal vehicles of heredity*.

Thus, the vegetative mutants differ from the wild type by possessing mitochondria which are incapable of synthesizing the respiratory enzymes. When the two types are crossed, the active and inactive mitochondria are mixed and the cell can respire. Since mitochondria are self-replicating, the daughter cells all receive sufficient active mitochondria. The *spontaneous occurrence of vegetative mutants* can be attributed to a *separation* of active and inactive mitochondria. Inactive mitochondria arise from active ones through spontaneous, irreversible mutations.

Mitochondria of the suppressive type of vegetative mutant are assumed to inactivate the normal mitochondria of the wild type. This effect is lost after several cell generations because of the high frequency of mutation of the suppressive to the neutral type.

A significant *clue to the functional relationship between extrachromosomal hereditary determiners* with respect to the "respiratory" character comes from the genetic analysis of another petite mutant. The respiratory defect of this type of mutant is physiologically the same as that of the vegetative mutant, but it is controlled by a nuclear gene. This "segregational" mutant, when crossed with the wild type, yields a 2:2 segregation in the ascus. The diploid cells formed through the asexual reproduction of the zygote are normal. Crosses with the vegetative mutants lead to normal diploid cells and the asci exhibit 2:2 segregations for normal and defective progeny (Fig. VII-2).

These results can be explained by assuming that the segregating mutant carries a recessive gene in the presence of which the normal mitochondria are not physiologically active. In contrast, the *vegetative mutants* possess the *dominant allele, R, but no normal mitochondria*. The normal growth rate of the diploid cells is determined by the physiologically inactive, but potentially normal mitochondria of the segregational

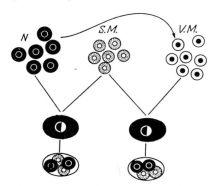

Fig. VII-2. Schematic representation of crosses between a segregational mutant (*S. M.*) and either a normal yeast (*N.*) or a vegetative mutant (*V. M.*). The nuclei and the plasmone of the normal strain are designed in black and the mutated nuclei of *S. M.* and the mutated plasmone of *V. M.* in white. The unmutated but nonfunctional plasmon of the *s.m.* is characterized by dots. The arrow indicates the relatively high mutation frequency of *N.* to *V. M.* (From EPHRUSSI, 1956)

mutant on the one hand, and the dominant allele of the nucleus of the vegetative mutant on the other. The inactive particles coming into the presence of the dominant allele attain their full activity (Fig. VII-2). After segregation of the heterozygous nucleus during ascus formation, half of the spores again loose their activity (Fig. VII-2).

The investigations on *S. cerevisiae* have shown the following: (1) *The formation of respiratory enzymes is controlled by an interplay of chromosomal and extrachromosomal factors.* (2) *The extrachromosomal factors are probably distinct self-replicating particles in the cytoplasm, namely mitochondria.*

The studies of the EPHRUSSI group on respiration-defective yeasts have been repeated and extended by a number of other investigators: BAUTZ (1954a, b, 1955), BAUTZ and MARQUARDT (1953a, b, 1954), BOGEN and KRAEPELIN (1964), HARTMAN and LIU (1954), KRAEPELIN (1964), MARQUARDT (1952), MARQUARDT and BAUTZ (1954), RAUT (1951, 1953, 1954), SHERMAN (1963, 1964), YCAS and STARR (1953).

Cytological investigation of normal and respiration-defective yeasts has shown that distinct structures in the cytoplasm, "grana with mitochondrial function" (BAUTZ and MARQUARDT), "mitochondria" (HARTMAN and LIU), carry the cytochrome system.

RAUT found a petite mutant following UV irradiation which is identical with the segregational mutant in its crossing behavior. The loss of respiratory capacity by this mutant is also determined by a gene, but not by an allele of *R*, nor does it have the same physiological effect. Although the mutant (*W-1*) lacks cytochrome *a* and *b*, the cytochrome *c* content is markedly reduced instead of increased. Depending upon external conditions the catalase activity of the mutant is significantly reduced or entirely absent (YCAS and STARR). These investigators demonstrated further that the mutant has a glycine deficiency which could be alleviated by the addition of the latter substance. From these and other experiments it was concluded that the gene responsible for the respiration-deficient mutant, *W-1*, disturbs another component of the cytochrome-producing system of yeast, namely the synthesis of prosthetic groups of certain respiratory enzymes.

SHERMAN obtained thirteen different mutants with partial cytochrome defects through UV or nitrite treatment. These could be assigned to six loci. They resemble the vegetative petite phenotypically. By crossing one of these mutants with the wild type, hybrids were obtained from which "true" petite cells with mutated plasmones arose at unusually high frequencies following meiotic as well as mitotic divisions. According to SHERMAN, the genes mentioned above may control the synthesis and transmission of a "cytoplasmic factor".

BOGEN and KRAEPELIN investigated the mechanism of the production of respiration-deficient mutants with acridine orange. They found that the dye is mutagenic in only certain developmental stages of the yeast cells. Under certain culture conditions, reverse mutations can be induced.

In this connection it should be pointed out that cytochrome differences between baker's and brewer's yeast had already been described by FINK (1932) and FINK and BERWALD (1933). Brewer's yeast can only ferment glucose and thus corresponds physiologically to the petite mutants. FINK was able to obtain brewer's yeast from baker's yeast and vice-versa by altering the culture conditions. Unfortunately these experiments were carried out in the absence of any genetic information.

II. *Poky* mutants of *Neurospora*

Extrachromosomal inheritance of the *poky* character in *Neurospora crassa* also appears to be connected with the mitochondria (MITCHELL and MITCHELL, 1952; MITCHELL et al., 1953; HASKINS et al., 1953; HARDESTY and MITCHELL, 1963). A *nucleocytoplasmic interaction* likewise exists in this case, since the *poky* phenotype can be suppressed by the *f* gene. The exact nature of the genome-plasmone relationship is not yet known (SILAGI, 1963).

Poky mutants, as the name implies, are characterized by a retarded growth of the mycelium. The phenomenon is maternally inherited, i.e. all spores arising from a perithecium of a *poky* strain produce *poky* mycelia. However, if *poky* conidia are used to fertilize wild type protoperithecia only wild type mycelia arise from the ascospores. The *poky* character is not determined by infectious particles (p. 444f.), since it cannot be transferred to wild type hyphae through hyphal fusions. Biochemically the *poky* strains when compared with the wild type are characterized by an altered complement of cytochromes and respiratory enzymes.

They form large amounts of cytochrome *c*. In young *poky* cultures cytochromes a and b are absent or only very small amounts are present. In older mycelia these amounts may reach about 2% of the cytochrome produced by the wild type. An analysis of the particulate fraction of cell-free enzyme extracts obtained by centrifugation revealed that the young *poky* cultures lack the respiratory enzymes succinate-oxidase and cytochrome oxidase. Although the succinate oxidase activity reaches the level of the wild type with increasing age of the *poky* culture, the cytochrome activity remains far below the normal amount. In the *poky* strains a cytochrome-degrading enzyme (cytochromase), which is absent in the wild type, has been demonstrated. The high cytochrome c content of *poky* is made possible by the presence of an unknown system which inhibits the cytochromase activity.

The Mitchell group isolated still another plasmone mutant *mi-3*, with similar growth characteristics and cytochrome content as those of *poky*. *Poky* and *mi-3* differ morphologically as well as in growth rate. These differences are also apparent in reciprocal crosses between *poky* and *mi-3*. A complementation of both defects in the heterokaryon as is generally the case with gene mutations, is not possible (GOWDRIDGE, 1956). The combination of *poky* and *mi-3* with similar, although gene-determined growth mutants (*C 115* and *C 117*) has shown that the characters inherited through the nuclei are not correlated with extrachromosomal characters.

These results lead to the conclusion that the *cytoplasms of poky and mi-3 possess an altered cytochrome system* which reproduces itself from generation to generation without becoming modified by nuclear genes. Since the cytochromes and the respiratory enzymes are associated with the mitochondria, as described in the case of the respiration-deficient yeasts, it may be assumed that *poky* and *mi-3* contain altered mitochondria. Moreover, these mutated *mitochondria* can be viewed as having genetic continuity, *representing extrachromosomal hereditary determinants* as in the petites of yeast.

Summary

1. The "respiratory" character in *Saccharomyces cerevisiae* is determined through an interplay of genome and plasmone. Formation of the essential enzymes of the cytochrome system is prevented by mutational changes in the plasmone as well as by single gene mutations in the genome. Thus respiration is in effect eliminated. The capacity of the plasmone to respire can only be realized, if the appropriate gene in the nucleus has not mutated.

2. Distinct self-replicating particles in the cytoplasm are involved in the plasmone mutation; these are very likely associated with the mitochondria.

3. Plasmone mutants in which the cytochromes and respiratory enzymes are altered are also known in *Neurospora crassa*. A nucleo-cytoplasmic relationship appears to be involved. It can be assumed that the mitochondria are also determinants of extrachromosomal heredity in *Neurospora*.

C. Infectious particles as hereditary determinants

I. Barrage phenomenon

If two fungal strains of the same species are placed on an agar medium the two mycelia generally intermingle in the zone of contact and form innumerable hyphal fusions. After a time the border between the two mycelia is hardly recognizable any longer. In contrast to such normal contact behavior, a series of cases are known in which two mycelia exhibit an antagonism in the zone of contact. The two types of hyphae form abnormal and often lethal fusions. The hyphal tips branch profusely. A clear line of separation in the zone of contact appears with increasing age of the culture. This phenomenon was first observed in hemi-compatible crosses (with common *B* factors) of Basidiomycetes exhibiting tetrapolar incompatibility and named "barrage" by VANDENDRIES (1932) (p. 59).

As we have already mentioned in the discussion of heterogenic incompatibility (p. 70f.), a barrage also occurs between strains of different races of the Ascomycete, *Podospora anserina* (RIZET, 1952, 1953a; ESSER, 1959). Mycelia of the same race show a normal contact reaction. The barrage only influences the sexual compatibility of two mycelia if the hyphal antagonism occurs in connection with heterogenic incompatibility. The barrage phenomenon itself involves vegetative incompatibility (p. 100). Barrage formation similar to that in *Podospora* has also been observed in another Ascomycete, *Gelasinospora tetrasperma* (DOWDING and BAKERSPIGEL, 1956).

In general the *formation of a barrage is determined by genes*. In the Basidiomycetes it is an expression of the homogenic incompatibility of identical *B* factors. On the other hand, *Podospora* and *Gelasinospora* only form barrages if two mycelia carry different genes; in this case the mechanism is heterogenic. In *Podospora anserina* a number of loci are known which produce this effect even if only one of them is represented by different alleles. The strains *S* and *s* constitute an *exception*; in these *plasmone differences* are responsible *for the barrage reaction* in addition to a single gene difference (RIZET, 1952; SCHECROUN, 1958a, b, 1959; RIZET and SCHECROUN, 1959; BEISSON-SCHECROUN, 1962). Before going into these nucleo-cytoplasmic relationships, we wish to describe briefly the development and appearance of the barrage in *Podospora*.

The barrage occurs in the zone of contact between two mycelia as a macroscopically recognizable, pigment-free region several millimeters wide (Fig. II-6). The hyphal tips in this region fail to produce melanin pigment in contrast to the remaining parts of the mycelium. The absence of pigment formation is a consequence of vegetative incompatibility, the same phenomenon as in *Neurospora crassa* (p. 98f.): Through hyphal fusions arising in the barrage region a cytoplasmic exchange occurs which leads to the death of the hyphal segments involved in a short time. The living segments adjacent to the cytoplasmic connections branch and also fuse with those of the other mycelium. In this way a hyphal pad arises which can readily be recognized under low magnification.

Thus the barrage is a macroscopically visible expression of vegetative incompatibility between S and s. A model of the way in which the genome and plasmone interact to produce such hyphal lethality has been developed through genetic and cytological analysis.

1. *Genetic constitution of the strain s.* The strains S and s differ by a single allele. In the F_1 of the cross, $S \times s$, a segregation of S and s mycelia does not occur as expected; rather S and a modified s strain, called s^S

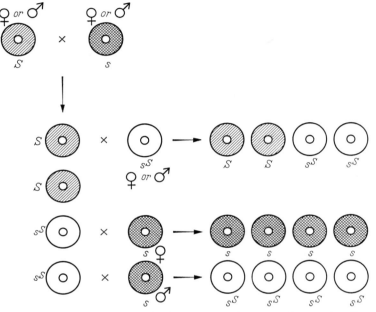

Fig. VII-3. Diagram of the analysis of a cross between the strains S and s of *Podospora anserina*. In crosses of $S \times s$, the progeny segregate $2S:2s^S$, with the total loss of the s phenotype, and appearance in its place of the new property s^S. In reciprocal crosses of $s^S \times S$, the progeny segregate $2:2$ normally. However, in crosses of $s^S \times s$, the F_1 progeny all resemble the female parent, thereby showing nonchromosomal inheritance of the difference between s and s^S. Chromosomal genes segregate normally in these crosses. (After RIZET et al., 1958; adapted from SAGER and RYAN, 1961)

are found. The latter does not give a barrage with either S nor s. It has been shown through appropriate backcrosses that although the s^S mycelia carry an unchanged s gene, the phenotype is determined by a cytoplasmic alteration (Fig. VII-3). The plasmone mutation apparently arises under the influence of the S gene in the zygote. The mutation shows a typical maternal inheritance and persists through multiple sexual generations of s^S strains.

A reversion of s^S to s at a frequency of about 10^{-7} per cell can be observed with prolonged vegetative reproduction. It spreads like an infection in the mycelium at a rate of about 7 cm per day. A reversion can also be produced as a result of cytoplasmic contact with s (hyphal fusion).

A single fusion between an s^S and an s mycelium is sufficient to induce a reversion. The exchange of cytoplasm can be clearly seen under the microscope. Nuclei do not migrate through the point of contact, as shown by using marker genes. If the cytoplasmic bridge is broken after a few hours, a stable s mycelium grows out of the s^S hypha.

The plasmone mutants s^S can be obtained at will (through crossing of s with S) as well as by reversion. Mutations in the direction of s to s^S do not occur.

RIZET and his coworkers based their interpretation of these results on the fact that the induced reversion occurs only after cytoplasmic contact, but without nuclear migration from one mycelium to the other. Therefore, they assumed that the reversion is induced by distinct particles which enter the s^S cytoplasm from s. Since the nuclei of the s and s^S hyphae are identical, the plasmone difference in the two strains consists of the presence and absence of such infectious particles. The regularly occurring spontaneous reversions are explained by the spontaneous activation of inactive particles or a *de novo* origin of new particles.

The Rizet group invoked a model for the origin of new particles which corresponds to that in phage for lysogeny and in bacteria for episomes (JACOB et al., 1960). The s nuclei contain a "pro-particle" which like the prophage can enter the cytoplasm spontaneously and there replicate itself as an infectious particle. Accordingly, such plasmone particles exist in two forms, one as an integrated component of the genome, and one as an infectious particle in the cytoplasm.

In summary it can be said that the *phenotype s* (barrage with S) is *determined by the gene s and specific infectious plasmone particles.* The *particles are eliminated or inactivated* in the presence of the *gene S* (in the zygotes $s \times S$ or in artificially produced heterokaryons). The *phenotype s^S* (no barrage with S) which *differs from s only in the cytoplasm, arises* in this way.

2. Genetic constitution of the strain S. Microsurgical experiments have indicated that the S mycelia also carry infectious cytoplasmic factors, which, in contrast to the s particles, are constantly regenerated by the nuclei.

This has been shown by the differences in the reciprocal crosses, $s \times S$. The male parent supplies only the nucleus in both crosses. If the male nucleus carries the S gene, all the S progeny receive the S factors. Conversely, the s mycelia have no s particles; their phenotype has been modified to s^S. A plasmone mutant of S comparable to s^S does not exist.

Since it is possible to consider the S cytoplasmic factors as diffusible gene products and not as self-replicating particles, the *reaction of the S strain cannot be explained with certainty as having an extrachromosomal basis.*

A choice in favor of a particulate basis would be indicated if a plasmone mutant of the genotype S, but lacking the S cytoplasmic factors were found and converted to the S phenotype through infection of the cytoplasm with a normal S hypha.

3. Interaction between the genomes and plasmones of the s and S strains.
Distinct particles or unknown cytoplasmic factors with a characteristic
behavior arise in homokaryotic mycelia under the influence of the *s* and *S*
genes. Genetic and microsurgical experiments have shown that these
determinants behave antagonistically. This leads to vegetative incompatibility as evidenced by the barrage. The antagonism is not limited to
the plasmones, but also involves the genomes.

The following combinations may serve as examples: 1. plasmone *s* /
gene *S*: both the formation of the cytoplasmic factor *S* and the reproduction
of the *S* nuclei are inhibited. 2. plasmone *S* / gene *s*: synthesis of the cytoplasmic particle *s* is inhibited. The rate of division of the *s* nuclei is not
affected.

All these interactions depend upon the quantities of the individual
components present. If any one of the two components predominates,
they are absent or occur only partially. The vegetative incompatibility
of genetically different mycelia expressed through *barrage formation*
results from an *incompatibility of different cytoplasms*.

BEISSON-SCHECROUN (1962), borrowing from a theoretical analysis
of extrachromosomal inheritance by L'HÉRITIER (1962), has recently
proposed another *explanation of the barrage phenomenon* based on the
fact that the cytoplasmic factors of the *S* strain cannot be considered as
particulate with certainty; they can equally well be viewed as non-
particulate products of the *S* gene. This explanation stands in contrast to
that held by the RIZET group until now (e.g. RIZET et al., 1958).

BEISSON-SCHECROUN assumes that not only the *plasmone determinants
of S, but also those of s, are products of the S and s alleles. Specific cytoplasmic conditions* in the sense of a steady state (DELBRÜCK, 1949) *either
permit or inhibit the activity of the S and s genes.* This model, which will
not be discussed in greater detail here, can only be viewed as a working
hypothesis in view of the absence of biochemical data.

II. Senescence syndrome

Senescence which regularly appears in higher plants and animals also
occurs in fungi. It has long been known that strains of fungi maintained
in culture collections through serial transfer of mycelia become senescent
in time. They generally first lose the capacity to reproduce sexually and
in the case of pathogens, their pathogenicity. This is often followed by
a degeneration of the mycelium which may lead to a complete cessation
of hyphal growth. Such senescence syndromes were previously viewed as
spontaneous mutational alterations of the genome. This explanation
does not hold generally, since in many fungi degeneration is nullified
through plasmogamy (sexual reproduction or heterokaryotization with
young strains). A definitive idea of the mechanism of senescence has
been developed through the investigations of RIZET and coworkers on
Podospora anserina (RIZET, 1953 b, c, 1957; RIZET and MARCOU, 1954;

Marcou, 1954a, b, 1957, 1958, 1961; Marcou and Schecroun, 1959). These investigators demonstrated that in this Ascomycete *senescence is determined by a plasmone mutation*.

All wild type strains of *P. anserina* show senescence after prolonged vegetative growth. The growth rate decreases, the hyphae become slender and undulating. The hyphal tips become vacuolated, swell up, and may burst. A few days after the appearance of the senescence syndrome growth ceases entirely and the mycelium dies. The time of onset of senescence depends upon the genotype and environmental conditions. Senescence begins after forty days at a temperature of 24° C in the wild type strains mostly utilized for these investigations.

A single gene difference is sufficient to prolong or shorten the juvenile phase. Lowering the temperature from 24° C, normally used for culturing *Podospora*, to 16° C lengthens the life of the culture. At a temperature of 3° C the mycelium will remain in the juvenile state for years. Higher temperatures accelerate the ageing process. The onset of senescence is favored by light and optimal nutrition.

Sexual reproduction of aged *Podospora* strains is not affected by the senescence syndrome. When a senescent strain is crossed with the juvenile one from which it arose (the latter held at 4° C) reciprocal differences appear (compare Fig. VII-1).

Perithecia, the asci of which yield only spores producing normal mycelia, arise from ascogonia of the juvenile strains which have been fertilized with spermatia from the senescent strain. The reciprocal cross (senescent ♀ × juvenile ♂) gives as high as 90% perithecia with "senescent" spores.

Thus *senescence* is *maternally inherited*. The senescent condition determines the degree of ageing of all spores formed in the perithecium. Further, since crossing experiments show that although genes determine the onset of senescence, they are not responsible for senescence itself, the extrachromosomal nature of the inheritance of this character may be considered proven.

Through microsurgical experiments similar to those used in analyzing the barrage phenomenon it was shown that the senescence syndrome could be transferred from a senescent to a juvenile hypha by the cytoplasmic bridge formed between them. A transfer of nuclei was excluded. Rizet and coworkers concluded from these results that *senescence was determined by infectious particles in the cytoplasm*. These particles, the nature of which remains unknown, must be *self-replicating*, because they multiply in the cytoplasm of a juvenile strain and migrate through the hyphal septa. Once infection of a juvenile strain has occurred, it leads to complete senescence of the particular mycelium in a short time. Since senescence occurs in every wild type or mutant mycelium sooner or later, one must assume that each strain contains at least precursors of the particle; or inactive particles which acquire their infective character only after an extended period of vegetative growth. Whether or not and to what extent such "pro-particles" are correlated with the genome cannot be established as yet. However, a nucleocytoplasmic relationship is indicated in that the time of onset of the syndrome is determined by the genome.

Senescent strains can be rejuvenated by various methods: storage of the cultures for several months at $3°$ C or under mineral oil, culturing on a growth-inhibiting medium, desiccation. When these methods are applied long enough, mycelia with a normal life span arise in all cases. The same strain can be repeatedly rejuvenated. The reversion, *senescent→ juvenile*, can be explained by assuming that the particles are destroyed by the treatment or that only those hyphae grow which do not contain particles.

A non-uniform distribution of particles in the mycelium is indicated by the results of the cross, senescent female × juvenile male (see above), since up to 10% of the ascogonia produce only juvenile offspring after fertilization. This can only result from the failure of the ascogonium to contain any senescent particles. The unequal distribution of the plasmone determinants is also indicated by another series of experiments: When a senescent mycelium is fragmented in a blendor and the homogenate diluted and plated, some juvenile strains regularly arise.

Extrachromosomally inherited senescent syndromes are also known in other fungi: *Podospora setosa* (RIZET, 1953a), *Aspergillus glaucus* (JINKS, 1957; SHARPE, 1958), *Helminthosporium victoriae* (LINDBERG, 1959), and *Pestalozzia annulata* (CHEVAUGEON and DIGBEU, 1960). Infectious plasmone particles as determinants of senescence were demonstrated in *P. setosa*, *A. glaucus* (only in the work of SHARPE) and *P. annulata*. The mechanism of extrachromosomal inheritance of senescence needs further investigation in the remaining species.

III. Morphological mutants

Plasmone mutations which *alter the habit of the fungal thallus frequently* lead to the *formation of sectors* in petri dish cultures. The sectors differ from the original mycelium in growth form of the hyphae, failure to form conidia, density of conidiophores, number of fruiting bodies, or other macroscopic characters. A mutational alteration of the plasmone for the origin of such sectors, can be assumed only if the latter occur with a frequency which greatly exceeds that of gene mutation. This applies to sectoring in *Aspergillus glaucus* (SHARPE, 1958), *Pestalozzia annulata* (CHEVAUGEON and LEFORT, 1960) and *Curvullaria pallescens* (CUZIN, 1961). If one disregards the developmental differences of these three species (*P. annulata* and *C. pallescens* are imperfects), the investigations of sectoring have led to essentially the same results:

1. Sectors which exhibit the same syndrome arise regularly in each culture. Sectoring is not dependent upon the genome.

Various gene-determined morphological mutants of both imperfect fungi form, with equal regularity, sectors which are similar in habit to those of the wild type.

2. The mycelial modification persists through asexual reproduction.

Both imperfect fungi can transmit the modification through hyphae but not through conidia. In *A. glaucus* it can be transmitted by both only if large numbers of conidia or large segments of mycelium are used. The variant always disappears after sexual reproduction.

3. The mycelial modification can be transferred to normal mycelium through cytoplasmic contact (hyphal fusion).

The mycelial modification spreads with a speed exceeding ten times the growth rate in the hyphae of the wild strain of *P. annulata*.

These results indicate that a *plasmone mutation is very likely involved in the sectoring and that self-replicating infectious particles determine it*. The expression of the plasmone mutation appears to depend upon a particular concentration of particles.

If a hyphal segment with only a few septa is used as inoculum, normal mycelium can be recovered from any mutated sector.

Further, one can assume that the concentration for each strain is specific. For example, it is greater in the two imperfect fungi than in *A. glaucus*, since the former do not transmit the particles through the conidia which have only a small amount of cytoplasm. On the other hand, in *A. glaucus* transmission of the plasmone mutation through the conidia is possible to a limited degree.

ARLETT (1957, 1960) has made the same observation. Plasmone-determined morphological mutants of *Aspergillus nidulans* were obtained by selection as well as with UV irradiation and treatment with acriflavine. Transmission of the stable mutant characters through the conidia did not occur in all cases. In two variants investigated in greater detail they seem to be determined through a heteroplasmic condition. As a matter of fact, the mutated plasmone is only viable when associated with the wild type plasmone (ARLETT et al., 1962; FAULKNER and ARLETT, 1964). It appears to consists of homologous subunits which can be inherited independently of one another (GRINDLE, 1964).

The fact that in each strain (wild type and mutant) the phenotypically identical plasmone mutation not only occurs regularly, but also appears with the same regularity after loss through sexual or asexual reproduction leads to the conclusion: Each mycelium bears a template for the formation or activation of infectious plasmone particles. Does this potential lie in the genome or in some other site within the cell? The same question was raised in connection with the senescence syndrome; in neither case it can be answered. The two cases of plasmone mutation, senescence and formation of morphologically distinct mycelial sectors, seem to have the same genetic basis: the spontaneous origin or activation of infectious cytoplasmic particles.

An example of a *nucleo-cytoplasmic interaction* in the expression of plasmone-determined morphological mutants is provided by the investigations of ROPER (1958) and by MAHONY and WILKIE (1962) on *Aspergillus nidulans*.

ROPER obtained stable morphological variants by transferring successively three genotypically different strains on an acriflavine-containing medium. All modified strains had lost the capacity to form conidia and to reproduce sexually. Through heterokaryon formation their characters could be transferred to the wild type. An "infection" of the variants by the wild type was not possible. The analysis of asci from the fruiting bodies of a heterokaryon always yielded a 3:1 segregation of variants to wild type. Nevertheless, the F_1 wild type strains were able to infect other wild strains with the characters of the variants. The 3:1 ratio is determined by a segregation of genes which act epistatically on the mycelial variants. One of these genes has been localized. The origin of stable morphological variants can thus be attributed to plasmone mutations the determinants of which are infectious particles.

MAHONY and WILKIE found 3 alleles which are responsible for the differential frequency of formation of sterile plasmone variants in different wild type strains. This defect can be compensated in heterokaryons with gene-determined sterile mutants, since one component has a non-mutant genome and the other a non-mutant plasmone.

These experiments show that *plasmone mutations* leading to morphological changes *can only produce their effect when the proper genetic information is present in the nucleus*. The parallel between this genome-plasmone relationship and that in the respiration-deficient yeasts and barrage formation in *Podospora* reinforces the view that an interaction between genome and plasmone is a general biological phenomenon.

Finally we wish to consider certain investigations on the extra-chromosomal inheritance of morphological characters carried out on members of the *Aspergillaceae* (JINKS, 1954, 1956, 1957; review 1958). Although an infectious nature could not be demonstrated for the particles responsible for the plasmone mutations, their mode of inheritance agrees in large part with those of the cases previously discussed.

Through serial transfer of isolated hyphal tips or conidia JINKS obtained mycelia with altered phenotypes from homo- and heterokaryons of *Penicillium cyclopium, Aspergillus nidulans*, and *A. glaucus*. These variants differ from each other and the wild type through germination rate of the spores, growth rate of mycelium, senescence, pigment formation, and fruiting body formation. Mycelia exhibiting the phenotype of the original strain could be recovered from all variants through a period of reverse selection. The extrachromosomal nature of these modifications can be demonstrated by two methods:

1. *Use of marker genes.* Segregation occurs for the markers but never for the thallus characters following sexual reproduction of homo- or heterokaryotic variants.

2. *Heterokaryon test.* Segregation of markers can be shown among conidia and ascospores following heterokaryon formation between variants and wild type. The variant characters fail to appear. Since microsurgical experiments similar to those in the *Podospora* work (p. 447) were not carried out, the existence of infectious particles in the wild type strain which migrate into the variant following cytoplasmic contact and cause them to revert has not been established. Nevertheless, such a possibility does not contradict the interpretation which JINKS gives for the extrachromosomal hereditary phenomena that he observed.

According to the idea of JINKS the origin of plasmone mutants can be attributed to a reduction in the number of cytoplasmic particles through selection. The equilibrium of cytoplasmic components disturbed in this way can be reestablished through a renewed accumulation of particles (by selection in the reverse direction).

One can explain in the same way the differential expression of the phenotype of a female sterile mutant of *Neurospora crassa* determined by two recessive genes (FITZGERALD, 1963). However, this variability has not been unequivocally shown to be controlled through cytoplasmic particles. It may also involve a differential activity of the two genes which adapt to the particular condition of the cell.

For the sake of completeness the study of MAHONY and WILKIE (1958) deserves a brief mention; this was published in connection with the previously cited investigations on extrachromosomal inheritance in the Aspergillaceae.

Through selection from a self-fertile strain of *Aspergillus nidulans* the authors obtained sterile variants which could reproduce vegetatively only by conidia. The following three categories of perithecia were obtained from crosses between fertile and sterile strains which carried marker genes: those which contained only spores of one or the other parental type and those in which only hybrid asci with spores of both parents or their recombinations were formed. On the basis of such crossing results the authors suggested the possibility that the asexual strains arise through a loss of a cytoplasmic factor. This does not seem to be correct because the behavior in crosses corresponds completely to the behavior of similar mutants of *Sordaria* (p. 44 f.). In the latter case three types of fruiting bodies also occur after crossing through an inductive effect, determined by compensation of non-allelic gene mutations.

Consideration of the investigations on the extrachromosomal inheritance of morphological characters (including the senescence syndrome), which have been carried out predominantly on the Aspergillaceae, permits certain generalizations. Above all, *the role of the cytoplasm in morphogenesis is clearly not limited to the determination of specific steps in differentiation the genetic information for which resides in the nucleus. The cytoplasm possesses a certain independence in its control of developmental processes. It can itself be the bearer of genetic information in the form of a self-replicating particle effective in morphogenesis. In such cases the genome assumes primarily an epistatic control function.*

Summary

1. Extrachromosomal inheritance in a series of Ascomycetes and related imperfect fungi is very likely determined by self-replicating cytoplasmic particles. These particles, the nature of which remains unknown, are infectious. They may migrate into other mycelia following cytoplasmic contact (hyphal fusion), multiply, and manifest themselves as a plasmone mutation.

2. These plasmone particles can be considered as autonomous determinants of genetic information. The nucleo-cytoplasmic relationship demonstrated in certain cases solely involves a control function for the genome, which on a higher level permits or prevents the development of the plasmone characters.

3. Examples of such an hereditary mechanism which have been discussed are the barrage phenomenon in *Podospora anersina*, the senescence syndrome which is widely distributed among Ascomycetes and imperfect fungi, and the plasmone-determined morphological mutants of the aspergillaceae.

D. Hereditary elements of unknown nature

In the previous sections we have mentioned only those extrachromosomal phenomena in which distinct particles in the cytoplasm have been demonstrated with more or less certainty as hereditary determinants of the plasmone. We now wish to review those cases of extrachromosomal inheritance in fungi in which the nature of the plasmone determinants remains unknown. Since these hereditary phenomena affect various characters in various organisms, a review organized on a conceptual basis is not possible. Thus the different experimental results are presented in an arbitrary sequence.

I. Recombination between genome and plasmone

Just as the expression of extrachromosomally inherited characters may depend upon the genome, genes of the genome may only manifest themselves in the presence of specific plasmone types. Such nucleocytoplasmic relationships have been observed repeatedly in the course of studies of extrachromosomal inheritance in fungi as well as flowering plants (survey of literature by CASPARI, 1948; and RHOADES, 1955; reviews by MICHAELIS, 1954; and STUBBE, 1959).

Particularly the extensive researches of MICHAELIS on the genus *Epilobium* and those of STUBBE on the complex-heterozygotes of *Oenothera* have shown that the phenotype is not only the product of the genotype and the environment, but is to a great extent dependent upon the interaction between genome and plasmone (and plastome respectively). Both investigators have demonstrated that the reciprocal dependence of chromosomal and extrachromosomal genetic information is an essential factor in speciation.

SRB (1958) undertook an experiment parallel to the investigations on *Epilobium* and *Oenothera to determine the relationships between genome and plasmone within the genus Neurospora* by obtaining recombinations between plasmone and genome of different species and races.

He used the species, *Neurospora crassa*, *N. intermedia*, and *N. sitophila* for his investigations. Three races of *N. crassa* of different geographic origins were available. A slow growing plasmone mutant (*SG*) was isolated following acriflavine treatment of the US race of *N. crassa*.

In order to replace the genome *SG* with another, ten successive backcrosses between *SG* and the wild type strain of *N. sitophila* were carried out. An *SG* strain from the progeny of the preceding cross was always selected as maternal parent and the same wild type strain of *N. sitophila* as paternal parent. An extensive replacement of the *N. crassa* genome by that of *N. sitophila* could be demonstrated particularly by the fact that after 8 generations a shift in the post reduction frequency of the mating type marker to the value typical of *N. sitophila* had occurred. The *SG* plasmone was combined with the genomes of *N. intermedia* and the two other *N. crassa* races in a similar way. The *SG* character was retained in all cases.

In contrast to this manifestation of a plasmone-determined mutant character apparently uninfluenced by the genome, the expression of two characters determined by nuclear genes was extensively affected by the plasmone.

The gene *ac* (weak conidial formation) can produce its effect in only the *N. sitophila* and not in the *N. crassa* plasmone. Similarly, the gene *S* leads to formation of minute colonies only if it is combined with the plasmone of one of the three races of *N. crassa*.

A variant which arose following acriflavine treatment in another experimental series differs completely in its hereditary behavior (MUNETA and SRB, 1959). This strain, characterized by a very slow growth rate, can be propagated vegetatively without change. After crossing with the wild type, it fails to appear among the offspring of either reciprocal cross. However, if this variant is crossed with the plasmone mutant, *SG*, reciprocal differences are found; slow-growing strains are found exclusively only in the cross, *SG* ♀×*variant* ♂. As soon as *SG* is the donor of the male nuclei, *SG* strains occur predominantly; only a few offspring with the variant phenotype appear. This is the *first case of paternal inheritance in fungi,* judging from the brief report so far published[1]. Additional experimental work is needed to formulate a hypothesis explaining this phenomenon.

The fact the SRB has only been able to demonstrate a dependence of the genome on the plasmone in the expression of phenotypic characters does not exclude the possibility that after further analysis of plasmone mutants a dependence in the reverse sense can be found. The significance of these experiments, which (as SRB notes himself) can only be considered as a beginning of a detailed genome-plasmone analysis, lies particularly in the fact that they show the fungi to be as suitable as the flowering plants for such studies. Thus one can utilize the advantages that fungi offer in contrast to the flowering plants for the investigation of extra-chromosomal inheritance, namely the short generation time, tetrad analysis, and seasonal independence.

II. Disturbance of meiosis through mutational alteration of plasmone

Following the crossing of two monokaryons of *Coprinus lagopus,* DAY (1959) showed reciprocal differences with respect to the number of spores produced per basidium. When *strain 68* was utilized as the cytoplasmic donor, all basidia possessed 4 spores. In the reciprocal combination (*strain 54* dikaryotized with nuclei from *strain 68*) the numbers of spores formed on the basidium varied between 2 and 6. Further, after one nuclear component was removed from the normal and from the abnormal dikaryons (see below), they retained their reaction when used in further crosses as a nuclear recipient. Thus the extrachromosomal character of this basidial anomaly is unequivocally established. Cytological investigation showed that the plasmone mutation induces irregularities in meiosis. These disturbances (failure of meiosis II to occur, bridges between chromosomes, formation of micronuclei etc.) result in basidia with a variable number of functional spores.

---

[1] Paternal inheritance through plastids in flowering plants has been known for some time (for literature see STUBBE, 1963).

III. Variability following removal of nuclei from dikaryotic mycelia

Forty years ago HARDER (1927a) attempted to determine the behavior of a nucleus in the cytoplasm of another cell. He hoped in this way to differentiate between the effects of genome and plasmone on the expression of phenotypic characters. He selected the dikaryotic mycelia of the Basidiomycetes *Pholiota mutabilis* and *Schizophyllum commune* as experimental material. With the aid of a microsurgical technique (HARDER, 1927b), he succeeded in preparing uninucleate cells from hyphal tips of dikaryons and in isolating monokaryotic mycelia which developed from such cells. These "neo-monokaryons" lost their typical dikaryotic character, the clamp connections, after a few cell generations and were no longer distinguishable in this respect from the original monokaryotic strains from which they had arisen. The neo-monokaryons were highly variable, since two morphologically different races had been used to prepare the dikaryon. The strains obtained by removing one of the nuclear components showed all gradations between the growth forms of the two parents and retained these characters through vegetative propagation. Oddly enough only those strains remained viable which possessed the nuclear type of a particular parent. Segregation of the mycelial characters did not occur following either the backcrossing of a neo-monokaryon to the other parent or the crossing of the two original strains with each other. In both cases only the full range of intermediates between the two parental forms arose (just as after the surgical removal of a nuclear component from the mycelium).

In interpreting his results, HARDER assumed that the cytoplasm of the two partners became mixed as a result of dikaryon formation. The variability which appeared within the experimentally produced monokaryons as well as in the mycelia produced from sexual reproduction was attributed to the differential mixing of the two parental cytoplasms, which are thus to be considered as carriers of heredity.

The questions raised by HARDER were not taken up by other investigators until much later. ASCHAN (1952), ASCHAN-ÅBERG (1960) on *Collybia velutipes*, FRIES and ASCHAN (1952) on *Polyporus abietinus*, and PAPAZIAN (1955) on *Schizophyllum commune* obtained similar results by using the Harder technique. These investigators, by applying the modern approaches of fungal genetics marked the original strains with genes and thus were able to identify with greater reliability the nuclei present in the neo-monokaryons. Moreover, ASCHAN observed a certain variability among the mycelia germinating from uninucleate oidia of the dikaryon; such mycelia can be considered as naturally-occurring neo-monokaryons. This variability, however, was appreciably less than among the experimentally produced neo-monokaryons.

The fact that after removal of one of its nuclear components a dikaryotic mycelium takes on a habit intermediate between that of the strains from which it originated can no longer be viewed as a fully reliable criterion of extra-

chromosomal inheritance in light of modern genetics (ASCHAN-ÅBERG, 1960).
The following considerations must be taken into account:

1. It is very unlikely that the cytoplasm of a dikaryon is a mixture of the cytoplasms of its two monokaryotic components.

As already mentioned earlier (p. 58), a dikaryon generally arises from a reciprocal exchange of nuclei between two sexually compatible mono-karyons. The migratory nucleus usually first induces clamp connection formation after it has moved through different cell septa from the point of fusion to the periphery of the mycelium. Any cytoplasmic components of the original mycelium which adhere to it are most likely lost during migration.

The cytoplasm of the dikaryon can at most be differentiated from that of the monokaryon through factors which have arisen *de novo* following dikaryotization.

2. Clear-cut crossing results are lacking.

The crosses carried out by ASCHAN-ÅBERG (1960) showed a Mendelian segregation for the markers, but no reciprocal differences for the morphological characters. This seems all the more strange, since in these experiments a migrating nucleus and recipient cytoplasm was identified with considerable reliability.

3. Variability was limited or failed to appear entirely in mycelia from oidia or cystidia which lost a nuclear component naturally (*Coprinus lagopus:* PAPAZIAN, 1956). FRIES and ASCHAN and also PAPAZIAN (1958) concluded from this that the morphological change in the neo-mono-karyons may be the result of a wounding stimulus.

4. Mitotic recombinations and spontaneous mutations cannot be excluded as possible explanations.

These considerations make it *difficult to decide at this time whether or not the altered characters of the neokaryotic mycelium are determined extrachromosomally.* In order to settle this question it is necessary to undertake reciprocal crosses between morphologically different strains which are adequately marked and to analyze their progeny genetically.

As the experiments of DAY (p. 457) have shown, such a procedure can be carried out with Basidiomycetes. It is unfortunate that the first attempt to gain an insight into the function of the plasmone in fungi has so far failed to yield significant results.

IV. Other evidence for extrachromosomal inheritance in fungi

In crosses between different races of the rust, *Puccinia graminis*, JOHN-SON (1946) observed reciprocal differences in pathogenicity. These characters were inherited unaltered through a number of generations. Analysis of crosses showed that genome differences are not responsible for these changes in pathogenicity.

QUINTANILHA and BALLE (1940) obtained a high frequency of dwarf mycelia as a result of plasmone mutations after germinating basidiospores of *Coprinus lagopus*.

The wooly growth of the "*fluffy*" mutant of *Coprinus macrorhizus* (DICKSON, 1936) likewise appears to be determined by a mutational alteration of the plasmone. Through a series of crosses between compatible mono- and dikaryons (p. 65) of *Coprinus macrorhizus*, KIMURA (1954) showed that the crossing reaction was probably influenced by autonomous cytoplasmic determinants.

Summary

1. Recombination between genome and plasmone of different races and species of *Neurospora* have shown that in this genus cooperation between chromosomal and extrachromosomal genetic material is essential for the expression of specific characters.

2. The plasmone in the Basidiomycete *C. lagopus* is partially responsible for the normal course of meiosis.

3. The frequently observed variability of neo-monokaryotic mycelia which arise after removal of one nuclear component from a Basidiomycete dikaryon cannot be attributed with certainty to a mutational change of the plasmone.

4. Other, less thoroughly analyzed examples of extrachromosomal inheritance are briefly described.

E. Concluding remarks

Study of extrachromosomal inheritance in the fungi has not only added examples of this phenomenon to those already known in higher plants and animals, but in particular has enhanced our knowledge of the mode of action of the plasmone. A survey of the relatively thoroughly analyzed examples leads to *two general points of view: 1. The genetic information of the plasmone is tied to particulate, self-duplicating hereditary determinants. 2. An interaction occurs between genome and plasmone.*

Although in two cases mitochondria are implicated as plasmone determinants, in all others the nature of the plasmone particles is unknown. Their infectious character, which has often been demonstrated, is reminiscent of the extrachromosomal *kappa* particles in *Paramecium* (SONNEBORN, 1959) and in *Drosophila* (CO_2 sensitivity: L'HÉRITIER, 1951). Alternatively it can be explained by the "lysogeny" or "episome" models (JACOB et al., 1960) of bacterial genetics. According to this working hypothesis proposed by RIZET and coworkers (1958) the *plasmone particles* exist in two different states, as *active, infectious, self-replicating components of the cytoplasm* and as *inactive, non-infectious "pro"-particles incapable of replication.* A great deal argues for the idea that these pro-particles can become integrated into the genome like lysogenic phage. A portion of the experimental data from fungi can be fitted to the episome model. Thus hereditary phenomena of eukaryotic and prokaryotic organisms can be viewed as having the same basic mechanism. It should be pointed out, however, that the existence of infectious particles is indicated only by indirect genetic evidence. The validity and generalization of this model depends to a critical degree on whether or not such particulate hereditary units can be identified and their constitution clarified in the future. It is conceivable that infection experiments with isolated cytoplasmic RNA may be helpful.

The clearly established *nucleo-cytoplasmic relationship* does not consist of a unilateral dependence of the plasmone on the genome; the genome on its part can only realize its potential if it is permitted expression by the genetic information of the plasmone. The *development of a character* occurs only *if both systems consisting of their autonomous determinants work together harmoniously*. Mutational changes in one of the two systems can either suppress the expression of a character entirely or lead to various abnormalities. This knowledge derived from the study of extrachromosomal inheritance is of basic significance for the understanding of morphology. It is generally assumed that all nuclei of an individual are identical genetically, whether they occur in differentiated or undifferentiated tissues. The differences in characteristics that somatic cells exhibit during development can be viewed at least partially as changes in the plasmone. This conception is confirmed by the fact that such changes can be maintained throughout extended vegetative propagation in tissue cultures. Thus further analysis of extrachromosomal inheritance in the fungi, which have a relatively simple organization, can be expected to lead to a model which may serve as a basis for corresponding experiments on differentiation in higher organisms.

Literature

ARLETT, C. F.: Induction of cytoplasmic mutations in *Aspergillus nidulans*. Nature (Lond.) **179**, 1250—1251 (1957).
— A system of cytoplasmic variation in *Aspergillus nidulans*. Heredity **15**, 377—388 (1960).
— M. GRINDLE, and J. L. JINKS: The "red" cytoplasmic variant of *Aspergillus nidulans*. Heredity **17**, 197—209 (1962).
ASCHAN, K.: Studies on dediploidisation mycelia of the basidiomycete *Collybia velutipes*. Svensk bot. T. **46**, 366—392 (1952).
ASCHAN-ÅBERG, K.: Studies on dedikaryotization mycelia and of F_1 variants in *Collybia velutipes*. Svensk bot. T. **54**, 311—328 (1960).
BAUTZ, E.: Beeinflussung der Indophenolblaubildung (Nadi-Reaktion) in Hefezellen durch Röntgenstrahlen. Naturwissenschaften **41**, 375—376 (1954a).
— Untersuchungen über die Mitochondrien von Hefen. Ber. dtsch. bot. Ges. **67**, 281—288 (1954b).
— Mitochondrienfärbung mit Janusgrün bei Hefen. Naturwissenschaften **42**, 49—50 (1955).
—, u. H. MARQUARDT: Die Grana mit Mitochondrienfunktion in Hefezellen. Naturwissenschaften **40**, 531 (1953a).
— — Das Verhalten oxydierender Fermente in den Grana mit Mitochondrienfunktion der Hefezellen. Naturwissenschaften **40**, 531—532 (1953b).
— — Sprunghafte Änderungen des Verhaltens der Mitochondrien von Hefezellen gegenüber dem Nadi-Reagens. Naturwissenschaften **41**, 121—122 (1954).
BEALE, G. H.: The genetics of *Paramaecium aurelia*. Cambridge (Engl.) 1954.
BEISSON-SCHECROUN, J.: Incompatibilité cellulaire et interactions nucleoplasmiques dans les phénomènes de «barrage» chez le *Podospora anserina*. Ann. Génét. **4**, 4—50 (1962).

Extrachromosomal inheritance

Bogen, H. J., u. G. Kraepelin: Induktion atmungsdefekter Mutanten bei *Saccharomyces cerevisiae* mit asynchroner und synchronisierter Sprossung. Arch. Mikrobiol. **48**, 291—298 (1964).

Bulder, C. J. E. A.: Induction of petite mutation and inhibition of synthesis of respiratory enzymes in various yeasts. Antonie v. Leeuwenhoek **30**, 1—9 (1964).

Caspari, E.: Cytoplasmic inheritance. Advanc. Genet. **2**, 1—66 (1948).

Catcheside, D. G.: Introduction to a discussion on the cytoplasm in variation and development. Proc. roy. Soc. B **148**, 285—290 (1958).

— Cytoplasmic inheritance. Nature (Lond.) **184**, 1012—1015 (1959).

Chevaugeon, J., et S. Digbeu: Un second facteur cytoplasmique infectant chez *Pestalozzia annulata*. C. R. Acad. Sci. (Paris) **251**, 3043—3045 (1960).

—, et C. Lefort: Sur l'apparition régulière d'un «mutant» infectant chez un champignon du genre *Pestalozzia*. C. R. Acad. Sci. (Paris) **250**, 2247—2249 (1960).

Correns, C.: Zur Kenntnis der Rolle von Kern und Plasma bei der Vererbung. Z. indukt. Abstamm.- u. Vererb.-L. **2**, 331—340 (1909).

— Nichtmendelnde Vererbung. In: E. Bauer u. M. Hartmann (Hrsg.), Handbuch der Vererbungswissenschaft, Bd. II, S. 1—159. Berlin 1937.

Cuzin, F.: Apparition régulière chez *Curvularia pallescenz*, d'une variation sectorielle contagieuse, non transmissible par les thallospores. C. R. Acad. Sci. (Paris) **252**, 1656—1658 (1961).

Day, P. R.: A cytoplasmatically controlled abnormality of the tetrades of *Coprinus lagopus*. Heredity **13**, 81—87 (1959).

Delbrück, M.: In: Unités biologiques douées de continuité génétique: Colloques internat. du CNRS, vol. 7, p. 33, Paris 1949.

Dickson, H.: Observations of inheritance in *Coprinus macrorhizus* (Pers.) Rea. Ann. Bot. **50**, 719—733 (1936).

Dowding, E. S., and A. Bakerspigel: Poor fruiters and barrage mutants in *Gelasinospora*. Canad. J. Bot. **34**, 231—240 (1956).

Ephrussi, B.: Quelques problèmes de la génétique des microorganismes. Arch. Klaus-Stift. Vererb.-Forsch. **26**, 403—425 (1951).

— The interplay of heredity and environment in the synthesis of respiratory enzymes in yeast. Harvey Lect. Series **46**, 45—67 (1952).

— Nucleo-cytoplasmic relations in microorganisms. Oxford (Engl.) 1953.

— Die Bestandteile des cytochrombildenden Systems der Hefe. Naturwissenschaften **43**, 505—511 (1956).

—, and H. Hottinguer: Cytoplasmic constituents of heredity. Cold Spr. Harb. Symp. quant. Biol. **16**, 75—85 (1951).

— — et A. M. Chimènes: Actions de l'acriflavine sur les levures. I. La mutation «petite colonie». Ann. Inst. Pasteur **76**, 351—364 (1949a).

— — et H. Roman: Sur le comportement des mutants a déficience respiratoire de la levure dans les croisements. Caryologia Vol. Suppl. 1112—1113 (1954).

— — — Supressiveness: A new factor in the genetic determinism of the synthesis of respiratory enzymes in yeast. Proc. nat. Acad. Sci. (Wash.) **41**, 1065—1071 (1955).

— — et J. Tavlitzki: Action de l'acriflavine sur les levures. II. Étude génétique du mutant «petite colonie». Ann. Inst. Pasteur **76**, 419—450 (1949b).

—, et P. P. Slonimski: La synthèse adaptive des cytochromes chez la levure de boulangerie. Biochim. biophys. Acta (Amst.) **6**, 256—267 (1950).

— — Yeast mitochondria, subcellular units involved in the synthesis of respiratory enzymes in yeast. Nature (Lond.) **176**, 1207—1209 (1955).

Esser, K.: Die Incompatibilitätsbeziehungen zwischen geographischen Rassen von *Podospora anserina (Ces.) Rehm*. II. Die Wirkungsweise der Incompatibilitätsgene. Z. Vererbungsl. **90**, 29—52 (1959).

Faulkner, B. M., and C. F. Arlett: The "minute" cytoplasmic variant of *Aspergillus nidulans*. Heredity **19**, 63—73 (1964).

FINK, H.: Klassifizierung von Kulturhefen mit Hilfe des Cytochromspektrums. Hoppe-Seylers Z. physiol. Chem. **210**, 197—219 (1932).

—, u. E. BERWALD: Über die Umwandlung des Cytochromspektrums in Bierhefen. Biochem. Z. **258**, 141—153 (1933).

FITZGERALD, P. H.: Genetic and epigenetic factors controlling female sterility in *Neurospora crassa*. Heredity **18**, 47—62 (1963).

FRIES, N., and K. ASCHAN: The physiological heterogeneity of the dikaryotic mycelium of *Polyporus abietinus* investigated with the aid of micrurgical technique. Svensk bot. T. **46**, 429—445 (1952).

GOWDRIDGE, B. M.: Heterocaryons between strains of *Neurospora crassa* with different cytoplasms. Genetics **41**, 780—789 (1956).

GRENSON, M.: A gene-induced cytoplasmic mutation in yeast. Proc. XI. intern. Congr. of Genetics, vol. 1, p. 202, The Hague 1963.

GRINDLE, M.: Nucleo-cytoplasmic interactions in the "red" cytoplasmic variant of *Aspergillus nidulans*. Heredity **19**, 75—95 (1964).

HAGEMANN, R.: Plasmatische Vererbung. Jena 1964.

— Extrachromosomale Vererbung. In: BÜNNING, E. et al. (Hrsg.): Fortschritte der Botanik, Springer: Berlin-Heidelberg-New York 1966, pp. 202—216.

HARDER, R.: Zur Frage nach der Rolle von Kern und Protoplasma im Zellgeschehen und bei der Übertragung von Eigenschaften. Z. Bot. **19**, 337—407 (1927a).

— Über mikrochirurgische Operationen an Hymenomyceten. Z. wiss. Mikr. **44**, 173—182 (1927b).

HARDESTY, B. A., and H. K. MITCHELL: The accumulation of free fatty acids in poky, a maternal inherited mutant of *Neurospora* crassa. Arch. Biochem. 100, 330—334 (1963).

HARTMAN, P. E., and C. J. LIU: Comparative cytology of wild type *Saccharomyces* and a respirationally deficient mutant. J. Bact. **67**, 77—85 (1954).

HASKINS, F. A., A. TISSIÈRES, H. K. MITCHELL, and M. B. MITCHELL: Cytochromes and the succinic acid oxidase system of poky strains of *Neurospora*. J. biol. Chem. **200**, 819—826 (1953).

JACOB, F., P. SCHAEFFER, and E. L. WOLLMAN: Episomic elements in bacteria. Microbiol. Genetics, X. Symp., pp. 67—91, London 1960.

JINKS, J. L.: Somatic selection in fungi. Nature (Lond.) **174**, 409—410 (1954).

— Naturally occurring cytoplasmic changes in fungi. C. R. Lab. Carlsberg, Sér. Physiol. **26**, 183—203 (1956).

— Selection for cytoplasmic differences. Proc. roy. Soc. B **146**, 527—540 (1957).

— Cytoplasmic differentiation in fungi. Proc. roy. Soc. B **148**, 314—321 (1958).

JOHNSON, T.: Variation and the inheritance of certain characters in rust fungi. Cold Spr. Harb. Symp. quant. Biol. **11**, 85—93 (1946).

JOLY, P.: Données récentes sur la génétique des champignons supérieurs (Ascomycètes et Basidiomycètes). Rev. Mycol. (Paris) **29**, 115—186 (1964).

KIMURA, K.: On the diploidization by the double compatible diploid mycelium in the hymenomycetes. Bot. Mag. (Tokyo) **67**, 238—242 (1954).

KRAEPELIN, G.: Normalisierung des Atmungsdefektes bei Hefe. Rückführung stabilisierter *RD*-Mutanten in voll atmungsfähige Normalzellen. Arch. Mikrobiol. **48**, 299—305 (1964).

L'HÉRITIER, P.: The CO_2 sensitivity problem in *Drosophila*. Cold Spr. Harb. Symp. quant. Biol. **16**, 99—112 (1951).

— Le problème de l'hérédité non chromosomique. Ann. Biol. **1**, 3—34 (1962).

LINDBERG, G. D.: A transmissible disease of *Helminthosporium victoriae*. Phytopathology **49**, 29—32 (1959).

LINDEGREN, C. C.: Cytoplasmic inheritance. Ann. N.Y. Acad. Sci. **68**, 366—379 (1957).

MAHONY, M., and D. WILKIE: An instance of cytoplasmic inheritance in *Aspergillus nidulans*. Proc. roy. Soc. B **148**, 359—361 (1958).

— — Nucleo-cytoplasmic control of perithecial formation in *Aspergillus nidulans*. Proc. ryo. Soc. B **156**, 524—532 (1962).

Marcou, D.: Sur la longévité des souches de *Podospora anserina* cultivées à divers températures. C. R. Acad. Sci. (Paris) **239**, 895—897 (1954a).
— Sur le rajeunissement par le friod des souches de *Podospora anserina*. C. R. Acad. Sci. (Paris) **239**, 1153—1155 (1954b).
— Rajeunissement et arrêt de croissance chez *Podospora anserina*. C. R. Acad. Sci. (Paris) **244**, 661—663 (1957).
— Sur la déterminisme de la sénescence observée chez l'ascomycète *Podospora anserina*. Proc. X. intern. Congr. of Genetics, vol. II, p. 179, Montreal 1958.
— Notion de longévité et nature cytoplasmique de déterminent de la sénescence. Ann. Sci. nat. Bot. **2**, 653—764 (1961).
—, et J. Schecroun: La sénescence chez *Podospora anserina* pourrait être due à des particules cytoplasmiques infectantes. C. R. Acad. Sci. (Paris) **248**, 280—283 (1959).
Marquardt, H.: Die Natur der Erbträger im Zytoplasma. Ber. dtsch. bot. Ges. **65**, 198—217 (1952).
—, u. E. Bautz: Die Wirkung einiger Atmungsgifte auf das Verhalten von Hefe-Mitochondrien gegenüber der Nadi-Reaktion. Naturwissenschaften **41**, 361—362 (1954).
Mather, K.: Nucleus and cytoplasm in heredity and development. Proc. roy. Soc. B **148**, 362—369 (1958).
Michaelis, P.: Interactions between genes and cytoplasm in *Epilobium*. Cold Spr. Harb. Symp. quant. Biol. **16**, 121—129 (1951).
— Cytoplasmic inheritance in *Epilobium* and its theoretical significance. Advanc. Genet. **6**, 287—401 (1954).
— Genetical interactions between nucleus and cytoplasmic cell constituents. Path. et Biol. **9**, 769—772 (1961).
Mitchell, M. B., and H. K. Mitchell: A case of "maternal" inheritance in *Neurospora crassa*. Proc. nat. Acad. Sci. (Wash.) **38**, 442—449 (1952).
— —, and A. Tissières: Mendelian and non-Mendelian factors affecting the cytochrome system in *Neurospora crassa*. Proc. nat. Acad. Sci. (Wash.) **39**, 606—613 (1953).
Muneta, J., and A. M. Srb: Paternal transmission of extra-chromosomal properties in *Neurospora*. IX. Intern. Bot. Congr., vol. II, p. 274, Montreal (Canada) 1959.
Nanney, D. L.: The role of cytoplasm in heredity. In: W. D. McElroy and B. Glass (edits.), The chemical basis of heredity, pp. 134—166. Baltimore 1957.
Oehlkers, F.: Neue Überlegungen zum Problem der außerkaryotischen Vererbung. Z. indukt. Abstamm.- u. Vererb.-L. **84**, 213—250 (1952).
— Außerkaryotische Vererbung. Naturwissenschaften **40**, 78—85 (1953).
Papazian, H. P.: Sectoring variants in *Schizophyllum*. Amer. J. Bot. **42**, 394—400 (1955).
— Sex and cytoplasm in the fungi. Trans. N.Y. Acad. Sci. **18**, 388—397 (1956).
— The genetics of basidiomycetes. Advanc. Genet. **9**, 41—69 (1958).
Quintanilha, A., et S. Balle: Étude génétique des phénomènes de nanisme chez les hymenomycètes. Boll. Soc. Brot. **14**, 17—46 (1940).
Raut, C.: Cytochrome deficient yeast strains. Genetics **36**, 572 (1951).
— A cytochrome deficient mutant of *Saccharomyces cerevisiae*. Exp. Cell Res. **4**, 295—305 (1953).
— Heritable non-genic changes induced in yeast by ultraviolet light. J. cell. comp. Physiol. **44**, 463—475 (1954).
Renner, O.: Die pflanzlichen Plastiden als selbständige Elemente der genetischen Konstitution. Ber. sächs. Akad. Wiss. Leipzig, math.-physik. Kl. **86**, 241—266 (1934).
Rhoades, M. M.: Interaction of genic and non-genic hereditary units and the physiology of non-genic inheritance. In: W. Ruhland (Hrsg.), Handbuch der Pflanzenphysiologie, Bd. I, S. 19—57. Berlin-Göttingen-Heidelberg 1955.

RIZET, G.: Les phénomènes de barrage chez *Podospora anserina*. I. Analyse génétique des barrages entre souches *S* et *s*. Rev. Cytol. Biol. végét. **13**, 51—92 (1952).
— Sur la multiplicité des mécanismes génétiques conduisant à des barrages chez *Podospora anserina*. C. R. Acad. Sci. (Paris) **237**, 666—668 (1953a).
— Sur l'impossibilité d'obtenir la multiplication végétative interrompue et illimitée de l'ascomycète *Podospora anserina*. C. R. Acad. Sci. (Paris) **237**, 838 —840 (1953b).
— Sur la longévité des souches de *Podospora anserina*. C. R. Acad. Sci. (Paris) **237**, 1106—1109 (1953c).
— Les modifications qui conduisent à la sénescence chez *Podospora* sontelles de nature cytoplasmique? C. R. Acad. Sci. (Paris) **244**, 663—665 (1957).
—, et D. MARCOU: Longévité et sénescence chez l'acsomycète *Podospora anserina*. Compt. rend. VIII. Congr. intern. Bot. Sect., vol. 10, pp. 121—128, Paris 1954.
— — et J. SCHECROUN: Deux phénomènes d'héridité cytoplasmique chez l'ascomycète *Podospora anserina*. Bull. Soc. Franc. physiol. végét. **4**, 136—149 (1958).
—, et J. SCHECROUN: Sur les facteurs cytoplasmiques associés ou couple des gènes *S*—*s* chez *Podospora anserina*. C. R. Acad. Sci. (Paris) **249**, 2392—2394 (1959).
ROPER, J. A.: Nucleo-cytoplasmic interactions in *Aspergillus nidulans*. Cold Spr. Harb. Symp. quant. Biol. **23**, 141—154 (1958).
SAGER, R., and F. J. RYAN: Cell heredity. New York and London 1961.
SCHATZ, G., H. TUPPY u. J. KLIMA: Trennung und Charakterisierung cytoplasmatischer Partikel aus normaler und atmungsdefekter Bäckerhefe. Ein Beitrag zur Frage der genetischen Kontinuität der Mitochondrien von *Saccharomyces cerevisiae*. Z. Naturforsch. **18**b, 145—153 (1963).
SCHECROUN, J.: Sur la réversion provoquée des souches *s*^{S} en souches *s* chez *Podospora anserina*. C. R. Acad. Sci. (Paris) **246**, 1268—1270 (1958a).
— Sur la réversion provoquée d'une modification cytoplasmique chez *Podospora anserina*. Proc. X. intern. Congr. Genetics, vol. II, pp. 252—253, Montrale (Canada) 1958b.
— Sur la nature de la différence cytoplasmique entre souches *s* et *s*^{S} de *Podospora anserina*. C. R. Acad. Sci. (Paris) **248**, 1394—1397 (1959).
SHARPE, S.: A closed system of cytoplasmic variation in *Aspergillus glaucus*. Proc. roy. Soc. B **148**, 355—359 (1958).
SHERMAN, F.: Respiration-deficient mutants of yeast. I. Genetics. Genetics **48**, 375—385 (1963).
— Mutants of yeast deficient in cytochrome C. Genetics **49**, 39—48 (1964).
SILAGI, S.: Interactions between a cytoplasmic factor and nuclear genes in *Neurospora crassa*. Proc. XI. intern. Congr. Genetics, vol. 1, pp. 202, The Hague 1963.
SLONIMSKI, P. P.: Action de l'acriflavine sur les levures. IV. Mode d'utilisation de glucose par les mutants «petite colonie». Ann. Inst. Pasteur **76**, 510—530 (1949a).
— Action de l'acriflavine sur les levures. VII. Sur l'activité catalytique du cytochrome C des mutants «petite colonie» de la levure. Ann. Inst. Pasteur **77**, 774—777 (1949b).
— Effet de l'oxygène sur la formation de quelques enzymes chez le mutant «petite colonie» de *Saccharomyces cerevisiae*. C. R. Acad. Sci. (Paris) **231**, 375—376 (1950).
— Recherches sur la formation des enzymes réspiratoires chez la levure. Thèse, Fac. des Sci. Paris 1952.
—, et B. EPHRUSSI: Action de l'acriflavines sur les levures. V. Le système de cytochromes des mutants «petite colonie». Ann. Inst. Pasteur **77**, 47—63 (1949).
— et H. M. HIRSCH: Nouvelles données sur la constitution enzymatique du mutant «petite colonie» de *Saccharomyces cerevisiae*. C. R. Acad. Sci. (Paris) **235**, 741—743 (1952).

Extrachromosomal inheritance

Sonneborn, T. M.: Kappa and related particles in *Paramecium*. Advanc. Virus Res. **6**, 231—356 (1959).

Srb, A. M.: Some consequences of nuclear cytoplasmic recombinations among various *Neurosporas*. Cold Spr. Harb. Symp. quant. Biol. **23**, 269—277 (1958).

Stubbe, W.: Genetische Analyse des Zusammenwirkens von Genom und Plasmon bei *Oenothera*. Z. Vererbungsl. **90**, 288—298 (1959).

— Extrem disharmonische Genom-Plastom-Kombinationen und väterliche Plastidenvererbung bei *Oenothera*. Z. Vererbungsl. **94**, 392—411 (1963).

Tavlitzki, J.: Action de l'acriflavine sur les levures. III. Étude de la croissance des mutants «petite colonie». Ann. Inst. Pasteur **76**, 497—509 (1949).

Vandendries, R.: La tétrapolarité sexuelle de *Pleurotus colombinus*. Cellule **41**, 267—278 (1932).

Wettstein, F. v.: Die genetische und entwicklungsphysiologische Bedeutung des Cytoplasmas. Z. indukt. Abstamm.- u. Vererb.-L. **73**, 349—366 (1937).

Windisch, S.: Über Bildung und Bedeutung plasmatischer Mutanten von Kulturhefen. Brauerei, Wiss. Beil. **11**, 3—7 (1958).

Ycas, M., and T. J. Starr: The effect of glycine and protoporphyrin on a cytochrome deficient yeast. J. Bact. **65**, 83 —88 (1953).

Yotsuyanagi, Y.: Mitochondria and refractive granules in the yeast cells. Nature (Lond.) **176**, 1209 (1955).

— Études sur le chondriome de la levure. II. Chondriomes des mutants à déficience respiratoire. J. Ultrastruct. Res. **7**, 141—158 (1962).

References

which have come to the authors' attention after conclusion of the German manuscript

A

Beale, G. H.: The role of the cytoplasm in heredity. Proc. roy. Soc. B **164**, 209—218 (1966).

Sager, R.: Mendelian and non-Mendelian heredity: a reappraisal. Proc. roy. Soc. B **164**, 290—297 (1966).

—, and Z. Ramanis: Recombination of nonchromosomal genes in *Chlamydomonas*. Proc. nat. Acad. Sci. (Wash.) **53**, 1053—1061 (1965).

Srb, A. M.: Extrachromosomal heredity in fungi. Reproduction: Molecular, subcellular and cellular, p. 191—211. New York 1965.

B I

Avers, C. J., and C. D. Dryfuss: Influence of added nucleosides on acriflavin induction of *petite* mutants in baker's yeast. Nature (Lond.) **206**, 850 (1965).

— C. R. Pfeffer, and M. W. Rancourt: Acriflavine induction of different kinds of "*petite*" mitochondrial populations in *Saccharomyces cerevisiae*. J. Bact. **90**, 481—494 (1965).

Ephrussi, B., et S. Grandchamp: Études sur la suppressivité des mutants à déficience respiratoire de la levure. I. Existence au niveau cellulaire de divers "degrés de suppressivité". Heredity **20**, 1—7 (1965).

— H. Jakob et S. Grandchamp: Études sur la suppressivité des mutants à déficience respiratoire de la levure. II. Étapes de la mutation grande en petite provoquée par le facteur suppressif. Genetics **54**, 1—29 (1966).

Horn, P., and D. Wilkie: Selective advantage of the cytoplasmic respiratory mutant of *Saccharomyces cerevisiae* in a cobalt medium. Heredity **21**, 625—635 (1967).

JAKOB, H.: Complementation entre mutants à déficience respiratoire de *Saccharomyces cerevisiae:* Établissement et régulation de la respiration dans les zygotes et dans leur proche descendence. Genetics 52, 75—98 (1965).

KRAEPELIN, G.: Zur Wirkung von 2,4-Dinitrophenol und Acridinorange auf *Saccharomyces cerevisiae* und dessen atmungsdefekte Mutante M_k. Arch. Mikrobiol. 50, 52—58 (1965).

— Der Einfluß der Vorkultur auf die Induktion atmungsdefekter Hefemutanten durch Acridinorange. Arch. Mikrobiol. 50, 59—62 (1965).

— Synchrone Vorgänge in asynchron sprossenden *Saccharomyces*-Kulturen: Mutationsauslösung und Abtötung durch Acridine. Arch. Mikrobiol. 50, 63—67 (1965).

MORITA, T., and I. MIFUCHI: Effect of methylene blue on the action of 4-nitroquinoline N-oxide and acriflavine in inducing respiration-deficient mutants of *Saccharomyces cerevisiae*. Jap. J. Microbiol. 9, 123—129 (1965).

MOUSTACCHI, E., et H. MARCOVICH: Induction de la mutation «*petite colonie*» chez la levure par le 5-flurouracile. C. R. Acad. Sci. (Paris) 256, 5646—5648 (1963).

—, and D. H. WILLIAMSON: Physiological variations in satellite components of yeast DNA detected by density gradient centrifugation. Biochem. biophys. Res. Commun. 23, 56—61 (1966).

SUGIMURA, T., K. OKABE, M. NAGAO, and N. GUNGE: A respiration-deficient mutant of *Saccharomyces cerevisiae* which accumulates porphyrins and lacks cytochromes. Biochim. biophys. Acta (Amst.) 115, 267—275 (1966).

TUPPY, H., E. HASLBRUNNER u. G. SCHATZ: Extranukleare Desoxyribonukleinsäure in aerob und anaerob gezüchteter normaler Bäckerhefe sowie in der atmungsdefizienten „*petite*"-Mutante. Mh. Chem. 96, 1831—1841 (1965).

WINTERSBERGER, E.: Synthesis and function of mitochondrial ribonucleic acid. In: J. M. TAGER, S. PAPA, E. QUAGLIARIELLO, and E. C. SLATER: (eds.): Regulation of metabolic processes in mitochondria. Amsterdam 1966.

—, u. H. TUPPY: DNA-abhängige RNA-Synthese in isolierten Hefe-Mitochondrien. Biochem. Z. 341, 399—408 (1965).

B II

HELLER, J., and E. L. SMITH: The amino acid sequence of cytochrome C of *Neurospora crassa*. Proc. nat. Acad. Sci. (Wash.) 54, 1621—1625 (1965).

McDOUGALL, K. J., and T. H. PITTENGER: A cytoplasmic variant of *Neurospora crassa*. Genetics 54, 551—565 (1966).

MUNKRES, K. D., and D. O. WOODWARD: On the genetics of enzyme locational specificity. Proc. nat. Acad. Sci. (Wash.) 55, 1217—1224 (1966).

SILAGI, S.: Interactions between an extrachromosomal factor, *poky*, and nuclear genes in *Neurospora crassa*. Genetics 52, 341—347 (1965).

SRB, A. M.: Extrachromosomal factors in the genetic differentiation of *Neurospora*. Symp. Soc. Experim. Biol. XVII, Cell Different. 1963, p. 175—187.

C III

ARLETT, C. F.: The radiation sensitivity of a cytoplasmic mutant of *Aspergillus nidulans*. Int. J. Radiat. Biol. 10, 539—549 (1966).

CHEVAUGEON, J.: Modification extra-chromosomique et âge du thalle chez le *Pestalozzia annulata*. C. R. Acad. Sci. (Paris) 255, 1980—1982 (1962).

— Conditions de la différenciation du mycélium modifié chez le *Pestalozzia annulata*. C. R. Acad. Sci. (Paris) 255, 3450—3452 (1962).

Extrachromosomal inheritance

CHEVAUGEON, J.: Une période d'incubation s'interpose entre deux événements impliqués dans la modification extra-chromosomique du *Pestalozzia annulata*. C.R. Acad. Sci. (Paris) **257**, 217—220 (1963).
— Modification expérimentale du phénotype normal chez le *Pestalozzia annulata*. C. R. Acad. Sci. (Paris) **261**, 2730—2733 (1965).
— Mise en évidence de mécanismes de répression de la variation extra-chromosomique chez le *Pestalozzia annulata*. C. R. Acad. Sci. (Paris) **263**, 120—123 (1966).
—, et L. CLOUET: Temps et lieu des deux événements aléatoires impliqués dans la modification extra-chromosomique du *Pestalozzia annulata*. C. R. Acad. Sci. (Paris) **256**, 4068—4071 (1963).
— L. CLOUET et G. MICHEL: Nutrition, croissance et modification extra-chromosomique du *Pestalozzia annulata*. C. R. Acad. Sci. (Paris) **261**, 517—520 (1965).

D IV

CROFT, J. H.: A reciprocal phenotypic instability affecting development in *Aspergillus nidulans*. Heredity **21**, 565—579 (1967).
GREGORY, K. F., and J. C. C. HUANG: Tyrosinase inheritance in *Streptomyces scabies*. I. Genetic recombination. J. Bact. **87**, 1281—1286 (1964).
— — Tyrosinase inheritance in *Streptomyces scabies*. II. Induction of tyrosinase deficiency by acridine dyes. J. Bact. **87**, 1287—1294 (1964).
KERR, S.: Disappearance of a genetic marker from a cytoplasmic hybrid plasmodium of a true slime mold. Science **147**, 1586—1588 (1965).

Appendix

1. Genetic Stock Centers

a) Europe

1. Centraalbureau voor Schimmelculture, Baarn (Netherlands), Java-laan 20.
A wide variety of fungus cultures are available. A charge is made for cultures.
2. The International Registry of Microbial Genetic Stocks was organized in 1966 by the International Cell Research Organization (sponsored by UNESCO). For information write:

> Dr. B. HAYES
> Microbial Genetics Research Unit
> Hammersmith Hospital, Ducane Road
> London W 12, Great Britain

The registry includes a list of available strains of bacteria, bacterio-phages, and fungi as well as the names and addresses of workers co-operating in this venture; the latter are grouped into eight geographical regions and are listed by countries within the regions. A table organized by geographical region gives the source of any particular genetic strain.

b) U.S.A.

The Committee on the Maintenance of Genetic Stocks of the Genetics Society of America (Chairman, Dr. R. W. BARRATT, Dept. of Biological Sciences, Dartmouth College, Hanover, New Hampshire 03755) has published a list of genetic stock centers in *Genetics* (volume 52 : 59—62, 1965). The following stock centers may be of interest to scientists working with fungi:

> American Type Culture
> 12301 Parklawn Drive
> Rockville, Maryland 20852

Species of bacteria, fungi including yeasts and plant rusts, human and animal viruses, human and animal cell lines, and bacteriophages. A charge is made for cultures. Partially supported by NSF and the Public Health Service.

> Fungal Genetics Stock Center
> Dr. R. W. BARRATT
> Department of Biological Sciences
> Dartmouth College
> Hanover, New Hampshire 03755

Appendix

Inbred, wild type, and mutant stocks of *Neurospora*; wild type and mutant strains of *Aspergillus nidulans*. Cultures distributed free. Supported by NSF. Stock list available free.

Yeast Genetic Stock Centers

Dr. R. K. MORTIMER
Donner Laboratory
University of California
Berkeley, California 94720

Genetic stocks of *Saccharomyces cerevisiae*

Dr. C. C. LINDEGREN
Biological Research Laboratory
Southern Illinois University
Carbondale, Illinois 62901

Genetic stocks of *Saccharomyces*

2. Newsletters and information services

The Committee on the Maintenance of Genetic Stocks has also compiled a list of newletters, publications containing stock lists, and information of special value to geneticists. Copies of this list are available from members of the Committee. The following information bulletins may be of interest to fungal geneticists:

Aspergillus Newsletter
Dr. J. A. ROPER, Editor
Department of Genetics
The University
Sheffield 10, England
(1 issue a year — free)

Neurospora Newsletter
Dr. BARBARA J. BACHMANN, Editor
Department of Microbiology
New York University School of
Medicine
550 First Avenue
New York, N.Y. 10016
(2 issues a year — free; supported
by NSF)

Microbial Genetics Bulletin
Dr. HOWARD ADLER, Editor
Division of Biology
Oak Ridge National Laboratory
Oak Ridge, Tennessee 37831
(2 issues a year — free; supported
by Oak Ridge National Laboratory)

Notes on Fungal Incompatibility
Dr. P. G. MILES, Editor
Department of Biology
State University of New York
Buffalo, New York 14214

Yeast Newsletter
Dr. HERMAN J. PHAFF, Editor
Department of Food Technology
University of California
Davis, California 95616
(2 issues a year; subscription
necessary)

A very valuable compilation of literature is:

Neurospora Bibliography and Index
Edited by B. J. BACHMANN and W. N. STRICKLAND
Yale University Press New Haven and London 1965

This booklet contains an author index and a general index.

Author index

Page numbers in *italics* refer to the literature citations

Subject index

Page numbers in *italics* refer to definitions

Rhizopus, sex hormones 90
— nigricans 49, 90
— sexualis 90
reproduction, sexual *48*
—, systems of sexual *48—91*
reproductive systems, division 52
reunion of chromatid arms 234—235
riboflavin, synthesis 362
ribose nucleic acid (RNA), genetic
    code 344—346
—, messenger *342—345*
—, structure 129
—, transfer *342—345*
ribosome protein synthesis 342—343

Saccharomyces, α-glucomelizitase 377
—, α-glucosidase 376
—, mating types 56
— cerevisiae 51, 87, 408
— —, abnormal mating behavior
    86—87
— —, affinity 144, 145, 146
— —, cell form 314
— —, chromatid interference 182,
    183, 184
— —, chromosome interference 188,
    189, 191, 192
— —, complementation 387
— —, diagnostic characters 30
— —, distribution of nuclei 138
— —, DNA content 220, 315
— —, DNA replication 176
— —, dose-effect curves 285
— —, enzymatic changes after
    mutation 374
— —, frequency of spontaneous
    reverse mutations 305
— —, influence on recombination
    frequencies 168, 176, 177
— —, intragenic complementation
    388
— —, isolation of auxotrophic
    mutants 273, 276
— —, linkage groups and chromo-
    some maps 167, 201, 208
— —, mitochondria and respiratory
    defects 442—444
— —, mitotic crossing over 206
— —, morphology 11
— —, mutagenic effect of chemical
    agents 288
— —, — of radiation 279, 282,
    284
— —, negative interference 217
— —, non-reciprocal recombination
    222, 223
— —, nutritional deficiency
    mutants 351
— —, "petite" colony mutants
    442—444

Saccharomyces cerevisiae, physiolo-
    gical dioecism 86—87, 91
— —, polyploidy 314, 315, 317, 318
— —, restoration of "petite" colony
    mutants 444
— —, respiratory defect mutants
    (see: "petite" colony mutants)
— —, sex reaction 91
— —, somatic recombination 94,
    204
— —, synthesis of amino acids 354
— —, transformation 290
— chevalieri 87
— ludwigii, lethal mutants 271
Salmonella, block mutations 312
Salpiglossis, non-reciprocal recombi-
    nation 223
Sapromyces reinschii, morphological
    dioecism 82
Schizophyllum 24, 42
— commune 45, 55, 58, 59, 60, 63,
    65, 66, 74, 78, 97
— —, diagnostic characters 31
— —, different recombination fre-
    quencies 168
— —, distribution of nuclei 138
— —, frequency of spontaneous for-
    ward mutations 304
— —, isolation of unordered
    tetrads 153
— —, mapping of genetic markers
    201
— —, mitotic recombination 208
— —, morphology 24
— —, mutability of incompatibility
    factors 74
— —, neo-monokaryons 458—459
— —, nutritional deficiency
    mutants 351
— —, physiology of incompatibility
    factors 80
— —, somatic recombination 95,
    204
— —, structure of incompatibility
    factors 64
Schizosaccharomyces pombe 87, 301
— —, chromosome interference 190
— —, diagnostic characters 30
— —, DNA content 220
— —, dose-effect curves 281
— —, frequency of spontaneous
    reverse mutations 305
— —, intragenic complementation
    387, 388, 389
— —, — recombination 220
— —, — specificity 295, 296, 299
— —, isolation of auxotrophic
    mutants 273
— —, mapping of genetic markers
    201

Subject index

Subject index

Universitätsdruckerei H. Stürtz AG Würzburg